# Toward a Unified Theory of Development

*Oxford Series in Developmental Cognitive Neuroscience*

Series editor: Mark H. Johnson,

*Centre for Brain and Cognitive Development, Birkbeck College, University of London, United Kingdom*

***Books in this series:***

**Neuroconstructivism, Volume One: How the Brain Constructs Cognition**

Denis Mareschal, Mark H. Johnson, Sylvain Sirois, Michael W. Spratling, Michael S. C. Thomas, and Gert Westermann

**Neuroconstructivism, Volume Two: Perspectives and Prospects**

*Edited by* Denis Mareschal, Sylvain Sirois, Gert Westermann, and Mark H. Johnson

**Toward a Unified Theory of Development: Connectionism and Dynamic Systems Theory Re-Considered**

*Edited by* John P. Spencer, Michael S. C. Thomas, and James L. McClelland

# Toward a Unified Theory of Development

## CONNECTIONISM AND DYNAMIC SYSTEMS THEORY RE-CONSIDERED

*Edited by*

JOHN P. SPENCER

MICHAEL S. C. THOMAS

JAMES L. McCLELLAND

OXFORD
UNIVERSITY PRESS
2009

OXFORD
UNIVERSITY PRESS

Oxford University Press, Inc., publishes works that further
Oxford University's objective of excellence
in research, scholarship, and education.

Oxford New York
Auckland Cape Town Dar es Salaam Hong Kong Karachi
Kuala Lumpur Madrid Melbourne Mexico City Nairobi
New Delhi Shanghai Taipei Toronto

With offices in
Argentina Austria Brazil Chile Czech Republic France Greece
Guatemala Hungary Italy Japan Poland Portugal Singapore
South Korea Switzerland Thailand Turkey Ukraine Vietnam



Published by Oxford University Press, Inc.
198 Madison Avenue, New York, New York 10016
www.oup.com



**Library of Congress Cataloging-in-Publication Data**

Toward a unified theory of development : connectionism and dynamic systems theory re-considered /
edited by John P. Spencer, Michael S. C. Thomas, and James L. McClelland.
p. cm.—(Oxford series in developmental cognitive neuroscience ; 2)
Includes bibliographical references and index.
ISBN 978-0-19-530059-8
1. Developmental psychology. 2. Cognitive psychology. 3. Connectionism. 4. Differentiable dynamical
systems. 5. Dynamics. I. Spencer, John P. II. Thomas, Michael S. C. III. McClelland, James L.
BF713.T69 2009
155—dc22
2008033827

9 8 7 6 5 4 3 2 1

Printed in China
on acid-free paper.

*To the memory and vision of Elizabeth Bates and Esther Thelen*

# ACKNOWLEDGMENTS

The preparation of this book was supported by funds from the University of Iowa Obermann Center for Advanced Studies and by the Center's Laura Spelman Rockefeller Fund.

The editors are sincerely grateful to the Obermann Center for its support of this project and for its dedication to the advancement of scholarship in the arts and sciences.

The preparation of this volume was also supported by NSF HSD 0527698 awarded to John P. Spencer, by UK MRC Grant G0300188 and European Commission Grant NEST-029088 awarded to Michael S. C. Thomas, and by NIMH Integrative Behavioral Science Center Grant MH64445 awarded to James L. McClelland.

# CONTENTS

## III. CONNECTIONISM: CASE STUDIES



## IV. HYBRID MODELS: CASE STUDIES



## V. REACTIONS FROM THE OUTSIDE



## VI. REACTIONS FROM THE INSIDE

# CONTRIBUTORS

**Frank D. Baughman**
Developmental Neurocognition Laboratory
School of Psychology
Birkbeck College
University of London
London, United Kingdom

**Richard P. Cooper**
School of Psychology
Birkbeck College
University of London
London, United Kingdom

**Daniela Corbetta**
Department of Psychology
University of Tennessee
Knoxville, Tennessee

**Evelina Dineva**
Department of Psychology
Iowa Center for Developmental and Learning Sciences
University of Iowa
Iowa City, Iowa

**Kurt W. Fischer**
Harvard Graduate School of Education
Harvard University
Cambridge, Massachusetts

**Jessica S. Horst**
Department of Psychology
University of Sussex
Brighton, United Kingdom

**Jeffrey S. Johnson**
Department of Psychology
University of Wisconsin
Madison, Wisconsin

**Timothy D. Johnston**
College of Arts and Sciences
University of North Carolina
Greensboro, North Carolina

**Heidi Kloos**
Department of Psychology
University of Cincinnati
Cincinnati, Ohio

**Robert Leech**
School of Psychology
Birkbeck College
University of London
London, United Kingdom

**Robert Lickliter**
Department of Psychology
Florida International University
Miami, Florida

**Yeou-Teh Liu**
Department of Kinesiology
Taipei Physical Education College
Taipei City, Taiwan (R.O.C)

**Denis Mareschal**
School of Psychology
Birkbeck College
University of London
London, United Kingdom

**Gottfried Mayer-Kress**
Department of Kinesiology
Pennsylvania State University
University Park, Pennsylvania

**James L. McClelland**
Department of Psychology
Center for Mind, Brain and Computation
Stanford University
Stanford, California

**Bob McMurray**
Department of Psychology
Iowa Center for Developmental and Learning Sciences
University of Iowa
Iowa City, Iowa

**J. Bruce Morton**
Department of Psychology
University of Western Ontario
Ontario, Canada

**Yuko Munakata**
Department of Psychology
University of Colorado
Boulder, Colorado

**Nora S. Newcombe**
Department of Psychology
Temple University
Philadelphia, Pennsylvania

**Karl M. Newell**
Department of Kinesiology
Pennsylvania State University
State College, Pennsylvania

**Lisa M. Oakes**
Department of Psychology
University of California
Davis, California

**Sammy Perone**
Department of Psychology
University of Iowa
Iowa City, Iowa

**Jodie M. Plumert**
Department of Psychology
Iowa Center for Developmental and Learning Sciences
University of Iowa
Iowa City, Iowa

**Maartje E. J. Raijmakers**
Department of Psychology
University of Amsterdam
Amsterdam, the Netherlands

**Fiona M. Richardson**
Developmental Neurocognition Laboratory
School of Psychology
Birkbeck College
University of London
London, United Kingdom

**Larissa K. Samuelson**
Department of Psychology
Iowa Center for Developmental and Learning Sciences
University of Iowa
Iowa City, Iowa

**Anna C. Schapiro**
Department of Psychology
Center for Mind, Brain and Computation
Stanford University
Stanford, California

**Matthew Schlesinger**
Department of Psychology
Southern Illinois University Carbondale
Carbondale, Illinois

**Gregor Schöner**
Institut für Neuroinformatik
Ruhr-Universität Bochum
Bochum, Germany

**Linda B. Smith**
Department of Psychological and Brain Sciences
Indiana University
Bloomington, Indiana

**John P. Spencer**
Department of Psychology
Iowa Center for Developmental and Learning Sciences
University of Iowa
Iowa City, Iowa

**Whitney Tabor**
Department of Psychology
University of Connecticut
Storrs, Connecticut

**Michael S. C. Thomas**
Developmental Neurocognition Laboratory
School of Psychology
Birkbeck College
University of London
London, United Kingdom

**Joseph C. Toscano**
Department of Psychology
University of Iowa
Iowa City, Iowa

**Gautam Vallabha**
Department of Psychology
Stanford University
Stanford, California

**Han L. J. van der Maas**
Department of Psychology
University of Amsterdam
Amsterdam, the Netherlands

**Paul van Geert**
Department of Clinical and
Developmental Psychology
University of Groningen
Groningen, the Netherlands

**Guy C. Van Orden**
Department of Psychology
University of Cincinnati
Cincinnati, Ohio

# Introduction

*John P. Spencer, Michael S. C. Thomas, and James L. McClelland*

The study of development has always been organized around big issues and overarching theoretical ideas. From William James to Sigmund Freud to Jean Piaget to B. F. Skinner, researchers (and parents alike!) have wondered how children move from the blooming, buzzing confusion of infancy, through the tumult of childhood and adolescence, into adulthood. Does development occur continuously over time or via dramatic stages? Is development driven by learning or biological maturational processes? What is the nature of experience, and how does it generate change? And what are the best ways to understand development; for instance, can we bring modern advances in neuroscience and modern theoretical approaches such as computational modeling to bear on these big questions? This theoretical ferment percolates into articles, books, symposia, and debates, both formal and informal.

In the mid-1990s, two new contenders weighed in, building on broad frameworks introduced a decade earlier. In 1994, Esther Thelen and Linda B. Smith published *A Dynamic Systems Approach to the Development of Cognition and Action*. These authors, inspired by earlier work on dynamical systems in physics, motor control, and psychology, proposed a radical view of development as fluid, multidetermined, grounded in perception and action, and free of mental structure. Two years later, Jeff Elman, Elizabeth Bates, Mark Johnson, Annette Karmiloff-Smith, Dominic Parisi, and Kim Plunkett followed with the publication of *Rethinking Innateness: A Connectionist Perspective on Development*. Here the inspiration for a new view came from earlier work on connectionist/parallel distributed processing models and from new work on early brain development.

Concurrent with publication of these landmark books, researchers from a variety of disciplines—motor development, cognitive development, socio-emotional development—began using dynamic systems and connectionist approaches in their developmental research. These efforts have led to profound insights into developmental process across a broad range of topics. Nevertheless, despite many commonalities, the diversity of these efforts has led to an associated diversity of views about what constitutes a dynamic systems approach and what constitutes a connectionist approach. Moreover, given the many shared concepts between these approaches—concepts of attractors, self-organization, emergence, soft assembly—it has become unclear exactly how these two approaches relate to one another.

As an initial exploration of these issues, John Spencer and Esther Thelen organized a symposium for the 2001 Biennial Meeting of the Society for Research in Child Development entitled Connectionism and Dynamic Systems Theory: Are They Really Different Approaches to Development? The purpose of the symposium was to bring together some of the leading proponents of the two views to illuminate both similarities and differences. Moreover, we attempted to situate these new contenders within the context of other overarching theories of child development. The symposium sparked intense discussion and resulted in a 2003 special issue of the journal *Developmental Science*. This collection of papers made important strides in identifying the many commonalities between the approaches; however, there was less agreement about the strengths and weaknesses of the approaches and only limited discussion about the direction in which the approaches might evolve.

Thus, in 2005, John Spencer and Jay McClelland organized a 3-day conference at the University of Iowa to move the discussion of connectionist and dynamic systems approaches to the next level. The conference had broad representation from a diversity of perspectives, including scholars with limited computational modeling experience and from different theoretical traditions. Day 1 focused on defining the approaches. Day 2 focused on the strengths and weaknesses of each approach as illustrated by concrete examples, including connectionist and dynamic systems approaches to two tasks that Piaget used to chart developmental change in infants' and children's behaviors (the A-not-B task and the balance scale problem). Day 3 focused on issues central to the future of both approaches, including cases where one approach has a clear advantage over the other.

The conference was extremely stimulating, and a set of crosscutting conceptual themes emerged from our discussions. These themes, summarized below, provided a core set of questions we asked the authors of the chapters in this volume to address:

- What changes across development? How do the connectionist and dynamical systems approaches explain/capture developmental stability and change?
- What commitments do the two approaches make to standard cognitive concepts such as resources and representations? What commitments do the two approaches make to dynamical concepts such as attractors, stability, bifurcation, self-organization, emergence, and so on? What commitments do the two approaches make to the dimensionality of cognition/(behavior or to the compression of dimensions)?
- What timescales are the two approaches usually focused on? How do the approaches conceptualize/realize the integration of timescales?
- Different approaches often focus on explaining certain phenomena while taking other things for granted; what are the things-to-be-explained versus the things-taken-for-granted for each approach? What is the level of description/explanation at which each type of model is typically pitched, and what are different types of models trying to capture?
- What empirical data are used to constrain assumptions of models within each approach? How strong are these constraints? What other types of evidence are used to constrain each type of model (e.g., neural constraints)?
- How are the types of models formally related (e.g., difference equations vs. differential equations)?
- Do the approaches amount to more than a collection of models? How do the approaches differ from other developmental theories in terms of the types of explanations they offer?

This book is organized as follows. Part I provides an introduction to the two approaches by McClelland and Vallabha (Chapter 1) and Schöner (Chapter 2). The next three parts offer case studies that illustrate how both approaches have been used to study development. Part II provides case studies of dynamical systems research highlighting approaches to motor development and embodiment by Corbetta (Chapter 3), to the A-not-B error and early language development by Smith (Chapter 4), to embodied cognition using dynamic neural fields by Spencer and colleagues (Chapter 5), and to integration of processes over multiple timescales by Newell and colleagues (Chapter 6). Part III provides case studies of connectionist approaches highlighting approaches to perseveration and the A-not-B error by Morton and Munakata (Chapter 7), to acquisition of grammar by Tabor (chapter 8), and to design of autonomous robots by Schlesinger (Chapter 9). Part IV offers a third set of case studies that illustrate hybrid connectionist/dymamical systems models of analogy by Mareschal and colleagues (Chapter 10) and of speech categorization and early word learning by McMurray and colleagues (Chapter 11).

The final two sections of this book offer a collection of commentaries by researchers outside and inside the connectionist/dymamical systems groups. In particular, Part V presents reactions from the *outside*, by Kloos and Van Orden (Chapter 12), who examine connectionism and dynamic systems theory in the context of mechanistic accounts of cognition by Oakes and colleagues, who present the view from so-called good old-fashioned cognitive development (Chapter 13), and by Johnston and Lickliter (Chapter 14), who adopt a developmental systems perspective. Part VI presents reactions from the *inside*, by van der Maas and Raijmakers (Chapter 15), van Geert and Fischer (Chapter 16), Thomas and colleagues (Chapter 17), and Spencer and colleagues (Chapter 18). The latter two chapters effectively present the conclusions of the editors regarding the central themes of this book, highlighting the strengths of each approach, the ways in which the approaches are complementary, and the promise the future holds for a synthesis and discussing the idea that connectionism and dynamical systems provide complementary views of the developmental process.

We are tremendously excited about the chapters in this volume—the scope and depth of the ideas and research reported here illustrate how far the field has come. The authors have tackled difficult developmental questions, making this volume a rich source of contemporary thinking about developmental process. We are indebted to the authors for their willingness to contribute to this volume and for their patience with the ensuing editorial process. We are also grateful to the graduate students who attended the 2005 conference, as well as the graduate students (both current and former students) and postdoctoral fellows from the University of Iowa and Birkbeck College who provided helpful reviews of the chapters as well as miscellaneous editorial assistance. We would particularly like to thank Frank Baughman, Aaron Buss, Evelina Dineva, Jeff Johnson, Chelsey Patterson, Sammy Perone, Fiona Richardson, Shannon Ross-Sheehy, Anne Schutte, and Vanessa Simmering. Finally, the editors extend a huge thank you to Valerie Vorderstrasse for her invaluable help in putting together the manuscript of the book.

In a very real sense, the overarching theories of development have a profound influence on daily life. These theories influence national policy and educational curricula, and they influence how parents think about their own children—they truly form a framework within which real children learn and grow. Thus, debate and discussion about these two new theoretical contenders—connectionism and dynamic systems theory—will have far-reaching influence. More locally within the field of developmental science, the time is ripe to raise the question of what connectionist and dynamic systems theorists might learn from one another. These theoretical approaches are now old enough to enable us to consider them in detailed ways using concrete examples, yet researchers have not become too entrenched to consider new challenges and new directions. This book can also play an important role for groups not directly affiliated with the approaches, offering a formal framework within which to think about the central concepts of our science. Finally, this book allows readers to assess a critical issue initially raised at the symposium in 2001—whether connectionist and dynamic systems approaches should be treated as a unified framework that serves as a new overarching theory of development.

This edited volume is unique in that it defines and refines two major theories of development and evaluates whether they may someday merge into one. We hope that this book will bring the ideas of connectionism and dynamic systems theory into sharper focus, and that we will inspire more developmentalists to see their usefulness and explanatory power. We view this book as an important stepping-stone for what we hope will be a long and fruitful process of further elaboration on these two approaches. Although the outcome has yet to be determined, we look forward to watching the unfolding of these two (or one?) theoretical approaches. This is truly an exciting time to be a developmentalist—an exciting time to grapple with the big issues of our science.

# PART I

# Introduction to the Approaches

# CHAPTER 1

# Connectionist Models of Development: Mechanistic Dynamical Models with Emergent Dynamical Properties

*James L. McClelland and Gautam Vallabha*

The symbolic paradigm of cognitive modeling, championed by Minsky and Papert (1969), Newell and Simon (1976), and other pioneers of the 1950s and 1960s, remains very much alive and well today. Yet an alternative paradigm, first championed in the 1950s by Rosenblatt (1958), in which cognitive processes are viewed as emergent functions of a complex stochastic dynamical system, has continued to attract adherents. A cooling of interest in such approaches in the 1960s did not deter Stephen Grossberg (1976, 1978a) or James Anderson (1973), and the approach emerged again full force in the 1980s. Thereafter, many others began to explore the implications of this paradigm for development (Elman et al., 1996; McClelland, 1989; Plunkett & Marchman, 1993; Schulz, Mareschal & Schmidt, 1994). A parallel and closely related movement, emerging from the physical sciences, also began to gain adherents during the same decade (Schöner & Kelso, 1988) and began to attract the interest of developmental psychologists in the early 1990s (Smith & Thelen, 1993; Thelen & Smith, 1994).

The present chapter represents our attempt to underscore the common ground between connectionist and dynamical systems approaches. Central to both approaches is the emergent nature of system-level behavior and changes to such behavior through development.

## Mechanistic Dynamics Versus Emergent Dynamics

We first make a key distinction between mechanistic and emergent dynamics, which parallels one made by Schöner and Kelso (1988) in their seminal article. We consider a mechanism to be a set of rules or equations that *stipulate* certain microscopic behaviors in a system. For example, the Hodgkin-Huxley equations stipulate certain relations between membrane potentials and neurotransmitter concentrations; these relations are considered to be causal (and hence mechanistic) because of the underlying biology and chemistry. When the stipulations are in the form of differential equations, the resulting system behaviors are *mechanistic dynamics*. Such behaviors are usually unsurprising since they have been stipulated to be just so. When several mechanistic subsystems interact, however, there is the possibility of surprising behaviors. We term such behaviors *emergent*, and they are typically at a different level of description than mechanistic dynamics. For example, lateral inhibition may lead to sharper tuning curves and heightened contrast, but the terms *tuning curve* and *contrast* are a step removed from the mechanistic level of membrane potentials and neurotransmitters. Such emergent properties may have their own (non-stipulated) temporal evolutions, and we term such evolutions *emergent dynamics*.

The above terminology allows a crisp characterization of connectionism and dynamical systems theory. Both approaches assume that (a) mechanistic dynamics are best described by differential equations that are rooted in biology and (b) psychological phenomena such as categorization and selection reflect the emergent dynamics of the system. The key theoretical issue is not just to describe the emergent dynamics, but also to explain *how* emergent dynamics arise from the underlying mechanistic dynamics. This last point is crucial, since it distinguishes connectionist and dynamical systems approaches from other approaches in which the focus is on a descriptive characterization of emergent dynamics.

A further characteristic of connectionist systems is a graded notion of emergent dynamics. Before elaborating on this, it is helpful to consider the four parts of a generic connectionist system. These parts are depicted schematically in Figure 1.1: (1) the activations of the units, with a set of differential equations stipulating how inputs to units are related to their activations; (2) the connection weights between the units, with a set of differential equations stipulating how weights are to be updated; (3) the environment of the system, which provides input stimulation; and (4) the mismatch between the system's behavior and the target behavior (also known as the error). The units are akin to pools of neurons, and the activations are akin to the population activity of these neurons. Consequently, the activations are influenced by the current weights and environment; the weight changes are influenced by the activations and the mismatch; and the mismatch itself is influenced by the environment and the activations (which constitute the ongoing behavior of the system).

There are several things to note about the generic connectionist system. First, the mechanistic dynamics of the system are simply those that result from the differential equations. These mechanistic dynamics can give rise to emergent dynamics. Examples include (a) the tuning curves and contrast that emerge from lateral inhibition and (b) the spatial organization that emerges from competitive topographic maps (e.g., Kohonen, 1982). Second, the weights can change over a variety of timescales, but the activations are assumed to only change over a fast timescale. This difference in timescales for the mechanistic dynamics leads to a difference in timescales for the emergent dynamics. Thus, there can be short-term emergent dynamics (in the range of milliseconds to seconds) that coexist with long-term emergent dynamics (in the range of minutes to years). The former are used to model activation processes that are assumed to be complete within the range of a second (e.g., the process involved in settling on a perceptual interpretation of a visual input such as a line drawing or a word). By contrast, long-term emergent dynamics are assumed to operate over a developmental timescale and may therefore also be termed *developmental dynamics*.

Since the focus of many connectionist models has been on change over a developmental

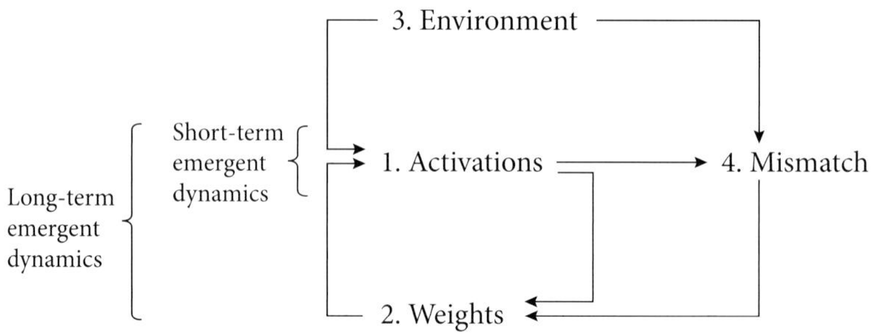


**Figure 1.1. Schematic overview of a connectionist system**
Activations and weight changes are specified in mechanistic terms (typically, differential equations). Arrows indicate direction of influence.

timescale, their emphasis tends to fall along the same time-evolution as connections. Within multilayered connectionist networks, time-evolution can have complex dynamical properties, undergoing both accelerations and decelerations, even though the weight-change mechanism itself does not undergo any change. We shall consider such issues in detail later, but for now the essential point is this: *Connectionist systems are dynamical systems, and like other dynamical systems, they can have complex emergent properties which themselves may be described by dynamical variables (rates of change with respect to time) and dynamical structures (attractors, transitions, and instabilities).*

Now we can elaborate the earlier observation that connectionist systems have a graded notion of emergent dynamics. On a short timescale, connection weights can be treated as fixed, as contributing (along with the architecture of the system and the mechanistic dynamics of the activations) to the emergent time course and outcome of a single act of information processing and/or behavior, such as reaching to one of two locations. Over a longer timescale (measured in months or years), changes in the connection weights lead to changes in the short-term emergent dynamics and may make possible new kinds of behavior (e.g., a 20-month-old toddler may remember and reach differently than a 10-month-old infant). Consequently, *emergence* is not sharply demarcated in time or scale—it occurs gradually and concurrently on different timescales, with long-term emergence shaping the short-term emergent behaviors and vice versa. Both the short- and long-term behaviors exhibit properties congruent with those postulated by dynamical systems theory (DST), as articulated by Schöner and Kelso (1988; see also Chapter 2 of this volume). What connectionism adds over and above DST is an explanation of how the different scales of emergence are related.

## Examples of Activation and Weight-Change Dynamics

The system in Figure 1.1 is very abstract, so it is helpful to consider a few concrete cases. One example of activation dynamics is the cascade model (McClelland, 1979), which describes the time-evolution of activation across the units in a linear feed-forward network.

$$\Delta a_i/dt = \lambda(net_i - a_i) \quad (1)$$

$$net_i = \Sigma_s w_{is}a_s \quad (2)$$

Here, $a_i$ is the activation of the $i^{th}$ unit, $net_i$ is its net input, and $w_{is}$ is the weight of the connection from unit $s$ to unit $i$. Essentially, the activation of a unit tends to change slowly toward its instantaneous net input. While some units receive direct stimulation from the outside world, most units receive their input from other units, and thus, what determines their activation is the matrix of values of their incoming connection weights. The cascade model demonstrated that changes in the parameters of equation (1) (the mechanistic dynamics) modulated the overall system behavior (the short-term emergent dynamics). For example, changes in the rate constants at two different layers of the cascaded feed-forward network had additive effects on reaction time at the system level.

More complex equations are often used in other models (e.g., Grossberg, 1978b):

$$\Delta a_i/dt = \alpha E(M - a_i) + \beta I(a_i - m) - \gamma(a_i - r) \quad (3)$$

Grossberg's formulation separates the excitatory input $E$ and the inhibitory input $I$ (where $E \geq 0$ and $I \leq 0$) and lets each drive the activation up or down by an amount proportional to the distance from the current activation to the maximum $M$ or minimum $m$. Also operating is a restoring force driving the activation of units back toward rest $r$; each force has its own strength parameter ($\alpha$, $\beta$, or $\gamma$). A similar function was used in the interactive activation model (McClelland & Rumelhart, 1981). Analogous to equation (2), $E$ and $I$ in equation (3) are essentially the sum of the excitatory input (mediated by excitatory connections) and inhibitory input (mediated by inhibitory connections). In his pioneering work in the 1970s, Grossberg examined in detail the conditions under which these mechanistic dynamics gave rise to stable emergent behaviors such as attractors and various other important emergent structures (Grossberg, 1978b).

In the above two examples, short-term emergent behavior arises from the mechanistic dynamics of the activation updates. However, weight-update dynamics may also lead to emergent behavior. For example, Linsker (1986) suggested that the update of the connection weight from a sending neuron *s* to a receiving neuron *r* might obey a Hebbian-type learning rule:

$$dw_{rs}/dt = \varepsilon(a_r - \Theta_r)(a_s - \Theta_s) - \beta w_{rs} \qquad (4)$$

Here, $a_r$ and $a_s$ are the activations of the receiving and sending units, $\beta$ is a constant regulating a tendency for weights to decay toward 0, $\varepsilon$ is the learning rate, and $\Theta_r$ and $\Theta_s$ are critical values that can have their own mechanistic dynamics. Linsker (1986) showed that equation (4) can lead to receptive fields with surprisingly complex structures, such as center-surround and orientation selectivity. In other words, the weight-update dynamics stipulated in equation (4) regulate the long-term emergent behavior of the system.

As with activation functions, many variations have been proposed for connectionist learning rules. What is crucial to this discussion is that the time-evolution of the connection weights (which are the parameters that determine the emergent dynamics and outcome of the short-term processing) are themselves defined in terms of differential equations. The results of these connection dynamics are emergent cognitive structures such as feature detectors, perceptual categories, and behavioral abilities, as well as the trajectories of change in such properties.

## Simplified Mechanisms

The discussion thus far has sidestepped a crucial issue. Equations (1)–(4), while relatively close to the neurophysiology, are still highly abstract. For example, equation (4) assumes that the receiving and sending units' activity is exactly synchronized, though recent work suggests a complex role for asynchrony (Roberts & Bell, 2002). So at what level should the mechanistic dynamics be stipulated?

The pragmatic answer is that it depends on the research problem at hand. In some cases, the focus is on short-term emergent dynamics, so the mechanics of the weight change are simplified. In a Hopfield net (Hopfield, 1982), for example, the weights may be calculated beforehand from the training ensemble, and thus there are no weight-change dynamics at all. Likewise, in the interactive activation model (McClelland & Rumelhart, 1981) or the model of the Stroop effect (Cohen, Dunbar, & McClelland, 1990), the weights are directly stipulated by the modeler to yield particular activation dynamics. The converse—where activation dynamics are simplified to focus on long-term emergent dynamics—can also happen. In the Kohonen map (Kohonen, 1982), the activation dynamics are simplified so that the unit with the maximum input is stipulated to be the winner and assigned an activation of 1.0. This allows the modeler to focus on the weight-change mechanisms and the developmental emergence of the topographic map. (Kohonen, 1993, examines how this simplification may be reconciled with more realistic activation dynamics.) Likewise, Carpenter and Grossberg (1987) and Rumelhart and Zipser (1985) use binary input and simplified activation to focus on unsupervised competitive learning.

Simplification of activation dynamics sometimes goes hand in hand with simplified weight-change dynamics. This is the case when a network is trained using back-propagation, which updates each weight in proportion to its influence on the overall error (Rumelhart, Hinton, & Williams, 1986). Back-propagation as numerically implemented is biologically implausible, and the mechanics of the weight update may be much more complex (O'Reilly & Munakata, 2000). However, back-propagation is a useful simplification that allows one to study how network architecture, statistical structure of input, and experience history interact to produce particular patterns of emergent, long-term behavior.

It is important to note that many connectionist models simplify the activation dynamics in ways that may mislead those unfamiliar with the framework and underlying theory. For example, in most connectionist models of past-tense verb inflection or single-word reading, an input pattern is propagated forward through one or more layers of connection weights to an output layer. Furthermore, learning and performance variability is usually introduced through initial randomization of weights and training schedule but not in propagation of activity during a single trial. Such models may lead to the misconception that connectionist theories *advocate* non-dynamic and non-stochastic processing. However, this is not the case. The final output of the network represents an asymptotic (i.e., stable or settled) state that would be achieved in an identical cascaded feed-forward network or even in a recurrent network. (See Plaut, McClelland, Seidenberg, & Patterson, 1996, for an explicit comparison of one-pass feed-forward and recurrent versions of a network for single-word reading.) Likewise, while intrinsic variability (such as due to neural noise) is often left out of specific models for simplicity, it is assumed to always be at work in processing (McClelland, 1991, 1993). Similar simplifications are often made in simple recurrent networks (Elman, 1990), which can be viewed as chunking time into relatively large-scale, discrete units and setting the state for the next discrete unit based on the state of the one before. While such networks look discrete, this method is an approximation that increases the computational feasibility of a simulation, since allowing a network to settle over many time steps multiplicatively scales the time it takes to run the simulation. For example, if there are ten steps per simulated time unit, the simulation can take ten times as much memory *and* take ten times as long to run (Williams & Zipser, 1995).

Another simplification often found in connectionist systems concerns the input representation, specifically, whether it is localist or distributed. In a localist input representation, only one input unit is active at a given time, whereas in a distributed representation several units may be concurrently active. Distributed representations are more plausible from both computational and biological points of view: they allow many more stimuli to be represented (an $N$-unit distributed representation can potentially represent $2^N$ patterns, whereas a localist representation can only represent $N$ patterns); they can capture the similarity structure between the input stimuli (due to the overlap between the distributed representations); and they are more in line with what is known about neural representations (e.g., Haxby et al., 2001). Hence, a localist representation is often a simplification of an underlying distributed representation. Such a simplification has some theoretical support. For example, Kawamoto and Anderson (1985) found that the competition between distributed representations in an attractor network mimicked that between two units. That is, in certain cases the interaction between distributed patterns (between different spatial modes) may be recast as an interaction between localist units.

In summary, then, connectionist systems are dynamical systems that can have both short- and long-term emergent behaviors. These emergent behaviors may themselves be described by dynamical variables, such as attractors and transitions, and are therefore similar in spirit to the models postulated in dynamical systems theory. In the following discussion, the points above will be elaborated through presentation of a series of models.

We begin with the model of activation dynamics recently proposed by Usher and McClelland (2001), showing how a simple architecture can give rise to interesting dynamical properties (including attractors and path dependence); these properties will be treated collectively as emergent response dynamics of the connectionist system. We will then consider a more complex model proposed by Vallabha and McClelland (2007), which embeds the Usher-McClelland activation dynamics within a network with simple weight-change dynamics; the complex model thereby learns in response to experience and exhibits interesting developmental dynamics in the domain of acquisition of phonological distinctions. Further emergent dynamic properties will be explored within this context. Finally we will consider a model by Rogers and McClelland (2004) that addresses the time-evolution of conceptual category representations. This model simplifies both the activation and weight-change dynamics (as in the past-tense verb inflection and reading models mentioned above) to address the gradual evolution of semantic category representations over the first ten years of life. Once again, simple learning rules give rise to complex, emergent dynamical properties, now seen at a developmental timescale. Overall, we hope to highlight the capacity of these models to instantiate many of the properties of the dynamical models proposed by other contributors to this volume.

## Short-Term Emergent Dynamics in Perceptual Classification

We begin with Usher and McClelland's (2001) model of the time course and accuracy of responding in speeded perceptual classification tasks. The goal of this model was to provide an account of classification responses, including time-accuracy tradeoffs and reaction-time distributions, as a function of classification difficulty. The architecture of this model is very similar to that used in parallel distributed processing (PDP) models of reading (McClelland & Rumelhart, 1981) and speech perception (McClelland, 1991; McClelland & Elman, 1986), and incorporates the basic principles of graded, interactive, and stochastic processing. At its simplest level, the Usher-McClelland network model consists of a layer of units, in which each unit corresponds to one response (Fig. 1.2A). Each unit $i$ receives excitatory activity $\rho_i$ from an input layer that represents the external evidence or "support" for that particular response. In addition, each unit has an excitatory connection to itself ($\alpha$) and inhibitory connections ($\beta$) to all other units in that layer. The key premise of the model is that each unit takes up the external evidence in a cumulative and noisy manner while competing with the other units. It is important to keep in mind that this is an abstract description of a system of mutual constraints that does not impose a specific division between perceptual and motor processing. For example, each "unit" may be instantiated as a spatially dispersed pattern of neural activity that encompasses both perceptual and motor systems.

The dynamics of this system are governed by a set of stochastic nonlinear differential equations:

$$dx_i = [\rho_i - \lambda x_i + \alpha f_i - \beta \Sigma_{i \neq j} f_j] \cdot dt + \xi_i \sqrt{dt} \quad (5)$$

Here, $\rho_i$ is the external input to unit $i$, $x_i$ is the unit's instantaneous net input (its *current*), $f_i$ is the unit's instantaneous output (its *firing rate*; $f_j$ is likewise the output of unit $j$), $\alpha$ is the unit's self-excitation factor, $\lambda$ is the spontaneous decay

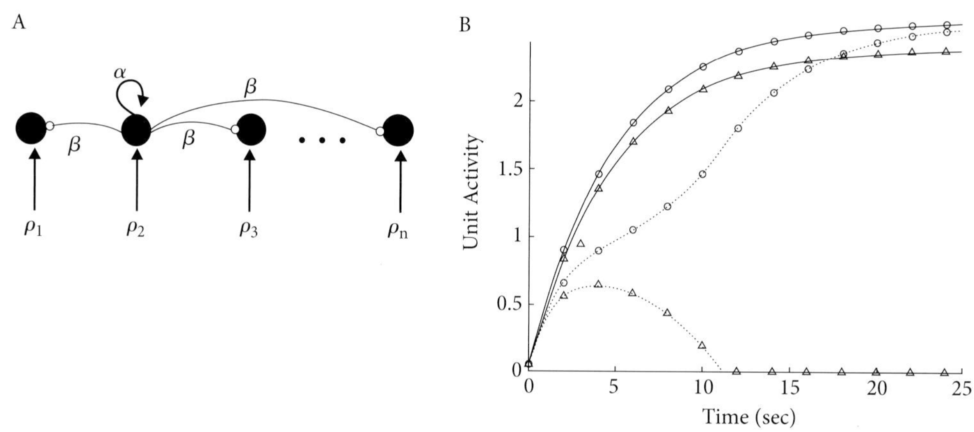


**Figure 1.2. Architecture of the Usher-McClelland model**
(A) For unit $i$, $\alpha$ and $\beta$ are the self-excitation and inhibition weights, respectively; $\rho_i$ is the external input to unit $i$. For clarity, inhibitory connections are only shown for a single unit. (B) Graph depicts the unit activities $x_1$ and $x_2$ (○ and Δ, respectively) when $\beta = 0$ (solid lines) and $\beta = 0.4$ (dotted line). In all cases, $k = 0.2$, $\rho_1 = 0.52$, $\xi_1 = \xi_2 = 0$, and $dt = 0.1$.

rate of the unit's net input, $\beta$ is the strength of the inhibitory connection between units, and $\xi_i$ is the standard deviation of the integration noise. For simplicity, Usher and McClelland assumed that $x_i = \max(x_i, 0)$, and that $f_i = x_i$ which reduces equation (5) to the following linear equation:

$$dx_i = [\rho_i - kx_i - \beta\Sigma_{i\neq j}x_j]\cdot dt + \xi_i\sqrt{dt} \quad (6)$$

Here, $k \equiv \lambda - \alpha$. It should be noted that while equation (6) is linear, the overall system is nonlinear because of the floor on the activity of $x_i$. The behavior of this system can be understood by considering a network with two units with the added constraint that $\rho_1 + \rho_2 = 1.0$. The state of the network then depends on three central parameters: $k$, the net amount of activity leakage; $\beta$, the interunit inhibition level; and $\rho_1$, the external input to unit 1. Figure 1.2B depicts the dynamics of the units in two sample regimes (both with $k = 0.2$ and $\rho_1 = 0.52$). With $\beta = 0.0$ there is no inhibition, so the two units find a balance between the leakage current and the sustained external input. This results in a slight advantage for $x_1$ because $\rho_1 > \rho_2$. With $\beta = 0.4$, the lateral inhibition results in competitive interaction that depresses both $x_1$ and $x_2$. However, once $x_1$ overcomes $x_2$ (at $t \approx 11$ s.), its rate of increase is similar to that in the $\beta = 0.0$ condition.

The main point to take from Figure 1.2B is that the system's dynamics tend toward a stable state, with the precise nature of that state being governed by the three central parameters. In order to map these stable states, we consider the dynamics of $\upsilon \equiv x_1 - x_2$. For each parameter setting, we numerically integrated equation (6) for 25 seconds ($\xi_1 = \xi_2 = 0$) with initial conditions ranging from $\upsilon = -5.0$ to $\upsilon = +5.0$. From these trajectories, we calculated $d\upsilon/dt$ and estimated the locations of the fixed points of $\upsilon$ for that parameter setting. Figure 1.3 depicts the resulting bifurcation diagrams for various values of $\rho_1$, $k$, and $\beta$. The diagrams show that the system has a wide variety of dynamics. For $k \leq 0$, there is no leakage, so both units' activities continually increase (the sign of $\upsilon$ is determined by the unit with the faster increase). As a result, the system only has unstable fixed points.

Once $k > 0$, however, the net leakage dominates the external input and restricts the overall levels of $\upsilon$. Furthermore, the final state of the system is sensitive to the initial condition $\upsilon_0$. If inhibition is weak, the input $\rho_1$ allows one unit to quickly overwhelm the other, even if the other unit has a slight initial advantage (see, for example, $k = +0.2$, $\beta \leq 0.2$). If the inhibition is strong, however, the initial state $\upsilon_0$ allows a unit to establish a decisive early advantage. For example, let $k = +0.2$, $\beta \geq 0.4$, $\rho_1 = 0.6$, and $\upsilon_0 = -2$. Unit 1 has a slight input advantage, but the $\upsilon_0$ allows unit 2 to suppress unit 1's activity and win the competition. On the other hand, if the $\upsilon_0 \geq 0$ (or some critical value $\upsilon^*$) then obviously unit 1 will win the competition. Thus, there is *bistability* in the dynamics. If the inhibition is even greater (e.g., $\beta = 0.6$), the early advantage is established more rapidly; for instance, an input $\rho_1$ is more likely to be overcome by an initial bias $\upsilon_0$. Consequently, the bistable regime becomes wider as the inhibition $\beta$ increases. The bistability also implies that the system is capable of *hysteresis*. Say $\rho_1$ is initially equal to zero, so that the system settles at $\upsilon^-$. If $\rho_1$ is then gradually ramped-up, the system will stay at $\upsilon^-$ until the end of the bistable regime, at which point it will snap to $\upsilon^+$. If $\rho_1$ is then gradually decreased, the system will stay at $\upsilon^+$ during the bistable regime, indicating path dependence in the system dynamics.

A key point to keep in mind here is that the Usher-McClelland model was not designed to produce the dynamics in Figure 1.3. Its purpose was to account for ensemble statistics of categorization performance using the principles of interactive, stochastic, and graded processing (Usher & McClelland, 2001); the dynamical properties fell out as one consequence of these principles. Furthermore, if equation (6) is augmented with an additional depression term (so that a unit is less able to compete if it has recently been active), its dynamics become remarkably similar to those of an explicitly dynamical model of categorization (Ditzinger, Tuller, Haken, & Kelso, 1997; Tuller, Case, Ding, & Kelso, 1994). This kinship suggests that the PDP approach (as instantiated in the Usher-McClelland model) and the dynamical approach (as instantiated through

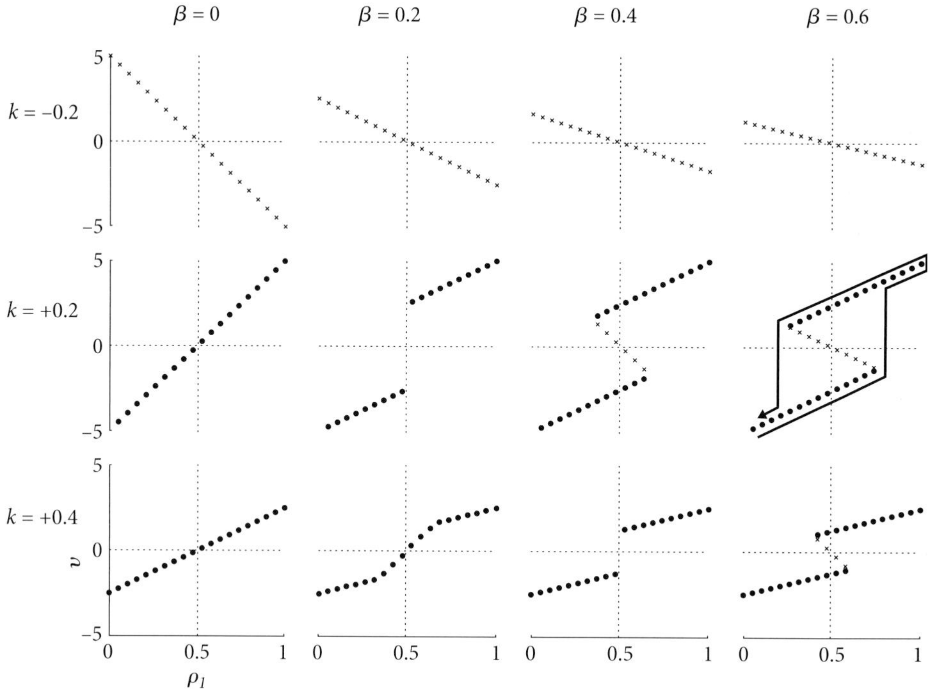


**Figure 1.3. Bifurcation diagrams for the Usher-McClelland model**
Bifurcation diagrams depict the stable and unstable fixed points (● and x, respectively) for the dynamics of $\upsilon$ ($\upsilon \equiv x_1 - x_2$). In all cases $\xi_1 = \xi_2 = 0$ and $dt = 0.1$. When there are two stable fixed points for a parameter setting, they are referred to as $\upsilon^-$ and $\upsilon^+$. The solid line for $k = +0.2$ and $\beta = 0.6$ depicts the system states as $\rho_1$ is ramped from 0 to 1 and then ramped back down to 0. The state at $\rho_1 = 0.5$ is $\upsilon^-$ on the upward ramp and $\upsilon^+$ on the downward ramp, indicating hysteresis.

the dynamic field or synergetic approaches, see Chapter 2) are in fact closely related approaches to the characterization of response dynamics.

## Developmental Emergent Dynamics I: Perceptual Learning

There are two broad ways in which developmental dynamics have been explored within the connectionist approach, one based on supervised error-driven training and the other based on unsupervised Hebbian learning. The power of the former method is that it can re-represent the input in sophisticated ways to match the task at hand (for example, a network may assign finer grain representations to salient input and coarser grain ones to unimportant or rare input). However, in many cases it is not plausible to assume that outcome information is consistently available, and unsupervised learning based on Hebbian learning can provide insight into such problems (Petrov, Dosher, & Liu, 2005; Rumelhart & Zipser, 1985). In addition, there is substantial evidence for the biological plausibility of Hebbian learning (Ahissar, Abeles, Ahissar, Haidarliu, & Vaadia, 1998; Syka, 2002). Finally, Hebbian learning with bidirectional weights typically results in symmetrical connections that facilitate interactive attractor dynamics (Grossberg, 1988).

The relation between Hebbian learning and developmental dynamics was explored in a model proposed by Vallabha and McClelland (2007). The motive for the model was to explore how perceptual attractors can help establish speech categories in infancy while impeding the acquisition of speech categories in adulthood. Here we present a simplified version of this model that focuses on the emergence of a *perceptual magnet effect* as a result of perceptual experience (Kuhl, 1991). The effect is marked by a decrease in discriminability between adjacent stimuli as they get closer to the category prototype, and the effect is developmental as it only shows up for sound categories that are distinguished in the native language (Iverson et al., 2003). We will focus on how exposure to native-like sound structure results in the formation of perceptual categories, and how these in turn affect discrimination, producing a perceptual magnet effect. Below we describe how the Vallabha-McClelland model addresses this issue.

The model consists of two layers of units: L1 and L2 (Fig. 1.4), with 80 units in L1 and 2 units in L2. Each unit has excitatory connections with every unit in the other layer and inhibitory connections with all other units in its own layer. The pattern of activity over L1 is taken to be the *perceptual representation*, and the activity in L2 is taken to be the *category representation*, with each unit standing for a distinct response category. The dynamics for the units, represented in equations (7) and (8), are similar to those for the Usher-McClelland model:

$$dx_i = \left[\rho_i - x_i + \Sigma_i wscale_{ij} \cdot w_{ij} f_j\right] \cdot dt + \xi\sqrt{dt} \quad (7)$$

$$f_i = \min(0, \tanh(x_i \cdot gain_{act})) \quad (8)$$

Here $w_{ij}$ is the weight to unit $i$ from unit $j$, $gain_{act}$ is the gain of the activation function, and $wscale_{ij}$ is a *weight scaling* parameter set to 5.0 for L2→L1 weights and to 1.0 for all other weights. [In equation (7), the parameter $wscale_{ij}$ simulates the effect of a group of similarly wired L2 units that act in concert.]

Each external input consists of a Gaussian bump of activity over the L1 units, specified by $\rho_i \equiv 0.8 \exp[-(i - x)^2/17]$. The center of the bump, $x$, is designated as the location of that particular input, and it can range from 1 to 80 (the number of units in L1). For current purposes, the input locations are drawn from two Gaussian distributions, $N(29,3)$ and $N(51,3)$, with 400 samples from each distribution. On each trial, one input location was chosen at random from the 800 samples and the corresponding bump of activity was presented to L1. The network was then allowed to settle for 30 time steps

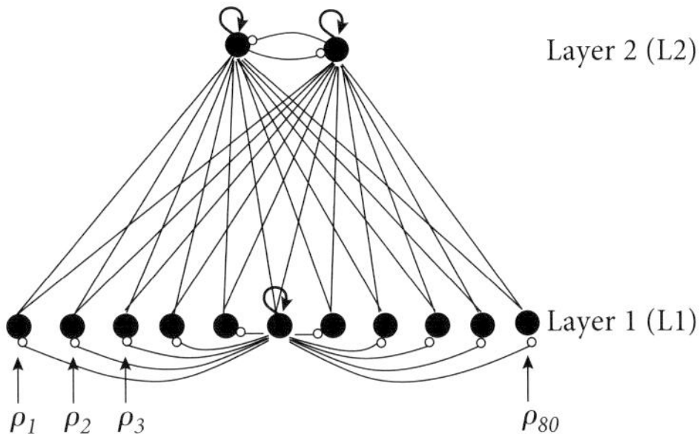


**Figure 1.4. Architecture of the Vallabha-McClelland model**
Units in L1 and L2 are fully connected to each other, with the L1→L2 weights initialized from a Uniform distribution between 0 and 0.03, and the L2→L1 weights initialized from a Uniform distribution between 0 and 0.0005. Each unit has a self-connection with a weight of +1.0. The within-layer inhibitory connections have a weight of −0.2 for L1 units and −2.0 for L2 units.

with $dt = 0.2$, $\xi = 0.04$, and $gain_{act} = 0.5$, with the input being presented the entire time. Once the settling was completed, the weights between L1 and L2 were updated using a Hebbian rule, $\Delta w_{ij} = \eta \cdot f_i \cdot f_j$. Following the update, the magnitude of the weight vectors was adjusted as follows. Let $W$ be the vector of weights from L1 to a particular unit in L2. At the start of training, $|W|$ (the absolute value of $W$) was allowed to grow freely, as $|W|$ increased, the increases in $|W|$ became smaller so that $|W|$ rose asymptotically toward a limiting value of 1.0. This *gradual normalization* allowed the network to start with small random weights and gradually discover a good solution while limiting the growth of the individual connection weights. In order to ensure that the L1→L2 and L2→L1 weights were approximately symmetric, the normalization was done over incoming weights for L1→L2 projections, and over outgoing weights for L2→L1 projections (Grossberg, 1988).

The above dynamics resulted in competitive learning between the units in L2 (cf., Carpenter & Grossberg, 1987; Rumelhart & Zipser, 1985). For example, one L2 unit developed strong connections to L1 units around input location 29 (the center of one of the input distributions), with input locations that are active more often getting higher connection strengths than locations that are only occasionally active. In doing so, the other L2 unit was inhibited from becoming sensitive to those same input locations. Now consider what happens when an input stimulus is presented to the network. The stimulus causes a small bump of activity on L1. Due to the within-L1 interactions, this bump coalesces and becomes more prominent. Concurrently, it activates the L2 unit that is sensitized to that input location. The L2 unit recurrently excites the input locations that are most frequently active for that category; for example, the L2 unit excites the prototypical L1 representation for that category. As a result, the L1 activity becomes skewed toward the category prototype. Thus, the categorization ability of the network (reflected in L2 activity) changes the perceptual representations (reflected in L1 activity) and shapes the discrimination of input stimuli.

Figure 1.5 illustrates the above process by showing the evolution of L1 activity at different stages of training or development. The key point to note is the *skew* of the L1 activity at the final time step of the settling. Before training, the skew is negligible. After 1500 weight updates, the final activities are noticeably skewed toward the center of the category. A consequence of this skew is that the final L1 activities for adjacent input stimuli (e.g., at input locations 25 and 27) become more similar to each other. If this similarity, as measured by overlap or by Euclidean distance, is used as a measure of discriminability, then we get the perceptual magnet effect. After extensive exposure to exemplars of a category, more prototypical stimuli are harder to distinguish than less prototypical ones (because the

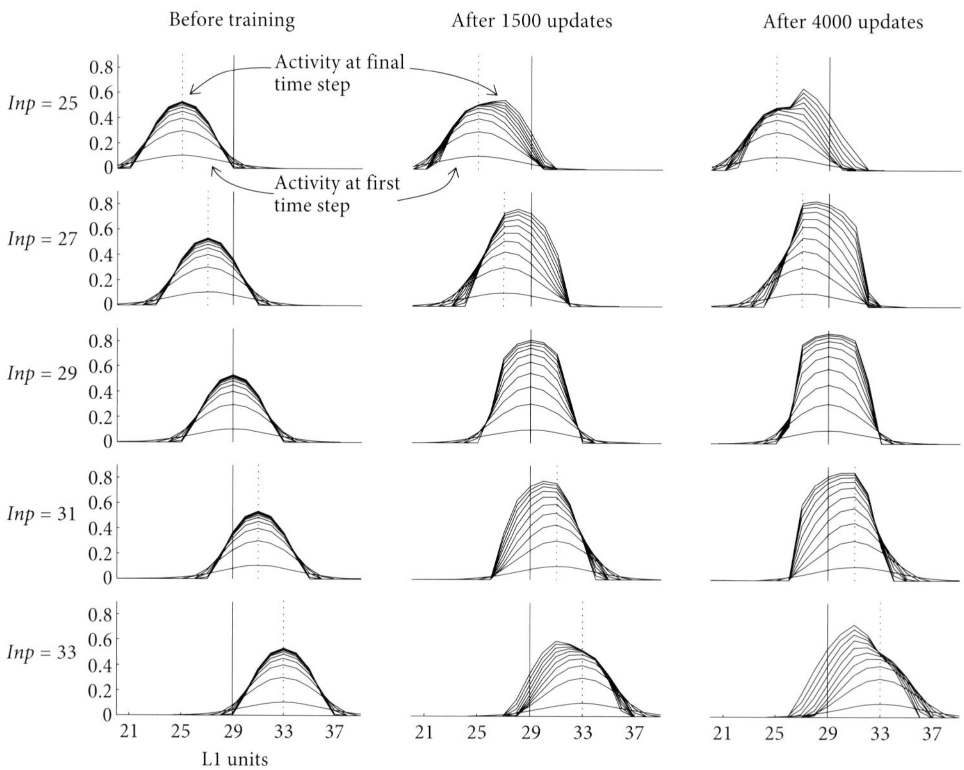


**Figure 1.5. Layer 1 activity at different stages of training in the Vallabha-McClelland model**
Each pane depicts the evolution of L1 activity for one input location. *Inp* = $x$ indicates that the input is centered on L1 unit $x$ (indicated by a vertical dotted line). A vertical solid line indicates the center of the category (input location 29).

amount of skew is greater with the former than with the latter).

The developmental dynamics of learning may be visualized through the relation between the input location and the amount of skew. We calculated the skew using equation (9). For each input location $k$, we presented an input centered at $k$ and ran the network for 30 time steps. Then we took the final L1 activity vector $y^{(k)}$ and calculated its *center of mass* $c_k$.

$$c_k = \sum_{i=1}^{80} i \cdot \left( y_i^{(k)} / \Sigma_j y_j^{(k)} \right) \qquad (9)$$

The amount and direction of skew are indexed by $c_k - k$ (rightward and leftward skews are indicated by positive and negative values, respectively). Figure 1.6A shows the skew over the input space at different stages of training. The centers of the distributions (input locations 29 and 51) function as dynamical attractors, and the center of the space (input location 40) functions as an unstable fixed point. In fact, if we use the number of updates as a control variable, we induce a pitchfork bifurcation over the representational space. Furthermore, if we calculate the skew in L1 representation at each time step (rather than just the final time step), an interesting pattern emerges. Figure 1.6B shows the skew for input location 26 as a function of processing stage (the number of time steps) *and* developmental stage (the number of weight updates). The effect of training is to accelerate the response dynamics: a skew of 0.5 took 16 time steps to develop after 1500 updates, 13 time steps after 2000 updates, and only 10 time steps after 4000 updates.

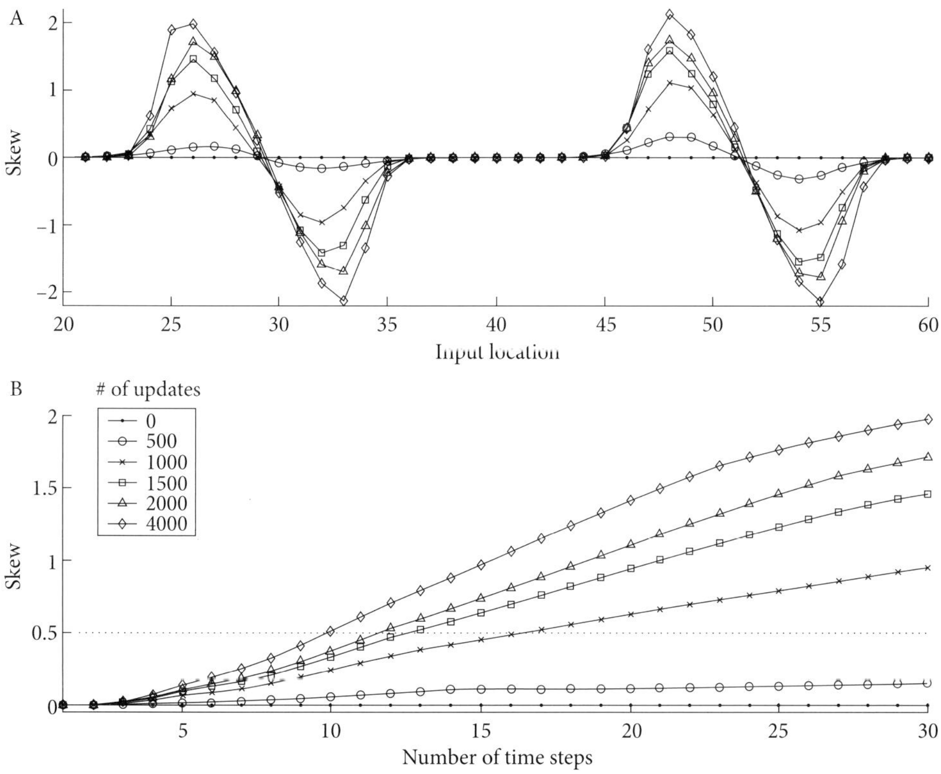


**Figure 1.6. Developmental "phase plot" for the Vallabha-McClelland model**
(A) For each input location $k$, skew is depicted at different stages of training (skew $\equiv x - c_k$; with $c_k$ as the center of mass at the final time step). See equation (9) and text for details. (B) The development of skew for input location 26 is depicted as a function of processing time (the number of time steps) and training (the number of weight updates). The horizontal dotted line illustrates that a particular level of skew is achieved more rapidly with training.

Figure 1.6B suggests how response dynamics and developmental dynamics may be linked. A system without learning (such as the Usher-McClelland model) displays a rich variety of short-term emergent dynamics (Fig. 1.3). Some of these dynamics, and correspondingly some values of $k$ and $\beta$ (the net amount of activity leakage and the interunit inhibition level), may be particularly relevant for accomplishing a task such as categorizing an input stimulus. The subsequent Hebbian learning facilitates just those task-relevant dynamics by entrenching the current values of the parameters. These entrenched parameters influence the outcomes of future trials (e.g., by determining the dominant attractor for a particular input); these outcomes shape subsequent learning and so on. Thus, response dynamics and developmental dynamics are different timescales of emergent activity that can coexist in the same system. Furthermore, the linkage between the two scales—the developmental effects facilitate certain response dynamics, which in turn shape further development—lead to a kind of reciprocal causality. One consequence of this linkage is that the developmental changes need not be imposed on the system (by stipulating a change in learning rate or in lateral-interaction process, for example), but rather can emerge from the operation of the system.

## Developmental Emergent Dynamics II: Semantic Learning

We now turn to a model of conceptual knowledge acquisition introduced by Rumelhart (Rumelhart, 1990; Rumelhart & Todd, 1993) and studied in detail by Rogers and McClelland (2004). This model simplifies the activation dynamics down to a single deterministic pass through a feed-forward network, and also uses the back-propagation algorithm to adjust the connection weights. Models such as these do have limitations that have perhaps impeded their acceptance by some dynamical systems theorists who see them as failing to incorporate certain key principles. However, we regard these limitations as simplifications that allow one to explore two abilities of such models: (1) learning structured representations that are sensitive to the structure of the training environment, and (2) demonstrating interesting dynamical properties as they move from an initially naïve unstructured state toward a state fully sensitive to the experiential structure.

Rogers and McClelland (2004) explored these issues in a domain initially explored by Rumelhart (1990). Rumelhart's initial goal was to study how a conceptual hierarchy like that in Figure 1.7 may be acquired through graded and distributed representations. The figure specifies sets of three-part propositions: for example, "living thing can grow" and "living thing is living." An animal has all the properties of a living thing: an "animal can grow" and an "animal is living". An animal has more properties besides: an "animal can move" and an "animal has skin." Note that widely different objects can have similar properties: for instance, a "sunfish is yellow" and a "daisy is yellow." The model only gets experience with the bottommost layers of this conceptual tree: with concrete facts such as "a sunfish is yellow" and "an oak is tall." Rumelhart was able to show that the hierarchical relations between the encountered objects can be acquired in quite a simplified multilayered network. Here we focus primarily on the developmental course of these changes, as explored by Rogers and McClelland (2004).

The network consists of a series of nonlinear processing units organized into layers and connected in a feed-forward manner as shown in Figure 1.8. It may be taken as a simplified model of experience with objects in the world and spoken statements about these objects. The *Item* layer is a simplified proxy for an input representation of an object as encountered in experience; the *Relation* layer is a simplified specification of the context in which the item is encountered (e.g., the *can* relation corresponds to a context in which the behaviors of the object might be observed); and the *Attribute* layer may be thought of as representing the consequences following from the occurrence of the object in the given context. When presented with a particular *Item* and *Relation* pair in the input, the network's task is to turn on the *Attribute* units in the output that correspond to valid completions

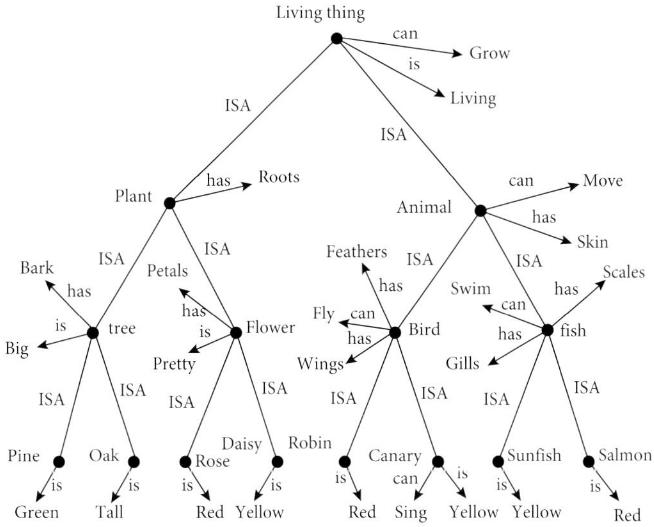


**Figure 1.7. Training data for the Rumelhart network**
Each line indicates a relation (e.g., "living thing can grow"). ISA relations indicate that the child node has all the relational attributes of the parent, allowing propositions such as "animal can grow," "bird can grow," and "bird can move."

of the proposition. For example, when the units corresponding to *canary* and *can* are activated in the input, the network must learn to activate the output units *move*, *grow*, *fly*, and *sing*. Patterns are presented by activating one unit in each of the *Item* and *Relation* layers (i.e., these activations are set to 1 and activations of all other input units are set to 0). Activation then feeds forward through the network, layer by layer. To update the activation of a unit, its net input is calculated [equation (2)] and transformed into an activation by the logistic function. Each target state consists of a pattern of 1s and 0s like the one shown for the input *canary can* in Figure 1.8—the target values for the black units are 1 and for all other units they are 0.

Initially, the connection weights in the network have small random values, so that the activations produced by a given input are weak and random in relation to the target output values. To find an appropriate set of connection weights, the model is trained with the backpropagation algorithm (Rumelhart et al., 1986). Training consisted of a set of epochs, each encompassing the presentation of every three-part proposition in the training set. On each trial, the *Item* and *Relation* are presented to the network as input, the resulting output states are compared to the target values, and the error information is "propagated" backward through the network. Each weight is adjusted slightly to reduce the error, with weights responsible for more error receiving larger adjustments. Overall the weights adapt slowly, yielding gradual evolution of the patterns of activation at each level of the network and gradual reduction of error. Crucially, the procedure also adjusts the weights from the *Item* layer to the *Representation* layer (see Fig. 1.8). Hence, each item is mapped to an internal representation, which is a distributed pattern of activity and which changes gradually over the course of learning. The manner and direction of this change provides an index of how the semantic knowledge is being acquired, and is therefore of central importance.

As the training progresses, the network gradually adjusts its weights to capture the semantic similarity relations that exist among the items in the training environment. Figure 1.9 shows the representation for eight-item input at three points in learning. Initially, the patterns representing the items are all very similar, with activations hovering around 0.5. At epoch 100, the patterns corresponding to various animal instances are similar

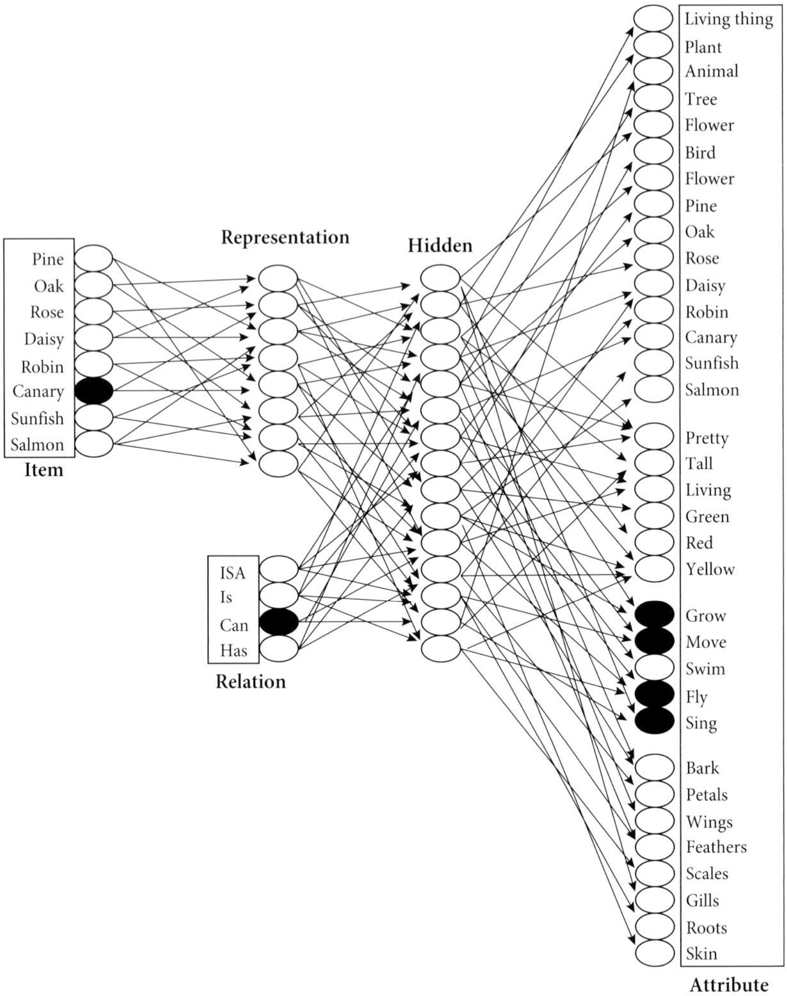


**Figure 1.8. The Rumelhart model**
This schematic depicts how the *Item*, *Relation*, and *Attribute* layers instantiate the propositions (e.g., "canary can grow," "canary can fly") within the network. A similar model was used for the simulations in Rogers and McClelland (2004).

to one another but are distinct from the plants. At epoch 150, items from the same intermediate cluster, such as *rose* and *daisy*, have similar but distinguishable patterns and are now easily differentiated from their nearest neighbors (i.e., *pine* and *oak*). Thus, each item develops a unique representation, but semantic relations are preserved in the similarity structure across representations.

In order to visualize the conceptual differentiation, Rogers and McClelland performed a multidimensional scaling of the representations for all items at ten equally spaced points during the first 1500 epochs of training in a replication of the above simulation. Specifically, the *Representation* vector for each item at each point in training was treated as a vector in an eight-dimensional space, and the Euclidean distances were calculated between all vectors at all points over development. Each vector was then assigned a two-dimensional coordinate such that distances between points in the two-dimensional space were as similar as possible to distances in the original eight-dimensional space. The solution is plotted in Figure 1.10. Note that the trajectories

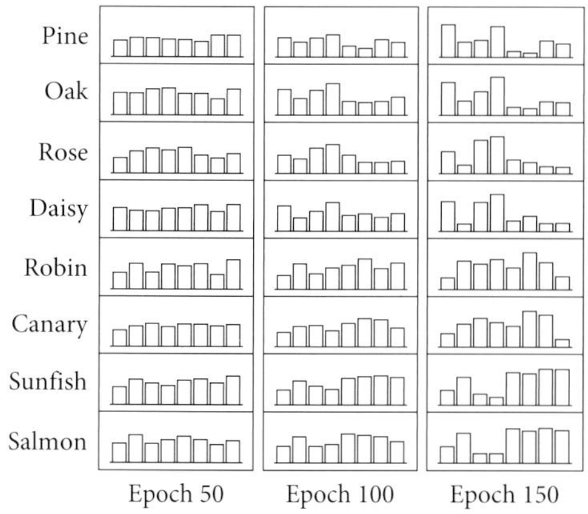


**Figure 1.9. Differentiation of the representations at three points in learning**
The panels shows the activity of the Representation layer after 50, 100 and 150 epochs of training. Each set of bars within a panel depicts the activations of the representation units in response to a particular Item input (*pine, oak,* etc.); the height of each bar depicts the activation level of a single representation unit.

are not straight lines. The items, which initially are bunched together in the middle of the space, first divide into two global clusters, one containing the plants and the other containing the animals. Next, the global categories split into smaller intermediate clusters, and finally the individual items are pulled apart. Thus, differentiation happens in relatively discrete stages, first occurring at the most general level before progressing to successively fine-grained levels.

Three aspects of the acquisition are pertinent to the issue of developmental emergence: the differentiation itself, the stagelike nature of the differentiation, and the relation between the differentiation and the statistical structure of the input. We will consider each in turn.

### Why Does the Differentiation Happen at All?

Consider how the network starts to learn about the following four objects: oak, pine, daisy, and salmon. Early in learning, when the weights are

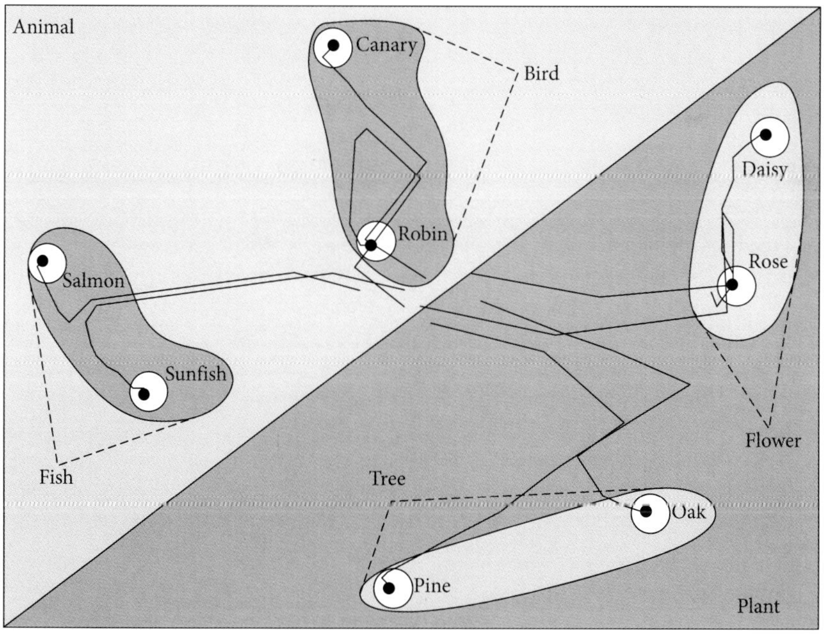


**Figure 1.10. Progress of differentiation during training**
The two-dimensional space is the projection of the eight-dimensional space of the *Representation* units. Lines trace the trajectory of each item through learning. The labeled end points of the lines represent the internal representations learned after 1500 epochs of training.

small and random, all inputs produce a similar meaningless pattern of activity throughout the network; in particular, the input for these four objects will produce similar representations, with only slight random differences. Since oaks and pines share many output properties, this pattern results in a similar error signal for the two items, and the weights leaving the *oak* and *pine* units move in similar directions. Because salmon shares few properties with oak and pine, the same initial pattern of output activations produces a different error signal, and the weights leaving the *salmon* input unit move in a different direction. What about the daisy? It shares more properties with oak and pine than it does with salmon, and so it tends to move in a similar direction as the other plants. Similarly, other animals tend to be pushed in the same direction as salmon. As a consequence, on the next pass, the pattern of activity across the representation units will remain similar for all plants, but will tend to differ between plants and animals.

This initial similarity of representations facilitates the subsequent learning. In particular, any weight change that captures shared attributes for one item will produce a benefit in capturing these attributes for other, related items. For example, weight changes that allow the network to better predict that a canary has skin and can move will improve the network's predictions for robin, sunfish, and salmon. On the other hand, weight changes that capture an idiosyncratic item attribute will tend to be detrimental for the other items. For example, two of the animals (canary and robin) can fly but not swim, and the other two (salmon and sunfish) can swim but not fly. If the four animals all have the same representation, what is right for half of the animals is wrong for the other half, and the weight changes across different patterns will tend to cancel each other out. In short, coherent covariation of attributes across items tends to accelerate learning (and change representations in the same direction), while idiosyncratic variation tends to hamper learning.

### Why Does the Differentiation Exhibit Stages?

The above description of differentiation suggests why the differentiation follows the category hierarchy: the attributes that distinguish high-level categories such as *animal* vary more consistently and are learned more quickly than attributes that distinguish low-level categories such as *fish* and *bird*. But this does not quite explain the stagelike learning where long periods of very gradual change are interleaved with sudden bursts of rapid change (Figs. 1.10 and 1.11). The key reason is that back-propagation of error is much stronger through weights that are already structured to perform useful forward-mappings.

This point is illustrated in Figure 1.11, which shows the network's output activity over the entire training, along with the magnitude of the error information reaching the *Representation* layer (see Fig. 1.8). Initially, there is little difference between the representations (Fig. 1.11C). The network first reduces error by modifying the *Hidden→Attribute* weights, and hence, little error information percolates down to the *Representation* layer (Fig. 1.11B). Furthermore, since the representations are initially all alike, the error information is not useful for developing fine distinctions; for example, the error for the attribute *fly* is likely to change the representations for pine and daisy as well. This situation is not a complete impasse, since the error information about the largest distinction—plants versus animals—does percolate through and accumulate at the *Representation* layer (see Fig. 1.8). (As noted above, this distinction is picked up first because of the coherent covariation of the attributes with *plant* and *animal*.) By around 800 epochs, the representations start to differentiate plants and animals, which allows them to better predict the plant versus animal attributes, which allows more error information to percolate down, which encourages further differentiation. Thus, there is an accelerated differentiation of the plant versus animal representations, leading to an increase in distance (Fig. 1.11C). After this reorganization, the error for the *swim* and *fly* attributes is usefully applied to modifying *animal* representations. Hence, the entire process (slow percolation of error to the *Representation* layer, followed by an accelerated differentiation)

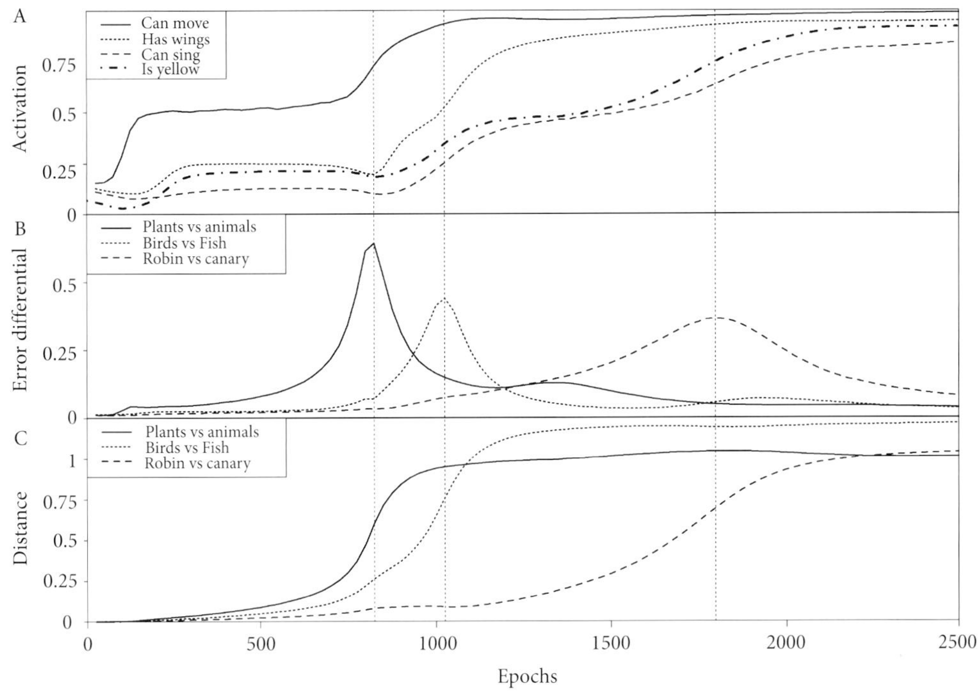


**Figure 1.11. Details of the differentiation process**
(A). Activation of the output units (*move, wings, sing,* and *yellow*) for the input item *canary*; (B). Mean error reaching the *Representation* layer from output units that reliably distinguish *plants* versus *animals*, *birds* versus *fish*, and *robin* versus *canary*; (C). Distance between the plants and the animals, between the birds and the fish, and between the two bird representations.

repeats for finer distinctions such as *bird* versus *fish*.

Figure 1.11 also indicates that the rapid learning of coherently covarying properties is not solely driven by frequency of occurrence. In this training corpus, the attribute *yellow* is true of three objects (canary, daisy, and sunfish), whereas the attribute *wings* is true of only two (robin and canary). Nevertheless, the network learns that the canary has wings more rapidly than it learns that the canary is yellow (Fig. 1.11A). The reason is that *wings* varies coherently with several other attributes (*bird, fly, feathers*), which allows learning about the robin to generalize to the canary (and vice versa), whereas *yellow* is more idiosyncratic.

## Relation between the Differentiation and the Statistical Structure of the Input

The timing of the developmental changes noted above are not "built into" the model, but rather are shaped by the higher-order covariation among the attributes of items in its environment. This pattern of covariation is indicated in Figure 1.12, which shows three eigenvectors of the property covariance matrix. The first eigenvector weights attributes, such as *roots* and *move*, which covary most robustly and discriminate *plant* versus *animal*. The second vector weights attributes that covary less robustly and distinguish *fish* versus *bird*, and the third vector picks out attributes that distinguish *tree* versus *flower*. Thus, the properties to which the model first becomes sensitive, and which organize its nascent conceptual distinctions, are precisely those that consistently vary together across contexts. Furthermore, each level of covariation only maps systematically to the system's internal representations after the prior stage of differentiation has occurred. Consequently, the timing of different waves of differentiation and the particular groupings

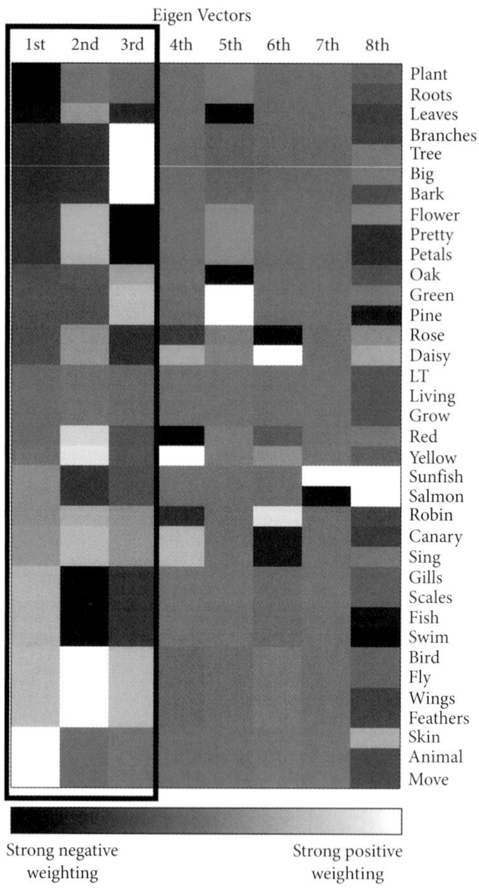


**Figure 1.12. Eigenvectors of the property covariance matrix**

Eigenvectors were formed by calculating the correlation of attributes across the different items in the model's training environment. Each column is one eigenvector. All eigenvectors were normalized to span the same range and ordered by amount of variance explained (left to right represents most to least). The first eigenvector distinguishes plants from animals, the second fish from birds, and the third trees from flowers. In the right-hand column, LT stands for the category *living thing.*

of internal representations that result are governed by the statistical structure of the training environment.

Some of the dynamical properties of the Rogers and McClelland simulation are also observable in other simulations, including McClelland's (1989) simulation of how children learn the two factors (weight and distance from the fulcrum) that affect the functioning of a balance scale. In that simulation, connections also exhibited a tendency to adapt very gradually at first, then to undergo a rapid acceleration, then to stop changing as error was finally eliminated—as with the Rogers and McClelland model, the connection weights had to be already partially organized in order to benefit from experience. The simulation also exhibited the differences in readiness to learn shown by children of different ages (Siegler & Klahr, 1982). Specifically, a network that had just become sensitive to the weight cue did not benefit when exposed to problems that involved distance, and in fact regressed in its performance. However, a network that had more experience with the weight cue benefited from the distance problems and progressed quickly. Thus, changes in connection weights affect both the subsequent processing and learning dynamics, leading to non-linearities in sensitivity to experience.

## Discussion

In the preceding sections, we reviewed three types of connectionist models. The first model shows instabilities and attractors emerging from a simple connectivity pattern (viz., interactive activation) and demonstrates the short-term emergent dynamics implicit in networks. The second model incorporates Hebbian learning into the above scheme and demonstrates that long-term developmental dynamics emerge naturally in such a system. These long-term dynamics are difficult to characterize analytically since they are shaped by the particular training history of the network; even so, they exhibit characteristics of dynamical systems (Fig. 1.6). The third model shows that error-correcting learning also leads to emergent developmental patterns. Specifically, the error correction does not lead to gradual and uniform changes in the representations, but rather results in waves of differentiation intimately tied to the statistical structure of the input (Figs. 1.11 and 1.12). Based on these results, how should one conceive of the relation between connectionist models and those proposed by dynamical systems theorists?

A first point to note is that we regard models of the above sort as *explorations*—examinations of the implications of certain sets of assumptions—rather than statements of belief. For example, the Usher-McClelland model explored how interactive activation may account for reaction time phenomena and the Vallabha-McClelland model explored how Hebbian learning may account for auditory learning. However, these explorations do not preclude the possibility that other mechanistic assumptions can give rise to similar system-level behavior.

With the above caveat firmly in place, we suggest that connectionist and other dynamical theories may be clustered into three overlapping classes. One class of theories, which may be exemplified by the work of van Geert (see Chapter 16), directly model cognitive phenomena using dynamical equations. Such theories do not show how the emergent dynamics arise from the mechanistic dynamics and, in fact, may leave out the mechanistic dynamics entirely. Consequently, they may be seen as a compact dynamical description of the phenomena. A second class of theories seek to explain how complex phenomena such as stability and entrainment arise from simple nonlinear interactions. However, this emergence is characterized at a short-term timescale, and developmental changes (if any) are imposed by the modeler rather than arising from within the system. In particular, there is no account of the interaction between short-term dynamics and developmental dynamics. Examples of such models include the work of Schöner and Kelso (1988), the dynamic field model of Schutte and colleagues (Schutte et al., 2003), and the reaction-time model by Usher and McClelland (2001). It is worth emphasizing that the focus on short-term dynamics is usually due to modeling expediency, and should not be taken as a theoretical claim that developmental dynamics are outside the scope of a dynamical theory (see Kelso, Jirsa, & Fuchs, 1999, for an instance of a more integrated dynamical account).

Theories of the third class attempt to show how mechanistic, short-term, and developmental dynamics interact with and influence each other. Many connectionist models, such as Vallabha and McClelland (2007) and the LEABRA model (O'Reilly & Munakata, 2000), are conceptually in this class. However, such an explicit linkage of levels is not present in every connectionist model; as noted earlier, some connectionist models simplify the mechanistic activation dynamics while others simplify the weight-change dynamics. One reason for the simplification is computational feasibility. Doing a continuous, recurrent back-propagation (as in Plaut et al., 1996, simulation 3) is computationally very expensive, and training proceeds much more slowly than in a nonrecurrent network of comparable size. A second and more important reason is that a detailed model of microdynamics can sometimes obscure the key characteristics of the problem. For example, a neurotransmitter-level model with several timescales of learning can obscure the point that stagelike learning does not require fluctuations in the learning rates, but can result from the error-based learning itself. In other words, the level of the model should be pitched to the problem under study. The key is to get the right dynamics in the simplification, and theoretical analyses are crucial in this regard. For example, Wong and Wang (2006) were able to analytically reduce the dynamics of 2000 spiking neurons in the monkey lateral intraparietal cortex to a simple two-variable model similar to that depicted in Figure 1.2A. It is worth noting that such reductions can also be applied to dynamical theories (e.g., Jirsa & Haken, 1997).

It should be noted that the use of simplifications may limit the ability to address some fine-grained aspects of the phenomena under consideration while increasing the ability to address other aspects at a coarser grain. For example, McClelland (1989) and Rogers and McClelland (2004) used a one-pass feedforward network trained by back-propagation to simulate stagelike patterns (e.g., accelerations and plateaus) seen on a long-term, developmental timescale. However, these networks failed to exhibit some characteristic hallmarks of stage transitions, namely increased variability in the region during the transitional period. We suggest that models of this type would naturally exhibit such characteristics if they used continuous, stochastic activation as well as recurrent connectivity, as in the models of Usher and McClelland (2001) and Vallabha and McClelland (2007). As we showed in our simulations, these models can exhibit considerable variability, especially when weak stimuli and weak connectivity support newly emerging attractors. The larger point here is that connectionist simplifications sometimes make it difficult to capture behavioral regularities that are central in dynamical accounts. Replacing these simplifications using models that capture the mechanistic dynamics in more detail would be a useful step toward integrating connectionist models with explicitly dynamical accounts of cognition. Such models may also address some of the criticisms of existing models (e.g., van der Maas, Chapter 15 of this volume).

In conclusion, in our view connectionist and dynamical approaches address a similar question: How do varied forms of cognitive organization emerge from the underlying physical substrate? The terminologies and the emphases are different—for example, dynamical system researchers tend to emphasize the mechanical constraints imposed by the organism's body, while connectionists tend to consider such constraints as properties of the physical elements within the nervous system (such as neurons and neural connections, or at least abstractions of their properties). Likewise, explicitly dynamical models address the constraint satisfaction using dynamical metaphors such as coupling and stability, while connectionist models address it using neuronal metaphors such as propagation of unit activity and weight change. However, these are differences of emphasis that should be seen as complementary rather than competing. For example, within connectionist approaches, consideration of dynamics at many scales is essential since this allows for gradual integration over constraints at various spatial and temporal scales (McClelland & Rumelhart, 1981; Spivey & Dale, 2005). Thus, we view the two approaches as converging on the same problem from different sides, and we hope that the insights of each approach will continue to inform the ongoing development of the other.

## Acknowledgments

Preparation of this chapter was supported in part by Program Project Grant MH 64445 (J. McClelland, Program Director). The authors thank the participants in the Meeting on Dynamical Systems and Connectionism (University of Iowa, July 2005) for useful discussions of the version of this work presented in spoken form at the meeting. Correspondence concerning this article should be addressed to Dr. McClelland, Department of Psychology, Center for Mind Brain and Computation, Stanford University, Palo Alto, CA.

## References

Ahissar, E., Abeles, M., Ahissar, M., Haidarliu, S., & Vaadia, E. (1998). Hebbian-like functional plasticity in the auditory cortex of the behaving monkey. *Neuropharmacology, 37*(4–5), 633–655.

Anderson, J. A. (1973). A theory for the recognition of items from short memorized lists. *Psychological Review, 80*(6), 417–438.

Carpenter, G. A., & Grossberg, S. (1987). A massively parallel architecture for a self-organizing neural pattern recognition machine. *Computer Vision, Graphics, and Image Processing, 37*(1), 54–115.

Cohen, J. D., Dunbar, K., & McClelland, J. L. (1990). On the control of automatic processes: A parallel distributed processing account of the Stroop effect. *Psychological Review, 97*(3), 332–361.

Ditzinger, T., Tuller, B., Haken, H., & Kelso, J. A. S. (1997). A synergetic model for the verbal transformation effect. *Biological Cybernetics, 77*(1), 31–40.

Elman, J. L. (1990). Finding structure in time. *Cognitive Science, 14*(2), 179–211.

Elman, J. L., Bates, E. A., Johnson, M. H., Karmiloff-Smith, A., Parisi, D., & Plunkett, K. (1996). *Rethinking innateness: A connectionist perspective on development.* Cambridge, MA: MIT Press–Bradford Books.

Grossberg, S. (1976). Adaptive pattern classification and universal recoding, I: Parallel development and coding of neural feature detectors. *Biological Cybernetics*, 1976, 23, 121–134. CITED in Chapter 11.

Grossberg, S. (1978a). A theory of human memory: Self-organization and performance of sensory-motor codes, maps, and plans. *Progress in Theoretical Biology*, *5*, 233–374.

Grossberg, S. (1978b). A theory of visual coding, memory, and development. In E. L. J. Leeuwenberg & H. F. J. M. Buffart (Eds.), *Formal theories of visual perception*. New York: Wiley.

Grossberg, S. (1988). Nonlinear neural networks: Principles, mechanisms, and architectures. *Neural Networks*, *1*, 17–61.

Haxby, J. V., Gobbini, M. I., Furey, M. L., Ishai, A., Schouten, J. L., & Pietrini, P. (2001). Distributed and overlapping representations of faces and objects in ventral temporal cortex. *Science*, *293*(5539), 2425–2430.

Hopfield, J. J. (1982). Neural networks and physical systems with emergent collective computational abilities. *Proceedings of the National Academy of Sciences, USA*, *79*(8), 2554–2558.

Iverson, P., Kuhl, P. K., Akahane-Yamada, R., Diesch, E., Tohkura, Y., Kettermann, A., et al. (2003). A perceptual interference account of acquisition difficulties for non-native phonemes. *Cognition*, *87*(1), B47–B57.

Jirsa, V. K., & Haken, H. (1997). A derivation of a macroscopic field theory of the brain from the quasi-microscopic neural dynamics. *Physica D: Nonlinear Phenomena*, *99*(4), 503–526.

Kawamoto, A. H., & Anderson, J. A. (1985). A neural network model of multistable perception. *Acta Psychologica*, *59*(1), 35–65.

Kelso, J. A. S., Fuchs, A., & Jirsa, V. K. (1999). Traversing scales of brain and behavioral organization: Part I. Concepts and experiments. In C. Uhl (Ed.), *Analysis of neurophysiological brain functioning* (pp. 73–89). New York: Springer.

Kohonen, T. (1982). Self-organized formation of topologically correct feature maps. *Biological Cybernetics*, *43*(1), 59–69.

Kohonen, T. (1993). Physiological interpretation of the self-organizing map algorithm. *Neural Networks*, *6*(7), 895–905.

Kuhl, P. K. (1991). Human adults and human infants show a "perceptual magnet effect" for the prototypes of speech categories, monkeys do not. *Perception and Psychophysics*, *50*(2), 93–107.

Linsker, R. (1986). From basic network principles to neural architecture: Emergence of spatial-opponent cells. *Proceedings of the National Academy of Sciences, USA*, *83*(19), 7508–7512.

McClelland, J. L. (1979). On the time relations of mental processes: An examination of systems of processes in cascade. *Psychological Review*, *86*, 287–330.

McClelland, J. L. (1989). Parallel distributed processing: Implications for cognition and development. In R. G. M. Morris (Ed.), *Parallel distributed processing: Implications for psychology and neurobiology* (pp. 8–45). New York: Oxford University Press.

McClelland, J. L. (1991). Stochastic interactive processes and the effect of context on perception. *Cognitive Psychology*, *23*(1), 1–44.

McClelland, J. L. (1993). Toward a theory of information processing in graded, random, and interactive networks. In D. E. Meyer & S. Kornblum (Eds.), *Attention and performance: Vol. XIV. Synergies in experimental psychology, artificial intelligence, and cognitive neuroscience* (pp. 655–688). Cambridge, MA: MIT Press.

McClelland, J. L., & Elman, J. L. (1986). The TRACE model of speech perception. *Cognitive Psychology*, *18*(1), 1–86.

McClelland, J. L., & Rumelhart, D. E. (1981). An interactive activation model of context effects in letter perception: Part 1. An account of basic findings. *Psychological Review*, *88*(5), 375–407.

Minsky M. L. & Papert S. A. (1969). *Perceptrons*. Cambridge, MA: MIT Press.

Newell, A., & Simon, H. A. (1976). Computer science as empirical enquiry: Symbols and search. Communications of the ACM, 19(3), 113–126.

O'Reilly, R. C., & Munakata, Y. (2000). *Computational explorations in cognitive neuroscience: Understanding the mind by simulating the brain*. Cambridge, MA: MIT Press–Bradford Books.

Petrov, A. A., Dosher, B. A., & Liu, Z.-L. (2005). The dynamics of perceptual learning: An incremental reweighting model. *Psychological Review*, *112*(4), 715–743.

Plaut, D. C., McClelland, J. L., Seidenberg, M. S., & Patterson, K. E. (1996). Understanding normal and impaired word reading: Computational principles in quasi-regular domains. *Psychological Review*, *103*, 56–115.

Plunkett, K., & Marchman, V. A. (1993). From rote learning to system building: Acquiring verb morphology in children and connectionist nets. *Cognition*, *48*(1), 21–69.

Roberts, P. D., & Bell, C. C. (2002). Spike timing dependant synaptic plasticity in biological systems. *Biological Cybernetics, 87*(5–6), 392–403.

Rogers, T. T., & McClelland, J. L. (2004). *Semantic cognition: A parallel distributed processing approach.* Cambridge, MA: MIT Press.

Rosenblatt, F. (1958). The perceptron: A probabilistic model for information storage and organization in the brain. *Psychological Review, 65*(6), 386–408.

Rumelhart, D. E. (1990). Brain style computation: Learning and generalization. In S. F. Zornetzer, J. L. Davis, & C. Lau (Eds.), *An introduction to neural and electronic networks* (pp. 405–420). San Diego, CA: Academic Press.

Rumelhart, D. E., Hinton, G. E., & Williams, R. J. (1986). Learning internal representations by error propagation. In D. E. Rumelhart, J. L. McClelland, & PDP Research Group (Eds.), *Parallel distributed processing: Explorations in the microstructure of cognition: Vol. 1. Foundations* (pp. 318–362). Cambridge, MA: MIT Press–Bradford Books.

Rumelhart, D. E., & Todd, P. M. (1993). Learning and connectionist representations. In D. E. Meyer & S. Kornblum (Eds.), *Attention and performance: Vol. XIV. Synergies in experimental psychology, artificial intelligence, and cognitive neuroscience* (pp. 3–30). Cambridge, MA: MIT Press.

Rumelhart, D. E., & Zipser, D. (1985). Feature discovery by competitive learning. *Cognitive Science, 9*(1), 75–112.

Schöner, G., & Kelso, J. A. S. (1988). Dynamic pattern generation in behavioral and neural systems. *Science*(4847), *239*, 1513–1520.

Schutte, A. R., Spencer, J. P., & Schöner, G. (2003). Testing the dynamic field theory: Working memory for locations becomes more spatially precise over development. *Child Development, 74*(5), 1393–1417.

Shultz, T. R., Mareschal, D., & Schmidt, W. C. (1994). Modeling cognitive development on balance scale phenomena. *Machine Learning, 16*(1–2), 57–86.

Smith, L. B., & Thelen, E. (Eds.). (1993). *A dynamic systems approach to development: Applications.* Cambridge, MA: MIT Press–Bradford Books.

Siegler, R, S, & Klahr. D, (1982). When do children earn? The relationship between existing Knowledge and the acquisition of new knowledge. In R. Glaser (Ed.), Advnaces in instructional psychology: Vol. 2. pp 121-211. Hillsdale, NJ: Lawerence Erlbaum Associates.

Spivey, M. J., & Dale, R. (2005). The continuity of mind: Towards a dynamical account of cognition. In B. Ross (Ed.), *The Psychology of Learning and Motivation* (Vol. 45, pp. 85–142). Amsterdam: Elsevier Academic Press.

Syka, J. (2002). Plastic changes in the central auditory system after hearing loss, restoration of function, and during learning. *Physiological Reviews, 82*(3), 601–636.

Thelen, E., & Smith, L. B. (1994). *A dynamic systems approach to the development of cognition and action.* Cambridge, MA: MIT Press–Bradford Books.

Tuller, B., Case, P., Ding, M., & Kelso, J. A. S. (1994). The nonlinear dynamics of speech categorization. *Journal of Experimental Psychology: Human Perception and Performance, 20*(1), 3–16.

Usher, M., & McClelland, J. L. (2001). The time course of perceptual choice: The leaky, competing accumulator model. *Psychological Review, 108*(3), 550–592.

Vallabha, G. K., & McClelland, J. L. (2007). Success and failure of new speech category learning in adulthood: Consequences of learned Hebbian attractors in topographic maps. *Cognitive, Affective, and Behavioral Neuroscience, 7*(1), 53–73.

Williams, R. J., & Zipser, D. (1995). Gradient-based learning algorithms for recurrent networks and their computational complexity. In Y. Chauvin & D. E. Rumelhart (Eds.), *Backpropagation: Theory, architectures, and applications* (pp. 433–486). Hillsdale, NJ: Erlbaum.

Wong, K.-F., & Wang, X.-J. (2006). A recurrent network mechanism of time integration in perceptual decisions. *Journal of Neuroscience, 26*(4), 1314–1328.

# CHAPTER 2

## Development as Change of System Dynamics: Stability, Instability, and Emergence

*Gregor Schöner*

The 2005 conference out of which this book grew was dedicated to Esther Thelen (1941–2004). Three themes that permeated Esther Thelen's work are also central to dynamical systems thinking (Smith & Thelen, 1993; Smith & Thelen, 2003; Thelen & Smith, 1994). The first theme is the conviction that cognition cannot be separated from its sensorimotor embodiment in a rich and changing environment. The second theme is the idea that cognition and other complex forms of behavior may emerge in suitable behavioral and environmental contexts. The soft assembly of such skills may be multicausally linked to diverse processes, none of which need be the single cause for the emergence of the skill. The third theme is that all behavior and development unfold over time. Behavioral history (in a task context) and developmental history matter in so far as when and how a skill emerges. Development occurs, therefore, through individual developmental trajectories.

These themes structure dynamical systems thinking at different levels. Esther Thelen emphasized how dynamical systems thinking can provide metaphors that help ask new questions, generate new experimental paradigms and measures, and lead to new kinds of explanations. In this chapter, I will review dynamical systems theory (DST) as a set of concepts that formalizes such metaphors and thus becomes a scientific theory of considerable rigor. While dynamical systems theory is sometimes viewed as a collection of certain kinds of models, this chapter will neither focus on any specific model nor review modeling work in general.

In fact, DST is not about modeling systems using differential equations or dynamic iterative maps. Instead, DST is actually a specific set of concepts. I will emphasize five of these concepts in this chapter. (1) Behavioral patterns resist change; that is, they are stable. This may be mathematically characterized by considering behavioral patterns as the attractor states of a dynamical system. (2) Behavioral change is brought about by a loss of stability. (3) Representations possess stability properties, as well, and can be understood as the attractor states of dynamic fields, that is, of continuous distributions of neuronal activation. (4) Cognitive processes emerge from instabilities of dynamic fields. (5) Learning occurs as changes in behavioral or field dynamics that shift the behavioral and environmental context in which these instabilities occur.

Below, I will walk the reader through each of these five conceptual steps to explain/illustrate the key ideas, evoke exemplary model systems that bring these ideas to fruition, and finally link the ideas to other approaches/problems.

### Dynamical Systems Theory

The central nervous system is tightly interconnected. As a result, any given state of the central nervous system is exposed to a range of influences, some of which are pushing to change the neural state. The interconnectedness of the central nervous system—along with the flexibility and multifunctionality of many of its components—make perturbations the norm rather than the exception. Only neural states that resist such perturbations will persist long enough to be observable, to influence downstream processes, and to induce behavioral and long-term effects. Stability, the capacity to resist perturbations, is thus a key property of the functional states of nervous systems. Stability is needed not only to protect functions against distractive internal couplings, but also to enable neural states to maintain sustained coupling to the external world through a continuous link to sensory information. Organisms are embodied nervous systems in that their perceptual, motor, and cognitive processes are intermittently coupled to sensory information from the sensory and motor systems. Given a complex environment and a complex body with many degrees of freedom—many more than recruited for any particular task—such couplings are sources of perturbation. Only stable functional states persist in the face of such perturbations.

Mathematically, stability is the constitutive property of attractors. Illustrated in Figure 2.1, an attractor is an invariant (i.e., unchanging in time) solution of a dynamical system toward which solutions converge (if they start near an attractor). If perturbations push the state away from an attractor, the dynamical system restores the attractor state. Attractors emerge from the dynamics of a system as points at which the forces pushing in opposite directions converge and balance. A stabilization mechanism is implied whenever the rate of change, $dx/dt$, of a dynamical state variable, $x$, depends on the curr ent state (as shown in Fig. 2.1). One may think of stability as an abstract generalization of the physical concept of friction. Imagine a ball moving through water. The velocity of the ball along a line is the state variable $x$. Friction reduces positive velocities by decelerating the ball. If the ball moves in the opposite direction, its velocity is formally negative, and friction changes that velocity to zero. Thus, the state of zero velocity is a stable state of this system. The figure and the analogy illustrate the concept of a

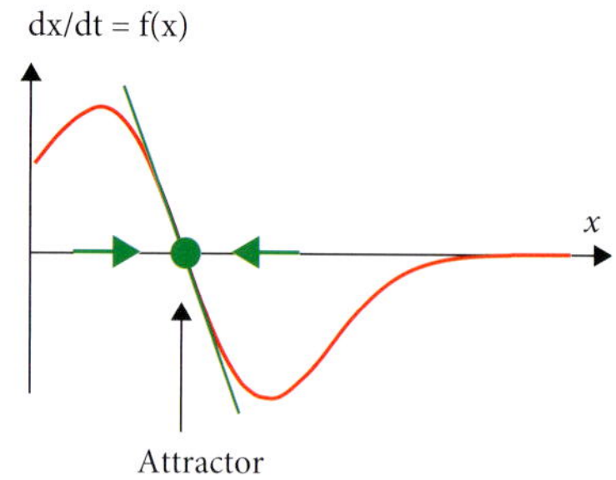


**Figure 2.1. Dynamical system with a fixed point attractor**
A differential equation model of a dynamical system is defined by how the rate of change, $dx/dt$, of the state variable, $x$, depends on the current state: $dx/dt = f(x)$. The present state thus determines the future evolution of the system. In the presence of an attractor, the system evolves by converging to the attractor as indicated by the arrows: A negative rate of change for values larger than the attractor state leads to a decrease in time toward the attractor; a positive rate of change for values smaller than the attractor leads to increase in time toward the attractor. The time-invariant attractor thus structures the temporal evolution of the state of the dynamical system in its vicinity.

fixed point attractor, that is, a stable state that is itself a constant solution of the dynamical system. Stability may be defined for more complex solutions as well, such as oscillatory solutions (limit cycle attractors), oscillatory solutions with multiple frequencies (quasi-periodic attractors), or solutions with complex time structure (strange attractors). In this review, I shall limit the discussion to fixed point attractors, a concept that is sufficient to describe most states of the nervous system.

Neurons and neural networks are naturally described as dynamical systems (Wilson, 1999) and provide through their intrinsic dynamics the mechanisms that stabilize attractors (see Hock, Schöner, & Giese, 2003, for a discussion). In that sense, stability comes for free from the neuronal dynamics prevalent in the central nervous system, although the sensorimotor periphery may also contribute (through muscle viscosity and elasticity, for instance). Once a stabilization mechanism is in place, there is

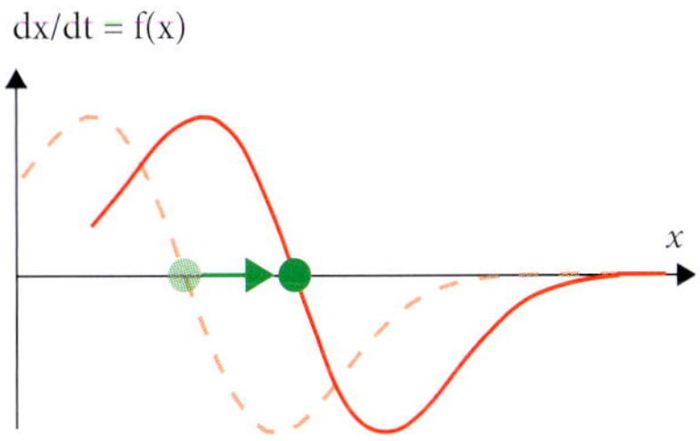


**Figure 2.2. Change in a dynamical system to a new attractor state**
When a dynamical system changes (from the form shown as a dashed line to the form shown as a solid line), the stability of the new attractor state leads automatically to an updating of the state through relaxation to the new attractor, that is, through a change of state until the rate of change again reaches zero.

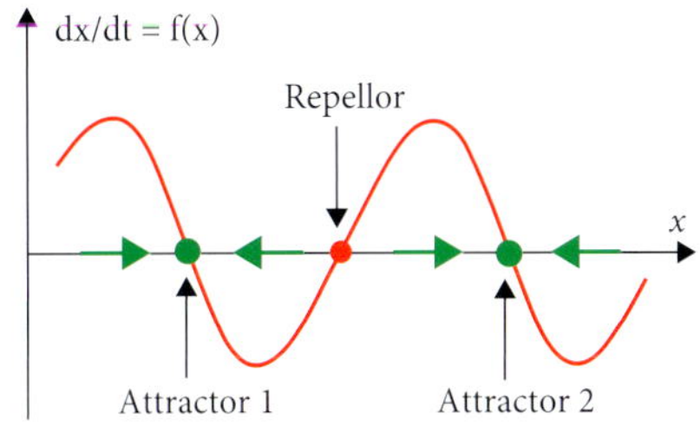


**Figure 2.3. Dynamical system with bistability**
This schematic depicts a nonlinear dynamical system with two attractors separated by a repellor, a time-invariant solution from which the system diverges.

no need for other computational mechanisms in order to determine the output of a dynamic neural network. In fact, the very forces that restore an attractor state following perturbations also help the system track any changes to the attractor state incurred as input to the network change (see Fig. 2.2). Such changing input moves the attractor state to a new location in the state space. The old attractor state is then no longer a constant solution (i.e., zero rate of change). It is instead associated with nonzero rates of change that drive the system toward the new attractor state.

The idea of a system tracking a stable state that moves as input varies is shared with the conceptual framework of cybernetics or control theory. In cybernetics, the attractor is called a set-point, and deviations from the set-point are called errors. The control system is designed to reduce such errors to zero. A conceptual step beyond this analogy is made, however, when multiple attractors coexist. The simplest case of such, bistability, is illustrated in Figure 2.3. Which of the two attractor states is realized depends on the prior history of the system. As long as the state of the system lies within the basin of attraction of attractor 1, the system tracks that state as the dynamics change. Note that it no longer makes sense to talk about the deviation from either attractor as an error—error relative to which of the two attractors? Thus, the function of the dynamical system is not to reduce error to zero. Instead, the selection of one attractor over another is a simple form of decision making, and the dynamics stabilize such decisions. On rare occasions, stochastic perturbations may shift the system sufficiently far that it crosses over into an alternative basin of attraction. This induces a stochastic switch of state.

Stochastic switches out of an attractor basin become more likely if a change of the dynamics reduces the stability of the attractor. Figure 2.4 illustrates the situation when changes of the dynamical system reach a critical point at which an attractor (attractor 1 on the left) loses stability. In this particular instance, the attractor on the left and the repellor in the middle move toward each other until they collide and then both disappear. No zero-crossing of the rate of change remains in the vicinity of the former attractor and repellor. If the system was originally in or near attractor 1, the system tracked this attractor as it moved toward the repellor. But when attractor and repellor collide and disappear, the system must switch to attractor 2 (if the system did not escape with the help of an earlier stochastic perturbation). Mathematically, such a change in the number and stability of attractors and/or repellors is a bifurcation. In simple dynamical systems such as the one shown in Figure 2.4, a bifurcation can only occur by collision of an attractor and a repellor with each

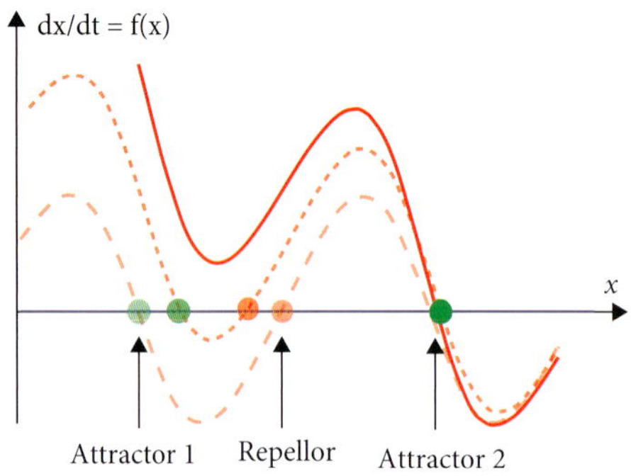


**Figure 2.4. An instability in a dynamical system** Changes of a bistable dynamical system (from dashed line via dotted line to solid line) lead to an instability. Attractor 1 collides with the repellor, leaving attractor 2 behind. This bifurcation thus leads to a change from bistable to monostable dynamics.

other. To envision this, visualize the graph of the functional dependence of the rate of change of the current state as a flexible rubber band. This graph and the associated function can be deformed by parametric changes of the dynamics, but not cut (because the rate of change must be a continuous function of the state). The only way to eliminate a zero-crossing of the function is thus to make two or more zero-crossings collide. The case shown in Figure 2.4 is the simplest such bifurcation in which the rubber band lifts off from the $x$ axis just as it becomes tangent to that axis. This so-called tangent bifurcation is one of four elementary bifurcations (the others being transcritical, pitchfork, and Hopf bifurcations), which are most likely to be observed in real systems because they make the smallest number of demands on the parameter values of the dynamical system (Perko, 1991).

Instabilities may thus lead to qualitative changes, as contrasted to the mere tracking of changing states. Changes are preceded by telltale signatures of instability such as increasing variability and increasing time needed to recover from perturbations. These properties can be exploited to detect instabilities and thus to distinguish qualitative from quantitative change (c.f. Schöner & Kelso, 1988; van der Maas & Molenaar, 1992). In nonlinear dynamical systems, instabilities arise naturally in response to changes, even relatively unspecific changes of the dynamics. In Figure 2.4, for instance, the changes to the dynamics are not specifically localized around the attractor 1 that is losing stability, but rather are essentially an increasing bias toward larger values of the state variable $x$.

This illustrates that attractors are not fixed entities. When they disappear, they are neither stored nor simply deactivated. Attractors emerge, seemingly out of nowhere, when the conditions (the dynamics) are right. This can be visualized by considering the scenario of Figure 2.4 in the reverse order: As the bias to larger values of $x$ is reduced, a new attractor may be formed spontaneously, virtually out of nowhere, and branching off an associated repellor, which forms a new boundary between two basins of attraction.

So far, I have talked about the *state* of a system in the abstract. What kind of variables $x$ would describe behaviors, patterns, decisions? Many readers may be familiar with now classical examples of DST in interlimb coordination, reviewed in Scott Kelso's (1995) book, for instance. In this work, the relative phase between two rhythmically moving limbs has been shown to be sufficient to characterize patterns of movement coordination. Each of the two common patterns of coordination can be characterized through a specific, constant value of the relative phase. For the *in phase* pattern of coordination (relative phase = 0°), homologous muscles co-contract. Typically, the *in phase* is the most stable pattern of coordination with a neuronal basis in shared descending input to the two limbs as well as excitatory coupling. In the *anti phase* pattern of coordination (relative phase = 180°), homologous muscles alternate contractions. The *anti phase* pattern is less stable. In the laboratory, one may push the *anti phase* pattern through instability by, for instance, increasing the frequency of the periodic limb motion. At a critical frequency, the stability of *anti phase* coordination is lost, leading to increased fluctuations of relative phase, to increased time needed to recover from perturbations, and (ultimately) to a switch to the *in phase* pattern of coordination (Schöner & Kelso, 1988).

Relative phase may be the only obvious example of a single variable that clearly describes a pattern of relative timing and that is independent of other details of how the pattern is generated (e.g., the movement amplitude vs. the exact trajectory shape). There are many other examples from the literature, but the variables describing the patterns are less obvious. For instance, oscillator models of coordination of rhythmic movement are formulated in terms of variables that describe the spatial position and velocity of the moving effector, although these variables are not identical to the associated physical quantities. When the limb is mechanically perturbed, for instance, the physical position and velocity are changed, but the oscillator driving the movement is not necessarily affected (Kay, Kelso, Saltzman, & Schöner, 1987). Conversely, the oscillator variables are not directly related to neuronal activations either (but see Grossberg, Pribe, & Cohen, 1997, for an account at the level of neuronal oscillators).

That dynamical systems models may require only a small number of variables (i.e., may be low-dimensional) is a central assumption of DST. What is the basis of this assumption? Why should it be possible to use a simple differential equation of one or two variables to describe what happens when a nervous system generates behavior—when millions of neurons engage sensory and motor processes and are coupled by feedback through the outer world? There is a mathematical answer to this question; I will briefly sketch a response, even if a full explanation goes beyond what can be achieved in a survey chapter.

The ensemble of neural processes and their coupling to the sensory and motor surfaces form a very high-dimensional dynamical system (Wilson, 1999). Stable states in such high-dimensional dynamics are the persistent macroscopic states which are observable at the behavioral level. Stability means that in all directions of the high-dimensional space, restoring forces secure the state against perturbations. An attractor becomes unstable when the restoring forces in one particular direction

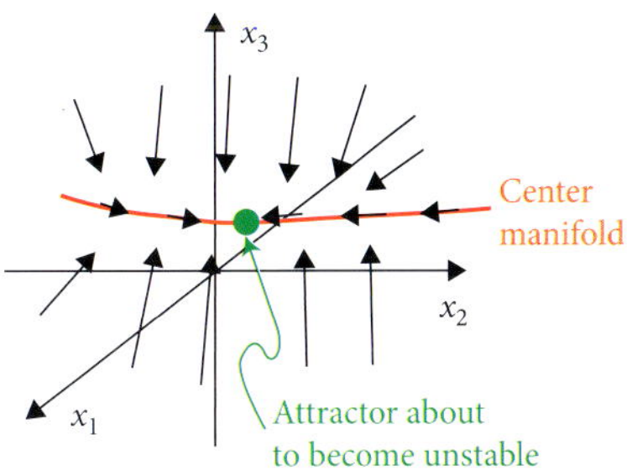


**Figure 2.5. Center manifold theorem**
When an attractor of a high-dimensional dynamical system (of which three dimensions are sketched here) becomes unstable, there is typically one direction in the high-dimensional space along which the restoring forces begin to fade (shorter arrows); while in other directions, the stabilization mechanism remains strong (longer arrows). The one direction along which forces fade spans the *center manifold*.

begin to fail (see Fig. 2.5). Only under exceptional circumstances caused by symmetries would stability fail in multiple directions at the same time. The direction along which the restoring forces become weak defines the low-dimensional *center manifold* (see Perko, 1991, for a textbook treatment).

The temporal evolution of the state of the system along the center manifold is slower than in any other direction of the high-dimensional state space. This is because the restoring forces are weaker in this direction, leading to slower movement toward the attractor state along this direction than along any other direction. Perpendicular to the center manifold, in contrast, restoring forces are strong and the system quickly moves from wherever it started out to some point on the center manifold. Thus, the long-term evolution of the system is essentially dictated by movement within the center manifold. This intuition is formalized in the center manifold theorem, which states that knowing how the system evolves along the center manifold uniquely determines how the system evolves in the original high-dimensional space. Thus, to capture the macroscopic states of the high-dimensional dynamics and their long-term evolution, it is sufficient to model the dynamics along those dimensions along which stability breaks down.

In Figure 2.4 for instance, the dimension, *x*, would correspond to the single direction in a high-dimensional state, along which the stabilization mechanism breaks down. The true dynamical system may have many more dimensions (e.g., activation levels of many neurons involved in stabilizing attractor 1). But when the bifurcation occurs, the system is still sitting in the attractor state along all the other dimensions except the one, shown in the figure, along which the instability occurs. The switch to attractor 2 arises from movement along the unstable direction. The other dimensions do not add anything qualitative to the dynamics, and can therefore, be left out of the description.

The center manifold theorem implies a huge reduction in the number of dimensions that need to be measured/modeled to understand macroscopic states and their changes. Although the theorem is mathematically true only exactly at a bifurcation, in practice the low-dimensional description provides a fair representation of the fuller dynamics whenever the system is near an instability (even if not exactly at the instability) (Haken, 1983). DST is based on the assumption that nervous systems are almost always near instability and can thus be described by low-dimensional dynamical systems most of the time.

Does this mean that the low-dimensional dynamical systems are purely descriptive while a fully mechanistic account must take place in the original, high-dimensional space? The answer depends on what is meant by a *fully mechanistic account*. If that means, literally, an account that captures the state of all neuronal processes, then, by definition, only extensive high-dimensional computational modeling will be satisfactory. If this means, however, that an account is sufficient to actually generate a behavior in a real system, then capturing the macroscopic, low-dimensional dynamics qualifies. A proof of sufficiency in that the latter sense has been provided, for instance, by generating simple robotic behaviors (such as target acquisition, target selection, obstacle avoidance, and so on) from low-dimensional attractor dynamics with appropriate bifurcations interfaced with simple sensory and motor systems (for an example see Bicho & Schöner, 1997; for a review see Schöner, Dose, & Engels, 1995). Such robotic implementations also prove that DST models are embodied/situated so that no new concepts are needed when dynamical systems models are acted out with real bodies moving in real environments based on real sensors. Such *acting out* in the real world is an interesting challenge to theoretical accounts/models of cognition. Accounts rooted in information processing have traditionally relied on relatively high-level interfaces with the sensory and motor processes necessary to act in the real world (e.g., world models and configuration space planning). These high-level interfaces have typically been difficult to put into practice (Brooks, 1991).

A related question is how abstract the low-dimensional dynamical descriptions of behavior end up being. The center manifold argument suggests that fairly abstract models may result—models that cut through the high-dimensional space describing the neural systems' supporting behavior in ways that depend on the task, the state studied, and the particular parametric/input conditions under which instability is observed. On the other hand, over the last few years, a closer alliance of dynamical systems models with neurophysiological principles has contributed much to reducing the gap between the low-dimensional dynamical descriptions and the neural networks that implement them. This will be a theme in the next section.

Given the abstract nature of DST accounts/models, why is DST often perceived to be primarily about motor behavior? Many of the exemplary model systems that influenced the development of dynamical systems ideas did come from the motor domain (reviewed in Kelso, 1995; see Hock, Kelso, & Schöner, 1993, for early work using DST in perception). On the other hand, much work since has shown that the ideas are not intrinsically tied to motor behavior. van der Maas and Molenaar (1992), for instance, applied these concepts to characterize a wide variety of developmental changes including cognitive development. Similarly, van Geert (1998) has used the abstract setting of DST to think quite generally about continuity versus discontinuity in development, using test scores in a broad variety of tasks to map the changes of dynamics over development. In many of these cases, however, the level of abstraction increased substantially when moving from simple motor skills to cognitive skills. This is due to a conceptual problem that must be confronted when stepping beyond the domain of motor behavior, and that I shall address now.

## Dynamic Field Theory (DFT)

It is easy to talk about the dynamical state of a motor system without excessive abstraction. For instance, the position of my arm, the values of the joint angles in my arm, their level of neuronal activation, and the frequency/phase of my arm's rhythmical movement are all perfectly good candidates for dynamical state variables that are not particularly abstract. They have well-defined values that evolve continuously over time.

When we move beyond pure motor control, we encounter problems with these variables, however. What value, for instance, does the phase of my arm's rhythmical movement have before I start the movement or after I have stopped moving? What values are appropriate for the movement parameters, *amplitude* and *direction*, before I select the target of my movement? Obviously, selection/initiation of motor acts—and also generation of perceptual patterns and memorization of perceptual/motor states—requires a different kind of variable than the kind used to describe motor control. These variables must capture the abstract state in which variables appear to have well-defined values at some times but not at all times. More fundamentally, we must understand how state variables may change continuously even during seemingly discrete acts such as initiation/termination of a movement.

The classical concept of activation can do this work for us. As a neural concept, activation is invoked in much of cognitive psychology and in all connectionist models. Activation may be mapped onto observable behavioral states by postulating that high levels of activation impact downstream systems (including the motor system), while low levels of activation do not. This captures the fundamental sigmoidal nonlinearity of neuronal function: activated neurons transmit to their projection targets, while insufficiently activated neurons do not. There are multiple mechanisms through which activation may be realized at the level of the neuron (e.g., through the intracellular electrical potential of neurons, through the firing rate of neurons, through the firing patterns of neural populations, or even through the amount of synchrony between the spike trains of multiple neurons; for detailed examples, see Dayan & Abbott, 2001). But the concept of activation does its work for DST independently of the details of its neural implementation.

The second concept that we will need is that of an activation field, that is, a set of many continuous activation variables, $u(x)$, defined over a continuous dimension, $x$, which spans a range of behaviors/percepts/plans. While the identification of psychologically meaningful dimensions is one of the core problems of cognitive science, there is little doubt that a wide range of perceptual, cognitive, and motor processes can be characterized by such dimensions (Shepard, 2001). Below I will argue that neuronal representations in the higher nervous system provide guidance in identifying relevant dimensions. Note that activation, $u$, is now taking on the role of the continuous state variable that was played by $x$ in the previous section. The dimension, $x$, now is a continuously valued index of multiple such state variables. In a moment I will show, that this shift of notation keeps the concepts of DST and DFT aligned.

Different states of affairs can be represented by activation fields (see Fig. 2.6). Localized peaks of activation (top panel) indicate two things: the presence of large *values* of activation means that the activation field is capable of influencing downstream structures/behavior; the locations of the peaks indicate the current values along the field *dimension* that are handed on to downstream structures. By contrast, flat patterns of low-level activation (middle panel) represent the absence of specific information about the dimension. Graded patterns of

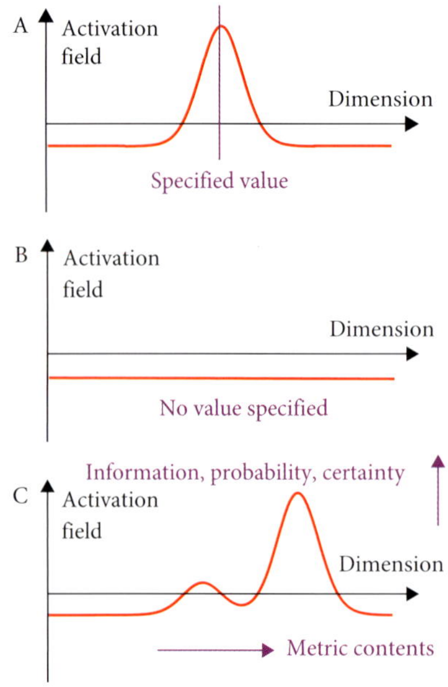


**Figure 2.6. Depicting a single activation peak over field dimension**

(A) An activation field defined over a continuous dimension, *x*, may represent (through a localized peak of activation) both the presence of information and an estimate of the specified value along the dimension. (B) Flat, low-level distributions of activation indicate the absence of information about the dimension. (C) Graded activation patterns may represent the amount of certainty about different values along the metric dimension.

activation may represent varying amounts of information, probabilities of input/responses, or how close the field is to bringing about an effect (bottom panel).

Conceptually, localized peaks are the units of representations in DFT. The location of a peak represents the value along the dimension that this peak specifies. The dimensional axis thus encodes the metrics of the *representation*—what is being prepared, perceived, or memorized and how different the various possible values along the dimension are. For instance, during the preparation of a goal-directed hand movement, a peak of activation localized along the dimension *movement direction* is a movement plan; the peak both specifies the direction in which the movement target lies and the readiness to initiate the movement (Erlhagen & Schöner, 2002). If the target is shifted during the preparation of the movement, the peak may shift continuously along the field dimension and thus provide an update of the movement plan to changed sensory information. In this situation of a continuously moving peak, we are back to the simpler picture of DST, in which the value of the state variable, *x*, now represented by the peak location, changes continuously. In this simpler description, the peak location, *x*, is the dynamical variable.

This form of representation is essentially the space code principle of neurophysiology (e.g., Dayan & Abbott, 2001); the location of a neuron in the neural network determines what the neuron encodes; the level of activation of a neuron represents how certain/important/imminent the information is that the neuron transmits. Feature maps (consisting of ensembles of neurons that code for the feature dimensions) are possible neuronal realizations of such activation fields,. Such maps conserve topology, so that neighboring neurons represent neighboring feature values. The activation field is the approximation of such a network by a continuum of neurons, justified because neural tuning curves overlap strongly.

In creating feature maps, the spatial arrangement of neurons along the cortical surface does not actually matter. The function of a neuronal network is solely determined by its connectivity, after all. The spatial arrangement is merely a matter of organizing the axons and dendritic trees in efficient ways (critical, no doubt, for growth processes). Neuronal fields capture organizational principles of feature maps: neurons exhibit broad, overlapping tuning curves and coupled through patterns of interactive connectivity that will be discussed below. Overlapping tuning curves imply that whenever a particular value of a feature dimension is specified, an entire population of neurons is active. Based on this insight, distributions of population activation may be constructed from the neural activation levels of many neurons, irrespective of where these neurons are located in a cortical or subcortical area (Erlhagen, Bastian, Jancke, Riehle, & Schöner, 1999). An example is

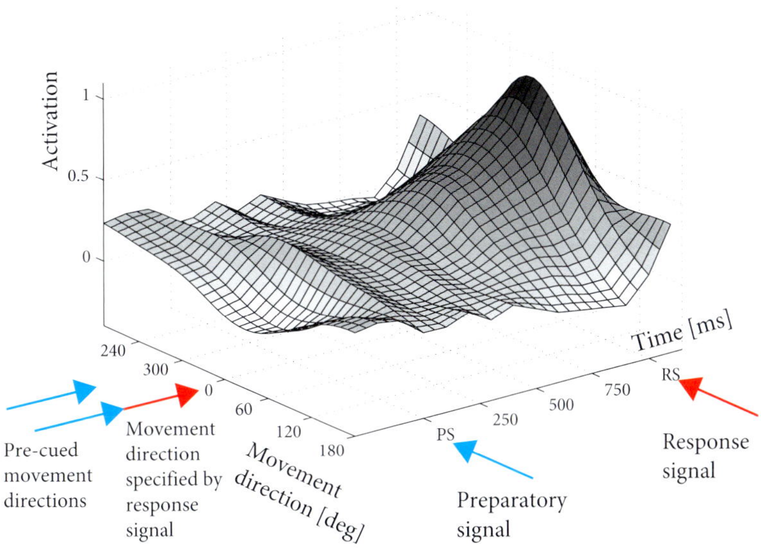


**Figure 2.7. Depicting population activation over field dimension over time**
A distribution of population activation over the dimension of "movement direction" evolves over time under the influence of a preparatory signal (which specifies two possible upcoming movement directions) and a response signal (which selects one of the two). The distribution is estimated from the tuning curves of about a hundred neurons in the motor cortex and their firing rates in 10ms-intervals. See text for details.

shown in Figure 2.7. About a hundred neurons in motor cortex were recorded while a monkey performed center-out hand movements toward visual targets. These neurons were tuned to movement direction and their tuning curves were weighted with the current firing rate of each neuron to define its contribution to the distribution of population activation (the data are from Bastian, Schöner, & Riehle, 2003; the representation is generated based on the optimal linear estimator method as described by Erlhagen et al., 1999). That distribution is essentially an estimate of the dynamic activation field that represents the planned movement direction and evolves over time reflecting the process of specification of a directed movement. In the example shown, a graded pattern of activation first arises when a preparatory signal (PS) indicates two possible movement targets. Low-level activation is centered over the two associated movement directions. One second later, a response signal (RS) indicates to the animal, which of the two targets must be selected. This leads to the generation of a peak of activation centered over the selected movement direction. The movement is initiated when activation reaches a critical level (Bastian et al., 2003, predicts response times from activation levels).

How can activation fields be endowed with stability and attractors? Naturally, the fields themselves are assumed to form dynamical systems, consistent with the physiology and physics of the corresponding neural networks. The temporal evolution of these *dynamic fields* is thus generated by forces that determine the rate of change of activation at each field site. Two factors contribute to field dynamics: inputs and interactions. Inputs are contributions to the field dynamics that do not depend on the current state of the field. The role of inputs may therefore be understood separately for every field site. The rate of change of activation, $du(x)/dt$, at a particular field site, $x$, and its dependence on input are illustrated in Figure 2.8. The fundamental stabilization mechanism originating with the biophysics of neurons is modeled by

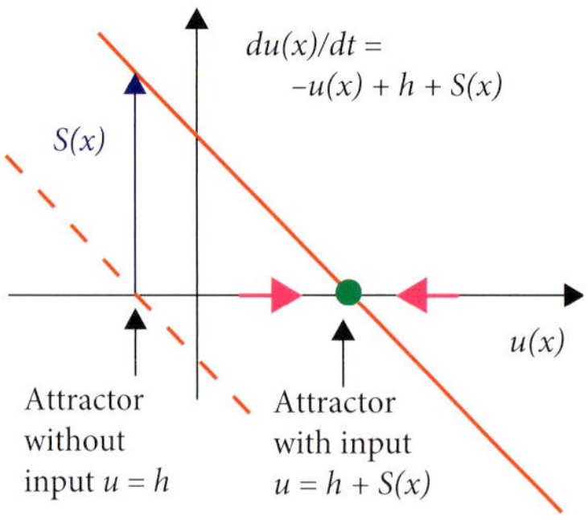


**Figure 2.8. Field dynamics factors: Excitatory inputs $S(x)$**
The generic dynamics of an activation variable, $u(x)$, is illustrated in the absence of interaction. The fundamental stabilization mechanism ($u$) generates an attractor state at the (negative) resting level $h$ in the absence of input (dashed line). Input, $S(x)$, shifts the rate of change upwards, generating a new attractor at $u = h + S(x)$ (solid line).

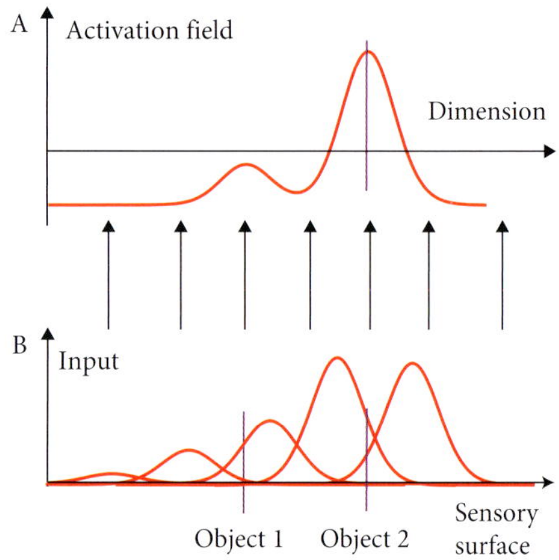


**Figure 2.9. Field dynamics factors: Input**
Input to a dynamic activation field (A) may derive from a sensory surface (B). The map from the surface to the field may involve feature extraction or complex transformation, although it is represented by a homogenous one-to-one mapping here. Sensory cells with broad tuning generate inputs to a range of field sites as illustrated here for a sample of only five sensor cells. In this example, the input arises from two objects on the sensory surface. The dynamic field selects object number 2 through interaction.

a monotonically decreasing dependence of the rate of change on the current level of activation. This leads to an attractor state at the level of activation at which this function intersects the activation axis (where the rate of change of activation = 0). In the absence of inputs, the attractor is at the resting level of the activation variable (assumed negative by convention although the units are arbitrary). Excitatory inputs, $S(x)$, shift the rate of change upwards by a constant amount. This shifts the attractor toward positive levels of activation.

Different field sites, $x$, receive different inputs, $S(x)$. For sensory representations, the way inputs vary along the field dimension, $x$, defines the meaning of that dimension. The typical idea is that inputs derive from the sensory surfaces, so that the input function varies continuously as a function of $x$ and neighboring sites receive overlapping inputs from any item on the sensory surface (Fig. 2.9). For instance, if $x$ represents retinal location, then inputs coming from the retina are determined by the receptive field of every field site. If $x$ represents orientation, then the neuronal wiring that extracts the orientation of edges generates a link between the retina and the associated locations in a direction field (Jancke, 2000). For motor fields, the meaning of the dimension is established by how the field projects onto the motor surface (that is, onto the space of all possible effector configurations). Connectionist network models are sometimes based on the same assumption of a continuous one-to-one mapping between sensory surfaces and the networks (e.g., when modeling cortical feature maps). In other instances, however, connectionist networks use much more complex input functions. These may lead to distributed activation rather than localized patterns of input that are not directly compatible with DFT.

Interactions are all contributions to the field dynamics that depend on the current activation level at any location of the field. Connectionists refer to networks with interactions as *recurrent* networks. DFT is based on the assumption that there is a universal principle of interaction—local excitation/global inhibition (Fig. 2.10, right panel). First, neighboring field sites which represent similar values of the dimension are assumed to interact in an excitatory fashion, that is, activation at either site generates positive rates of change at the other site. This form of interaction stabilizes peaks of activation against decay and thus contributes to the stability of peaks. Second, field sites at any distance are assumed to interact in an inhibitory fashion, that is, activation at either site generates negative rates of change at the other site. This form of interaction prevents peaks from broadening through lateral diffusion and thus also contributes to the stability of peaks. This pattern of interaction is ubiquitous in cortex and many subcortical structures. Only field sites with a sufficient level of activation contribute to interactions (Fig. 2.10, left panel)—a principle of neural function described by sigmoid transmission functions and the source of the fundamental nonlinearity of neural dynamics (e.g., Grossberg, 1973).

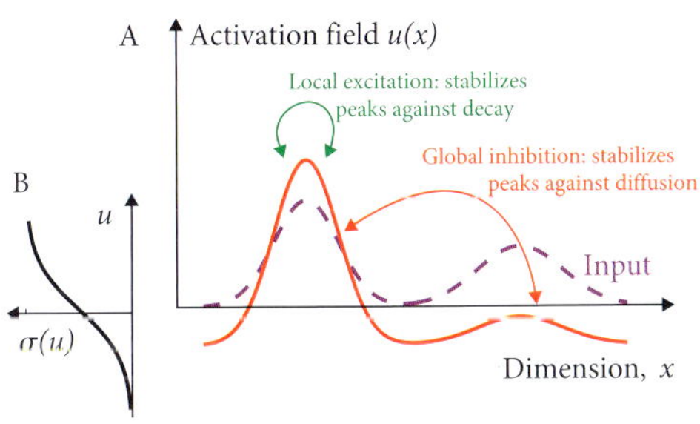


**Figure 2.10. Interaction—local excitation/global inhibition**
(A) Locally excitatory and globally inhibitory interactions enable dynamic fields to generate localized peaks of activation (solid line, peak on left) centered on locations of maximal input (dashed line, peak on left), as well as to suppress activation at competing locations (solid line, small hump on right). The activated site on the left, for instance, strongly suppresses activation at the site on the right (see difference between input and activation levels). The much less activated field site on the right has only very little inhibitory influence because its activation level is low. (B) Interaction passes through a sigmoid function, $\sigma(u)$, so that only activated sites contribute {where $u(x)$ is sufficiently large and hence $\sigma[u(x)] > 0$}.

When interaction is strong it may dominate over inputs so that the attractor states of the dynamic field are no longer dictated by input. In such a regime, dynamic fields are not described by input-output relationships. Instead, strong interactions support decision making. This may be demonstrated even for the simplest response of a dynamic field, the detection of a single localized object on the sensory surface (Fig. 2.11, left panel). For weak input strength, the field simply reproduces the input pattern as its stable activation pattern. This is the input-driven regime in which most neural networks operate in connectionist modeling. When input becomes stronger, this state becomes unstable because local excitatory interactions begin to amplify the stimulated field site (Fig. 2.11, right panel). The field relaxes to the other available stable state in which the activated peak is *self-stabilized* by the combination of local excitation and global inhibition (see Amari, 1977, for seminal mathematical analysis).

Under appropriate conditions, a self-stabilized peak of this kind may persist as an

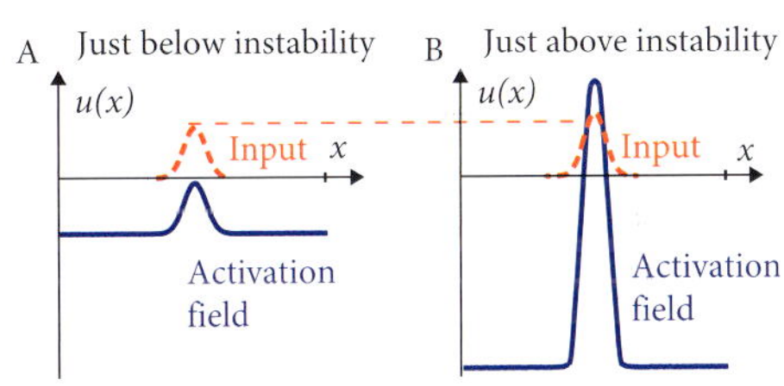


**Figure 2.11. Detection instability—local excitation/global inhibition**
(A) In response to a single localized input (fat dashed line), a dynamic field generates an input-defined pattern of activation (solid line) while the input is below the *detection instability.* (B) When the input is increased beyond a critical level (the thin dashed line is a visual guide to enable comparison with the input level in panel A), the field goes through an instability, in which the input-defined pattern becomes unstable and the field relaxes to an interaction-dominated peak. This peak is stabilized by local excitatory interaction, which pulls the activation peak higher than specified by input, and global inhibition, which strongly suppresses all locations outside the peak.

attractor of the field dynamics in the absence of input. It may then serve as a form of *self-sustained* activation, supporting the function of working memory (Schutte, Spencer, & Schöner, 2003; Spencer & Schöner, 2003; Thelen, Schöner, Scheier, & Smith, L. B., 2001; for neurocomputational models see, for example, Durstewitz, Seamans, & Sejnowski, 2000, and Wang, 1999). Such self-sustained peaks may coexist with input-defined patterns of activation, forming a bistable dynamical system (Fig. 2.12, right panel) with two accessible attractor states. By lowering the resting level, the self-sustained solution may be made unstable, returning the system to a monostable, input-driven state (Fig. 2.12, left panel) and *resetting* working memory. The detection instability, conversely, "sets" the memory by making the input-driven, memoryless state unstable.

The capability of dynamic fields to make decisions by selecting one of multiple sources of input emerges similarly from instability (Fig. 2.13). When the multiple sources are metrically close, the resulting *fused* peak positioned over an *averaged* location is monostable (Fig. 2.13, top panel). When the multiple sources are far from each other in the metric of the dimension, *x*, the field *selects* one of the sources and suppresses activation at the other locations (Fig. 2.13, bottom panel). How does the transition from averaging to selection occur? In related research trying to account for the transition from averaging to selection in visually guided saccades (i.e., rapid eye movements), similar transitions/instabilities have been successfully described in considerable detail as behavioral and neural features of saccade initiation (Kopecz & Schöner, 1995; Trappenberg, Dorris, Munoz, & Klein, 2001; Wilimzig, Schneider, & Schöner, 2006). The basic mechanism by which dynamic fields make selection decisions is the same as that in competitive activation networks (e.g., Usher & McClelland, 2001).

Why is it important that representations are endowed with stability? And what role do instabilities play? The computational metaphor is fundamentally timeless, conceiving cognition as the computation of responses to inputs. The inputs are given at some point in time; the responses are generated after a delay often called "latency" and assumed to reflect the amount of computation involved. An embodied/situated

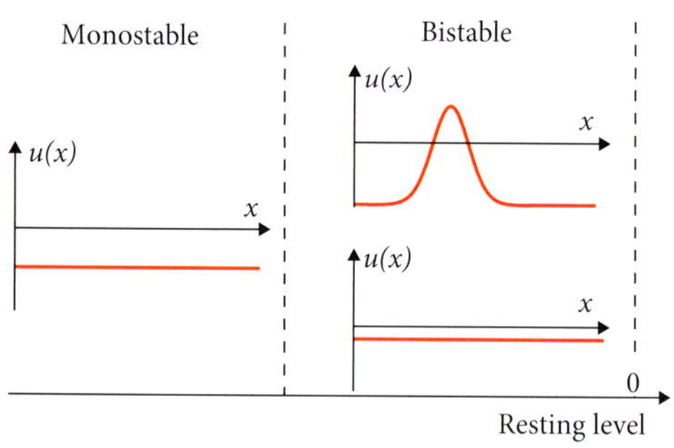


**Figure 2.12. Self-sustained activation can give rise to a bistable regime**

For sufficiently large values of the resting level (right), a *self-sustained* activation peak (top right) and a flat activation pattern near resting level (bottom right) can coexist as attractors creating a bistable regime. For smaller values of the resting level (left), the self-sustained activation mode becomes unstable and a monostable regime results.

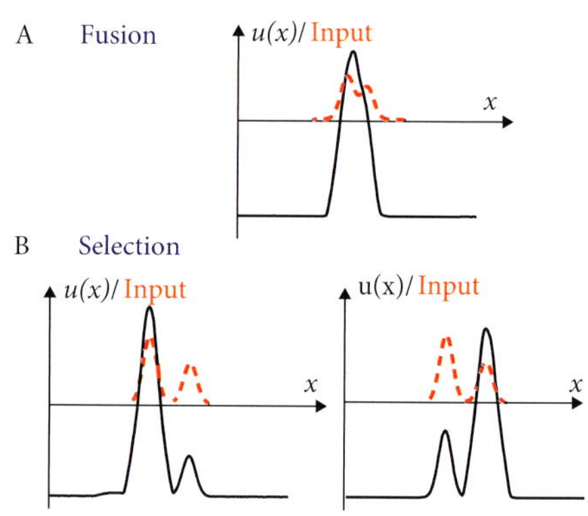


**Figure 2.13. Selection of one peak from multiple sources**

(A) Two closely spaced peaks in the input (dashed line) may generate a monostable, fused activation peak (solid line) positioned over an averaged location. Averaging (a fused solution) is due to local excitatory interaction. (B) When the two input peaks are increasingly more separate, a fused solution becomes unstable; the fused attractor bifurcates into two attractors. In each, one input peak is selected while activation at the other is suppressed.

cognitive system, by contrast, acts under the continuous influence of inputs. New processes are started on a background of ongoing processes. To provide for any kind of behavioral coherence, cognitive processes must be stabilized against the continuous onslaught of new sensory information, against sensory feedback from ongoing action, and against internal interactions from parallel (possibly competing) processes. The localized activation peaks of DFT are instances of stable states that support representation and resist change through self-stabilization induced by interaction. Resistance to change induces the reverse problem of achieving flexibility—being capable of releasing one process from stability to give way to a new process. In DFT, stable peaks of activation must be destabilized to allow for the generation of new peaks; instabilities play this crucial role. Moreover, through instabilities, cognitive properties emerge such as sustained activation and capacities to select among inputs, to fuse inputs, to couple time-varying inputs, or to suppress such coupling.

Stabilities/instabilities thus form the foundation for embodied cognition because they enable the emergence of cognitive states while retaining close ties to the sensory and motor surfaces, which are situated in rich, structured, and time-varying environments. Within DFT, the mechanisms for creating stability and instability are consistent with the organiztion of neuronal interaction found in many cortical and subcortical structures. Finally, the concepts of dynamical systems (both in the classical form of DST and in the expanded form of DFT) open up pathways toward a better understanding of learning and development.

## Learning and Development

In DST, learning means changing the dynamics of the system. Although easily stated and seemingly obvious, this insight has far-reaching consequences. Research on how people learn new patterns of interlimb coordination may serve to illustrate the ideas (Schöner, 1989; Schöner, Zanone, & Kelso, 1992; Zanone & Kelso, 1992).

Participants practiced a new pattern of bimanual coordination, with a 90° phase relationship, over five sessions in as many days. They were provided with knowledge of results after each trial. Before and after each session, participants performed other phase relationships, sampling the range from *in phase* (0°) to *phase alternation* (180°) in 10 steps. These required relative phases were invoked by presenting two metronomes with the requested phase relationship. Figure 2.14 highlights the main results of

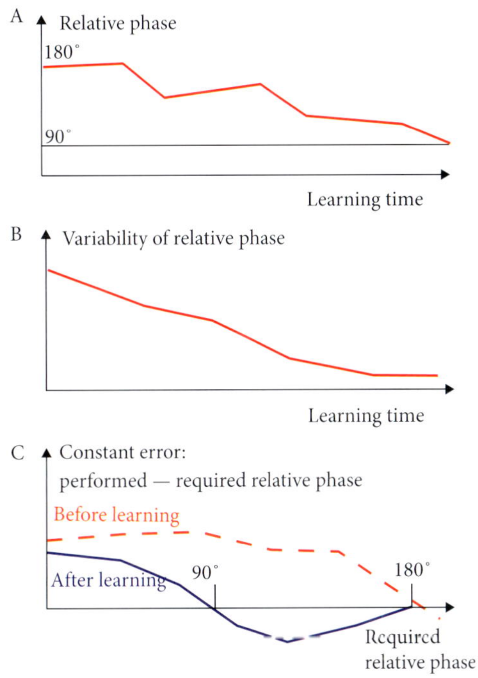


**Figure 2.14. Schematic summary of results from Zanone-Schöner-Kelso experiment**
The Schöner, Zanone, and Kelso (1992) experiment investigated the learning of a bimanual coordination pattern of 90° relative phase. Over learning, the mean relative phase approached correct performance (A) at 90° from an initial bias toward phase alternation (180°), while the variability of relative phase decreased (B). (C) The constant error of the relative phase (at a range of required relative phases) before (dashed line) and after learning (solid line) are compared. Note (a) the bias toward phase alternation before learning (i.e., constant error is positive for required relative phases below 180°) and (b) the bias toward 90° after learning (i.e., constant error is positive below 90°, negative above 90 degrees). Constant error is reduced after learning at all patterns, not only at the practiced 90° pattern.

the Schöner-Zanone-Kelso experiment (1992). During learning, both constant error and variable error decrease, consistent with increasing stability of the practiced pattern (a 90° relative phase pattern in this experiment). Before learning, performance of other patterns of relative timing is systematically biased toward phase alternation, one of the intrinsically stable patterns of coordination. After learning, performance not only improves at the practiced 90° pattern, but also for all patterns tested. There is, after learning, a systematic bias toward 90°, reflecting the changed dynamics of coordination, which have acquired a new force attracting toward 90° relative phase.

Learning a motor skill thus amounts to generating new forces which stabilize the new patterns. To see what such learning processes look like in neural language, let us examine the simplest learning processes. Patterns of activation may be stabilized simply by leaving a memory trace of ongoing activity (Erlhagen & Schöner, 2002; Schöner, Kopecz, & Erlhagen, 1997; Thelen et al., 2001). Such a memory trace, in effect, pre-activates and thus *pre-shapes* the activation field when new stimuli or new task demands arise. The memory trace thus biases neural representations toward previously activated patterns. The concept of a memory trace has a long history in psychology, dating back at least to William James (1899, chapter IV on habit formation) and Gestalt psychology and also playing a role in modern accounts of human memory (Baddeley, 1997). Memory traces may be generated through a variety of neuronal mechanisms. In particular, Hebbian strengthening of those inputs that have been successful in inducing an activation peak would be a simple neural mechanism for the functionality of a memory trace. The combined mechanisms of a memory trace and its role of pre-shaping the field around previously activated locations provide an elementary form of learning, that is, experience-dependent change of the dynamics of an activation field. Because a memory trace mechanism is formally a dynamical system, as well, we sometimes refer to this learning mechanism as the *pre-shape dynamics of a dynamic field* (Erlhagen & Schöner, 2002; Thelen et al., 2001).

Support for the concept of pre-shaping and pre-shape dynamics comes from accounts of a wide range of phenomena, including how the probability of choices influences reaction times (Erlhagen & Schöner, 2002), how motor habits are formed and then influence motor decisions in the *A-not-B* paradigm (Thelen et al., 2001), and how spatial memory is biased toward previously memorized locations (Schutte et al., 2003; Spencer, Smith, & Thelen, 2001). As an illustration, consider perseverative reaching in the A-not-B paradigm (reviewed in Wellman, Cross, & Bartsch, 1986). In this task setting, infants are shown two reachable locations, *A* and *B* (typically two wells within a small box). While the infant is watching, a toy is hidden either in the A well (on an initial set of trials called the A–trials) or in the B well (on subsequent set of trials called the B-trials). After a short delay of a few seconds, the box is pushed within reach of the infant, who will typically then make a motor decision of reaching either toward the A or the B location. Young infants below 11- to 12-months of age make the A-not-B error of reaching toward the A location on B-trials, thus perseverating in the action they had performed on the preceding A-trials. Smith, Thelen, Titzer, and McLin (1999) showed that perseverative reaching is observed in a toyless version of the paradigm, in which the experimenter merely attracts the attention of the infants to either the A or B location rather than actually hiding a toy there. In an account based on DFT, Thelen and colleagues (Thelen et al., 2001) described the rich phenomenology of this paradigm. An activation field representing the planned movement receives input from the attention-getting stimulation, but also from the perceptually marked A and B locations and from a memory trace built up during the first few reaches to the A location. On the first B-trial, the activation induced by the attention-getting stimulus at the B location competes with the activation induced by the memory trace at the A location. In young infants, the field is largely dominated by these inputs. For sufficiently long delays, the attention-induced activation decays and the pre-shape induced by the memory trace promotes reaching to the A location instead of the B location, thus generating error. In older infants, in contrast, the field is dominated by interaction which stabilizes the initial decision to reach to the B location against the competing influence of the memory trace.

Recent work by Evelina Dineva (Dineva, 2005, unpublished; Dineva, & Schöner, Thelen, 2008, in preparation) has elaborated how the behavioral history in the task predicts performance. Figure 2.15 illustrates the evolution of dynamic fields representing reaching directions and pre-shaping over six A-trials and two B-trials. Reaches to A leave a memory trace, which pre-shapes the field at A and thus promotes further reaches to A, inducing the A-not-B error in the B-trials (Fig. 2.15, left panels). In contrast, spontaneous errors (i.e., reaches to B on A-trials) leave a memory trace at B, which promotes further reaches to B and supports correct reaching to B on B–trials, reducing the A-not-B error (Fig. 2.15, right panels). Later chapters in this book examine the A-not-B paradigm in more detail.

Thus, DFT accounts for perseverative reaching by postulating that over the course of the few reaches performed during the experiment, infants acquire a motor habit (i.e., learn to reach to the A location). This creates an error when the cued location switches but also stabilizes their reaching behavior in an unchanging environment. Does learning always have a stabilizing effect? In the domain of motor learning, we have already seen that learning may destabilize patterns that were stable before learning. Learning a new pattern of coordination may reduce the stability of old patterns of coordination (see Schöner, 1989, and Schöner, Zanone, & Kelso, 1992, for accounts of how instabilities may arise during motor learning). According to DFT, a form of learning that destabilizes previously stable patterns accounts for *habituation* (Schöner & Thelen, 2006). Habituation is the gradual reduction of sensitivity to a repeated form of stimulation and is observed across a wide range of behaviors in many species (Thompson & Spencer, 1966). Habituation plays an important role in developmental psychology as a probe of infant perception and cognition (Kaplan, Werner, & Rudy, 1990; Spelke, Breinlinger, Macomber, & Jacobson, 1992).

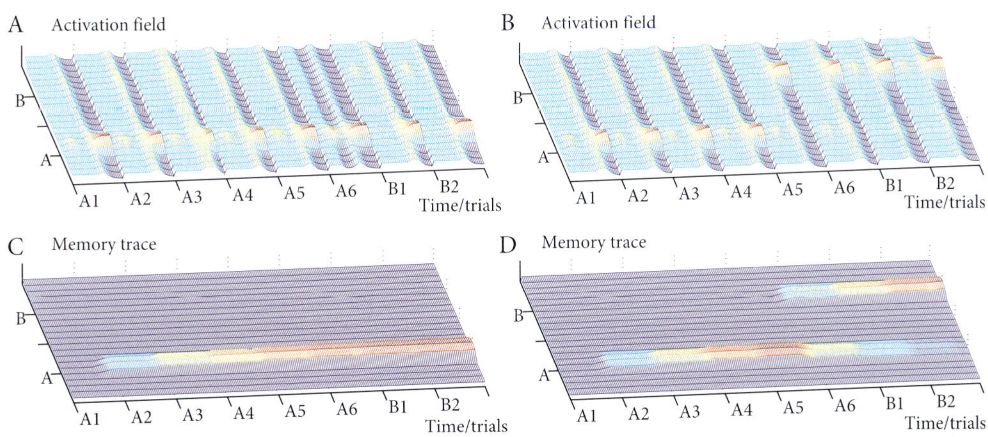


**Figure 2.15. Depiction of fields in the A-not-B task**
Dynamic fields representing the reaching direction in the A-not-B task are shown here (A, B) together with the associated pre-shape fields (C, D). Along the time/trials axis, the time structure of the task is reflected by a sequence of six A-trials (stimulation at A) and two B-trials (stimulation at B). On each trial, a peak is generated either at A or at B and leaves a matching memory trace, which in turn pre-shapes the field on the next trial. This process generates a tendency for perseveration both in the conventional sense (perseveration to A after a number of reaches to A on A-trials [left panels]) as well as in a new sense (correctly reaching to B after a number of reaches to B on A-trials [right panels]).

To illustrate how habituation is used to this end, I will briefly describe the influential *drawbridge paradigm* (Baillargeon, Spelke, & Wasserman, 1985). Infants are presented with a visual stimulus, which consists of a flap moving repeatedly through a rotation of 180° (as illustrated in the top left panel of Fig. 2.16); an infant observes this stimulus from the right. The infants habituate to this stimulus, that is, over time they tend to look less at the stimulus. Infants are then tested with two other stimuli. The *impossible stimulus* (middle left panel of Fig. 2.16) involves the same 180° flap motion (in that respect, a familiar stimulus); however, initially a block is visible that would appear to block the path of the flap (in that respect, violating expectancy if the infants' visual representations are structured by such knowledge about the world). In the *possible stimulus* (bottom left panel of Fig. 2.16), the flap motion is stopped at an angle of 112° (in that respect, a novel stimulus); this is consistent with the initially visible block remaining in place and preventing further motion of the flap (in that respect, not violating expectancy). A novel stimulus is predicted to elicit more renewed looking on the test trials than a familiar stimulus. When instead, the familiar but impossible stimulus arouses more looking, this observation is then used to infer that infants are, in fact, building expectations about the perceived scene based on knowledge they have about the world (such as the knowledge that objects cannot penetrate each other).

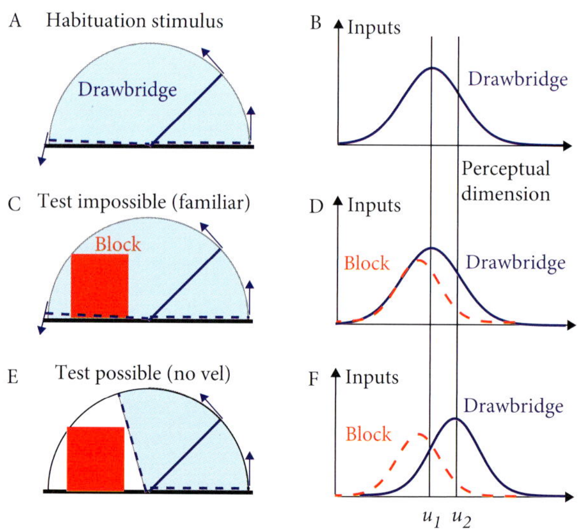


**Figure 2.16. Influential drawbridge paradigm: Schematic**
(Left column) These three drawings depict the three stimuli presented to the infant, who observes the stimuli from the right in this figure. Panel A depicts the moving flap alone during habituation. Panel C depicts the same flap motion with a block on its movement path; actually, the block is quickly removed through a trapdoor as the flap occludes/conceals the block (i.e., the impossible test stimulus). Panel E depicts the reduced flap motion with the stationary block (i.e., the possible test stimulus). (Right column) These stimuli are shown in a perceptual dimension that spans the visual depth at which the flap and the block are seen (from the right). The moving flap provides broad input (solid line) across this spatial dimension; this input is centered at a larger mean visual depth for the 180° motion (top two diagrams, B and D) than for the blocked motion (bottom diagram, F). The block provides added input (dashed line) at a larger depth (bottom two panels) since it is farther from the infant. The two marked locations (vertical bars) along the perceptual dimension are represented by two activation variables ($u_1$, $u_2$) as in the dynamic model of Schöner and Thelen (2006).

DFT accounts for this pattern of looking in a way that makes no use of knowledge but relies instead on the metrics of the perceptual events induced by this sequence of stimuli. The three stimuli are embedded into a perceptual dimension, for instance, the spatial location (in depth) at which both flap motion and block are perceived (right column of Fig. 2.16). In this embedding, it is obvious that the impossible test stimulus overlaps more with the habituation stimulus than with the possible test stimulus.

These stimuli provide input to an activation field that controls looking behavior (see Fig. 2.17). Levels of activation larger than a looking threshold lead to looking at the stimulus, levels below that threshold lead to looking away from the stimulus. The activation field drives a second, inhibitory field, which projects back onto the activation field as inhibition. During habituation, this accumulated inhibition in the inhibitory field gradually reduces the amount of activation a given stimulus induces. This destabilizes the positive peak of activation induced by the stimulus and promotes looking away from the stimulus. The coupled dynamics of activation and inhibition thus account for the typical pattern of looking time during the habituation phase (bottom right panel of Fig. 2.17).

When a new stimulus is presented during the test phase, the amount of dishabituation depends on how strongly the test stimulus overlaps with the habituation stimulus (see Fig. 2.18). A strongly overlapping stimulus (such as the familiar/impossible stimulus in the drawbridge paradigm) activates field sites that have previously been activated and are thus initially at a higher level of activation; this leads to more looking (top right panel of Fig. 2.18). A less overlapping stimulus (such as the novel/possible stimulus in the drawbridge paradigm) activates

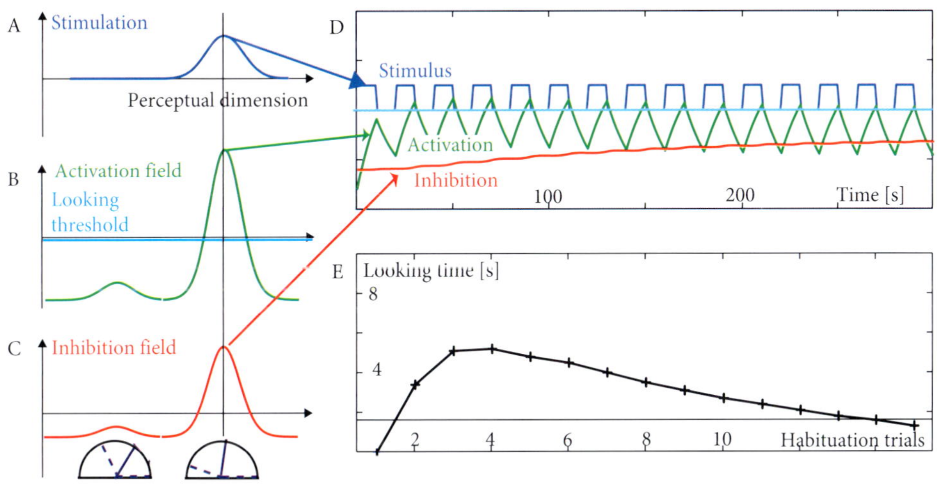


**Figure 2.17. Influential drawbridge paradigm: Habituation trials**
(A–C) A dynamic field model of habituation consists of two layers: (1) an activation field (B), which receives sensory input (A) and represents the propensity to look at the stimulus, and (2) an inhibition field (C), which is driven by the activation field which it in turn inhibits. The fields span the relevant perceptual dimension; here the visual depth of a stimulus in the drawbridge paradigm as schematically indicated by the diagrams at the bottom. The stimulus (the 180° drawbridge motion of the flap; marked by a vertical line) is presented periodically during the habituation phase. (D, E) The stimulus (square trace), activation field (zigzag trace), and inhibition field (increasing trace) at that marked location of the field are shown as a function of time (D). The looking time is shown across trials as predicted by this model (E). For each stimulus presentation, looking time is computed as the amount of time that activation is above the looking threshold (0 s in this model). The habituation criterion (i.e., half of the average looking time during the first three habituation trials) is indicated by the thin horizontal line.

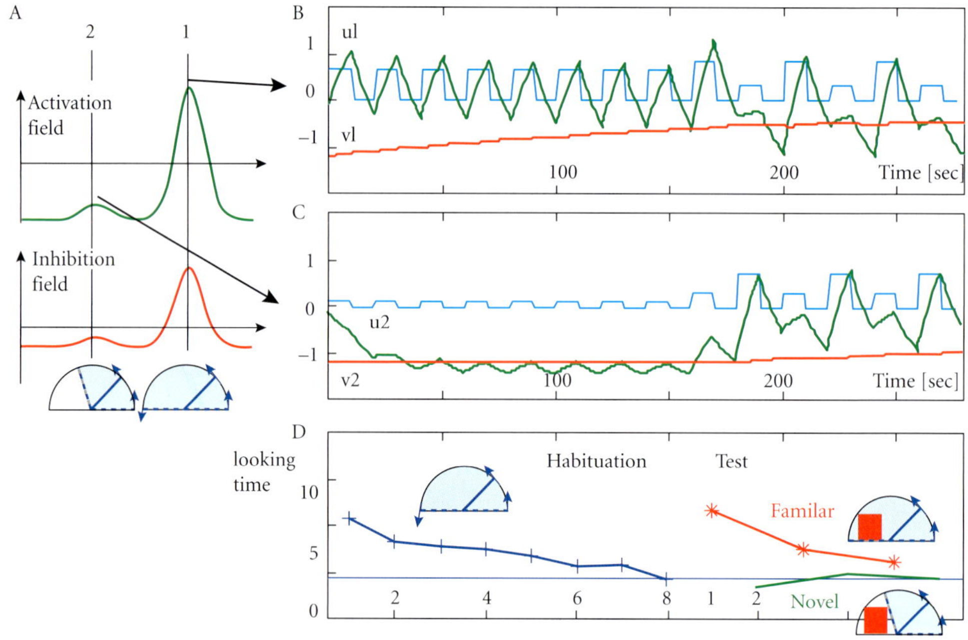


**Figure 2.18. Influential drawbridge paradigm: Dishabituation trials**
(A) Two locations in the field are marked by vertical lines. Location 1 (on the right) corresponds to the familiar/impossible test stimulus of the drawbridge paradigm with 180° motion. Location 2 (on the left) corresponds to the novel/possible test stimulus with reduced motion of the flap. (B, C) Activation and inhibition at these two locations are shown as functions of time. The traces are identifiable as in Figure 2.17. Panel C shows the looking time as a function of stimulus presentations (i.e., trials). During the habituation phase, activation is relatively stable and inhibition builds (A) leading to a decrease in looking time (C); looking approaches the habituation criterion indicated by the thin horizontal line (half of the average looking time [in seconds] during the first three habituation trials). During the test, the block provides a boost of input to both variables, leading to renewed looking (dishabituation). There is more dishabituation to the familiar/impossible stimulus than to the novel/possible stimulus when the familiar is presented first, as shown here.

field sites that were previously at rest, leading to initially lower rates of looking (middle right panel of Fig. 2.18). The pattern of looking also depends on the temporal structure of visual experience. Only the looking times on the first few test trials are determined by the level of activation at the beginning of the test phase. After these first few presentations of test stimuli, differences in initial activation have been washed out (bottom right panel of Fig. 2.18). At that point, differences in the level of inhibition determine differences in activation and the model predicts larger looking times for novel stimuli. In the experimental literature, this time dependence of looking on tests shows up as an interaction between the novelty versus the familiarity of stimuli and the order of presentation. The preference for the familiar/impossible stimulus arises only when that stimulus is presented first during the test, not when it is presented second (Baillargeon, 1987).

What about development? Central to the dynamical systems approach to development is the postulate that development depends on experience and is, therefore, in large part a learning process (see Thelen & Smith, 1994, for a detailed argument). Similar to the learning of motor skills as just described, what changes during the developmental process are the neuronal dynamics supporting a particular function.

In the DFT model for perseverative reaching, for instance, a change of the field dynamics from largely input-driven to largely interaction-driven accounts for the subsidence of perseverative reaching with increasing age. This would explain a range of age-related changes, including the tolerance of increasing delays and the resistance to distractor stimuli (Thelen et al., 2001). John Spencer and his colleagues have generalized this approach to the processes supporting spatial representations in what they call the *spatial precision hypothesis* (Schutte et al., 2003; Spencer, see chapter 5). These investigators have shown that a shift from weak and spatially diffuse interaction to stronger and spatially focused interaction in dynamic fields accounts for the time dependence and variance of performance in tasks requiring spatial working memory. This approach has implications for a range of other behaviors involving spatial representations such as spatial long-term memory, spatial discrimination, and the capacity to maintain object-centered reference frames (Simmering, Schutte, & Spencer, 2008). In all of these behaviors, the spatial precision hypothesis predicts correlated changes during development.

That development means change of the neuronal dynamics gives a new and concrete meaning to the idea of emergence. If it is the dynamics that change/develop, then developmental processes must be analyzed by investigating the causes of behavioral patterns, not just the patterns themselves. Thus, at a particular stage of a developmental process, the system may have a certain propensity to generate patterns of sustained activation supporting working memory, but whether or not such patterns are actually stable may depend on the perceptual context, the short-term behavioral history, or the total amount of stimulation (Schöner & Dineva, 2007).

Similarly, DFT accounts for habituation by postulating that what changes during development is the neuronal dynamics of activation and inhibition (Schöner & Thelen, 2006). The activation fields of older infants are assumed to have larger resting levels, so that less stimulation is required for them to reach the point at which inhibition starts to build. This accounts for both the faster rates of habituation in older infants and their increased preference for novel stimuli under equivalent stimulus conditions. Recent work by Perone, Spencer, and Schöner (2007) showed that this account can be linked to a generalization of the spatial precision hypothesis (space in this instance refers to the space of visual features).

The postulation that development amounts to a change in dynamics has also been made in slightly different form by van der Maas and Molenaar (1992). These investigators have examined more complex cognitive tasks (e.g., Piaget's task of judging the amount of water in differently shaped containers), which were assessed by overall performance scores. The level of performance at any particular point during development is interpreted as reflective of an attractor state of a dynamical system. It is not entirely clear what the status of that dynamical system is. Its attractors do not describe real-time behavior, which is a complex sequence of actions in many of the cognitive skills to which this model refers. These attractors characterize something closer to an overall state of competence. Time and stability have an uncertain status, as well, but may refer to factors such as how reproducible successful displays of a competence are. The main point of the approach is, however, that graded and continuous changes of the parameters of such attractor dynamics may lead to categorical and temporally discrete changes of the number and nature of the attractors of such a system. Using catastrophe theory—with noise added, so that the account is closer to the theory of nonequilibrium phase transitions of Haken (1983) and others—van der Maas and Molenaar (1992) describe various signatures of such qualitative changes of competence dynamics. These include the critical fluctuations and critical slowdown for which direct evidence has been provided in the context of movement coordination (Schöner, Haken, & Kelso, 1986; Schöner & Kelso, 1988). Here, however, fluctuations and resistance to perturbations are looked at in a different way. These adjustments occur within a general competence, so that performance in a cognitive task may vary—vary from one test to another, for instance.

But what of the developmental processes through which neuronal dynamics change? To date, the DFT approach has not been used to address these processes—not with the specificity/detail that would be comparable to that achieved in capturing the behavioral dynamics. This is clearly a task for the future. Knowing what changes during development is, of course, a critical prerequisite to providing a process account for such change.

In contrast, Paul van Geert (1991, 1998) and colleagues have used dynamical systems thinking to approach development—primarily the timescale of development. They postulate that the process of development can be characterized as some sort of dynamical system, evolving on the slower timescale over which development unfolds. Specifically, they propose to think of development as a growth process. Qualitatively different patterns of growth can be modeled; these include continuous growth, variable growth rates (e.g., S-shaped growth [from slow to fast to slow]), and oscillatory growth rates. Moreover, coupling among various growth processes may induce complex patterns of growth, such as stagelike alternations between slow and fast rates of change.

The concrete mathematical models that are used to illustrate these ideas encounter the same kind of problems discussed above for the account of van der Maas and Molenaar (1992). The state variable that reflects a particular stage in development is not directly linked to real-time behavior. Again, it represents something like a general level of competence, assessed through measures such as the size of vocabularies or the frequency of use of certain competences. In fact, one version of the model (van Geert, 1998) explicitly introduces a large ensemble of possible competences, which are given weights that may reflect how likely certain competences are to be displayed. These weights are the dynamical variables that evolve over developmental time. There is clearly a gap between these abstract, disembodied variables and the concrete neuronal processes that generate the behavior that is used to assess competences. Moreover, the competences are preformed and only need to be activated by increases in their respective weight values. These limitations make clear that this approach does not yet link developmental processes to the stream of behavioral experience. Implicitly, the model seems to assume that there is an intermediate level at which the developing nervous system takes stock of its current set of skills and updates the probability of using these skills. How the skills are implemented in the first place, where they come from, how they are employed to deal with incoming sensory information, and what sort of sensory information drives the learning process; such questions remain open. The strength of the model consists, instead, of a demonstration that qualitative changes may emerge from developmental processes that are governed by fairly simple dynamical laws.

Consistent with use of these types of models, Paul van Geert and his colleagues emphasize that dynamical systems thinking may be powerful at the level of metaphor (see van Geert & Steenbeek, 2005, where the relation between the Bloomington and Groningen approaches is discussed in some detail). Dynamical systems thinking as metaphor has been an important source of new ideas, new methods of analysis, and new questions—by both approaches. One instance is the emphasis on variability as a measure of underlying stabilization mechanisms; this is important on the timescale of behavior (Thelen & Smith, 1994), on the somewhat indeterminate timescale of expression of competence (van der Maas & Molenaar, 1992), and on the developmental timescale (van Geert, 1998).

## Conclusion

In this overview, I have shown how four concepts—attractors, their instabilities, dynamic activation fields, and the simple learning mechanism of a memory trace—together may enable an embodied theory of cognition. This is a theory that (a) takes seriously the continuous linking of cognitive processes to the sensory and motor surfaces and (b) takes into account that cognition emerges when systems are situated in richly structured environments. The theory is open to an understanding of learning and is based on neuronal principles.

Connectionist thinking shares many of the same ambitions and core assumptions, in particular, the commitment to using neuronal language to provide accounts for function and learning. In some instances, connectionist models have used the exact same mathematics as dynamical systems models (e.g., Usher & McClelland, 2001), and many connectionist networks are formally dynamical systems. Connectionism is really about change, about how (under the influence of a universe of stimuli) statistical regularities may be extracted by learning systems (Elman et al., 1996). These properties may be reflected in the representations that such learning systems evolve (Elman et al., 1996). Signatures of that process include accounting for learning curves, so that connectionist models speak to the evolution of systems on the timescale of development. On the other hand, some connectionist models speak to the flow of behavior on its own timescale, the generation of responses to inputs provided in a given task setting. Connectionism thus straddles the two timescales that the two approaches of dynamical systems thinking have studied, for the most part, separately.

The detailed comparison of dynamical systems thinking and connectionism is not the goal of this overview (see Smith & Samuelson, 2003, and much of this book). Maybe I can say just this much. The problem of bringing dynamical systems thinking to bear on the process of development itself may be the most constructive point of convergence. While connectionist networks have shown how the statistics of stimulation drive the development of representations, the models to date have not taken into account how infants and children control their own stimulation through their own behavior, how individually different developmental trajectories may emerge from that fact, and how the space-time constraints of behavior determine what can be learned and when. DST provides the framework within which online linkage to the environment can be modeled and demonstrates how cognitive functions emerge from instabilities. By combining the structured account for real-time behavior of DST with the neural substrate for learning provided by connectionism, maybe we will be able to understand not only what develops but also how individuals (through their own behaviors) generate the experiences on which development is based.

## Acknowledgments

Support of this work from the Ministry of Science and Research of the state of North-Rhine Westfalia, Germany as well as from the Federal Ministry of Education and Research, Berlin, Germany through a Bernstein Group Grant is gratefully acknowledged. I appreciate discussion with Evelina Dineva and John Spencer.

## References

Amari, S.-I. (1977). Dynamics of pattern formation in lateral-inhibition type neural fields. *Biological Cybernetics, 27*(2), 77–87.

Baddeley, A. D. (1997). *Human memory: Theory and practice.* London: Taylor & Francis–Psychology Press.

Baillargeon, R. (1987). Object permanence in 3 1/2– and 4 1/2–month-old infants. *Developmental Psychology, 23*(5), 655–664.

Baillargeon, R., Spelke, E., & Wassermann, S. (1985). Object permanence in 5-month-old infants. *Cognition, 20*(3), 191–208.

Bastian, A., Schöner, G., & Riehle, A. (2003). Preshaping and continuous evolution of motor cortical representations during movement preparation. *European Journal of Neuroscience, 18*(7), 2047–2058.

Bicho, E., & Schöner, G. (1997). The dynamic approach to autonomous robotics demonstrated on a low-level vehicle platform. *Robotics and Autonomous Systems, 21*(1), 23–35.

Brooks, R. A. (1991). New approaches to robotics. *Science, 253*(5025), 1227–1232.

Dayan, P., & Abbott, L. F. (2001). *Theoretical neuroscience: Computational and mathematical modeling of neural systems.* Cambridge, MA: MIT Press.

Dineva, E. (2005). *Dynamical field theory of infants reaching and its dependence on behavioral history and context.* Unpublished doctoral dissertation, Ruhr–Universität Bochum, Bochum, Germany.

Dineva, E., Schöner, G., & Thelen (2008). Behavioral history matters: A dynamic field account of spontaneous and perseverative errors in infant reaching. Manuscript in preparation.

Durstewitz, D., Seamans, J. K., & Sejnowski, T. J. (2000). Neurocomputational models of working memory. *Nature Neuroscience, 3*(Suppl.), 1184–1191.

Elman, J. L., Bates, E. A., Johnson, M. H., Karmiloff-Smith, A., Parisi, D., & Plunkett, K. (1996). *Rethinking innateness: A connectionist perspective on development.* Cambridge, MA: MIT Press.

Erlhagen, W., Bastian, A., Jancke, D., Riehle, A., & Schöner, G. (1999). The distribution of neuronal population activation (DPA) as a tool to study interaction and integration in cortical representations. *Journal of Neuroscience Methods, 94*(1), 53–66.

Erlhagen, W., & Schöner, G. (2002). Dynamic field theory of movement preparation. *Psychological Review, 109*(3), 545–572.

Grossberg, S. (1973). Contour enhancement, short term memory, and constancies in reverberating neural networks. *Studies in Applied Mathematics, 52*, 217–257.

Grossberg, S., Pribe, C., & Cohen, M. A. (1997). Neural control of interlimb oscillations. I. Human bimanual coordination. *Biological Cybernetics, 77*, 131–140.

Haken, H. (1983). *Synergetics—an introduction: Nonequilibrium phase transitions and self-organization in physics, chemistry, and biology* (3rd ed.). New York: Springer.

Hock, H. S., Kelso, J. A. S., & Schöner, G. (1993). Bistability and hysteresis in the organization of apparent motion patters. *Journal of Experimental Psychology: Human Perception and Performance*(1), *19*, 63–80.

Hock, H. S., Schöner, G., & Giese, M. A. (2003). The dynamical foundations of motion pattern formation: Stability, selective adaptation, and perceptual continuity. *Perception and Psychophysics, 65*(3), 429–457.

James, W. (1899). *Principles of psychology* (Vol. i). New York: Henry Holt.

Jancke, D. (2000). Orientation formed by a spot's trajectory: A two-dimensional population approach in primary visual cortex. *Journal of Neuroscience, 20*(14), U13–U18 (RC86:1–6).

Kaplan, P. S., Werner, J. S., & Rudy, J. W. (1990). Habituation, sensitization, and infant visual attention. In B. Hopkins, S. Johnson (Series Eds.), C. Rovee-Collier, & L. P. Lipsitt (Eds.), *Advances in Infancy Research* (Vol. 6, pp. 61–109). Norwood, NJ: Ablex.

Kay, B. A., Kelso, J. A. S., Saltzman, E. L., & Schöner, G. (1987). Space-time behavior of single and bimanual rhythmical movements: Data and limit cycle model. *Journal of Experimental Psychology: Human Perception and Performance, 13*(2), 178–192.

Kelso, J. A. S. (1995). *Dynamic patterns: The self-organization of brain and behavior.* Cambridge, MA: MIT Press–Bradford Books.

Kopecz, K., & Schöner, G. (1995). Saccadic motor planning by integrating visual information and pre-information on neural, dynamic fields. *Biological Cybernetics, 73*(1), 49–60.

Perko, L. (1991). *Differential equations and dynamical systems.* New York: Springer.

Perone, S., Spencer, J. P., & Schöner, G. (2007). A dynamic field theory of visual recognition in infant looking tasks. *Proceedings of the International Conference of the Cognitive Science Society* (pp.580–585). Nashville, TN.

Schöner, G. (1989). Learning and recall in a dynamic theory of coordination patterns. *Biological Cybernetics, 62*(1), 39–54.

Schöner, G., & Dineva, E. (2007). Dynamic instabilities as mechanisms for emergence. *Developmental Science, 10*(1), 69–74.

Schöner, G., Dose, M., & Engels, C. (1995). Dynamics of behavior: Theory and applications for autonomous robot architectures. *Robotics and Autonomous Systems, 16*(2–4), 213–245.

Schöner, G., Haken, H., & Kelso, J. A. S. (1986). A stochastic theory of phase transitions in human hand movement. *Biological Cybernetics, 53*(4), 247–257.

Schöner, G., & Kelso, J. A. S. (1988). Dynamic pattern generation in behavioral and neural systems. *Science, 239*(4847), 1513–1520.

Schöner, G., Kopecz, K., & Erlhagen, W. (1997). The dynamic neural field theory of motor programming: Arm and eye movements. In G. Stelmach (Series Ed.), P. G. Morasso, & V. Sanguineti (Vol. Eds.), *Advances in Psychology: Vol. 119. Self-organization, computational maps, and motor control, psychology series* (pp. 271–310). Amsterdam: Elsevier-North Holland.

Schöner, G., & Thelen, E. (2006). Using dynamic field theory to rethink infant habituation. *Psychological Review, 113*(2), 273–299.

Schöner, G., Zanone, P. G., & Kelso, J. A. S. (1992). Learning as change of coordination dynamics: Theory and experiment. *Journal of Motor Behavior, 24*(1), 29–48.

Schutte, A. R., Spencer, J. P., & Schöner, G. (2003). Testing the dynamic field theory: Working memory for locations becomes more spatially precise over development. *Child Development, 74* (5), 1393–1417.

Shepard, R. N. (2001). Perceptual-cognitive universals as reflections of the world. *Behavioral and Brain Sciences, 24*(4), 581–601.

Simmering, V. R., Schutte, A. R., & Spencer, J. P. (2008). Generalizing the dynamic field theory of spatial cognition across real and developmental time scales. In S. Becker (Ed.), Computational cognitive neuroscience [special issue]. *Brain Research*(C), *1202*, 68–86.

Smith, L. B., & Samuelson, L. K. (2003). Different is good: Connectionism and dynamic systems theory are complementary emergentist approaches to development. *Developmental Science, 6*(4), 434–439.

Smith, L. B., & Thelen, E. (Eds.). (1993). *A dynamic systems approach to development: Applications.* Cambridge, MA: MIT Press–Bradford Books.

Smith, L. B., & Thelen, E. (2003). Development as a dynamic system. *Trends in Cognitive Sciences, 7*(8), 343–348.

Smith, L. B., Thelen, E., Titzer, R., & McLin, D. (1999). Knowing in the context of acting: The task dynamics of the AnotB error. *Psychological Review, 106*(2), 235–260.

Spelke, E. S., Breinlinger, K., Macomber, J., & Jacobson, K. (1992). Origins of knowledge. *Psychological Review, 99*(4), 605–632.

Spencer, J. P., & Schöner, G. (2003). Bridging the representational gap in the dynamic systems approach to development. *Developmental Science, 6*(4), 392–412.

Spencer, J. P., Smith, L. B., & Thelen, E. (2001). Tests of a dynamic systems account of the AnotB error: The influence of prior experience on the spatial memory abilities of two-year-olds. *Child Development, 72*(5), 1327–1346.

Thelen, E., Schöner, G., Scheier, C., & Smith, L. B. (2001). The dynamics of embodiment: A field theory of infant perseverative reaching. *Behavioral and Brain Sciences, 24*(1), 1–34, commentary 34–70, authors' response 70–80.

Thelen, E., & Smith, L. B. (1994). *A dynamic systems approach to the development of cognition and action.* Cambridge, MA: MIT Press–Bradford Books.

Thompson, R. F., & Spencer, W. A. (1966). Habituation: A model phenomenon for the study of neuronal substrates of behavior. *Psychological Review, 73*(1), 16–43.

Trappenberg, T. P., Dorris, M. C., Munoz, D. P., & Klein, R. M. (2001). A model of saccade initiation based on the competitive integration of exogenous and endogenous signals in the superior colliculus. *Journal of Cognitive Neuroscience, 13*(2), 256–271.

Usher, M., & McClelland, J. L. (2001). The time course of perceptual choice: The leaky, competing accumulator model. *Psychological Review, 108*(3), 550–592.

van der Maas, H. L. J., & Molenaar, P. C. M. (1992). Stagewise cognitive development: An application of catastrophe theory. *Psychological Review, 99*(3), 395–417.

van Geert, P. (1991). A dynamic systems model of cognitive and language growth. *Psychological Review, 98*(1), 3–53.

van Geert, P. (1998). A dynamic systems model of basic developmental mechanisms: Piaget, Vygotsky, and beyond. *Psychological Review, 105*(4), 634–677.

van Geert, P., & Steenbeek, H. (2005). Explaining after by before: Basic aspects of a dynamic systems approach to the study of development. *Developmental Review, 25*(3–4), 408–442.

Wang, X.-J. (1999). Synaptic basis of cortical persistent activity: The importance of NMDA receptors to working memory. *Journal of Neuroscience, 19*(21), 9587–9603.

Wellman, H. M., Cross, D., & Bartsch, K. (1986). Infant search and object permanence: A meta-analysis of the AnotB error. *Monographs of the Society for Research in Child Development, No. 214, 51*(3), 1–67.

Wilimzig, C., Schneider, S., & Schöner, G. (2006). The time course of saccadic decision making: Dynamic field theory. *Neural Networks, 19*(8), 1059–1074.

Wilson, H. R. (1999). *Spikes, decisions, and actions: The dynamical foundations of neuroscience.* New York: Oxford University Press.

Zanone, P. G., & Kelso, J. A. S. (1992). The evolution of behavioral attractors with learning: Nonequilibrium phase transitions. *Journal of Experimental Psychology: Human Perception and Performance, 18*(2), 403–421.

# PART II

# Dynamical Systems Theory: Case Studies

# CHAPTER 3

## Brain, Body, and Mind: Lessons from Infant Motor Development

*Daniela Corbetta*

> *...One cannot simply "peel away" the body to understand the nervous system's role in adaptive behavior.*
>
> *Hillel J. Chiel & Randall D. Beer (1997, p. 554)*

---

The body is the host of our thoughts. It is with our body that we think, express emotions, reach states of mind, and formulate intentions. Although these mental processes are usually seen as the direct product of brain activity, they are inseparable from the body in which they reside and with which they are actualized. These embodied expressions of brain activity occur continuously, moment by moment, whether we are letting our thoughts wander while strolling along a pathway, or whether we are thinking intensely while sitting still in front of a computer. The body is also the crucial pathway through which our language, gestures, and actions are actualized and perceived by others. It is in concert with the body and through the mastery of its intrinsic physical/physiological properties that we learn to express, direct, perceive, receive, feel, and continuously exchange information with the environment. In sum, the body, with its physical properties, is the vital liaison between the mind and the outer world; this liaison is a constant throughout the life span as we grow and interact with the environment.

Can we ignore the body when studying the development of the brain and the mind? Is it possible to conceive a mind, pure and autonomous, that would not take into account, would not be influenced by, or would function independently from the body in which it resides? Could our thinking and reasoning, even in their most abstract form, grow and evolve without such a fundamental, bodily connection to the outer world? Despite the fact that body and mind are continuously and tightly interconnected in time, cognitive sciences have largely neglected the former to focus mainly on the later. Body and mind have been mostly defined hierarchically. The mind guides, commands, decides, discovers, and organizes, while the body executes, expresses, or simply reflects the commands or states dictated by the mind. This neglect in studying the body has become even more extreme today, as theoretical and technological advances have made it possible to study the brain—the site of the mind—more closely. Computer models simulating complex neural networks have emerged as a powerful way to understand the mind and its formation; however, in many instances, such models study a mind denudated from its body.

The goal of this chapter is to focus on the role that the body plays in the process of brain and mind formation. To do so, I borrow heavily from work on motor development that was performed using a dynamic systems approach. Dynamic systems and connectionist perspectives share common interests in their attempts to understand brain, behavior, and mind; however, compared to connectionist approaches, dynamic systems approaches have been more prone to consider the body—its physiological and biomechanical properties—as a crucial component in the formation of behavior and thought. I begin this chapter by reviewing briefly some theoretical concepts about embodiment and the dynamic systems approach. (For more extended tutorials on the dynamic systems approach, see other chapters in this book or other resources [i.e., Kelso, 1995; Smith & Thelen, 1993; Thelen & Smith, 1994].) Then, I concentrate on three examples of infant motor development that illustrate the strong coupling existing between the body and the mind in development and that highlight the usefulness of accounting for the body and its properties when investigating a developmental question. In the conclusion section, I extend the theme of this chapter to situations that do not involve motor activities per se, to illustrate that even such activities can be embodied. Overall, the theme of this chapter is to show that development is a process marked by continuous perceptual and sensorimotor experiences, which involve a brain with a body that can act and sense. I will argue that those perceptual and sensorimotor experiences are crucial for brain, mind, and behavioral development, and that they create the foundation of many forms of thinking in development.

## Dynamic Systems and the Embodied Brain: A Brief Overview

In 1997, Chiel and Beer published a wonderful paper in *Trends in Neurosciences* in which they offer a contrasting view to the commonly held view that the body is a mere acting device, subordinated to the commands of the mind. In that paper, the authors make a very simple, seemingly obvious, however extremely fundamental point; namely, the brain is completely embedded in the body, and the body, in turn, is completely embedded in the environment. Because of this, the body is an obligatory pathway between the mind and the environment, and therefore, understanding the brain should be done by including the body, in which it resides. Chiel and Beer (1997) also argue that the body, like the brain and the environment, is a complex and highly structured dynamical system that has unique, intrinsic mechanical and musculoskeletal properties. These properties, inherent to patterns of sensory and muscle activation, constitute a fundamental influence on the process of neural activity, and change over time. Finally, Chiel and Beer (1997) argue that brain, body, and environment are continuously coupled to one another through time, and that adaptive behavior emerges from the interaction of all three coupled components.

Chiel and Beer's view of the embodied brain is fully compatible with the dynamic systems approach. From a dynamic systems perspective, brain, body, and mind are not dissociable from one another. In fact, as discussed by Thelen (2000), there are no points in time when the mind is not embedded in the body, and there are no points in time when the brain does not receive information from the body. What changes with development and time is the nature of the coupling between brain, body, and environment. Thus, mind formation and the creation of more abstract forms of thinking over time can be seen as the product of this continuous change in coupling.

From a dynamic systems perspective, the body and its musculoskeletal properties can be important contributors to the process of change. The body, with its particular perceptual and motor capabilities, fully participates in the process of formation of mind and behavior, and actively contributes to the reorganization of the brain connections that support those behaviors. Furthermore, because change is the product of continuous, intricate, and multileveled interactions among brain, body, and environment, one cannot establish a hierarchical order between these elements, nor is there one element that is more privileged than the others in driving change (see, for example, Schöner, 1995; Steinhage & Schöner, 1997). The brain does not hold a particular or privileged status; rather, brain, mind, and behavior are constantly mapped onto one another, in concert with the body and the environment. Just as a symphony emanates from the cooperative activity of the many musicians and their instruments in the orchestra, behavior and thought emerge from the active interactions of the many parts and components of the organism in its environment.

For researchers interested in development, studying change and capturing the processes of change are of primary importance. Change occurs continuously, no matter the timescale or magnitude of change. Behaviors, thoughts, or states of mind do not have an enduring, definitive quality; they evolve continuously, form, dissipate, and eventually reform as a function of time and interactions between elements of a system. Those changes are also closely linked to the experiences associated with having a body with a particular set of perceptual and motor capabilities (Thelen, 2000). Those experiences, their behavioral/cognitive outcomes, and the subsequent developmental changes are not fully independent from the ways the body and its perceptual/motor systems are being organized over time.

In sum, brain, body, and environment all play central and intertwined roles in the process of developmental change. Whether one loses a limb, gains weight, discovers how to use computers, carries a heavy load, falls in love, or acquires the meaning of a new word—all these examples have the potential to dramatically alter the ways individuals, their bodies, brains, and minds perceive, act, and interact with their environment.

In the following sections, I use examples from research on motor development to illustrate specifically how the body—its characteristics and specific sensorimotor experiences emanating from using the body in particular ways—can play an integral part in the process of mind/behavior formation and can impact on neural organization. The behaviors I present and discuss hereafter relate to different aspects of the development of reaching in infants during the first year of life. These behaviors relate to the emergence of voluntary reaching around 3–4 months of age, to reaching in the context of the A-not-B task around 8–10 months of age, and to fluctuation in patterns of reaching throughout the first year of life. These three aspects of the development of reaching have been heavily studied, but the explanations provided in the past to account for the observed developmental changes in reaching are consistent and more in line with traditional, non-embodied views of cognition. Those views assume, for example, that an infant's reaching pattern is guided by vision and the mind, directly reflects the infant's state of reasoning or understanding of a particular situation at a particular point in time, or is the product of inborn/autonomous maturational processes in the brain. In the following sections, I will revisit those assumptions/interpretations and show that when we acknowledge the roles that the body and its sensorimotor characteristics play in the observed behavioral phenomena, another level of understanding emerges. In all three cases, I begin by presenting the behavioral facts and their classic interpretations; then I present the corresponding alternate interpretation from a dynamic systems–embodied brain perspective.

## Learning to Reach

The question of how young infants learn to reach for objects located in the nearby environment dates back to the first observations reported by Piaget (1952). How do infants, who have never before reached for objects voluntarily, figure out how to move their arms to the location of a desired toy? What are the characteristics of movement on their first reaching attempts? And how do infants progress over time? Piaget (1952) contended that early reaching emerges from seeing the hand and the toy simultaneously in the same visual field. From that moment, reaching develops in a process of gradual guidance of the hand toward the toy; in this process, the infant glances successively and alternately between hand and toy to bring the hand step-by-step closer to the toy. Piaget's observations set the stage for the long-lasting interpretation that reaching develops primarily under the guidance of vision. In subsequent studies that examined the development of reaching in finer detail (by analyzing movement kinematics), investigators found that during the few months following the onset of reaching infants produce highly meandering, indirect, zigzagging, discontinuous hand paths to the toy (Fetters & Todd, 1987; Mathew & Cook, 1990; Von Hofsten, 1979, 1991). These indirect trajectories were assumed to reflect the stepwise process of visual guidance of the hand, where the infant reassesses the position of the hand relative to the toy's position after each move and makes successive corrections to bring the hand progressively closer to the toy. This early period of discontinuous hand trajectories was later labeled *visually guided* (for reviews, see Bushnell, 1985; Streri, 1991/1993); this interpretation was very much in line with a cognitive approach to development where the mind commands the body.

The interpretation that early reaching is visually guided came into question in the late 1980s and early 1990s because of a number of studies performed by Rachel Clifton and her collaborators (Clifton, Muir, Ashmead, & Clarkson, 1993; Clifton, Rochat, Robin, & Berthier, 1994; Perris & Clifton, 1988). These researchers demonstrated that young infants do not really need to see their hands to find the toy; young infants can attain a glowing target in the dark or direct their hand to the target in the dark on the basis of auditory information alone. This work raised a critical new question. If young infants do not need to visually guide their hands to reach for the toy, why is their movement trajectory to the toy so convoluted? Why do infants show so many corrections in hand trajectory before attaining the target?

Answers were proposed by another group of researchers who discovered that the deviations in arm trajectory that are typical of infants' first attempts to reach arise from the interactions of the infants' own level of activity and the intrinsic biomechanical properties of the body and arm at work (Thelen et al., 1993; Thelen, Corbetta, & Spencer, 1996). Esther Thelen and her colleagues observed that infants who are more active and thus move their arms toward the target faster tend to generate more discontinuous movements; infants who are calmer and move their arms more slowly display much straighter trajectories to the toy. To understand this link between activity level and movement discontinuity, Thelen and colleagues analyzed the movement speed and trajectory patterns of the reaches in relation to the inertial forces generated by the infants' movements as they directed their hands toward the toy. These investigators found that fast movements are associated with high motion-dependent torques, which tend to drag the arm away from the intended goal location. Thus, the active infants who produce these high-speed motion patterns must take into account these intrinsic motion-generated forces to attain the toy. They need to use their muscles not only for initiating movement, but also for breaking these deviations in trajectory and repeatedly redirecting their arms toward the toy. (This is assumed to be done by matching the proprioceptive feeling of the arm with the sight of the toy as the arm moves closer to the toy.) These fast movements ultimately caused convoluted hand paths to the toy. In contrast, the slower movements of the quieter infants appeared much more direct, simply because they did not generate disruptive torques during the movement.

Thus, from this work, it appeared that the discontinuous trajectories of the arm to the toy were not the result of stepwise visual guidance of the hand, but rather the product of the amount of energy infants were bringing to the task of reaching for the objects. High energy levels generated high-speed movements and high motion-dependent torques which interacted with the infants' intents to direct their hands to the toy. Alternatively, low energy levels required mainly generating muscle strength to lift the arm and maintain it against gravity. Hence, learning to reach, attaining the toy, and improving movement directness over time required that infants discover unique movement solutions that were relative to their individual levels of energy; active infants needed to slow down their movements and quiet infants needed to speed up their movements in order to attain the toy more readily. This process of scaling the movement speed to the goal involved discovering the biomechanical properties of the arm in motion. This could only be done through repeated sensorimotor experience as infants produced and practiced a wide range of movement solutions—some fast, some slow—and matched their respective outcomes to the goal.

In this scenario, the brain and specific neural networks involved in the formation of reaching trajectories are not solely responsible for monitoring the trajectory of the hand; they are also involved in encoding and mastering the specific motion-dependent dynamics of the arm as it moves towards the goal. Those biomechanical, motion-dependent properties of the body and the arm are integral parts of the larger process of learning to reach for an object; the brain cannot ignore them when directing an arm to a specific goal location. When the mind plans an action to a goal, the characteristics of the tool involved in the action—the body—must be integrated with the resultant plan. In this context, body and mind are inseparable. Consistent with this scenario, Thelen and her colleagues, who followed the development of individual infants over time, were able to show that within a few weeks after the onset of reaching infants learned to adjust their movements to achieve more direct reaches: the active infants learned to slow down their movements, while the slow infants were able to increase movement speed. In sum, infants learned to reach and map their actions to their intentions through discovery of the properties of their bodies, more specifically through discovery of the appropriate pattern of movement dynamics that allowed them to attain the target more directly. This developmental change in action patterns involved coupling the activities of the mind, brain, body, and environment.

## Reaching in the Context of the A-not-B Task

The ability of infants to represent and retrieve objects hidden in particular locations in space is at the core of the second example used here. Piaget's (1954) interest in the origins of intelligence lead him to devise one of the most heavily used tasks in early cognitive and motor development—the A-not-B task. In this task, infants are presented with a surface containing two wells located a few inches apart; both wells are covered with an identical cloth or lid. The infant first watches an experimenter hide a toy in one of those wells (location A). After a short delay, the infant is invited to retrieve the toy. This hiding/retrieving event to location A is repeated a few times. The experimenter then hides the toy in the second well (location B) as the infant watches. Infants 8-10 months of age typically make an error on the B-trial; they search for the toy in the A location despite having witnessed the experimenter hide the toy at location B. According to Piaget (1954), infants fail to search at location B because they have a weak or incomplete representation of where the object is located.

Since Piaget, this error in the A-not-B task has been replicated across very large number of studies involving many variations of the task. In those studies, interpretations of the error were almost always of a cognitive nature, neglecting the role that action may have played in this task (see Chapter 7, for an example of disembodied approach). Interpretations of the A-not-B error stressed the limited memory capacity of infants, their lack of object permanence, or their inability to coordinate information in time and space (see Diamond, 1990; Smith, Thelen, Titzer, & McLin, 1999; Thelen, Schöner, Scheier, & Smith, 2001; Wellman, Cross, & Bartsch, 1986, for reviews).

Curiously, the A-not-B error in infancy is not limited to the context of a disappearing object. Some studies found that infants continue to make the error even when the object placed in location B remains fully visible (Butterworth, 1977), or when the task does not involve any hiding of the toy at all, but simply the repetition of reaches for a visible toy in an identical previous location (Munakata, 1997; Smith et al., 1999). Thus, from these studies it appears that the interpretations based solely on location memory or object representation are not sufficient to explain why infants continue to make the error even when the object is perfectly visible at location B. Why then do infants make the A-not-B error?

An alternative explanation involving the body and infants' inability to control their arms flexibly during reaching was provided by a group of researchers who looked closely at the spatiotemporal characteristics of the movement (Diedrich, Thelen, Smith, & Corbetta, 2000). These researchers recorded the kinematics of the arm movement while infants were reaching to location A for 6 consecutive trials and then when cued to location B for 2 additional trials. They observed that at each new reaching trial to location A, the speed profiles of arm movements to the target became more alike, as if infants were progressively forming a motor template or *motor habit* of a specific reaching pattern to one location. Infants who displayed such a growing pattern of similarity in movement speed as a function of the repeated A-trials were also more likely to continue to reach to location A during B-trials—thus making the A-not-B error. Those infants who made the error to reach to location A on B-trials appeared to lack movement flexibility to break the motor habit that had formed as a result of the previous reaches to location A (see also Clearfield & Thelen, 2001).

Other studies from the same group of researchers demonstrated that such motor habits and the associated likelihood of making the A-not-B error did not arise when infants performed a fewer number of reaches to location A before the object was switched to location B (Smith et al., 1999). Furthermore, when a infant's arm was loaded with a weight, increasing movement inertia and making it more difficult for the infant to change movement trajectory, infants made the error even more frequently (Smith, Clearfield, Diedrich, & Thelen, 2008, in preparation). Formal modeling of the A-not-B task and the perceptual-motor response associated with it have confirmed that the number of prior reaches to location A can contribute to forming a memory of the motor pattern, where brain and arm are a coupled system. This memory of the motor pattern can be strengthened or weakened depending on the number of previous reaches to location A, the motor experience of the child, the perceptual layout of the task, and other contextual factors; hence, the likelihood of making the A-not-B error can increase or decrease depending on how these factors interact (Spencer, see Chapter 5; Thelen et al., 2001). Thus, here again, the response and errors of infants are not solely the product of infants' decisions or lack of mental representations. The mechanical and musculoskeletal properties of the body, the lack of movement flexibility required to *break* a forming reaching habit, and the number of previous trials to location A, all contributed to the observed erroneous response on B-trials. Again, mind, body, and environment are tightly coupled in time in the formation of the response.

## Developmental Fluctuations in Hand Preference in Infant Reaching

The last example examines the development of hand preference in infant reaching. Hand preference is linked to specific brain asymmetries (Corballis, 1991; Hellige, 1993). Therefore, this example is useful to explore forms of coupling between brain asymmetries, body, and behavior and their evolution in early development. One strong characteristic of hand preference is that 75%–90% of the human population display a preferred right hand use when achieving tasks requiring a certain level of dexterity, such as writing or drawing. Furthermore, this preferred hand use is quite manifest and steady in adults. One question that has puzzled scientists for many decades relates to how such a manual preference forms. This has prompted a significant number of developmental studies and interpretations (for reviews, see Corbetta & Thelen, 2002; Provins, 1997). These interpretations have largely boiled down to two main views: one claims that hand preference is an inborn behavioral trait, the other contends that hand preference develops during childhood.

Because right-handedness is such a strong and widespread characteristic among many humans, the view that hand preference is inborn has a strong predominance (see, for example, Annett, 1985; McManus, 1985). A number of developmental studies have supported this view. One of the most convincing pieces of evidence comes from studies that examined newborns' spontaneous head turn (Coryell, 1985; Michel, 1981; Viviani, Turkewitz, & Karp, 1978). These studies found that most newborns prefer to rotate their heads to the right and that this right preferred head turn correlates with a preferred right hand use later in early childhood. However, the connection between preferred spontaneous head turn at birth and later hand preference is not so obvious. Following the assumption that newborns' spontaneous head turn reflects an inborn behavioral bias that sets the direction of hand preference early on, one would expect to see continuity in hand use as infants begin to reach and manipulate objects consistent with the early bias. Yet, studies that followed the development of hand preference during infancy did not find such consistent bias in hand use; rather, they found that infants' hand preference fluctuated many times during the first year of life, where periods of right hand use alternated with periods of left hand, two-hand, or no preferred hand use (Corbetta & Thelen, 1999; Gesell & Ames, 1947; Ramsay, 1985). Some authors proposed that these fluctuations were part of the general process of brain lateralization and neuromotor maturation (Gesell, 1956; Ramsay, 1984). Nevertheless, if one assumes that hand preference is determined from birth or genetically driven, as has been suggested (Annett, 1985; McManus, 1985), the question remains as to why hand use fluctuates so much during early development. This also raises the question of how stable hand preference emerges from such a background of fluctuating movements.

I became interested in these developmental questions while working with Esther Thelen and her collaborators on a longitudinal study on infant reaching in which 4 infants were followed at close time intervals over the first year of life (see Thelen et al., 1993; Thelen, Corbetta, & Spencer, 1996). In that study, we observed many aspects of the motor development of those infants, including which hand they used for reaching and when they developed new posture-related skills. All infants displayed alternating periods of one- versus two-handed reaching (Corbetta & Thelen, 1996) accompanied by periods of right, left, or no preferred hand use in reaching (Corbetta & Thelen, 1999). Interestingly, we noticed that developmental transitions in hand use for reaching occurred at particular times when the infants were beginning to develop new locomotor skills. We observed the subsequent developmental pattern. During the early period following the emergence of reaching, infants displayed a preferred right-hand bias. This right bias dissipated toward the middle of the year when infants began to self-locomote on hands and knees. From that time, infants alternated hands frequently, trial after trial, and week after week. Finally, toward the end of the first year, one infant began to perform his first upright independent steps. The emergence of independent walking in that infant coincided with the strong return of a two-handed pattern in reaching. This was a particularly surprising finding because at the end of the first year all infants had developed fine reaching skills, the objects were always small and easily graspable with one hand, and they were always presented at midline. Plus, all infants were highly familiar with the task and knew exactly what to expect—each had months of practice reaching for those small objects with one hand. Thus, why did this return to two-handed reaching at the end of the first year occur? Nothing had really changed in the task, other than the infant beginning to walk upright.

These early observations pointed to a developmental coincidence between the emergence of self-produced locomotion and change in pattern of hand use. The question arose as to whether the acquisition of new levels of postural control, as infants learn to self-locomote, entail successive reorganizations of the upper-limb system and change the way infants prefer to reach for and manipulate objects. That question was examined in follow-up studies with a larger subject pool and varied tasks.

The first study investigated whether underlying preferred lateral biases in hand use would dissipate when infants began to crawl (Corbetta & Thelen, 2002). Two groups of infants were followed over several weeks as they learned to retrieve an attractive toy concealed in a box with a lid. One group of infants was tested weekly from the age of 6 1/2 months, and the other group was tested weekly from the age of 8 months. In the younger group, infants were not crawling; in the older group, infants had just began or were beginning to crawl at the onset of the study. All infants were tested with the same task; they were required to learn how to coordinate their hands to lift/open the lid of the box with one hand while retrieving the toy from the box with other hand (for more detail, see Corbetta & Thelen, 2002).

The findings from this study were quite clear. The younger infants, who were not crawling, displayed a stunningly stable division of labor as they learned to solve the task over successive weeks. In particular, many infants systematically and consistently used the same hand to open/hold the lid and used the other hand to retrieve the toy. This consistent division of hand roles to retrieve the toy from the box was very steady, trial after trial and week after week—to the point that most infants displayed either no or little variability in response over the whole period of testing. In contrast, the older infants, who were crawling, displayed extremely changeable patterns of right and left hand use to retrieve the toy from the box. They did so both within and across sessions to the point that it was impossible to discern one preferred pattern of hand use (Corbetta & Thelen, 2002). These findings were consistent with our prior longitudinal observations on reaching, namely, that preferred biases in hand use present before the onset of crawling disappear when infants learn to locomote using four limbs.

The second study looked at the development of walking and investigated whether a return to two-handed reaching was associated with learning to walk (Corbetta & Bojczyk, 2002). In this study, only one group of infants was followed longitudinally. Infants were seen weekly over the transition to upright locomotion, from the time they were 8 months of age (and unable to walk), to about 2 months post upright locomotion. Every week, infants were presented with small and large objects at midline to test their one- or two-handed reaching tendencies, and they solved an object retrieval task identical to the one described above for the crawling infants. Additionally, the infants' postural and locomotor progress was tested every week to identify the week in which they took their first independent steps without support. After infants began to walk, improvement in walking was monitored by indexing the height at which they kept their arms during walking. (When infants are learning to walk, they keep their arms in a high guard position to control balance, but as they practice walking and gain better upright balance, they lower their arms alongside the body.)

The results from this study were again very clear. During the *pre-walking* period, infants reached for the large objects mainly with two hands and used mainly one hand for the small ones. They also learned to open the lid of the box with one hand and coordinate the activity of their two arms to retrieve the toy out of the box without difficulty, well before they began to walk. *When the infants began to perform their first independent steps, all these patterns changed.* Infants began to reach more with two hands for the small objects; they further increased their rate of two-handed reaching for large objects and began to open the lid of the box with two hands. All of these increases in two-handed performance were statistically significant (Corbetta & Bojczyk, 2002). Again, this study supported our prior observations that infants return to two-handed reaching when they are learning to walk.

These studies confirm that the development of early hand preference is not necessarily a behavioral trait set at birth, nor does it grow progressively and steadily over time. More importantly, these studies reveal that fluctuations in hand use do not happen randomly or autonomously; rather, they occur as a function of the successive locomotor skills infants develop over the course of the first year of life. The remaining problem was to elucidate why patterns of hand use fluctuate with infants' progression in locomotion. This remained unclear because in both studies infants were always reaching/retrieving objects from the box while they were fully supported in an infant seat; they were not crawling/walking while they were reaching, and there was no balance threat that could have prompted a change in manual response. Furthermore, the same seat was used for the same tasks across infants before they learned to walk, thus the postural support provided during reaching remained unchanged. So, why was the emergence of crawling associated with a disappearance of preferred hand biases and the emergence of walking associated with a return to two-handed reaching?

An answer to these questions was suggested by some data from the study on walking and reaching (Corbetta & Bojczyk, 2002). We looked at how long the period of two-handed reaching lasted after infants began to walk independently. Interestingly, we found that the rate of two-handed reaching declined significantly a few weeks after infants began to lower their arms alongside the body during walking. In other words, coupling between arms in reaching while sitting dissolved at about the same time infants began to decouple and lower their arms while walking—the return and decline in two-handed reaching seemed to mirror the emergence of and subsequent progress in independent walking. It is this particular finding that suggested to us the mechanism that may have tied the observed developmental shifts in reaching with the development of infants' posture and locomotion.

To explain this mechanism, consider the following findings from the neuroscience literature. There are now several classic studies on brain plasticity showing that specific and novel sensorimotor experiences entail a functional reorganization of the sensorimotor cortex (to cite a few, Jenkins, Merzenich, Ochs, Allard, & Guic-Robles, 1990; Karni et al., 1998; Kleim, Barbay, & Nudo, 1998; Merzenich, Allard & Jenkins, 1990; Merzenich & Jenkins, 1993; Petersen, van Mier, Fiez, & Raichle, 1998). These studies are not limited to rats, but extend to nonhuman primates and human adults. A similar cortical reorganization was reported in human infants, in particular, following the emergence of crawling (Bell & Fox, 1996). These cortical reorganizations are closely related to the task that is learned and practiced and, hence, to the specific limbs, body parts, and sensory organs that are used to achieve these tasks. Particularly relevant to our developmental story is the finding revealing that the acquisition of bimanual skills increases coupling of the cortical sensorimotor areas (Andres et al., 1999). Likewise, the use of one arm more than the other can entail a larger, upper-limb representation in the hemisphere contralateral to the hand used (Nudo, Milliken, Jenkins, & Merzenich, 1996). Finally and importantly, it has been shown that such experience-dependent neural reorganizations can, in turn, immediately shape motor performance (Dorris, Paré, & Munoz, 2000). Thus, together these studies show that there is a constant mapping between brain and behavior as new sensorimotor skills are learned and practiced. Moreover, such mapping occurs as these tasks are performed intentionally, that is, with a goal in mind.

What do these data imply regarding the development of infants' hand preferences? My colleagues and I hypothesized that a similar kind of mapping was taking place between reaching and crawling/walking; this mapping was taking place in the brain during critical periods of motor skill learning. When infants begin to crawl or walk, they develop a new behavior that was not present in their motor repertoire. This new behavior involves using the body in a new way, by coordinating patterns of muscles differently, in order to achieve a new goal—locomotion. When infants discover how to perform these new skills, they practice them a lot (Adolph & Berger, 2006). We believe it is the practice of these novel motor skills that temporarily *transfers* to reaching. Figures 3.1 and 3.2 illustrate how this transfer may occur. We describe the case of learning to crawl first (Fig. 3.1). When infants are learning to crawl they need to figure out how to alternate hands, arms, and legs to locomote. As an infant crawls, both arms acquire an identical status/role since they are now both used equally to support/move the body. Note that despite similarity of roles, the activities of the two hands remain uncoupled since they are used in alternation. Our data on hand preference reveal that such uncoupled, alternated (but similar) hand roles become the predominant mode of response in reaching, right at the time when infants begin to practice crawling (Corbetta & Thelen 1999, 2002). Hand preference declines because infants are beginning to use their hands equally to reach for the target.

The same scenario can be applied to the transition to upright locomotion. When infants are learning to perform their first independent steps, they raise their arms in a high-guard position, coupling the arms tightly

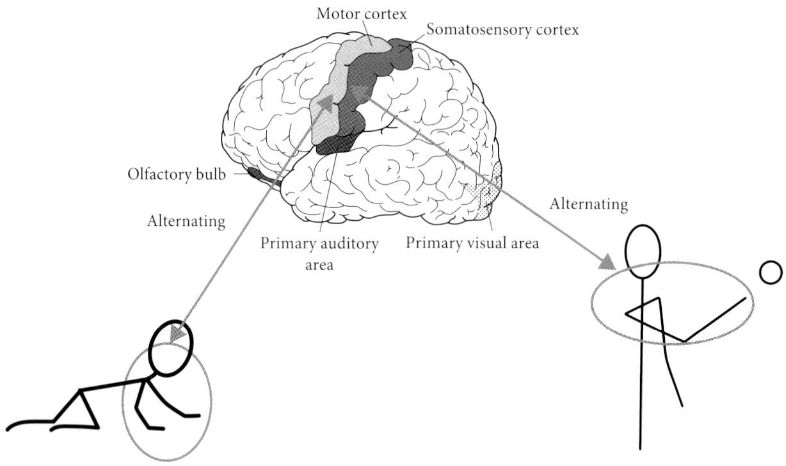


**Figure 3.1. Transfer of learning at the emergence of crawling**
When infants begin to self-locomote on four limbs, they learn to alternate arms and legs to move forward. This new, alternated functional use of the arms for locomotion *transfers* to reaching via brain functional reorganization, yielding a decline in preferred-hand use in reaching tasks.

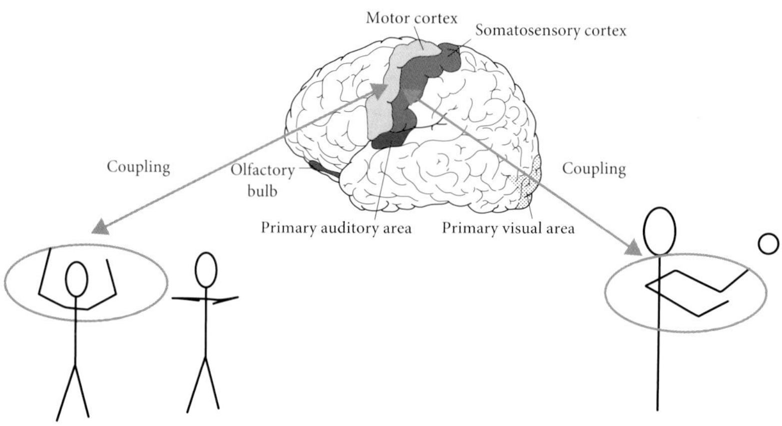


**Figure 3.2. Transfer of learning at the emergence of upright locomotion**
When infants begin to walk upright, they keep their upper arms coupled in a high-guard position. This upper-arm coupling performed during walking temporarily *transfers* to reaching via brain functional reorganization, yielding an increase in coupled, two-handed responses in reaching tasks.

to control their precarious balance during walking (Fig. 3.2). Again, when this new skill emerges, infants practice it intensely with their arms up and tightly coupled. We observed that infants who were beginning to walk also began to reach for objects at midline with two hands, suggesting that the tight arm coupling adopted and practiced during walking *transfers* to reaching. Furthermore, infants maintain this coupling in reaching tasks as long as they couple their arms during walking. When infants improve upright balance control, lower their arms, and thus decouple them, then two-handed reaching declines, as well (Corbetta & Bojczyk, 2002).

Recently, we obtained further support for our interpretation on the transfer of learning by studying the reaching patterns of two young infants who adopted unique forms of self-produced locomotion (Corbetta, Williams, & Snapp-Childs, 2006). One child, who began to locomote by scooting on his buttocks while in a sitting posture, also began to couple his arms during reaching over the same period. The coupling in reaching occurred, presumably, as a result of the arm coupling performed to generate the scooting motion. Another infant, who preferred to crawl on his belly and drag his body on the floor by using the same steady lateral pattern between hands and legs continued to maintain a strong right hand use for reaching. Unlike infants who alternated arms for crawling and displayed a disappearance in hand preference, this infant maintained a strong right bias as a result of never alternating arm movements during belly crawling.

Together, these examples on the development of hand preference in relation to the emergence of different forms of locomotion reveal that brain, body, and mind maintain a tight match over time depending on the type of locomotor pattern adopted by the child. Furthermore, this work reveals that developmental fluctuations in hand preference are intimately linked to the sensorimotor experiences that infants develop in the course of their first year. Thus, hand preference in early development is not necessarily a rigid, preestablished behavioral trait that is solely dictated by genes or an asymmetrical brain. Rather, early hand preference can be highly malleable and sensitive to the sensorimotor experiences and reorganizations of the postural system as infants learn to sit, crawl, and walk. Brain asymmetries may be subject to similar malleability. Finally, this work shows that even highly practiced skills, such as reaching at the end of the first year which is performed with a goal in mind and with prior knowledge of the actions involved and characteristics of the target object, can be sensitive to other patterns that are developed by the body over the same developmental period. Here again, the brain and the mind do not simply command the body but are constantly influenced by new patterns of activity and sensations received from the body.

## Conclusion

The work reviewed above shows that the body is not simply an instrument at the service of the mind or merely responding to the commands of the brain. The body possesses its own physical and biomechanical characteristics, which need to be assembled, integrated, and mastered as part of the processes of development and interaction with the environment. Discovering one's own body properties is a fundamental aspect of behavioral development as infants, children, and even adults engage in and learn new tasks. The brain is not foreign to this process, nor is it in greater command. Rather, the brain is involved in this process of behavioral learning as it fully partakes, in concert with the body and the environment, the goal of achieving a particular task. In this sense, the body, brain, and environment are continuously meshed together in time, and it is through changes in their mutual coupling that behavior and mind evolve in development.

All the examples used in this chapter focused on perceptual and motor development and relied on tasks that required specific actions from the body to be solved. These examples were chosen because they demonstrate how the body plays a role in the formation of behavior and learning. But what if the task did not involve bodily actions? One could argue that thinking does not always require the active participation of the body and that some forms of representation, mental images, or mental manipulations of concepts/symbols do not need the body to be evoked. Scientists can form theories, architects can envision buildings, physicists/mathematicians can reason about sophisticated laws. These are all activities that rely on very abstract mental processes and manipulate abstract symbols. Can we think of those mental activities as disembodied? Here, I will address this question two ways: First, by discussing how the body and the brain are still connected and are still communicating with each other while these abstract forms of mental activities are taking place; and second, by expanding this argument to the developmental/

learning domain and addressing the question of the origins of the mind.

The brain and the body are always connected. The brain uninterruptedly *listens* to the body and is continuously *tuned* to sensations and feelings coming from the body even when the body lies still. Thus, even when we assume that the only brain activity actually taking place is happening in our minds, we still think in concert with our bodies. Some examples of self-generated mind activities, such as meditating or concentrating on a problem while sitting, are not independent from bodily sensations. In appearance, these mind activities do not require active body involvement, yet, the brain, occupied by the thinking process, still continues to receive information about posture, body orientation, or even background sounds from the surrounding. Any sound from the environment, any feeling or sensation arising from our body, can suffice to disrupt the level of concentration or stream of thought during mental activity. The fact is that we learn throughout development to think with our body and within our body. If an uncomfortable posture is disrupting our train of thoughts or meditative activity, we learn to find a better seat, better position, or better location so the constraints arising from our own bodies won't interrupt our mental efforts. What we practice and consider as *pure* mental activities are, in reality, mind activities intimately coupled with our bodies. Other, more individualized forms of mind-body coupling during the so-called pure thinking process can be seen in the frowning of the forehead, shaking of the foot, tapping of the pencil, or any other forms of activities that people engage into with their bodies to stimulate thinking.

Recent research examples provide beautiful illustrations of mind-body coupling in tasks requiring perceptual judgment, recognition, or choice-making activity but no actual motor activity. They demonstrate that reasoning or cognitive processing in these tasks can be affected by physiological states and can be intimately associated with underlying sensorimotor or neural activity patterns arising from prior individual experiences. For example, Dennis Proffitt (see Proffitt, 2006, for a recent review of his work) has performed a series of experiments asking adults to estimate perceptually the slant of hills. In most tasks, participants were standing at the bottom of a hill and were asked to visually judge the steepness of the hill while looking at it (no actual climbing was involved). Proffitt and his collaborators found that the explicit slant judgments of the participants were scaled to the effort they would have had to make to climb the hill, even though they were not asked to climb the hill. Furthermore, participants overestimated the inclination of the slant if they were physically tired or had a weight added on their shoulders at the time they were making their judgments. Proffitt's work clearly demonstrates that perception of the world and the cognitive interpretation that derives from it are embodied. Perceptual judgment tasks are not simply determined by the cognitive ability to analyze and interpret the physical characteristics of surroundings (light conditions, geometry, perspectives, etc.) but are also influenced by the body, its physiological states, and potentials for action/interaction with the surroundings even though actions are not necessarily a requirement for the task.

Another example comes from the work of Günther Knoblich and his collaborators (for a recent review, see Knoblich & Sebanz, 2006) revealing how our basic understanding of the social world is embodied. Knoblich's argument relies on the mirror system (and mirror neurons in the brain) and indicates that social cognition is often constructed on the recognition of patterns of the perceptions/actions of others; this is especially relevant for patterns that match the patterns experienced by the observer. By identifying patterns of behaviors or actions in other people and recognizing them as one's own, one can make clear predictions about other's actions. This mechanism of identification is considered by Knoblich as a crucial building block of social understanding and social interaction.

Finally, a last example comes from a recent study by Beilock and Holt (2007) showing how preferences and choices can be unconsciously driven by our motor system and prior experience. These investigators asked expert and novice typists to simply choose between two neutral pairs of consonant letters that carried no meaning. One pair (*CJ*), if typed, would require the use of two fingers and would be easier to perform. The other pair (*FV*), on the other hand, would require the use of the same finger twice to be typed, which would be more likely to create interference. Note that there was no mention of typing in the study. The expert typists chose above chance the pair that would have required two fingers to be typed, although, there was no explicit reason to prefer that pair, whereas the novice typists showed no particular preference for one pair. These findings brought the authors to conclude that preference judgments, even of abstract, nonemotionally charged stimuli, can be driven by covert sensorimotor experience.

Together, these examples reflect that some patterns of thinking, even if they do not require actual motor activity, are the product of body-mind coupling, and this body-mind coupling is deeply grounded in the sensory and motor experiences of the thinker.

This leads to the second point which deals with the developmental origins of thought and embodied cognition. Where does the mind come from? How are thoughts generated? And how do memories or mental representations form? The answers to these questions are profoundly rooted in the developmental history of the individual, as infants, children, and adults act on and interact with the environment. Piaget (1967) considered mental representations, memories, and problem-solving abilities to be products of prior internalized patterns of perceptions and actions. Dynamic systems approaches are compatible with the view that it is through the history of prior perceptions and actions in interaction with the task that mental forms (such as memories, cognition, and patterns of behavior) emerge (see Smith, Chapter 5; Smith, 2005). Many cognitive skills are, in fact, the product of a long developmental history involving sustained interactions between brain, body, and the environment through perceptions and actions. The ability to count without needing our fingers or without needing pencil and paper, for instance, is the product of many repeated efforts at mapping the mind with the body pointing at targets or writing numbers. Language also is deeply grounded in the perceptual and motor systems. Very early in development, infants fine-tune their hearing to sounds specific to their native languages (Werker & Tees, 1984, 1992), and the production of sounds and speech develops as the child learns to control and master the complex speech apparatus over time (Goffman, Heisler, & Chakraborty, 2006). Spatial representations have also been shown to form from perceiving and acting in the environment from a very early age (Campos et al., 2000; Kermoian & Campos, 1988). To use yet another example, even highly skilled chess players, who can remember the position of each piece on the checkerboard and anticipate several potential moves ahead, have reached that level of mental performance after years of practice at a game grounded in perception and action. Each planned move reflects a potential for action that is deeply embodied. Highly skilled chess players have developed those forms of mental skills from hours of prior perceptions and actions involving the brain, the body, and the environment as the pieces were moved on the checkerboard and direct consequences of those particular moves were observed. These early skills then evolved to become more sophisticated, where multiple moves and their outcomes could be envisioned by the player without direct action. In sum, it is hard to think of a disembodied mind, especially when we realize that all the senses the mind relies on to connect with the world are deeply embodied. Our senses of sight, touch, smell, hearing, and taste are intimately coupled with specific motor activities such as orienting the head, moving the fingers, activating our mouth and tongue, or turning the eyes to detect, capture, and encode information from the environment.

The embodied view sees the ability to develop abstract forms of thinking from the continuous body and mind mapping that takes place over developmental time. Just as infants manage to learn over the years to reach for objects in an endless number of varied situations without effort and without having to think intently about the dynamics of their movements, we have learned to think, reason, count, and generate solutions in our minds without needing to activate or rely on our bodies. The fact that we can reach for objects seemingly automatically or think apparently without using our bodies is not evidence of a disembodied mind, but rather the indication that brain, body, and environment have reached a different level of coupling. In that sense, acquiring knowledge, becoming competent, and developing social skills are not simply the product of a cognitive achievement; they are the product of a developmental real-world process of mutual interactions between brain, body, and environment.

## References


Adolph, K. E., & Berger, S. E. (2006). Motor development. In W. Damon & R. M. Lerner (Series Eds.) & D. Kuhn & R. S. Siegler (Vol. Eds.), *Handbook of child psychology: Vol. 2. Cognition, perception, and language* (6th ed., pp. 161–202). New York: Wiley.

Andres, F. G., Mima, T., Schulman, A. E., Dichgans, J., Hallett, M., & Gerloff, C. (1999). Functional coupling of human cortical sensorimotor areas during bimanual skill acquisition. *Brain, 122*(5), 855–870.

Annett, M. (1985). *Left, right, hand and brain: The right shift theory.* Hillsdale, NJ: Erlbaum.

Beilock, S. L., & Holt, L. E. (2007). Embodied preference judgments: Can likeability be driven by the motor system? *Psychological Science, 18*(1), 51–57.

Bell, M. A., & Fox, N. (1996). Crawling experience is related to changes in cortical organization during infancy: Evidence from EEG coherence. *Developmental Psychobiology, 29*(7), 551–561.

Bushnell, E. W. (1985). The decline of visually guided reaching during infancy. *Infant Behavior and Development, 8*(2), 139–155.

Butterworth, G. E. (1977). Object disappearance and error in Piaget's stage IV task. *Journal of Experimental Child Psychology, 23*(3), 391–401.

Campos, J. J., Anderson, D. I., Barbu-Roth, M. A., Hubbard, E. M., Hertenstein, M. J., & Witherington, D. (2000). Travel broadens the mind. *Infancy, 1*(2), 149–219.

Chiel, H. J., & Beer, R. D. (1997). The brain has a body: Adaptive behavior emerges from interactions of nervous system, body and environment. *Trends in Neurosciences, 20*(12), 553–557.

Clearfield, M. W., & Thelen, E. (2001). Stability and flexibility in the acquisition of skilled movements. In C. A. Nelson & M. Luciana (Eds.), *Handbook of developmental cognitive neuroscience* (pp. 253–266). Cambridge, MA: MIT Press.

Clifton, R. K., Muir, D. W., Ashmead, D. H., & Clarkson, M. G. (1993). Is visually guided reaching in early infancy a myth? *Child Development, 64*(4), 1099–1110.

Clifton, R. K., Rochat, P., Robin, D. J., & Berthier, N. E. (1994). Multimodal perception in the control of infant reaching. *Journal of Experimental Psychology: Human Perception and Performance, 20*(4), 876–886.

Corballis, M. C. (1991). *The lopsided ape: Evolution of the generative mind.* New York: Oxford University Press.

Corbetta, D., & Bojczyk, K. E. (2002). Infants return to two-handed reaching when they are learning to walk. *Journal of Motor Behavior, 34*(1), 83–95.

Corbetta, D., & Thelen, E. (1996). The developmental origins of bimanual coordination: A dynamic perspective. *Journal of Experimental Psychology: Human Perception and Performance, 22*, 502–522.

Corbetta, D., & Thelen, E. (1999). Lateral biases and fluctuations in infants' spontaneous arm movements and reaching. *Developmental Psychobiology, 34*(4), 237–255.

Corbetta, D., & Thelen, E. (2002). Behavioral fluctuations and the development of manual asymmetries in infancy: Contribution of the dynamic systems approach. In S. J. Segalowitz (Ed.), *Handbook of neuropsychology: Vol. 8. Child neuropsychology, Part 1* (2nd ed., pp. 309–328). Amsterdam: Elsevier Health Sciences.

Corbetta, D., Williams, J., & Snapp-Childs, W. (2006). Plasticity in the development of handedness: Evidence from development and early asymmetric brain injury. *Developmental Psychobiology, 48*(6), 460–471.

Coryell, J. (1985). Infant rightward asymmetries predict right-handedness in childhood. *Neuropsychologia, 23*(2), 269–271.

Diamond, A. (1990). The development and neural bases of memory functions as indexed by the AB and delayed response tasks in human infants and infant monkeys. In A. Diamond (Ed.), *Annals of the New York Academy of Sciences: Vol. 608. The development and neural bases of higher cognitive functions* (pp. 267–317). New York: New York Academy of Sciences.

Diedrich, F. J., Thelen, E., Smith, L. B., & Corbetta, D. (2000). Motor memory is a factor in infant perseverative errors. *Developmental Science, 3*(4), 479–494.

Dorris, M. C., Paré, M., & Munoz, D. P. (2000). Immediate neural plasticity shapes motor performance. *Journal of Neuroscience, 20*(1), RC52, 1–5.

Fetters, L., & Todd, J. (1987). Quantitative assessment of infant reaching movements. *Journal of Motor Behavior, 19*(2), 147–166.

Gesell, A. (1956). The ontogenesis of infant behavior. In L. Carmichael (Ed.), *Manual of child psychology* (2nd ed., pp. 295–331). New York: Wiley.

Gesell, A., & Ames, L. B. (1947). The development of handedness. *Journal of Genetic Psychology, 70*, 155–175.

Goffman, L., Heisler, L., & Chakraborty, R. (2006). Mapping of prosodic structure onto words and phrases in children's and adults' speech production. *Language and Cognitive Processes, 21*(1–3), 25–47.

Hellige, J. B. (1993). *Hemispheric asymmetry: What's right and what's left.* Cambridge, MA: Harvard University Press.

Jenkins, W. M., Merzenich, M. M., Ochs, M. T., Allard, T., & Guic-Robles, E. (1990). Functional reorganization of primary somatosensory cortex in adult owl monkeys after behaviorally controlled tactile stimulation. *Journal of Neurophysiology, 63*(1), 82–104.

Karni, A., Meyer, G., Rey-Hipolito, C., Jezzard, P., Adams, M. M., Turner, R., et al. (1998). The acquisition of skilled motor performance: Fast and slow experience-driven changes in primary motor cortex. *Proceedings of the National Academy of Sciences, USA, 95*(3), 861–868.

Kelso, J. A. S. (1995). *Dynamic patterns: The self-organization of brain and behavior.* Cambridge, MA: MIT Press–Bradford Books.

Kermoian, R., & Campos, J. J. (1988). Locomotor experience: A facilitator of spatial cognitive development. *Child Development, 59*(4), 908–917.

Kleim, J. A., Barbay, S., & Nudo, R. J. (1998). Functional reorganization of the rat motor cortex following motor skill learning. *Journal of Neurophysiology, 80*(6), 3321–3325.

Knoblich, G., & Sebanz, N. (2006). The social nature of perception and action. *Current Directions in Psychological Science, 15*(3), 99–104.

Mathew, A., & Cook, M. (1990). The control of reaching movements by young infants. *Child Development, 61*(4), 1238–1257.

McManus, I. C. (1985). Handedness, language dominance and aphasia: A genetic model. *Psychological Medicine. Monograph Supplement, 8*, 1–40.

Merzenich, M. M., Allard, T. T., & Jenkins, W. M. (1990). Neural ontogeny of higher brain functions: Implications of some recent neurophysiological findings. In O. Franzén & P. Westman (Eds.), *Information processing in the somatosensory system* (pp. 193–209). London: MacMillan Press.

Merzenich, M. M., & Jenkins, W. M. (1993). Reorganization of cortical representations of the hand following alterations of skin inputs induced by nerve injury, skin island transfers, and experience. *Journal of Hand Therapy, 6*(2), 89–104.

Michel, G. F. (1981). Right handedness: A consequence of infant supine head-orientation preference? *Science, 212*(4495), 685–687.

Munakata, Y. (1997). Perseverative reaching in infancy: The role of hidden toys and motor history in the AB task. *Infant Behavior and Development, 20*(3), 405–416.

Nudo, R. J., Milliken, G. W., Jenkins, W. M., & Merzenich, M. M. (1996). Use-dependent alterations of movement representations in primary motor cortex of adult squirrel monkeys. *Journal of Neuroscience, 16*(2), 785–807.

Perris, E. E., & Clifton, R. K. (1988). Reaching in the dark toward sound as a measure of auditory localization in infants. *Infant Behavior and Development, 11*(4), 473–491.

Petersen, S. E., van Mier, H., Fiez, J. A., & Raichle, M. E. (1998). The effects of practice on the functional anatomy of task performance. *Proceedings of the National Academy of Sciences, USA, 95*(3), 853–860.

Piaget, J. (1952). *The origins of intelligence in children* (M. Cook, Trans.). Madison, CT: International Universities Press.

Piaget, J. (1954). *The construction of reality in the child* (M. Cook, Trans.). New York: Basic Books.

Piaget, J. (1967). *La psychologie de l'intelligence* [The psychology of intelligence]. Paris: Armand Colin.

Proffitt, D. R. (2006). Embodied perception and the economy of action. *Perspectives on Psychological Science, 1*(2), 110–122.

Provins, K. A. (1997). Handedness and speech: A critical reappraisal of the role of genetic and environmental factors in the cerebral lateralization of function. *Psychological Review, 104*(3), 554–571.

Ramsay, D. S. (1984). Onset of duplicated syllable babbling and unimanual handedness in infancy: Evidence for developmental change in hemispheric specialization? *Developmental Psychology, 20*(1), 64–71.

Ramsay, D. S. (1985). Fluctuations in unimanual hand preference in infants following the onset of duplicated syllable babbling. *Developmental Psychology, 21*(2), 318–324.

Schöner, G. (1995). Recent developments and problems in human movement science and their conceptual implications. *Ecological Psychology, 7*(4), 291–314.

Smith, L. B. (2005). Cognition as a dynamic system: Principles from embodiment. *Developmental Review, 25*(3–4), 278–298.

Smith, L. B., Clearfield, M. W., Diedrich, F. J., & Thelen, E. (2008). Piaget was right: Thoughts close to the sensory-motor surface. Manuscript in preparation.

Smith, L. B., & Thelen, E. (Eds.). (1993). *A dynamic systems approach to development: Applications.* Cambridge, MA: MIT Press–Bradford Books.

Smith, L. B., Thelen, E., Titzer, R., & McLin, D. (1999). Knowing in the context of acting: The task dynamics of the A-not-B Error. *Psychological Review, 106*(2), 235–260.

Steinhage, A., & Schöner, G. (1997). Self-calibration based on invariant view recognition: Dynamic approach to navigation. *Robotics and Autonomous Systems, 20*(2–4), 133–156.

Streri, A. (1993). *Seeing, reaching, touching: The relations between vision and touch in infancy* (T. Powell & S. Kingerlee, Trans.). Cambridge, MA: MIT Press. (Original work published 1991)

Thelen, E. (2000). Grounded in the world: Developmental origins of the embodied mind. *Infancy, 1*(1), 3–28.

Thelen, E., Corbetta, D., Kamm, K., Spencer, J. P., Schneider, K., & Zernicke, R. F. (1993). The transition to reaching: Mapping intension and intrinsic dynamics. *Child Development, 64*(14), 1058–1098.

Thelen, E., Corbetta, D., & Spencer, J. P. (1996). Development of reaching during the first year: Role of movement speed. *Journal of Experimental Psychology: Human Perception and Performance, 22*(5), 1059–1076.

Thelen, E., Schöner, G., Scheier, C., & Smith, L. B. (2001). The dynamics of embodiment: A field theory of infant perseverative reaching. *Behavioral and Brain Sciences, 24*(1), 1–34, commentary 34–70, authors' response 70–80.

Thelen, E., & Smith, L. B. (1994). *A dynamic systems approach to the development of cognition and action.* Cambridge, MA: MIT Press–Bradford Books.

Viviani, J., Turkewitz, G., & Karp, E. (1978). A relationship between laterality of functioning at 2 days and at 7 years of age. *Bulletin of the Psychonomic Society, 12*(3), 189–192.

Von Hofsten, C. (1979). Development of visually directed reaching: The approach phase. *Journal of Human Movement Studies, 5*(3), 160–178.

Von Hofsten, C. (1991). Structuring of early reaching movements: A longitudinal study. *Journal of Motor Behavior, 23*(4), 280–292.

Wellman, H. M., Cross, D., & Bartsch, K. (1986). Infant search and object permanence: A meta-analysis of the A-not-B error. *Monographs of the Society for Research in Child Development, No. 214, 51*(3), 1–67.

Werker, J. F., & Tees, R. C. (1984). Cross-language speech perception: Evidence for perceptual reorganization during the first year of life. *Infant Behavior and Development, 7*(1), 49–63.

Werker, J. F., & Tees, R. C. (1992). The organization and reorganization of human speech perception. *Annual Review of Neuroscience, 15*, 377–402.

# CHAPTER 4

# Dynamic Systems, Sensorimotor Processes, and the Origins of Stability and Flexibility

*Linda B. Smith*

Human intelligence reveals both remarkable stability and nimble flexibility. Stability emerges from the incorporation of the past into the present. Flexibility, in turn, requires an abandoment of (or selection among) past ways, a shifting of responses to meet new circumstances. All of cognitive science, one way or another, is concerned with processes of stability and/or flexbility (see Samuelson & Smith, 2000). These are also the core issues in this volume, a volume onstensibly about connectionist versus dynamic systems approaches to cognitive development. My charge in writing this chapter was to consider infant perseveration in the A-not-B task from a dynamic systems point of view. And so I shall. But my main point is not an argument for dynamic systems over connectionism, as that seems not to be an argument much worth having (see Samuelson & Smith, 2000). Much more interesting are the phenomena themselves. Furthermore, both connectionist and dynamic systems accounts of the A-not-B error provide insight into the multilayered and nested dynamics that make up human intelligence. And our intelligence, for the sake of *coherence*, always leans a bit to the *just-previous past* and is thus susceptible to preseveration—stable patterns or *stability*. This is a main point of this chapter. A second point is that human intelligence is also always open to sensory input and is thus always being perturbed. These perturbations can yield either variability or new stable patterns. It is the latter result that yields *flexibility*. The dynamic systems approach to the A-not-B error has provided new insights into the phenomenon itself, into the larger issues of stability and flexibility, and into the multilayered dynamics of the *sensorimotor processes* that underlie all of these.

In this chapter, I first present an overview of a dynamic systems account of the A-not-B error in infants. Second, I present evidence on how the processes that give rise to perseveration in infants are also processes that, in a quite positive way, create coherence and stability. Third, I present evidence on how flexibity emerges in this same system. I then consider the role of sensorimotor processes and implications of this work.

## Activation of a Motor Plan: The A-not-B Error

Jean Piaget (1963) viewed cognitive development as the progressive differentiation of internal representations from sensorimotor processes. Advanced forms of cognition were those that were separate from the here and now of perceiving and acting. The A-not-B task, along with the *object concept* this paradigm sought to measure, played a central role in then-current theorizing. The object concept was defined as the belief that objects persist in space and time independent of one's own perceptual and motor contact with them. The finding that infants search for a hidden object where they have just previously acted rather than where they have last seen the object dissappear was taken as indicative of representations that are sensorimotor in nature.

The A-not-B task begins with the experimenter hiding an enticing toy under a lid at location A. After a delay (typically 3–5 s), an infant is allowed to reach, and most infants reach to location A and retrieve the toy. This A-trial is repeated several times. Then, there is the crucial switch trial: the experimenter hides the object at location B as an infant watches. Where does an infant reach? After a delay, infants 8–10 months of age make the characteristic A-not-B error: that is, an infant reaches back to location A where the object had been found previously. Infants older than 12 months of age usually reach correctly on the crucial B-trials (see Wellman, Cross, & Bartsch, 1986). Piaget suggested that this pattern indicates that older infants, but not younger ones, know that objects can exist independently of their own actions. There has, of course, been much debate about this conclusion, and many relevant experiments tested a variety of alternatives (Acredolo, 1979; Baillargeon, 1993; Bremner, 1978; Diamond, 1998; Munakata, 1998). Indeed, the A-not-B task is now widely understood as one of a class of *executive control tasks* that require flexibility in the face of a past, stable response (Diamond, 1991; Munakata, 1998; Zelazo et al., 2003).

The dynamic field account (Smith & Thelen, 1993; Smith, Thelen, Titzer, & McLin, 1999; Spencer, Smith, & Thelen, 2001; Thelen, Schöner, Scheier, & Smith, 2001; Thelen & Smith, 1994) focuses on the sensorimotor processes that underlie a reach to a location in space and, in this way, returns to Piaget's original idea that the error is sensorimotor in its origins and reveals the nature of sensorimotor represesentations. The dynamic field account begins with a task analysis of these sensorimotor processes as illustrated in Figure 4.1. In the A-not-B task, infants look at objects at particular locations, and then—repeatedly—reach to those locations. Thus, the infant first *watches* a series of events: the toy being put into a hiding location (a well) and then covered with a lid. From this, the infant must *formulate* a *motor plan* to reach and must *maintain* the plan over the delay, and then *execute* the plan. Motor plans are themselves abstract in the sense that they are not descriptions of particular muscle movements (or even plans about specific movements using a particular body part). Motor plans can be *represented* with respect to multiple spatial reference frames (with repect to objects in the visual field/world,

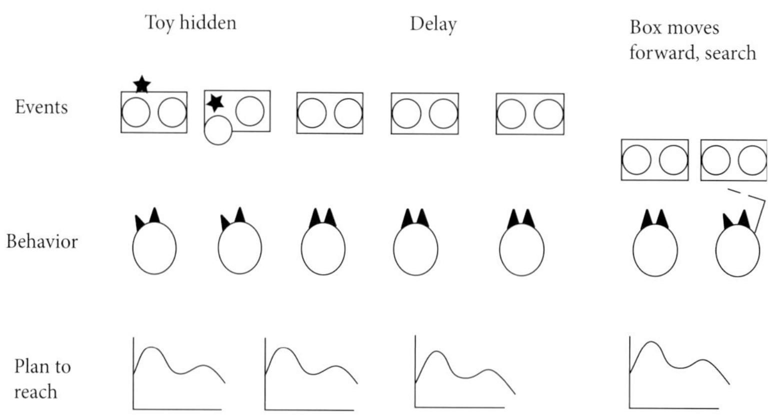


**Figure 4.1. Task analysis of the A-not-B task: Typical hiding event at location A**
The box and the two hiding wells (location A and location B) constitute the continually present visual input. The specific or transient input is the hiding of the toy in a well (location A in the event depicted). A delay is imposed between hiding the toy and allowing the infant to reach. During these events, the infant looks at the objects in view, remembers the cued location, and undertakes a planning process leading to the activation of reach parameters, followed by reaching itself.

to eyes, to the head, to effectors such as arms/hands, and to processes that transport them from one place to the other); but ultimately, motor plans must specify the relation between the body, its effectors, and the location of the target (see Cohen & Andersen, 2002). Such a motor plan is necessary in *any* account of an infant's actual performance of this task and, in and of itself, constitutes a *representation* of the existence of an object.

Thelen, Schöner, Scheier, and Smith's (2001) formal account of these processes using the dynamic field model is illustrated in schematic form in Figure 4.2. The central figure depicts the activation of a motor plan that codes the direction of a reach. Three dimensions define this motor planning field: the spatial direction of the reach (*x* axis), the activation strength (*y* axis), and time (*z* axis). Importantly, the activation strength must pass some threshold in order for a reach to actually be executed (this chapter returns to this point in the discussion of flexibilty). According to the dynamic systems model, activation in this field is driven by three inputs (three panels at left in Fig. 4.2). The first input is the continually present, sensory activation due to the covers/lids atop the two wells; this input drives activation (generally below a reaching threshold) to those two locations. The second input is the hiding event, the specific cue that instigates a rise in activation at one location when the object is hidden. It is this activation from the specific cue that must be maintained over the delay if the infant is to reach correctly after the delay (see Fig. 4.1). The third input is the relatively long-term memory of the previous reaches, which can perturb the evolving activation in the field, pulling it in the direction of previous reaches (i.e., a perservative response). The activation for reaching in the motor planning field (central panel in Fig. 4.2) evolves over time as a function of the sensory events, the dynamics of the activations those events drive, the memories of past reaches, and the

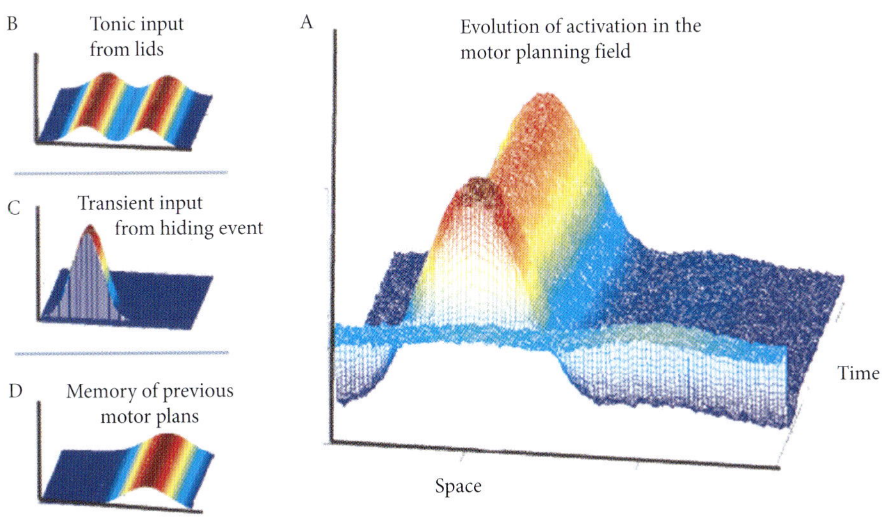


**Figure 4.2. Dynamic field model of the A-not-B task: Overview**
For the A-not-B task used with infants, the motor planning field is depicted and defined in three dimensions. The x axis indicates the spatial direction of the reach to the right or left; the y axis indicates the activation strength; the z axis is time. As depicted in three panels on the left, activation in the motor planning field is driven by the tonic input of the hiding locations (two wells covered by lids), the transient hiding event, and the memories of prior reaches. The central panel depicts a sustained activation to a hiding event on the left despite recent memories of reaching to the right (that is, a non-perservative response). See text for further explanation of events.

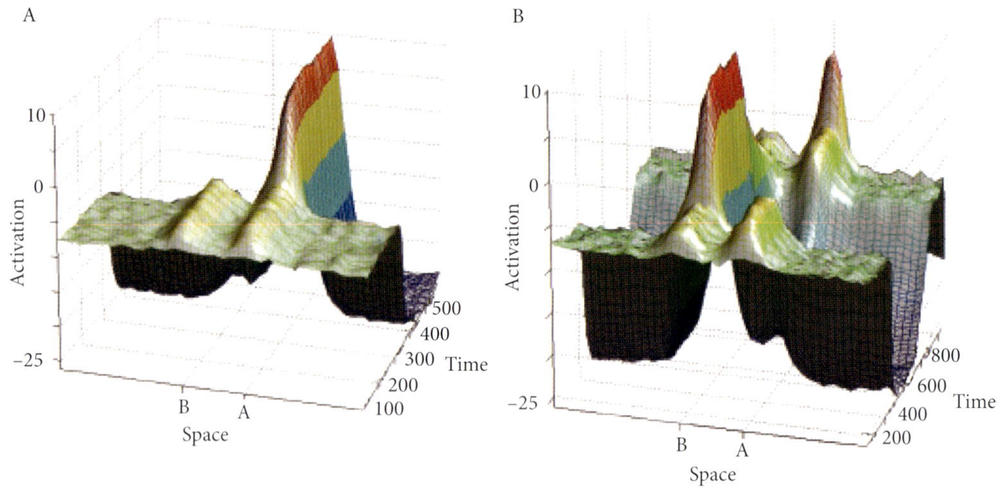


**Figure 4.3. Dynamic field model of the A-not-B task: Time-evolution**
(Panel A) The time-evolution of activation in the motor planning field is depicted for the initial A-trial. The activation rises for A as the object is hidden and, due to self-organizing properties in the field, is sustained during the delay. (Panel B) The time-evolution of activation in the motor planning field is depicted for the subsequent, first B-trial. There is heightened activation at A prior to the hiding event due to memory for prior reaches. As the object is hidden at location B, activation rises at B; but as this transient cueing event ends, activation at B decays and memory of previous motor plans dominate the field. Due to the memory properties of the field, activation increases at A, generating the A-not-B error. (Panels A and B) In these simulations of the model, time is depicted in seconds (100–800 s).

field's own internal dynamics. The central panel in Figure 4.2 depicts a sustained activation to a hiding event on the left despite recent memories of reaching to the right (i.e., a non-perservative response).

Figure 4.3 shows results from simulations of the model as a dynamic sensorimotor representation. Figure 4.3A illustrates the evolution of activation in the motor planning field on the first A-trial. Before the infant has seen any object hidden, there is low activation in the field at both A and B; this activation is generated from the perceptual input of the two hiding covers/lids. As the experimenter directs attention to location A by hiding the toy, this perceived event produces high transient activation at A. The field evolves and maintains a planned reaching direction to A. The evolution of a sustained activation peak that can drive a reach even after a delay, even when the object is hidden, is a consequence of the self-sustaining properties of the dynamic field. Briefly, the points within the motor planning field provide input to one another such that a highly activated point will exert a strong inhibitory influence over the points around it, allowing an activation to be maintained in the absence of external input.

This dynamic field, the plan to reach, is continuously informed by the sensory input and continuously driven by that input. But the activation in the field is also driven by—and constrained by—memories of previous reaches. Thus, on the second A-trial, there is increased activation at A because of the previous activity there. This combines with the hiding cue to produce a second reach to location A. Over many trials to location A, a strong memory of previous actions builds up; each trial embeds the history of previous trials. Figure 4.3B illustrates the consequence of this on the critical/first B-trial. The experimenter provides a strong cue to B by hiding the object at location B. But as the activation from that specific cue decays, the lingering memory of the actions at A begins to dominate the field, and indeed, over the course of the delay through the self-organizing properties of the field itself, activation shifts back to the site of habitual activation at A. The model predicts that the A-not-B error is time dependent: there is (all other things being equal) a brief period immediately after the hiding event in which infants should reach correctly (given that the hiding event was a sufficiently strong, specific cue). Past research shows that without a delay, infants typically do reach correctly in the standard version of the A-not-B task (Wellman et al., 1986).

The model makes a number of predictions that have been tested in a variety of experiments (see Thelen et al., 2001). Because reaching is continuously tied to immediate input, visual events at hiding, at the moment of the reach, and indeed even after the reach has begun can drive a different solution and push the reach to A or to B; this has been experimentally demonstrated (e.g., Clearfield, Dineva, Smith, Diedrich, & Thelen, 2008; Diedrich, Highlands, Spahr, Thelen, & Smith, 2001; Smith et al., 1999). Indeed, simulations from the model can be used to design experimental manipulations that cause infants at 8–10 month of age to search correctly on B-trials or that cause children at 2–3 year of age to make the A-not-B error (Schutte, Spencer, & Schöner, 2003; Spencer, Smith, & Thelen, 2001). Such effects can be achieved by changing the delay, heightening/lessening the attention-grabbing properties of covers/lids or hiding event, or by increasing/decreasing the number of prior reaches to A (Diedrich et al., 2001; Smith et al., 1999). All these effects show how the motor plan is dynamically connected to sensory events and motor memories. Because one can make the error appear or disappear in these ways over a broad range of ages (8–30 months), we know that the relevant processes cannot be tightly tied to one developmental period. Instead, changes in A-not-B task performance may be reflective of *general* processes that govern spatially directed action. The dynamic field model explicitly incorporates this idea by showing how seemingly qualitatively distinct patterns—perseveration, non-perseveration—emerge through small changes in the parameters.

Most critically for the discussion in this chapter, the processes that underlie the behavior—the activations in the dynamic field—are conceptualized as motor plans, plans that take the hand from its current location to the target. Memories for previous reaches—the memories that create the perseverative response/error—are also motor plans. That is, by this account, the processes that create the error reside close to the sensorimotor surface and are embedded in, not separate from, the processes of perceiving and acting. By this account, and as Piaget originally proposed, the A-not-B error arises from *sensorimotor representations*.

There is strong experimental support for this view. In particular, manipulations that distort the body's posture erase the memory for prior reaches and thus erase the cause of perseveration, leading to correct reaches on B-trials (Lew, Hopkins, Owen, & Green, 2007; Smith et al., 1999; Smith, Clearfield, Diedrich, & Thelen, 2008, in preparation). In one experiment, infants were in one posture (e.g., sitting) on A-trials and were then shifted to another posture (e.g., standing) on B-trials. This posture shift between A- and B-trials (but not other kinds of distractions) caused even 8- to 10-month-old infants to search *correctly*, supporting the proposal that the relevant memory is a motor plan. To be effective, motor plans must be in the coordinates of the body's position, that is, tied to the *current* relation of the body to the physical world. Thus, it makes sense that these memories for prior motor plans are not activated unless the body is in the proper position with respect to the targets in the world.

These results, and the conceptualization of the relevant processes as motor plans, fit well with Piaget's (1963) original conceptualization of A-not-B errors as representations tied to one's sensorimotor interactions with objects. The posture-shift results, in particular, fit with Piaget's idea that, for the young infant, the mental object and its location are inseparable from bodily action.

Such sensorimotor representations were considered by Piaget to be a limitation of an immature cognitive system. But the dynamic field account views these representations as general mechansisms underlying visually guided reaching at all ages. Might this form of sensorimotor representation that is so ill-suited to Piaget's A-not-B task be, in other contexts, a positive aspect of human intelligence? Might the processes that create the stability that leads to preseveration in the A-not-B task also create beneficial coherence in other tasks?

## Stability and Coherence: The Baldwin Task

By the dynamic systems account, infants keep track of objects in the A-not-B task by forming a plan for action, a motor plan of how to move the hand to the object's location. Because the motor plan necessarily specifies a target location, that plan represents the object's existence as a location. Activation of one component of this representation can give rise to activation of other components. Thus, in the A-not-B task, activation of the memory of the object and a goal to reach (through the hiding event at B) activates the *prior* motor plan and yields a reach to A. But this is just one possible consequence of the binding of an object to a location through an action plan. Activation of the action plan (and past location as a goal to reach), should also work to call forth the memory of the non-present object.

This section describes a series of recently completed studies that show how the same processes that create the A-not-B error (i.e., an inappropriate stability, given the task) also create coherence in other tasks, enabling the appropriate connection of an immediate sensory event to the *correct* contents in the just-previous past. The new task concerns how older infants (18–24 months of age) map names to referents when those names and referents are separated in time and occur in a stream of events with multiple objects and multiple shifts in attention among those objects. The experiments use an ingenious task that was created by Baldwin (1993) to study early word learning, but which is, in many ways, a variant of the classic A-not-B task.

The stream of events in Baldwin's task is illustrated in Figure 4.4A. The experimenter

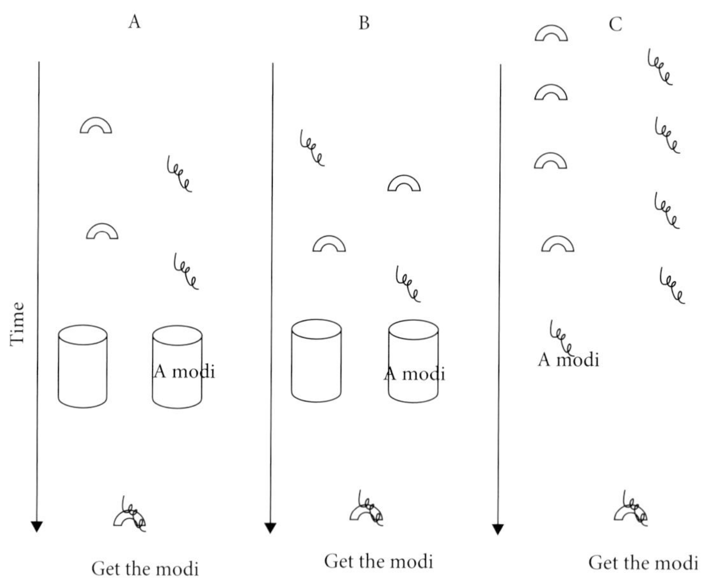


**Figure 4.4. The Baldwin task: Events in three different conditions**
In each condition, two novel, unnamed objects are repeatedly presented (2 or 4 times) with the respective side of presentation specified by the procedure. A verbal label is then provided for one object without the object in view. Subsequently, the objects are presented in a novel location and on top of each other (so that there are no spatial cues), and the child is asked which one is the referent of the label provided earlier. See text for further clarification.

sits before a child at a table and presents the child with first one unnamed object on one side of midline and then with a second unnamed object on the other side. This is repeated. Out of the sight of the child, the two objects are then put into containers, and the two containers are placed side by side on the table. The experimenter looks into one container and says, "I see a *modi* in here." The experimenter does not show the child the object in the container. Later the objects are retrieved from the containers and presented in a new location; the child is asked which one is a *modi*. Notice that the name and the object were never jointly experienced. Baldwin showed that young children nonetheless connect the object name to the correct object even though that object is not in view when the name is heard.

Figure 4.5 illustrates the A-not-B task and the Baldwin task showing their similar surface structures. In the A-not-B task, children turn to look at and reach to an object at a particular

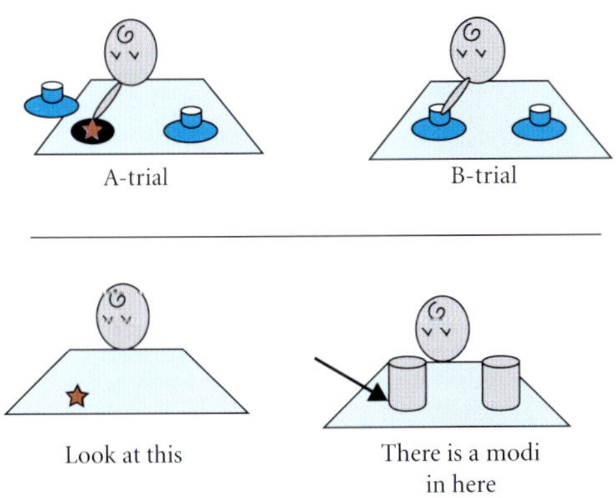


**Figure 4.5. Parallel steps in the A-not-B task and the Baldwin task**

In both the A-not-B task and the Baldwin task, children repeatedly look and reach to locations to interact with objects. (Top panel) In the A-not-B task, an object is presented at location A causing motor planning memory biased to location, and in this way, binding the object to location A. (Bottom panel) In the Baldwin task, two objects are presented at specific locations and—through remembered motor plans and attentional plans—remain bound to the respective locations. Children can then use this binding of objects to locations to link a name to a nonpresent object.

location. By hypothesis, this binds the object to the action plans of looking and reaching to that location. Subsequently, the goal of reaching for that object activates that plan and the old target location. In the Baldwin task, children turn attention to two different objects at two different locations and again by hypothesis bind these objects to spatially specific action plans of looking and reaching. If this analysis is correct, then, just as the goal to reach for the one object in the A-not-B task may call up the spatially specific action plan, then so could activation of the action plan—a look to a specific location—activate the memory of the specific object associated with that plan. In the Baldwin task, this association would lead the child to remember the correct object at the correct time, the moment the experimenter offered that name, enabling the child to associate the object and name. Could this parallel account of perseveration in the A-not-B task and successful mapping of a name to a nonpresent thing in the Baldwin task possibly be correct? Is success in the Baldwin task due to the same processes that yield failure in the A-not-B task? A series of recently completed experiments support this idea (Smith, Samuelson, & Spencer, 2008, in preparation).

The first experiment in the series sought to show that a common direction of spatial attention was critical to children's successful linking of the name to the object in the Baldwin task. As in the original Baldwin study, the participants were young children, 18–24 months of age. The experiment replicated the original Baldwin method in one condition, and in a second condition sought to disrupt children's word-object mappings by disrupting the link between the object and a single direction of attention and action, making spatial location more variable and less predictive of specific objects. The method in this new condition, shown in Figure 4.4B, is the same as that in Figure 4.4A *except* that the two objects were each, prior to the naming event, presented once on the right and once on the left. Thus each object was associated with both directions of attention prior to the naming event. If children keep track of objects through their action plans, and if they link a heard name to a remembered object by common direction of action, then this inconsistency in an object's location prior to naming should disrupt the mapping of the name to the thing in the container. This is what occurred. When each object was consistently linked to one direction of attention, Baldwin's original result was replicated, and the children chose the target object 73% of the time. In constrast, when the objects were not consistently linked to one direction of attention, children chose the target object only 46% of the time (which did not differ from a performance level due to chance). Clearly, children are learning *in the task* about the relation between objects and their locations and using this relation to coherently connect two events—presentation of a name and presentation of a thing—that are experienced at different moments during the task.

One difference between the A-not-B task and the Baldwin task is that the goal in the A-not-B task is to reach to and obtain the toy. Thus, it seems reasonable that the relevant memory in that task—a memory that represents the existence of the object when it is out of view—might be tied to a *motor plan*. But is this reasonable for the Baldwin task? Is turning attention to a location also correctly conceptualized as a motor plan? As Allport (1990) put it, attention is for action; we attend to locations in preparation for possible action, and actions are performed toward objects within the focus of attention. Thus, direction of attention, like the direction of a reach, may be coded through sensorimotor representations tied to the body and action. Accordingly, in the next experiment, the quesion was whether it is the body's direction of attention—and not a specific location in space—that enables children to access the correct object when the experimenter provides the name. If our analysis of the direction of attention as a motor plan for orienting the body in space is correct, then one should be able to activate the memory for one object or the other simply by shifting the child's orientation in one direction or the other. For instance, pulling attention generally to the left during the naming event should activate memories for the object seen on the left and then the name should be linked to that object. The method in this new condition (not shown) was similar to that illustrated in Figure 4.4A *except* that at the moment of naming, there was just an empty table top—no containers, no hidden objects. With the experimenter looking straight into the child's eyes, but with one hand held an arm's length to the side, the experimenter clicked her fingers and said, "modi, modi, modi." The clicking fingers and the outstretched hand directed children's attention to that side, so that when the child heard the name, the child was looking to that side. Later when children were asked to indicate which object was the modi, their responses indicated that the directional orientation at the moment of naming influenced their mapping of the name to the nonpresent object. Children chose the object spatially linked to the clicking fingers 68% of the time. This suggests that the direction of bodily attention is bound to the object. These results also highlight how the solution to connecting experiences separated in time may be found in the child's sensorimotor engagement in the task.

If the relevant link between objects and locations in the Baldwin task are through motor plans and thus close to the sensory surface, are they also disrupted by shifts in the body's postures? The bodily direction of attention, like a reach, is necessarily egocentric. Just how one shifts one's direction of gaze or turns one's body to bring an object into view depends on the body's *current* position in relation to that object. Accordingly, the next experiment altered children's posture. One posture setting consisted of the child sitting low and close to the table such that looks right and left (and reaches left and right) required lateral bodily moves. The second posture setting consisted of the child standing on the edge of the table so that the child looked down with a bird's eye view of the locations. The method was the original Baldwin task with containers as in Figure 4.4A. More specifically, in one condition, the child sat (as in the previous experiments) when the objects were first presented (unnamed) and attention and reaching was individually cued to each object at a different location. By the present theoretical analysis, this should associate the two objects with opposing directions of attention. Then during the naming event, the child stood up. If the memory of a previously experienced object is strongly linked to a plan for action, then this posture shift—by necessarily causing a reorganization of that plan—should disrupt the memory. Recall that in the A-not-B task, such disruption caused a lessening of perseveration and, thus, more correct responses. The prediction, here, is that the posture shift before the naming event should disrupt retrieval of the target object, and children should not be as successful in mapping the name to the object. The full experiment included four conditions: sit-sit, with a visual distraction before the naming event; stand-stand, with a visual distraction before the naming event; sit-stand; and stand-sit. In the visual distraction (no posture shift) conditions, children chose the target (spatially-linked) object on 70% of the test trials; in the posture-shift conditions, they did so 50% of the time (thus, no differerance from a performance level due to chance). These results strongly suggest that the memory for previously experienced objects is in or indexed through processes tightly tied to the body's current orientation in space. These are memories in spatial coordinates tied to the body's position. Again, these are representations close to the sensorimotor surface.

These sensorimotor representations give rise to children's ability to successfully map a name to a physically nonpresent object. But these representations appear fundamentally the same as those that lead even younger children to reach perseveratively in Piaget's classic A-not-B task. In one case these processes create *coherence* in the cognitive system, appropriately linking the correct object in the just-previous past to the immediate input (the naming event), enabling the child to keep track of objects as referents in a conversation and stream of events with attention shifting from one location to another. In the other task, the processes create *perseveration*, an inappropriate sticking to the past when the immediate cues call for a shift to a new response. Cognitive coherence in the Baldwin task is a good form of perseveration—one that yields a positive outcome. If this analysis is correct, can we alter the Baldwin task to create a task-inappropriate *sticking* to the just-previous past—a maladaptive stability? To this end, we attempted to create an A-not-B effect within the Baldwin task.

The reasoning behind this version of the experimental task is this: If attentional direction activates memories associated with motor plans that orient the body, then these activated memories should compete with an ongoing sensory event if that sensory input originates from the same place. This is how the experiment tested this prediction: Prior to the naming event, objects were presented four times, one always on one side of the table and the other always on the other side, in order to build up a strong link between a specific direction of bodily attention and each object. During the naming event, the experimenter showed the child one object, pointed to it and named it, *but did so with the object (and thus the child's attention) at the side associated with the other object* as illustrated in Figure 4.4C. This should set up a possible competition between the just previously experienced object at this location and the present one that is being named—a competition that, like that in the A-not-B task, might not be resolved in favor of the new cue. The results support this. Children selected the actual object they had seen named only 42% of the time, despite the fact that it was in view and pointed to when named. The prior experience of seeing one object in a particular location disrupted linking the name to a physically present object at that same location. This pattern strongly supports the idea that *where one looks selects and activates memories from the just-previous past.* This not only creates interference in our task but also creates interference in linking an audible word to an unseen object in the standard Baldwin task.

All in all, these results point to the power of sensorimotor intelligence. These results also fit with emerging ideas about so-called cheap solutions (cf. O'Regan & Noë, 2001), about how higher cognitive ends may be realized through sensorimotor processes and the continuous coupling of the mind to the physical world through the body. Young children's solution to the Baldwin task is a cheap solution in the sense that it does not require any additional processes other than those that must already be in place for perceiving and physically acting on objects in the world.

## Beyond Sensory Fields to Association Fields

These results discussed above align with Piaget's original idea that the A-not-B error was the result of sensorimotor representations, although Piaget missed the positive contributions these representations offer as a means of keeping track of nonpresent things. The Baldwin-task experiments also suggest that Piaget was correct about the emergence of representations not so tightly tied to sensorimotor processes. The present analysis views children's success in the Baldwin task (as well as infants' perseveration in the A-not-B task) as a form of sensorimotor intelligence—an intelligence bound to the here and now of perceiving and acting. However, not all aspects of children's performance in the Baldwin task fit this idea. Critically, once children map the name to the object, their knowledge of that mapping does not appear to be spatially fixed. This is seen at testing when the child is provided with the name and asked to choose to which of two objects it refers. In all the tasks, the objects are presented at a new location, and as shown in Figure 4.4, are overlapping and on top of one another. Thus, the direction of prior attention and action associated with the name and with the object can not be used to determine the intended target. This is important: *The processes that children use to learn the name and the processes that are available to them once the name has been learned are different.* When children form the link between the object and the name, they appear to use the spatial orientation of their body to retrieve from memory the nonpresent object. But once this mapping is made, they apparently no longer need any spatial correspondence with past experience to direct attention to the object. The course of events in this experiment is thus reminiscent of Piaget's (1963) grand theory of cognition in which he saw development as a progression from sensorimotor intelligence to representations that were freed from the binds of the here and now. In the Baldwin task, children use sensorimotor processes to keep track of things in space and mind, but naming brings a new index to memory that does not involve the body's disposition to act.

John Spencer, Gregor Schöner, Larissa Samuelson, and I are currently working on an extension of the dynamic field model to explain this. Figure 4.6 provides a schematic of the main ideas. At the top are two sensory fields, one field for *objects* plus the spatial direction of attention (Object-Space) and one field for *sounds* plus the spatial direction of attention (Word-Space). Within this theory, these are sensorimotor fields because they are driven by and continuously coupled to the sensory input and because they specify an action plan directing attention. As in the original dynamic field model of the A-not-B error, these fields also are driven by memories of their own recent activation. What is different in our new model is that these sensory fields are coupled to each other and to a new kind of field, an association field, which has no direct sensory input and is not a plan for action. Instead, the Word-Object *association field* has inputs only from the Object-Space field and the Word-Space field; this new association field represents the associations between words and objects in a manner that is unconnected to the spatial context of the original experience. It is, in this formulation, the association field that frees the mapping of words and objects to spatially directed action plans.

This general idea is similar to a proposal by Simmons and Barsalou (2003) concerning the origins of higher cognition in multimodal sensory processes and their associations to each other. Figure 4.7 illustrates this general idea. Consistent with well-established ideas in neuroscience, there are modality-specific and feature-specific areas of sensory and motor (and emotional) representations. These feed into a hierarchical system of association areas. At lower levels, association areas exist for specific modalities, capturing activation states within a single modality. At higher levels, cross-modal association areas integrate activations across

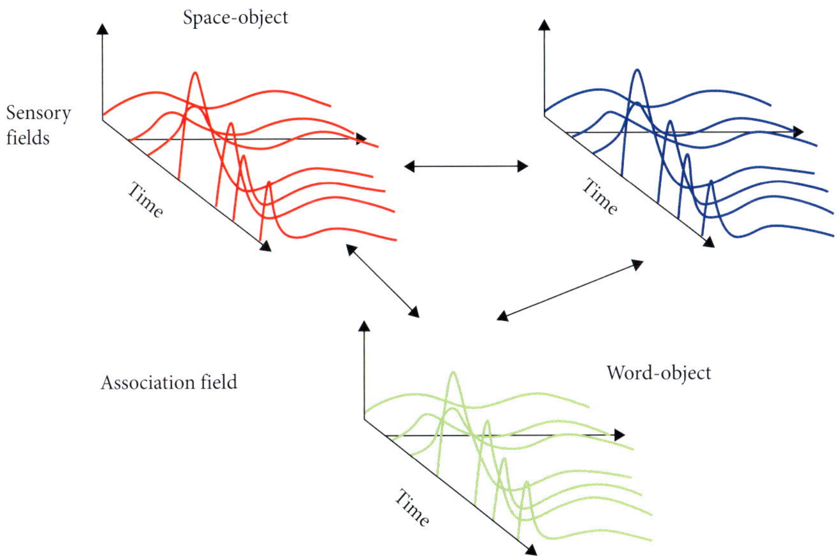


**Figure 4.6. Schematic model of an association field: Creating a word-object association field**
Two sensorimotor fields may be coupled to feed into an association field that no longer requires spatial links. This schematic depicts (1) a sensory field mapping objects with the spatial direction of attention (Object-Space), (2) a sensory field mapping sounds with the spatial direction of attention (Word-Space), and (3) the resulting association field mapping words with objects (Word-Object). In the coupling of the Object-Space sensory field with the Word-Space sensory field, the resulting Word-Object association field maps words to objects without representing the spatial links of those words and objects.

modalities giving rise to higher order regularities that are more abstract and that transcend modality-specific representations, but that are nonetheless built from them. The role of the association areas, then, is to capture modality-specific states for later representational use. In this way, they create stabilities that do not lie so close to the sensory surface and are not continuously coupled to sensory input and to action. The extension of the dynamic field model and the proposal of a word-object association field fits this general idea of a multimodal outside-in architecture.

The evidence reviewed in this chapter on the role of sensorimotor processes in the A-not-B task and in the Baldwin task suggests the following: (1) The processes that create perseverative reaching and the A-not-B error are a truly sensorimotor form of stability, realized in self-sustaining motor plans linked to objects and locations. (2) This stability creates what would seem to be a deficiency in infant cognition but is probably more profoundly about the sensorimotor bases of coherence and how those processes connect the events in the just-previous past to those in the present. (3) This sensorimotor intelligence is also a stepping-stone to representations distinct from current sensorimotor representations.

## Flexibility: Specific Cues

Stabilities are essential to human cognition, otherwise our mental life would flit from one thing to another with every change in sensory input. But too much stability—as the A-not-B error shows—can be a problem. Sometimes *a change in sensory events* requires *a change in response*. Where in this conceptualization is the origin of *flexibility*? It is in the sensory event, in the cue to shift. Figure 4.2, in its schematic presentation of the dynamic field theory, provides an answer: the strength of the specific cue and the subsequent generation

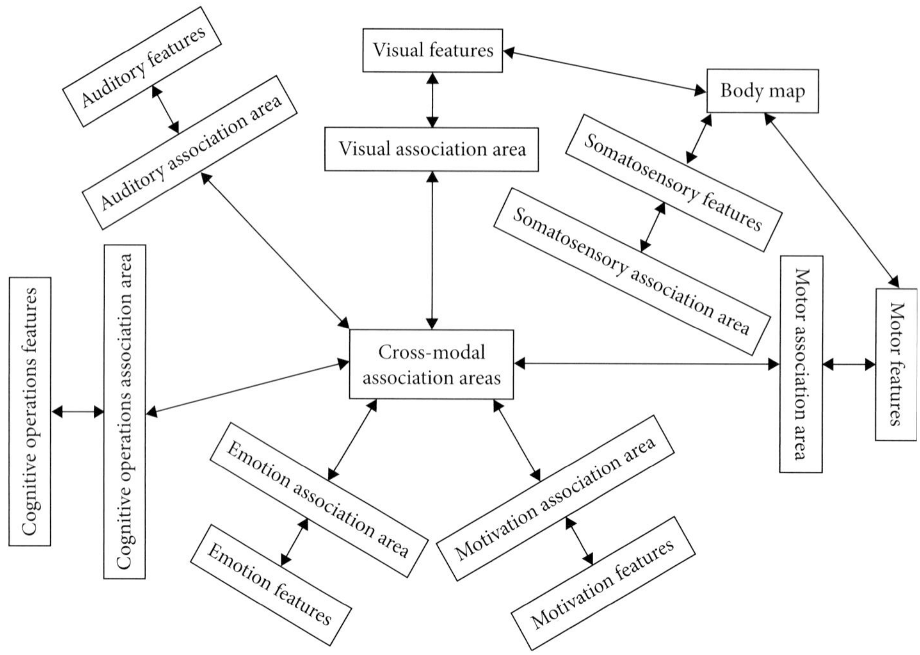


**Figure 4.7. Schematic conceptualization of the hierarchical associations in cognition**
This schematic illustrates the hierarchical architecture proposed by Simmons and Barsalou (2003) to conceptualize the associations leading to cognition. Major features include modality-specific and feature-specific areas of sensory, motor, and emotional representations. Modality-specific sensory and motor areas interact creating modality-specific association areas. Cross-modal interactions on a higher level create multimodal association areas.

of a sustained peak in a new direction point to flexibility. Many factors are relevant as to whether a peak will form in a new direction (Smith et al., 1999; Thelen et al., 2001). But, critically, since there is one final pathway—for example, the movement of the arm in the A-not-B task—all of these factors must be integrated into a single decision. More specifically, within the model, such a decision—to reach at all—depends on the activation surpassing some threshold, and thus releasing interactive processes among neurons in the field to create a single self-sustaining peak. Critically, then, the key question for flexibility is not simply whether the cue to shift is stronger than past memories, but whether it is strong enough to pass a threshold for self-sustaining cooperative interactions in the field.

In the A-not-B task, after a brief delay is imposed, the activation induced by this specific cue is no longer available and the motor planning peak may decay. However, before the reach, at the end of the delay, comes the signal to reach (the pushing of the box forward into the infant's reaching space). Within the model, the end of the delay and the signal that the infant can now reach is modeled by an increased activation to the field as whole, a *generalized excitation* in the motor planning field. This boost in the field's total activation is crucial in that it allows an activation peak to pierce threshold and form a self-sustaining, singular peak of activation that is a decision to reach. Whenever the activation in the field reaches a certain threshold, neural interactions—local excitations and global inhibitions—are set in action, which in turn allow for a single localized peak to form. There is *no* decision to reach unless activation passes a threshold and a single localized peak, specifying the direction, forms and sustains itself. This means that activation level matters. Just where this peak forms—and whether it forms in alignment with the most recent cue—determines whether the infant reaches *flexibly* or *perseveratively*.

In a recent paper, Clearfield, Dineva, Smith, Diedrich, and Thelen (2008) examined two factors relevant to the formation of the activation peak in a reaching task with infants; the salience of the specific cue calling for a shift in reaching direction and a delay between the cue and when the infant is allowed to reach. By dynamic field theory, flexibility and/or perseveration can occur as a function of the integrative response of the field to both the immediate input and the memory from previous reaches. Altering the relative strengths of these inputs to the decision to reach systematically alters the response of the field. Although perseveration is likely for young infants, (e.g., Clearfield et al., 2006; Munakata, 1998; Piaget, 1954; Smith et al., 1999; Thelen et al., 2001), flexibility is also possible. Perseveration and flexibility—even in an infant system with hard to achieve and rapidly decaying memory for specific input—should be systematically related to the strength of the specific cue and the delay period over which its influence on field dynamics must last. Simulations specifically suggested that with no delay, if cue salience is low enough, there should be perserveration (see Clearfield et al., 2008). This is a new prediction, one critical to the idea that total activation in the field and passing the threshold for the formation of a self-sustaining peak are critical. As the cue strength increases, the perseveration rate should drop as the generated activation is enough to pierce the threshold. With a delay, these same differences in cue intensity, will lead to different effects. Specifically, the simulations indicate that with a delay, perseveration rates depend nonlinearly on the cue strength. In particular, perseveration remains high for a wide range of cue strengths and drops fairly suddenly only at very intense, specific cues to shift to B. Thus, the dynamic field theory emphasizes the critical influence of time as well as the singularity of a decision to act.

Clearfield and colleagues tested these predictions in an A-not-B reaching task with infants at 8–10 months of age. Unlike the traditional A-not-B task, there was no hidden object. Instead, two objects were always in view and the experimenter cued the infant to reach to one or the other by holding the object up and waving it. More specifically, infants were presented with pairs of identical reaching targets. Across three between-subjects conditions, these targets varied in their intrinsic attention-getting properties. In all conditions, the targets were placed on a uniform brown presentation box from which they were picked up and cued. In all conditions, the targets, while sitting on the brown presentation, were the same uniform brown color; thus, the task input was the same in all conditions. However, the undersides of the targets—what the infants saw briefly when the experimenter lifted and presented the cue—differed. In the low-salience condition, the underside and thus transient display aspect of the targets were painted the same brown. In the medium-salience condition, the undersides were painted red, yellow, and blue with stripes and polka dots. In the high-salience condition, the undersides presented multicolored flashing lights. Infants were tested in one of two delay conditions (1 s or 3 s).

The stimulus differences are all strong enough to generate correct reaches on the A-trials (no shift). However, the stimulus differences led to dramatic differences in perseveration as shown in Figure 4.8. With no delay, infants perseverated only when the transient cue signaling a need to shift the direction of the reach was relatively weak, but not when the cue was strong. In contrast, with a delay, infants only succeeded in shifting correctly to B when the cue was very strong. These results tell us that infants—in this task—can be flexible or perseverative, both with and without delays; in this way, the infants, in effect, tell us where flexibility comes from. Flexibility originates in—is in the system's response to—perturbations from sensory events. This push from the sensory event is just the first step. Those sensory events must give rise to a new self-sustaining stability if there is to be a new behavioral decision and not just variability. Thus, just as stability can arise in the nested dynamics of sensorimotor processes, so does flexibility.

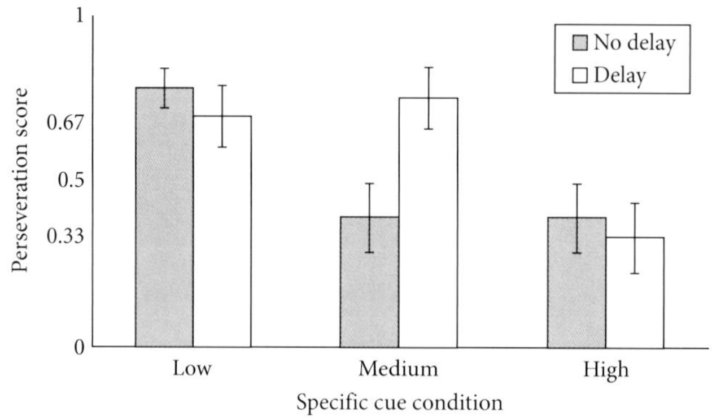


**Figure 4.8. Perseveration under varying cue conditions in the A-not-B task**
The bar chart depicts the perseverative responses of infants (8–10 months of age) under three specific cue conditions (low, medium, or high salience) under delay (3 s lag time) or no delay (1 s lag time) in the A-not-B task. The perserveration score is the percentage of infants exhibiting the A-not-B error during the B-trial in a modified A-not-B task, as described in the text.

## SENSORIMOTOR PROCESSES

At any moment in time, the complex system that is human cognition is constrained by the just-recent past, each moment preceding out of the just-previous moment. But at each moment, ongoing sensory input also impinges on and perturbs these internal dynamics. Therein lies the grist for the emergence of stability and flexibility. As noted at the start of this chapter, all accounts of the A-not-B error (and of executive control in general) recognize this tension between the ongoing activity generated by the just-previous past and the current input. Further, because this tension is *everywhere* evident in neural processes at many different timescales and in different regions of the brain, this tension between stability and flexbility is played out in many different systems. The dynamic field account—with its emphasis on sensorimotor processes—and the experiments it generates provide a number new insights; four insights on sensorimotor processes are discussed here.

### A Sensorimotor Object Concept

By the dynamic field account, objects and locations are bound together—and internally represented—via motor plans, which are themselves tightly tied to the current position of the body. These are dynamic representations, a kind of sensorimotor object concept, that provide a means of keeping track of nonpresent things and events over time. The embedding of representations in specific sensorimotor processes suggests that there is no *single* object concept anywhere. Classic theories of cognition focused on how stabilitites were maintained in the context of varying input. These classic theories explained stability by separating cognition from perception and action. The system was conceptualized as linear and unidirectional, from *sense* to to *act*, with the think part (as in most rule-based and connectionist models) segregated from perception and action, from input and output. The dynamic field model, along with other embodied approaches to cognition (cf. Barsalou, Breazeal, & Smith, 2007; Smith & Breazeal, 2007), views cognition and representation as being embedded in sensorimotor processes.

### Perseveration as a Source of Coherence

Sensorimotor representations tied to the body may seem maladaptive in the context of the A-not-B task (a laboratory task with little in common with real-world tasks), but these representations create coherence as the cognitive system moves from one moment to the next in the real-world task of learning object names. Thus the processes that create the A-not-B error are not primarily about a cognitive deficiency (or immaturity), but rather are about one important form of stability that enables the cognitive system to link cognitive contents through the body's disposition in space to ongoing events in the world (see also Ballard, Hayhoe, Pook, & Rao, 1997). The adaptive value of these processes within the human cognitive system is clearly seen in the Baldwin task and in children's ability to map names to nonpresent things. In a sense, current research on the A-not-B error concentrates on flexibility and ignores the value of stability, even though a system that is *too* variable (in the sense of shifting with every new stimulus) would not be adaptive, just as a system that is *too* stable is not. The cognitive achievement that underlies the development of perseverative tendencies in infancy was recently demonstrated in a longitudinal study of the A-not-B error (Clearfield et al., 2008). In that study, young infants (6 month of age) began flexibly following the cue on B-trials, then *became* perseverative as they grew older. Thus, perseveration in the A-not-B task is a developmental achievement requiring the building of memories of past reaches. (A similar point with a predicted curvilinear developmental trend has also been made by Munakata, 1998).

### Flexible Responding in a Stable System Requires Sustained Activation

The experiments manipulating the salience of the cuing event in the A-not-B task show how flexibility is, in part, a function of the strength of the stimulus cue signaling a shift (and, although not tested, a function of how strongly that new cue activates *appropriate* memories of prior reaches). The importance of the strength of the specific cue is consistent with other accounts of the A-not-B error, including the PDP account. However, one new idea in the current version of the dynamic field model (see also Clearfield et al., 2008; Schöner & Dineva, 2007) is that transient *active processes* from the specific cue are not enough; instead, the key issue is the ability of the system to stabilize an activation peak that can be sustained by the neural interactions within the system without further input. This is the key point of the experments showing perseveration even with no delay. No previous account has ever predicted that infants may perseverate without delay because *read-out* (i.e., response generation) always reflects the instantaneous maximum right after stimulation. This allows for a relatively weak cue to have a direct and thus relatively strong impact, always leading to a correct response without a delay. However, a response will not happen even without a delay unless a threshold that leads to sustained internally generated activation is pierced. Previous studies have shown that infants do not perseverate without a delay between the specific cue and the signal to reach (Marcovitch & Zelazo, 1999; for reviews, see Wellman et al., 1986). This is typically interpreted in terms of perseveration arising from the need to *represent* the target over a delay (Diamond, 1985; Munakata, 1998)—no delay implies no need to maintain a required representation and thus no perseveration arises. According to dynamic field theory, this is not precisely correct. Perseveration arises because of insufficient activation from the specific cue to the new target. This insufficient activation at the time of the signal to reach may arise either from a weak specific cue or from the decay of activation over a long delay. Flexibility requries not only a strong cue but also a system that can generate sufficient activation to recruit the self-sustaining cooperativity among local elements that is a decision to reach.

### Representations Without Direct Links to the Sensorimotor Surface

Results in the Baldwin task suggest that sensorimotor processes may give rise to representations that are not linked directly to sensory input (as in the association field depicted in Fig. 4.6). These representations—which also provide input to the motor planning fields—create both a new form of stability and a new way to augment specific cues; thus, these internal activations lead to increased flexibility and appropriate shifting with new cues (see Brown, Reynolds, & Braver, 2007, for research on executive control and attention shifting that fits this conceptualization).

## What Develops during Development?

The key aspects of the developmental pattern of infant perseveration are these: Very young infants do not perseverate but follow the cue (Clearfield et al., 2006). Infants 8–10 months of age typically perseverate in this task, under both relatively brief and long delays (Wellman et al., 1986). Infants 10–12 months of age will also perseverate, but only under increasing delays between the hiding event and the reach (Diamond, 1985). After 12 months of age, infants do not perseverate, even after very long delays (Wellman et al., 1986). But none of these response patterns are fixed. Instead, the behavioral outcome depends on the strength of the memories of past reaches, on the activation level in the field as a whole, on the decay rates, on the internal dynamics and connectivity in the field, and on the posture shifts. *All* of these are likely to have some developmental course of their own and may contribute to the overall developmental trend.

Current simulations of the model, however, have concentrated on resting levels of the motor planning field, modeling younger infants and older infants with lower and higher resting levels (Dineva, 2005, unpublished; Dineva, Schöner, & Thelen, 2008, in preparation; Thelen et al., 2001). The higher the resting level, the easier it is to enter the cooperative regime. A strong cue can compensate for the resistance (low resting level), enabling activation to pass the cooperativity threshold. Importantly, what the model predicts given an immature system (low resting level) plus a very strong specific cue versus a mature system (high resting level) plus any specific cue is fundamentally the same: activation reaches a threshold and feeds back on itself to suppress competing activation; this yields a sustained peak and, thus, a behavioral decision. For a younger infant, the amount of activation provided by the specific cue is more crucial in getting to a sustained regime than for an older infant; this is due to the low resting activation level in the field in the immature system of the younger infant. That is, in a more developed system, a sustained regime comes more easily.

The developmental question that needs to be answered is why this resting level goes up as the infant develops. This could be due to fundamental changes in the neural system, including, perhaps, the involvment of larger pathway systems, frontal areas of the brain, or association areas that amplify signals via learned associations and acquired cues (including language). These developmental changes may also be due to specific experience, with the curvilinear trajectory in the A-not-B task being driven by experience in goal-directed reaching itself (see Thelen et al., 2001). The apparently central role of the activation level may also help us understand the vulnerability of compromised systems to perseveration. Such perseveration is a hallmark of psychopathology and brain injury, including injury to the prefrontal cortex (e.g., Luria, Pribram, & Homskaya, 1964). The immaturity of the prefrontal cortex has also been implicated in young infants' failure to reach correctly on A-not-B tasks (e.g., Diamond & Goldman-Rakic, 1989). Often, variants of the A-not-B task are also used as diagnostic tools in the assessment of neurological disorders (e.g., Diamond, 2001; Gazzaniga, Ivry, & Mangun, 2002). By the present analysis, the core commonality in immaturity and in neurological damage may be decreased activation, making it more difficult for the system to get above the threshold for cooperativity. Further, damage to the system may result not in loss of some specific functionality, but rather in an unspecific decrease in overall activation levels (Kimberg & Farah, 2000; Farah & McClelland, 1991). For instance, a lesion may cut through a long-distance excitation-inhibition loop between different brain regions, thus reducing cooperative interactions within the system as a whole.

## Conclusion: It's the Brain in a Body in a World that Matters

The dynamic field theory's account of the A-not-B error—with its emphasis on the sensorimotor processes involved in keeping track of and reaching to objects—makes a unique contribution to the understanding of flexiblity and stability (and their development) by

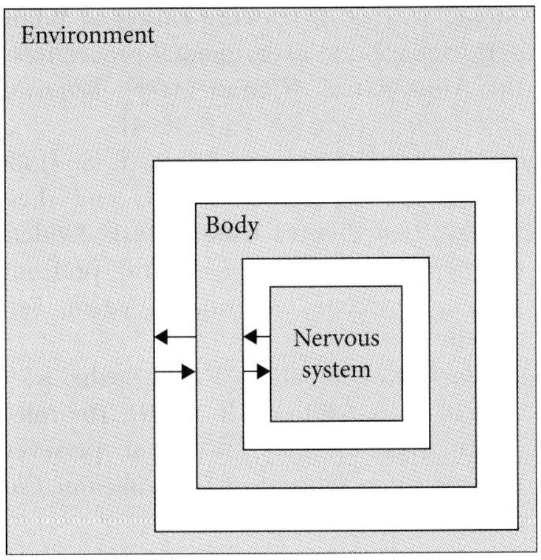


**Figure 4.9. Coupling of the brain to the body and to the world**
This diagram illustrates the concept that the brain is coupled to the body and, thus, to the world. The diagram appears simple; the implications for the real world are complex.

showing how these outcomes are the products of a complex dynamic system with many components that is *continually coupled to the world through the body*. This emphasis on processes close to the sensory surface, on the embedding of cognition in a body in a physical world is the most fundamental difference between the dynamic systems approach and a connectionist approach. Figure 4.9 shows a diagram often used by Randy Beer (Beer, 1995; Beer, in press; Chiel & Beer, 1997), a roboticist, who argues (and demonstrates) that complex systems of intelligence can only be understood through an integrated perspective on environment, body, and brain. He notes that the common way neural and cognitive scientists speak of, for example, "the neural control of behavior" reveals our bias of thinking of the body as a mere puppet responding to the demands of the nervous system (Beer, in press).

Internal representation—the nervous system—is crucially important to integration of information and to the range of behaviors in which the system can stably engage and among which the system can flexibly switch. However, the nervous system, the body, and the environment each possess their own complex intrinsic dynamics, and these are in continuous interaction with each other—always. The child's mapping of names to things in the Baldwin task, for instance, is built on a spatially ordered world (and spatial biases on the part of the social partner). This spatially ordered world is made up of a perception-action system (that must be tied to the body's current position in physical space), memories, and integrative processes. In turn, these integrative process are at work on multiple timescales and over many different sensory, motor, and internal sources of information. Human cognition is what it is because of this system as a whole. Evolution selects the behavioral efficacy of the *whole* package. Thus a complete explanation of stability and flexibility, of development, and of human cognition in general will require an explanation in terms of the complete brain-body-environment system as illustrated in Figure 4.9. The explanation cannot properly be found in any individual component of this coupled system.

## References

Acredolo, L. P. (1979). Laboratory versus home: The effect of environment on the 9-month-old infant's choice of spatial reference system. *Developmental Psychology, 15*(6), 666–667.

Allport, A. (1990). Visual attention. In M. I. Posner (Ed.), *The foundations of cognitive science* (pp. 631–682). Cambridge, MA: MIT Press–Bradford Books.

Baillargeon, R. (1993). The object concept revisited: New directions in the investigation of infants' physical knowledge. In C. Granrud (Ed.), *Visual perception and cognition in infancy. 23rd Carnegie Mellon Symposia on Cognition* (pp. 265–315). New York: Taylor & Francis–Psychology Press.

Baldwin, D. A. (1993). Early referential understanding: Infants' ability to recognize referential acts for what they are. *Developmental Psychology, 29*(5), 832–843.

Ballard, D. H., Hayhoe, M. M., Pook, P. K., & Rao, R. P. N. (1997). Deictic codes for the embodiment of cognition. *Behavioral and Brain Sciences, 20*(4), 723–742, commentary 743–758, authors' response 758–767.

Barsalou, L. W., Breazeal, C., & Smith, L. B. (2007). Cognition as coordinated non-cognition. *Cognitive Processing, 8*(2), 79–91.

Beer, R. D. (1995). Dynamical systems perspective on agent-environment interaction. *Artificial Intelligence, 72*(1–2), 173–215.

Beer, R. D. (in press). Beyond control: The dynamics of brain-body-environment interaction in motor systems. In D. Sternad (Ed.), *Advances in Experimental Medicine and Biology: Vol. 629. Progress in Motor Control V: A Multidisciplinary Perspective.* New York: Springer.

Bremner, J. G. (1978). Egocentic versus allocentric spatial coding in nine-month-old infants: Factors influencing the choice of code. *Developmental Psychology, 14*(4), 346–355.

Brown J. W., Reynolds, J. R., & Braver, T. S. (2007). A computational model of fractionated conflict-control mechanisms in task-switching. *Cognitive Psychology* 55*(1)*, 37–85.

Chiel, H. J., & Beer, R. D. (1997). The brain has a body: Adaptive behavior emerges from interactions of nervous system, body and environment. *Trends in Neurosciences, 20*(12), 553–557.

Clearfield, M. W., Diedrich, F. J., Smith, L. B., & Thelen, E. (2006). Young infants reach correctly in A-not-B tasks: On the development of stability and perseveration. *Infant Behavior and Development, 29* (3), 435–444.

Clearfield, M. W., Dineva, E., Smith, L. B., Diedrich, F. J., & Thelen, E. (2008). Cue salience and infant preseverative reaching: tests of the dynamic field theory. *Developmental Science.* Published online: Sepember 25, 2008. doi:10.1111/j.1467-7687.2008.00769.x

Cohen, Y. E., & Andersen, R. A. (2002). A common reference frame for movement plans in the posterior parietal cortex. *Nature Reviews. Neuroscience, 3*(7), 553–562.

Diamond, A. (1985). Development of the ability to use recall to guide action, as indicated by infants' performance on AB. *Child Development, 56*(4), 868–883.

Diamond, A. (1991). Neuropsychological insights into the meaning of object concept development. In S.Carey, & R. Gelman (Eds.), *The epigenesis of mind: Essays on biology and cognition.* Hillsdale, NJ: Erlbaum.

Diamond, A. (1998). Understanding the A-not-B error: Working memory vs. reinforced response, or active trace vs. latent trace. *Developmental Science, 1*(2), 185–189.

Diamond, A. (2001). Looking closely at infants' performance and experimental procedures in the A-not-B task [Commentary]. *Behavioral and Brain Sciences 24*(1), pp. 38–41.

Diamond, A., & Goldman-Rakic, P. S. (1989). Comparison of human infants and rhesus monkeys on Piaget's A-not-B task: Evidence for dependence on dorsolateral prefrontal cortex. *Experimental Brain Research, 74*(1), 24–40.

Diedrich, F. J., Highlands, T. M., Spahr, K. A., Thelen, E., & Smith, L. B. (2001). The role of target distinctiveness in infant perseverative reaching. *Journal of Experimental Child Psychology, 78*(3), 263–290.

Dineva, E. (2005). *Dynamical field theory of infant reaching and its dependence on behavioral history and context.* Unpublished doctoral dissertation, Ruhr-Universität Bochum, Bochum, Germany.

Dineva, E., Schöner, G., & Thelen, E. (2008). Behavioral history matters: A dynamic field account of spontaneous and perseverative errors in infant reaching. Manuscript in preparation.

Farah, M. J., & McClelland, J. L. (1991). A computational model of semantic memory impairment: Modality specificity and emergent category specificity. *Journal of Experimental Psychology: General, 120*(4), 339–357.

Gazzaniga, M. S., Ivry, R. B., & Mangun, G. R. (2002). Executive functions and frontal lobes. In M. S. Gazzaniga, R. B. Ivry, & G. R. Mangun, *Cognitive neuroscience: The biology of mind.* (2nd ed., pp. 499–536). New York: W. W. Norton.

Kimberg, D. Y., & Farah, M. J. (2000). Is there an inhibitory module in the prefrontal cortex? Working memory and the mechanisms underlying cognitive control. In S. Monsell & J. Driver (Eds.), *Attention and performance: Vol. XVIII. Control of cognitive processes* (pp. 739–751). Cambridge, MA: MIT Press–Bradford Books.

Lew, A. R., Hopkins, B., Owen, L. H., & Green, M. (2007). Postural change effects on infants' AB task performance: Visual, postural, or spatial? *Journal of Experimental Child Psychology, 97*(1), 1–13.

Luria, A. R., Pribram, K. H., & Homskaya, E. D. (1964). An experimental analysis of the behavioral disturbance produced by a left frontal arachnoidal endothelioma (meningioma). *Neuropsychologia, 2*(4), 257–280.

Marcovitch, S., & Zelazo, P. D. (1999). The A-not-B error: Results from a logistic meta-analysis. *Child Development, 70*(6), 1297–1313.

Munakata, Y. (1998). Infant perseveration and implications for object permanence theories: A PDP model of the AB task. *Developmental Science, 1*(2), 161–184, commentary 184–204, author's response 205–211.

O'Regan, J. K., & Noë, A. (2001). A sensorimotor account of vision and visual consciousness. *Behavioral and Brain Sciences, 24*(5), 939–973, commentary 973–1011, authors' response 1011–1021.

Piaget, J. (1954). *The construction of reality in the child* (M. Cook, Trans.). New York: Basic Books.

Piaget, J. (1963). *The origins of intelligence in children* (M. Cook, Trans.). New York: W. W. Norton. (Original work published 1953.)

Samuelson, L. K., & Smith, L. B. (2000). Grounding development in cognitive processes. *Child Development, 71*(1), 98–106.

Schöner, G., & Dineva, E. (2007). Dynamic instabilities as mechanisms for emergence. *Developmental Science, 10*(1), 69–74.

Schutte, A. R., Spencer, J. P., & Schöner, G. (2003). Testing the dynamic field theory: Working memory for locations becomes more spatially precise over development. *Child Development, 74*(5), 1393–1417.

Simmons, W. K., & Barsalou, L. W. (2003). The similarity-in-topography principle: Reconciling theories of conceptual deficits. *Cognitive Neuropsychology, 20*(3–6), 451–486.

Smith, L. B., & Breazeal, C. (2007). The dynamic lift of developmental process. *Developmental Science, 10*(1), 61–68.

Smith, L. B., Clearfield, M. W., Diedrich, F. J., & Thelen, E. (2008). Evidence for embodiment: Memory tied to the body's current position. Manuscript in preparation.

Smith, L. B., Samuelson, L. K., & Spencer, J. P. (2008). Binding names to things. Manuscript in preparation.

Smith, L. B., & Thelen, E. (Eds.). (1993). *A dynamic systems approach to development: Applications.* Cambridge, MA: MIT Press–Bradford Books.

Smith, L. B., Thelen, E., Titzer, R., & McLin, D. (1999). Knowing in the context of acting: The task dynamics of the A-not-B error. *Psychological Review, 106*(2), 235–260.

Spencer, J. P., Smith, L. B., & Thelen, E. (2001). Tests of a dynamic systems account of the A-not-B error: The influence of prior experience on the spatial memory abilities of two-year-olds. *Child Development, 72*(5), 1327–1346.

Thelen, E., Schöner, G., Scheier, C., & Smith, L. B. (2001). The dynamics of embodiment: A field theory of infant perseverative reaching. *Behavioral and Brain Sciences, 24*(1), 1–34, commentary 34–70, authors' response 70–80.

Thelen, E., & Smith, L. B. (1994). *A dynamic systems approach to the development of cognition and action.* Cambridge, MA: MIT Press–Bradford Books.

Wellman, H. M., Cross, D., & Bartsch, K. (1986). Infant search and object permanence: A meta-analysis of the A-not-B error. *Monographs of the Society for Research in Child Development, No. 214, 51*(3), 1–67.

Zelazo, P. D., Müller, U., Frye, D., Marcovitch, S., Argitis, G., Boseovski, J., et al. (2003). The development of executive function in early childhood. *Monographs of the Society for Research in Child Development, 68*(3), 1–137.

# CHAPTER 5

## Dynamic Field Theory and Embodied Cognitive Dynamics

*John P. Spencer, Sammy Perone, and Jeffrey S. Johnson*

The goal of this chapter is to explain some of the core concepts of dynamic field theory (DFT) and how this theory provides a formal framework for thinking about embodied cognitive dynamics.

DFT is now 15 years old. In 1993, Gregor Schöner and his colleagues published a paper (in the proceedings of an international conference) that presented a theory addressing how eye movements are planned; the authors also addressed the neural bases for their theory (Kopecz, Engels, & Schöner, 1993). Since that time, DFT has been extended to a range of topics including the planning of reaching movements (Bastian, Riehle, Erlhagen, & Schöner, 1998; Bastian, Schöner, & Riehle, 2003; Erlhagen & Schöner, 2002), the development of motor planning (Smith & Thelen, 1993; Thelen & Smith, 1994), the perception of motion (Hock, Schöner, & Giese, 2003; Jancke, Erlhagen, Schöner, & Dinse, 2004), the processes that underlie habituation in infancy (Schöner & Thelen, 2006), the control of autonomous robots (Bicho, Mallet, & Schöner, 2000; Steinhage & Schöner, 1998), the processes that underlie visuospatial cognition and spatial language (Lipinski, Spencer, & Samuelson, in press; Spencer, Simmering, Schutte, & Schöner, 2007), the development of visuospatial cognition (Simmering, Schutte, & Spencer, 2008), the processes that underlie visual working memory and change detection (Johnson, Spencer, Luck, & Schöner, in press), the fast learning of object labels and other aspects of word learning (Faubel & Schöner, 2008; Samuelson, Schutte, & Horst, in press), and the development of executive function (Buss & Spencer, 2008).

Despite the extensive application of DFT, discussion at the 2005 conference whence this edited volume emerged suggested that aspects of DFT are poorly understood. Pockets of researchers are familiar with one model or another, but there is little consensus regarding the broader theoretical concepts that underlie this theory. One barrier to understanding the underlying concepts is that there have been relatively few reviews of DFT pitched squarely at the central concepts. The present chapter takes up this charge.

We begin with some context, describing the milieu in which the theory emerged and the problems it was initially designed to tackle. In particular, we focus on a central problem in cognitive science: How do cognitive systems interface with real-time sensorimotor systems? That is, how can real-time cognitive dynamics be *embodied*? We discuss our answer to this question within DFT where we use neural dynamics and stable peaks of activation as the basic unit of cognition. Peaks can be continuously coupled to sensorimotor systems in real time. But peaks move far beyond a basic feed-forward picture, allowing for a host of cognitive properties that come for free in this theoretical framework. We highlight these properties and also show how peaks provide critical links to two other central challenges in the cognitive developmental sciences: the integration of processes over multiple timescales and the origins of behavioral flexibility.

## Toward Embodied Cognition: The Dynamic Approach to Motor Control and Development

To understand some of the central innovations of DFT, it is useful to describe the context in which this approach emerged (for reviews, see Johnson, Spencer, & Schöner, 2008; Spencer & Schöner, 2003). In the 1980s and 1990s, several researchers were interested in applying the concepts of *dynamical systems theory*[1] (Braun, 1994) to aspects of human behavior (Port & van Gelder, 1995). Much of this work was initially conducted to investigate motor control (Kelso & Tuller, 1984) and motor development (Thelen, 1995), in part because the continuous, real-time nature of movement mapped nicely onto continuously evolving dynamical systems.

As researchers probed the link between dynamical systems theory and motor behavior more deeply, a set of central concepts emerged (see Spencer & Schöner, 2003). First, action reflects a dynamic balance among stability (stably reaching for a cup of coffee despite a bump on the arm), instability (grabbing the cup a little too vigorously and spilling several drops), and flexibility (managing to drop the cup in order to catch the computer about to fall onto the concrete sidewalk!). Second, behavior is *softly assembled*[2] from multiple component processes—neural plans to move the hand to a particular location in space, oscillatory processes that regulate the timing of movement, forces generated by the contraction of muscles to move the hand through space, and so on (Schöner, 1995). Third, such processes are often nonlinear in nature, where small changes in one component (movement speed at the end of a reach) can lead to big changes in behavior (knocking the cup over). Finally, the dynamics of action change over multiple timescales, from the millisecond-to-millisecond unfolding of a single reach, to a sequence of actions in context—reach to a coffee cup, then to a computer—to the myriad reaches to objects that mount up over weeks, months, and years.

These four concepts have been extended to a variety of phenomena probing processes that underlie complex action systems and a host of topics in motor development (e.g., Corbetta & Thelen, 1996; Kelso, 1995; Schöner, 1995; Thelen & Ulrich, 1991). But even early in the development of these concepts a central issue arose: What about cognition?

Early explorations of this question led to a strong, antirepresentationalist stance and to a central challenge—to rethink cognition without using static/rigid symbols (Townsend & Busemeyer, 1995; van Gelder et al., 1998). In this sense, the dynamical systems community had much in common with connectionist ideas taking hold at the time (Rumelhart, McClelland, & PDR Group, 1986). In addition, there was a move to think of cognition as *embodied*, that is, always intimately tied to sensorimotor systems. This line of thinking resonated with fundamental issues within developmental science regarding the origin of cognitive systems (Smith & Thelen, 1993; Thelen & Smith, 1994). Recent work has demonstrated that embodiment plays a central role in adult cognitive activity, as well, including the neural bases of thought (Barsalou & Wiemer-Hastings, 2005; Damasio & Damasio, 1994; Spivey, 2007).

Although some researchers have made strong arguments in defense of embodiment (Glenberg, 1997) and dynamical systems approaches to cognition, there was, at the time, no formal theory for how such interactions could arise. The reason: There are some fundamentally challenging issues that lie at the heart of this appealing idea that perception, action, and cognition are all intimately intertwined.

The initial challenge is that sensorimotor systems evolve continuously in real time, but cognition can *jump* from one state to another, that is, from one thought to another. Consider both sides of this challenge. First, in what sense

[1] Note that *dynamic* systems theory and *dynamical* systems theory are used interchangeably in this volume.

[2] Behavior is characterized as softly assembled (or soft assembled) if the behavior is assembled to suit a current and specific task/environment. This term is used to distinguish such behavior from that existing prior to the current environment and independent of the specific task.

can't a sensorimotor system jump? When I reach my hand out to pick up a coffee cup, my hand obligatorily must move through the space that intervenes between the computer keyboard and the cup. Although I might move my hand quite quickly, there is no getting around this physical fact. By contrast, what would it mean for a cognitive system to jump? Motor plans in my brain can actually jump from a plan to, say, point to a key on my keyboard at 0° movement direction to a plan to reach 120° to the right to retrieve a cup. Although motor plans can evolve continuously (as is the case in some forms of mental rotation, see Georgopoulos, Lurito, Petrides, Schwartz, & Massey, 1989), they don't have to. Note that claims about *jumpy cognition* do not imply that motor plans and other forms of cognitive activity are somehow removed from continuous, real-time neural processes. The point here is that what these neural processes *represent* in the moment can jump in content.

This creates some serious issues. If cognition is jumping around, but heads, hands, fingers, and legs all move continuously through space and time, how does cognition remain connected? How does cognition remain embodied?

The second challenge is a challenge of *representation*. Clearly there must be something *stored* over time, some form of memory. But what is the nature of that memory? Work within the framework of information processing and artificial intelligence led investigators to hypothesize that cognition happens upon a bed of static, symbolic representations (Newell & Simon, 1976). If we are going to jettison static symbols, as early proponents of dynamical systems theory argued we should (van Gelder et al., 1998), what are we going to replace them with? What types of representational format can effectively interface with real-time sensorimotor systems, yet also do the job of cognition?

Again, there are layers of subtlety here. Consider one example from the domain of spatial cognition. According to one prominent theory of spatial memory, people remember locations over the long-term relative to spatial *prototypes* (Huttenlocher, Hedges, Corrigan, & Crawford, 2004; Huttenlocher, Hedges, & Duncan, 1991). For instance, if I want to remember where my computer screen is located, I might remember that it is on the right half of my office desk, roughly near the center of the right category, that is, near the right spatial prototype. But what happens when I'm sitting at my desk, and I turn my head to gaze longingly out the window? To keep track of my computer screen, I need to know where the desk is located relative to my current position in the world, and I need to know my orientation in space (so I can reconstruct what *right* means). Things become even more complicated when I get up and leave my office, only to return later. In this case, I need to make sure I enter the correct office, I need to reorient myself in this space, and then I can discover (much to my annoyance) that my computer screen was stolen the night before.

What's the point? The point is that forming an association between the right spatial prototype on my office desk and my computer screen *is only useful if that abstract representation is linkable in context to my real-time sensorimotor activity*. And it's not entirely clear how to make this link robust; particularly because sensorimotor systems rely on continuous, metric representations of space, but prototypes strip away much of this detail (Bridgeman, Gemmer, Forsman, & Huemer, 2000; Jackendoff, 1996). What system decides which details to throw away and which details to keep? And once those details are thrown away, how do we reconnect the abstract representation to the continuous, metric information provided by our senses? Of course, the alternative—storing all the details of daily experience—seems cumbersome, as well. Thus, what is the appropriate middle ground that effectively integrates sensorimotor and cognitive systems yet provides a useful level of abstraction for *higher cognition* (for a discussion of related issues, see Mareschal et al., 2007)?

## Setting the Stage: Grounding Cognition in Neural Process

In the early 1990s, Schöner, Thelen, Smith, and colleagues began thinking about how the dynamical systems approach to motor control and development could be extended to capture the dynamics of cognitive systems. The starting point for this approach was to consider several facts about neural systems. Neural systems are noisy, densely interconnected, and time dependent; they pass continuous, graded, and metric information to one another; and they are continuously coupled via both short-range and long-range connections (Braitenberg & Schüz, 1991; Constantinidis & Steinmetz, 1996; Edelman, 1987; Rao, Rainer, & Miller, 1997). These neural facts raise deep theoretical challenges (for background, see Skarda & Freeman, 1987). How can a collection of neurons *encapsulate* some bit of information (e.g., a static symbol) amidst near-constant bombardment by other neural signals? And how can a collection of neurons integrate information from multiple systems in a stable, reliable way?

A number of neurophysiologists and theoretical neuroscientists have argued that the densely interconnected and dynamic nature of the brain does not map cleanly onto the concepts of information processing. Thus, rather than imposing classic views of cognition onto the brain, the challenge is to take inspiration from what we know about neural reality to rethink cognition (Braitenberg & Schüz, 1991; Brooks, 1991). This perspective has taken hold within the field of theoretical neuroscience where theoreticians have embraced the use of complex, dynamic neural networks to capture brain-behavior relations. Although neural networks have architectures that can be depicted as separate systems, they are—at their core—complex, reentrant, densely interconnected, complex systems that violate the separability assumptions of information processing (Palmer & Kimchi, 1986).

One of the dominant approaches in theoretical neuroscience takes a biophysical approach to the study of brain-behavior relations: theoreticians attempt to build neurally realistic models of single neurons that capture key elements of neuronal firing rates, neurotransmitter action, and so on (Durstewitz, Seamans, & Sejnowski, 2000; Salinas, 2003). These neural models are then coupled together into populations, and populations are coupled together based on evidence of, for instance, cortical-cortical connections. The goal of such models is two-fold: (1) to model the behavior of different cell types within a cortical region, and (2) to capture aspects of behavioral performance via population dynamics.

Although the biophysical approach has led to new insights into brain function and neuronal dynamics, there's a problem: such models do a great job capturing the behavior of neurons, but they do *not* do a great job capturing the behavior of people. This is due, in part, to concrete limitations of such models: they can be very complex, they can have immense computational demands, and they can be difficult to analyze (Haskell, Nykamp, & Tranchina, 2001). Moreover, it is hard to get biophysical networks to behave in realistic ways. Such networks can generate behaviors (e.g., reaction times); however, there have been few attempts to create biophysical networks that are able to generate multiple types of behavior and to flexibly switch behaviors as task demands vary. Thus, while this approach to rethinking cognition is very promising, biophysical network models do well at the level of neurons, but poorly at the level of behavior (Finkel, 2000).

The dynamic neural field framework we have developed shares the same goal of many neural network approaches—to think about cognition as the emergent product of complex, contextually specific neural dynamics—but accomplishes this in a way that is more tightly linked to behavior. To achieve a stronger link to behavior, we focused our analysis at the level of population dynamics (see also Spivey, 2007). That is, rather than building networks that start from a set of, say, integrate-and-fire neurons that retain a great deal of biophysical detail, we have chosen to focus on the emergent product of the dynamics at the neural level—attractors at the level of the neural population.

To situate this work, consider classic research by Georgopoulos and colleagues (Georgopoulos, Kettner, & Schwartz, 1988; Georgopoulos, Taira, & Lukashin, 1993) that established how one could directly link population-based neural activity with behavior. These researchers examined the neural signals in motor cortex that specified the direction of an up-coming arm movement in a simple pointing task. Nonhuman primates were trained to point toward a lighted target from a central starting location at a *go* signal. Simultaneous neural recordings revealed that each neuron fired maximally when the monkey moved in a *preferred* direction, and this firing profile gradually tapered off as the arm moved in less preferred directions. To construct a population response, Georgopoulos and colleagues transformed the real-time pattern of neural activity from each neuron into a vector. The direction of each neuron's vector pointed in its *preferred* direction, and the magnitude of the vector was scaled by the strength of the neuron's current firing rate. These individual vectors could then be summed to produce a population vector. Results showed that this neural population vector accurately predicted the direction of an up-coming movement 160 ms before movement onset. More recent work has verified the predictive nature of the population vector. In a particularly vivid case, population vectors were shown to rotate in real time as nonhuman primates performed a mental rotation task, and the rotation time correlated with the response time (Georgopoulos et al., 1989).

Work by Georgopoulos and others (see, for example, Dayan & Abbott, 2001) suggests that stable patterns of activation at the neural population level directly give rise to behavior. Since this discovery in the 1980s, work has moved forward on two fronts. One line of investigation has tried to understand the link between population dynamics and single cell neurophysiology (Amit & Brunel, 1997; Compte, Brunel, Goldman-Rakic, & Wang, 2000; Durstewitz et al., 2000). The question here is how the activation of individual neurons gives rise to stable activation at the population level. A second line of investigation has tried to understand how population dynamics give rise to behavior. DFT emerged out of this second line of research. In the section that follows, we describe the central concepts that underlie DFT, including how this framework is linked to neurophysiology.

## The Real-Time Integration of Cognitive and Sensorimotor Systems: DFT Core Concepts

Figure 5.1 shows a particular implementation of DFT in the form of a one-dimensional, two-layer

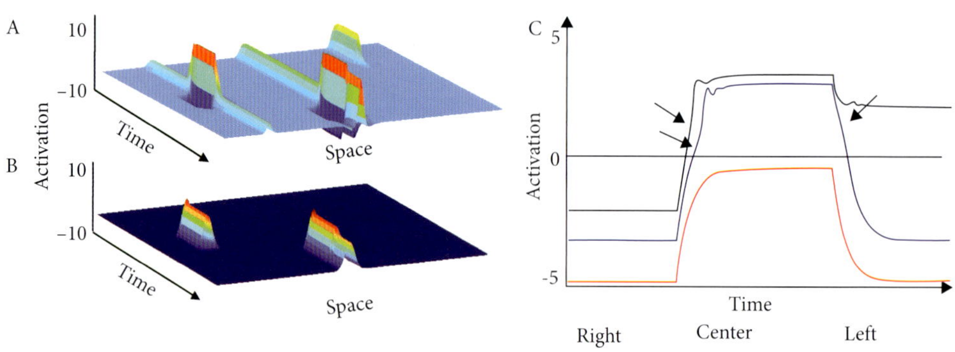


**Figure 5.1. Simulation illustrating a one-dimensional, two-layer dynamic neural field**
This implementation of the basic model consists of a field of excitatory neurons (panel A) reciprocally coupled to a field of inhibitory neurons (panel B). Neurons in the top layer (panel A) interact via *local excitatory connections* and drive activation in the inhibitory layer (panel B), which, in turn, projects broad inhibition (*longer range inhibition*) back to the excitatory layer. Panel C shows the activation profile through time at the spatial location associated with each of the three peaks in panel A: red trace = right hump (apex is yellow and may be difficult to see); black trace = central peak; and blue trace = left peak. Note that the activation profile for the three peaks are superimposed in time to highlight differences in activation strength. Arrows highlight joints in the activation curve at which the field is moving through a bifurcation from one attractor state to another.

dynamic neural field (DNF). DNFs were first developed by Amari (1977) and are part of a class of bistable attractor neural networks (Camperi & Wang, 1998; Wilson & Cowan, 1972). DNFs use the well-known construct of *activation* from mathematical psychology (Churchland & Sejnowski, 1992), which captures the certainty with which some bit of information is currently known to the nervous system—the greater the activation, the greater the certainty. Activation in DNFs is distributed over continuous dimensions (i.e., *fields*). These dimensions can be spatial locations, movement direction, a continuous color dimension, and so on.

A central issue for a dynamical systems approach to neural networks is to understand how patterns of activation change through time and the stability characteristics of these patterns. Activation in DNFs evolves continuously through time as a function of the current inputs impinging on the network, as well as the network's own *intrinsic* dynamic neural interactions. DNFs use a particular form of neural interaction based on the space code principle from neurophysiology. According to this principle, neurons that fire maximally to similar stimuli excite one another. These neurons are considered *close* along the represented dimension (e.g., close along a spatial dimension). By contrast, neurons that fire maximally to very different stimuli inhibit one another. These neurons are considered *far* along the represented dimension. This form of neural interaction gives rise to *local excitatory interactions* and *lateral or longer range inhibitory interactions.*

Note that in Amari's (Amari, 1977) initial work, neural fields reflected activation on the cortical surface because neurons in, for instance, the early visual cortex are lined up in this way (i.e., neurons close together on the cortical surface *code* for similar spatial locations). But other work by Georgopoulos and colleagues (e.g., Georgopoulos & Massey, 1988) revealed that one could construct a type of *functional topographic mapping*. Specifically, neurons in the motor cortex are not arranged topographically: two neurons next to one another in the motor cortex might prefer very different movement directions. Nevertheless, neural activation at the population level actually obeys topographic relations. That is, even though two neurons might be far apart on the cortical surface, they can still prefer the same movement direction and be mutually excitatory. Similarly, these same neurons will inhibit other neurons in the population that prefer very different movement directions. Critically, such interactions give rise to stable *peaks* of activation, even though the peaks are effectively distributed across the cortical surface. One can demonstrate this directly by lining up neurons in a field, not by their positions in the cortex, but rather, by their preferred movement directions (see Bastian et al., 1998; Bastian et al., 2003).

In Figure 5.1, the excitatory interactions all occur among neighboring neurons within the top layer of excitatory neurons (Fig. 5.1A). If these neurons are stimulated such that a local *hump*[3] of activation forms and goes above-threshold (i.e., above 0), the excitatory neurons begin to drive up the activation of neurons in the second layer of inhibitory interneurons (Fig. 5.1B). These inhibitory neurons, in turn, project broad inhibitory feedback to the excitatory layer. As excitation and inhibition cycle back and forth in the network, a stable *peak* of activation can form. The basic unit of cognition in DFT is a stable peak of activation.

At this point, it is important to note that only positive activation—*above-threshold activation*—contributes significantly to neural interactions within and between fields. That is, only field sites with activation greater than zero transmit their activation to other neurons and thus contribute to the evolving patterns of activation across the network. This premise is captured by the sigmoidal nonlinearity characteristic of neuronal dynamics (Grossberg, 1973); the same premise holds for all simulations

[3] We use the term *hump* here in contrast to the term *peak*. Humps refer to small, sub-threshold patterns of excitation that don't by themselves engage strong interaction. Peaks, by contrast, are supra-threshold activations that do engage strong neural interaction.

described herein. The presence of sub-threshold (i.e., below 0) patterns of activation can still exert a pronounced effect on activation patterns arising within the network, as discussed below. Note also, that all simulations begin with the excitatory and inhibitory neurons in a *resting state*, that is, stably sitting at a negative resting potential (i.e., a negative activation value).

One of the central contributions of the original work by Amari (1977)—and one of the primary reasons we build from this starting point—is that he analyzed the qualitatively different types of behaviors (or attractor states) this class of neural networks can exhibit. Amari identified five attractor states, three of which have been central to DFT: the resting state, the self-stabilized state where peaks of activation are stable but require input for their continued existence, and the self-sustaining state in which peaks of activation can remain stably *on* even in the absence of input.[4] These three states have a particular relationship to one another because DNFs are bistable. Under one parameter setting, the resting and self-stabilized states coexist; under other settings, the resting and self-sustaining states coexist.

The three attractor states are depicted in Fig. 5.1A. The first hump of activation—the one to the far right—depicts the resting state. The increase in activation around the right location is driven by the onset of a stimulus at the right location. When this input is presented, activation in the excitatory layer builds and comes close to the zero *activation threshold* (see red activation trace in Fig. 5.1C). However, the stimulus strength is not sufficient to create a peak and the field stays in the resting state. This can be observed directly by looking at activation in the inhibitory layer during this period of stimulation; the inhibitory layer is never engaged during the presentation of this first stimulus (see Fig. 5.1B). Consequently, once the input is turned *off*, the excitatory layer near the right location relaxes back to the negative resting level.

The second attractor state is depicted to the left (Fig. 5.1A). The same stimulus input was used; now, however, a self-stabilized peak builds. This occurs because the field has a small hump of activation present from the start of the trial. This small hump—which might reflect a contribution from long-term memory (see below)—elevated the resting level of the neurons near the left location (see blue line in Fig. 5.1C). Consequently, the target input moved the system through a dynamic instability and into a new attractor state. (Evidence of the instability can be seen in the small *joints* in the activation function indicated by the arrows in Fig. 5.1C.) As can be seen, the neural population remains in an *on* (i.e., above 0 activation) state while the stimulus input is present. Once the input is turned off, however, activation around the left location goes through another instability (see arrow in Fig. 5.1C) as the system moves from the self-stabilized state back to the resting state (for further discussion, see Spencer & Perone, 2008).

The final attractor state is shown around the central location in Figure 5.1A—the self-sustaining state. As can be seen in the figure, the central location in the field has a stronger hump of activation present at the start of the trial (see Fig. 5.1C). As with the peak around the left location, the peak at the center location grows with the onset of the stimulus. Once the input is turned off, however, the peak remains in the *on* state (note the above 0 level of the black line at the end of the simulation in Fig. 5.1C). In this case, neural interactions are strong enough to keep a memory of the stimulus around, even when the stimulus is no longer present. Note that we did not do anything special to the field to move it into this state. Rather, the stronger input was sufficient to create this shift from the self-stabilized attractor around the left location

[4] One of the attractor states Amari identified was a state in which the entire field becomes excited. Although this provides a potential mapping to cases of seizure, we avoid this state for obvious reasons. A fifth attractor state is a *spatially periodic* state, where the period varies with the average stimulation to the field. We suspect this state is involved in extracting the axes of symmetry we have probed in our work on the development of spatial recall (see, e.g., Schutte et al., 2003). This state might also have ties to the pattern of *grid cells* observed in the entorhinal and perirhinal cortex (Hafting, Fyhn, Molden, Moser, & Moser, 2005).

to a self-sustaining attractor state around the center location.

In the sections that follow, we provide a more detailed survey of the characteristics of self-stabilized and self-sustaining peaks. This survey highlights the many characteristics of peaks which *come for free* from the neural population dynamics within DNFs. It also provides grist for our claim that peaks provide an appropriate unit of cognition.

We will introduce these characteristics in the context of simple examples from spatial cognition, for instance, actively tracking the location of a moving object on a table or remembering the object's location during short-term delays (e.g., 10 s). We focus on these examples because they are conceptually simple and well-grounded in empirical observations. That said, the properties of fields that we illustrate could operate over any continuous, metric dimension such as object color or orientation (for examples, see Johnson, Spencer, Luck, & Schöner, in press; Johnson et al., 2008). Moreover, these properties can be extended to multidimensional fields, as well as to the coupling or *binding* of multiple cortical fields together (an issue we return to in the conclusion section; see also, chapter 4).

Another reason we focus on examples from spatial cognition is that they provide a link to formal demonstrations that dynamic field concepts are *embodied*. In particular, the characteristics we discuss below have proved useful in autonomous robotics scenarios in which robots make real-time, cognitive decisions (i.e., form peaks within DNFs); robots are able to sense features of the surrounds through sensory channels (such as cameras or infrared sensors that provide input to DNFs) and control their own navigation in real time via motors coupled to DNFs (e.g., Bicho et al., 2000; Faubel & Schöner, 2008; Iossifidis & Schöner, 2006; Steinhage & Schöner, 1997, 1998). We highlight a few specific robotics examples below. More generally, it is important to keep in mind that even though we use simulated fluctuations in input, the field concepts we highlight have proved critical to realizing autonomous and embodied links with perception, cognition, and action.

Finally, we want to emphasize that the simulations below use the same neural architecture except in a few cases where noted. Thus, even though we highlight different attractor states of DNFs, the only difference is typically a small change in the parameters of the model (e.g., weaker or stronger excitatory interactions among neurons). We begin with a series of simulations with short durations (1000 ms) to highlight the characteristics and detailed dynamics of each attractor state. Then we move to longer simulations with the types of variations one might experience during an experimental trial (e.g., across a 10-s memory delay). Later, we add long-term memory dynamics and look at the construction of memories on a trial-to-trial timescale (i.e., across thirty 10-s trials). Note that we have fixed these different timescales by mapping the features of DNFs onto empirical observations (e.g., Schutte & Spencer, in press).

### Self-Stabilized Peaks as a Basic Form of Perceptual Encoding

Self-stabilized peaks are stabilized in the sense that they are robust to noise—they don't go away even though activation fluctuates. Critically, however, they only stay around provided input is present. As Figure 5.1 shows, once the input is removed, the system relaxes back to the resting state.

Given that peaks in the self-stabilized state are dependent on input for their survival, it is tempting to think of them as *input-driven* or as a pattern of activation within a simple feed-forward network. Although these peaks are input-driven, they have some unique features that move them beyond the feed-forward picture.

The first property of self-stabilized peaks is that they are robust to variations in input—they are stable! In Figure 5.2, we use the same two-layer architecture previously shown in Figure 5.1; however, for simplicity, only the excitatory layer is shown (Fig. 5.2B). In this example, we present a single stimulus input at a central location (Fig. 5.2A); we used two types of fluctuations in the input. First, the amplitude

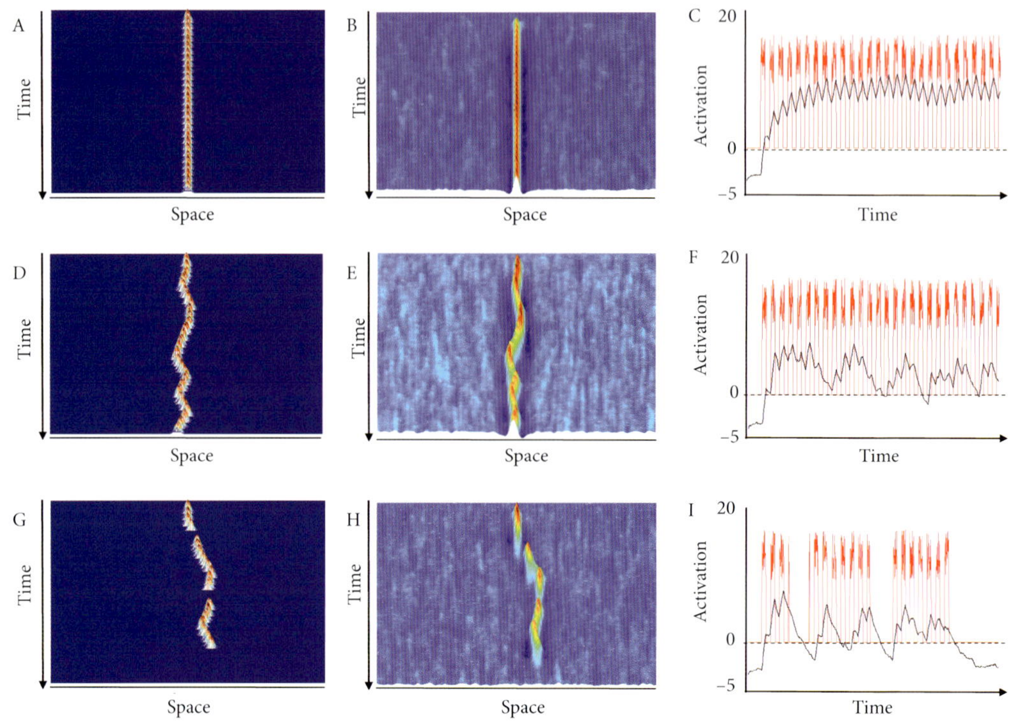


**Figure 5.2. Simulations of dynamic neural fields: Self-stabilized peaks**

This figure depicts three simulations of activation (panels B, E, and H) in a field with the two-layer architecture described for Figure 5.1. For each simulation, there is a single input (panels, A, D, and G) that fluctuates in amplitude and/or spatial location over time. **(Top row of panels)** Despite the fluctuations in input (panel A; colors indicate fluctuating/pulsating input), the field activation (panel B) remains stable in the *on* state throughout the trial; the black activation trace in panel C depicts the field activation through time at the spatial location of the peak in panel B. By contrast, the red line in panel C depicts the fluctuating/pulsating strength of the input through time at the spatial location of the input in panel A. **(Middle row of panels)** Even with strong amplitude fluctuations and spatial variation in the input (panel D and red trace in panel F), the activation peak remains *on*, tracking the input with smooth motion (panel E). **(Bottom row of panels)** When longer interruptions in the input are introduced (panel G and red trace in panel I), the activation peak dies out (panel H and black trace in panel I); but the field quickly reacquires the stimulus when it reappears after each occlusion. The time course of each simulation was 1 s (1000 ms).

of the input varies randomly; note the fluctuations in the red trace in Figure 5.2C. Second, the input is turned off for brief durations as occurs regularly in subjects during eye blinks. As can be seen in the figure, the peak stays in the *on* state (i.e., above 0 activation) throughout the trial (Fig. 5.2B), although the amplitude of the peak fluctuates with the input strength and as a function of the brief occlusions (see black trace in Fig. 5.2C).

This story gets more interesting if we look at what happens when we vary the input in both amplitude and spatial location over time. Recall that one of the goals when we started to use DNFs was to understand the link between perception-action and cognition—to keep cognition close to the sensory surface. Given that DNFs evolve continuously in time and across the continuous dimensions of space, color, and so on, they can provide a real-time connection to the outside world. The simulation in Figures 5.2D–F, begins with the presentation of a stimulus which builds a self-stabilized peak. Over the course of the simulation, however, the input takes a random walk (panel D) and shows dramatic amplitude fluctuations in real time (red trace in panel F). Despite great variation, the activation peak tracks the input in space and time (panel E).

Interestingly, we can make the situation even more complex by adding another common aspect of real-world neural signals—occlusions of longer durations. This occurs routinely when you move your eyes to scan a visual scene, when you turn your head, when you move your torso, and when you get up to walk around. What happens to peaks when these types of interruptions occur?

Because peaks in the self-stabilized state are *input-driven*, one might think longer interruptions would wreak havoc. The simulation in Figures 5.2G–I, shows that this is not the case—provided the interruption isn't too long. As can be seen in the figure, the self-stabilized peak tracks the input just as before, and it dies out during each long occlusion (see black trace in panel I). But the field is able to quickly reacquire the stimulus when it reappears.

Interestingly, we can make the input pattern even more complex to highlight yet another feature of self-stabilized peaks. Although the input pattern used in Figure 5.2 was noisy in multiple ways, it was still a far cry from the complexity of even simple real-world inputs. What happens to our peak picture when we add complexity in the form of *multiple* inputs appearing and disappearing over a longer timescale (such as a 10-s trial)?

In Figures 5.3A and B, we introduced new random, transient inputs at different distances from the central *target* input. (For simplicity, we focus on the dynamics within the fields shown in Figure 5.3 and do not show the input pattern directly.) The self-stabilized peak starts out tracking the input just fine. However, a few milliseconds into the simulation, a second *distractor* input appears. Although this second input is just as strong, the peak actively *suppresses* the distractor. This occurs via the lateral inhibitory interactions generated by the peak. About a third of the way through the simulation, there is a long occlusion. During the occlusion, a new distractor appears to the left and the field builds a peak at this location. Soon thereafter, the central target reappears along with a second distractor to the right. It takes several hundred milliseconds for the central peak to reach full strength again due to the active competition with the distractor items (see the dip in activation in Fig. 5.3B). Over time, however, the field reacquires the central target and tracks it for the remainder of the trial, suppressing several new distractors toward the end of the simulation. In summary, then, *peaks provide stability against transient local distractions.*

In the simulation depicted in Figure 5.3A, new distracting inputs were generally *on* for only a short period. What happens when distractors are far from the central target peak, sufficiently strong, and remain *on* for a significant duration—that is, when these distractions are discriminably different, sustained, and reliable patterns of input? Figure 5.3C depicts that a new type of behavior emerges: the field enters a multi-peak state and effectively tracks all three input patterns. The possibility for tracking multiple inputs opens up new types of functionality. This makes sense conceptually since the distractions depicted in Figure 5.3C can be considered interesting new inputs to *encode*.

At this point, it should be clear that peaks have some very interesting features that make them more than merely input-driven patterns of activation. They have important stability properties, tolerating a lot of noise in the input including fluctuations in amplitude, fluctuations along the represented dimension, fluctuations in the presence/absence of the signal, and the introduction of new transient inputs. Moreover, DNFs can operate in a mode in which they track multiple peaks simultaneously, provided that the input signals supporting these peaks are sufficiently robust (e.g., are strong enough, *on* for a sufficient duration, and so forth). Critically, the *decision* about which signals are robust and which are transient is emergent from the real-time dynamics of the system (e.g., excitatory and inhibitory interactions), the characteristics of the input,

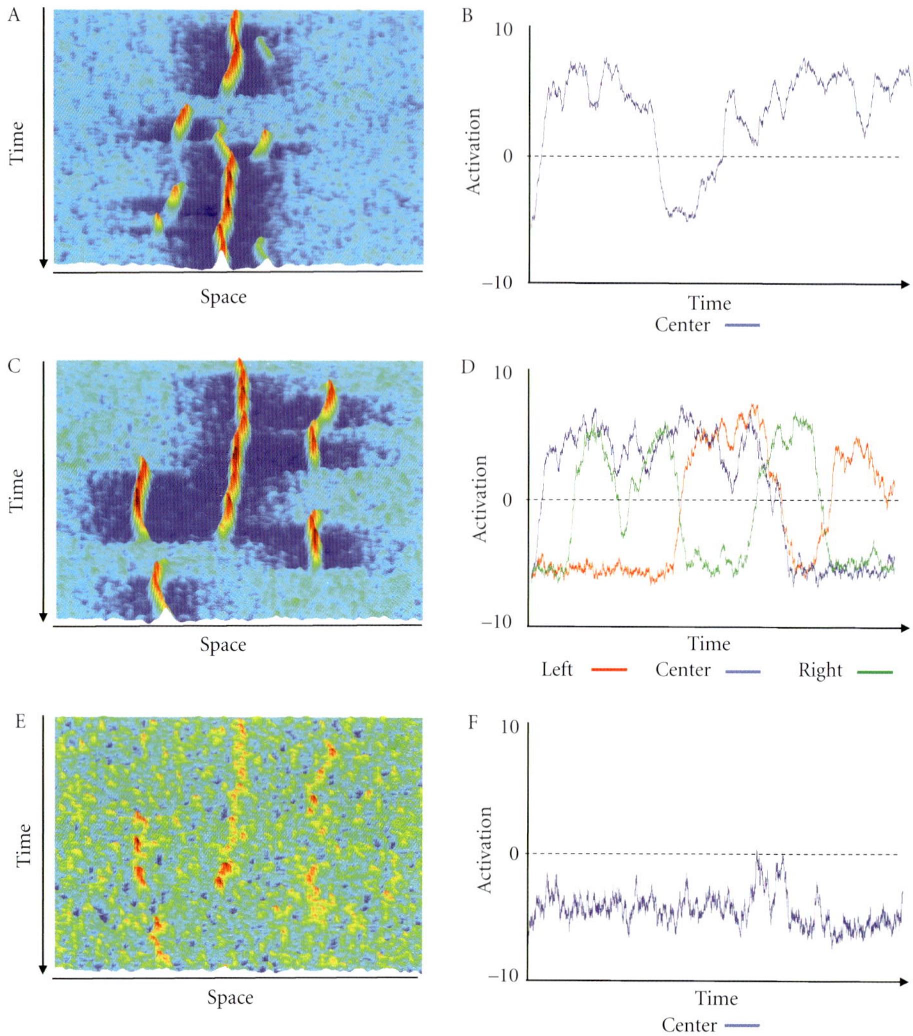


**Figure 5.3. Simulations of noisy dynamic neural fields: Self-stabilized peaks**

This figure depicts three simulations of activation (panels A, C, and E) in a field with the two-layer architecture as described for Figure 5.1 and depicted in Figure 5.2. The field activations depicted arise in response to target inputs (center peak) in the presence of one or more *distractor* inputs. Despite the presence of nearby distractors, the center activation peak in panel A remains above threshold (see red activation values in panel A) as long as the target input remains *on*. When the target input is *off* and a distractor appears, the center peak briefly goes below threshold (see blue trace in panel B) before reacquiring the target. When distractor inputs are farther away from the target (see left and right activation peaks in panel C), multiple self-stabilized peaks are formed in the field despite substantial variations in amplitude (see traces in panel D) and in spatial location. When neural interactions in the field are too weak to enter the self-stabilized state (panel E), the field functions as a low-pass filter on the same inputs presented in panel C. In this case, activation at the center location never consistently passes above threshold (see panel F). The time course of each simulation was 10 s.

and the *just-previous* history of activation in the field. These properties of self-stabilized peaks come for free. All of these properties allow peaks to actively maintain a stable, real-time link to the outside world (or to a complex inside world as in the case of multiple, coupled cortical fields).

We think the characteristics of peaks discussed thus far—the characteristics of stable patterns of activation at the level of a neural population—are critical for the nervous system. Imagine you are a neural field sitting in the posterior parietal cortex. You have just received input from early visual areas that a spot of light has been detected at a particular location on the retina and you have dutifully built an associated pattern of activation to represent this event. At the same time, you begin to receive input from inferior temporal areas that the feature values of a face might be visible. Frontal areas are telling you to keep focused on that computer screen and ignore the people walking around in your periphery at the coffee shop. You are also receiving input from somatosensory areas linked to the stomach saying that the vanilla latte you are drinking tastes really yummy, while your motor cortex wants to tell you about the lovely oscillation of your foot it has started up in response to the influx of caffeine. In short, your little cortical field is being bombarded with noisy inputs—some are relevant, some are irrelevant—and without the stability provided by peaks, you would be in sorry shape.

How sorry? Figure 5.3E depicts a field with the same input pattern used in Figure 5.3C, but with neural interactions too weak to enter the self-stabilized state. As can be seen in this figure, there is a hint of three reliable inputs, but only a hint. Fortunately, the neural dynamics in your parietal cortex operate more like the pattern in Figure 5.3C than in Figure 5.3E... which allows you to stay focused and keep working...

The simulations thus far have highlighted that self-stabilized peaks have some advantageous characteristics, creating stable perceptual decisions that an input is sufficiently strong and reliable to build a peak and subsequently tracking the details of the stimulus in real time. But there are also more cognitive aspects of the self-stabilized state worth highlighting.

Figure 5.4 depicts the neural field used in all simulations discussed thus far; it can track up to six inputs simultaneously (panel A). Once the number of inputs reaches seven, however, the field *selects* a subset of five inputs to track (panel C). Thus, fields operating in the self-stabilized state have a maximum *capacity*: only so many peaks can be simultaneously built (for discussion, see Johnson et al., 2008). What causes this capacity limit? In panel B, the red line depicts the summed activation in the *excitatory* layer shown in panel A (with 6 inputs); the blue line in panel B depicts the summed activation from panel C (with 7 inputs). Panel D depicts the corresponding summed activation in the *inhibitory* layer from each simulation. As can be seen in these panels, excitation (panel B) and inhibition (panel D) quickly reach their maximal levels when six inputs are presented (red lines). When seven inputs are presented (blue lines), excitation and inhibition increase as before, but the presence of an extra input tries to push excitation too high (the total excitatory strength cannot exceed the total inhibitory strength or activation in the network will grow without bounds). As a consequence, the excitatory and inhibitory strengths settle into a level that is lower than with 6 peaks, but where the difference between these strengths is very small—just large enough to keep inhibition reliably stronger than excitation.

In summary, the simulations in this section illustrate that self-stabilized peaks within DNFs provide a neural basis at the population level for (a) robust perceptual encoding and simple forms of perceptual decision making (e.g., should I build a peak or not?), (b) tracking/updating inputs despite variations in the input signal, (c) simple forms of distractor suppression when multiple inputs are present, and (d) multi-input tracking/updating when a capacity-limited set of reliable input signals are present. These properties of self-stabilized peaks could also be useful in cases in which it is necessary to track multiple moving objects and/or landmarks in the task space.

We think self-stabilized peaks provide a wonderful first step for linking simple forms of

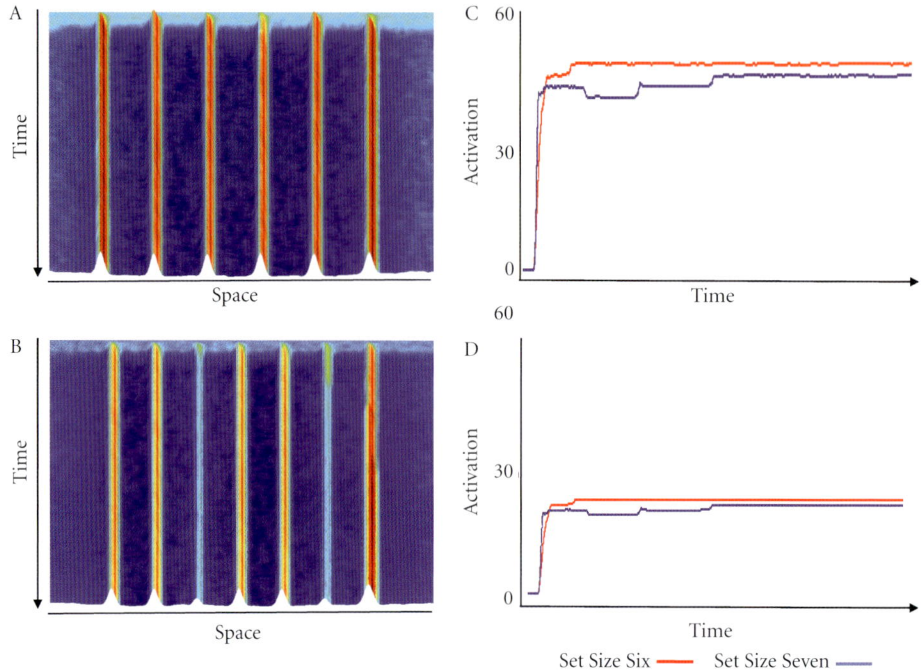


**Figure 5.4. Simulations depicting capacity limits of a field in the self-stabilized state**
This figure depicts a two-layer field operating in the self-stabilized state. The field successfully tracks up to six simultaneous inputs in panel A. When a seventh input is added in panel C, the field *selects* a subset, forming self-stabilized peaks for five out of the seven inputs. (Blue activation humps are subthreshold, that is, below 0.) Line plots show summed excitatory (panel B) and inhibitory (panel D) activation, respectively, with six (red line) versus seven (blue line) inputs. The time course of simulations was 10 s.

cognition to low-level sensory inputs to obtain states/behaviors that are robust over space and time from inputs that are noisy and transient. But peaks in this self-stabilized state have a fundamental limitation: they must be anchored to input. Once the input is removed for a sustained period, this state is destabilized and the field relaxes back to the resting state. In the next section, we show that this does not always have to occur; DNFs can operate in a different attractor state—the *self-sustaining state.*

## Self-Sustaining Peaks and Working Memory

By employing the simulator we've been using in all the simulations thus far and simply making the field a bit more excitable overall, a new, qualitatively different state emerges. Figure 5.5B depicts a simulation of a *self-sustaining peak*. This simulation uses the same input profile as in Figure 5.2G (see Fig. 5.5A), *except* that the resting level of the *inhibitory field* has been moved from −5 to −12. (The inhibitory field is not shown here; refer to Figure 5.1B for a depiction of an inhibitory field.) Note that the change in inhibition effectively increases the excitation in the system—less inhibition allows more excitation. This change is evident in the higher activation level in Figure 5.5C (black trace) relative to Figure 5.2I (black trace). The other critical difference is in how the peak behaves over time when the input is occluded for longer durations: rather than returning to the resting state during the three longer occlusion events, the peak remains in the *on* state throughout the simulation (Fig. 5.5C).

We have referred to this self-sustaining behavior as a form of *working memory* because it shows how a cortical field can actively maintain information about a stimulus during a memory delay (i.e., when the stimulus is no longer present). There is general agreement that some form of recurrent activation is the neural basis for working memory (Fuster, 1973; Goldman-Rakic, 1995), and DNFs provide one implementation of this concept.

There is also an important link between self-sustaining peaks and the original proposal by Baddeley (1986) that working memory is a form of active memory that can be used in the service of tasks that require operations on information. In our view, using information in a working memory store in the service of some cognitive operation *requires stability*. That is, the cortical system actively maintaining the information to be used must keep that neural representation stable to support operations on this information conducted by, for instance, some other cortical system. This requires stability in the face of neural noise, stability in the face of potentially *distracting* inputs, and so on. In short, we contend that the working memory construct requires the types of stability we have already shown that peaks have! Of course, the simulations in the previous section showed the stability properties of self-stabilized peaks. What are the stability properties and unique features of self-sustaining peaks?

As can be seen in Figure 5.5B, self-sustaining peaks can still handle variations in the amplitude, spatial location, and presence/absence of an input. Indeed, in some respects, self-sustaining peaks handle these variations more effectively. For instance, self-sustaining peaks can remain in the *on* state during long occlusions (Fig. 5.5C). Self-sustaining peaks are also very good at tracking multiple inputs with several intervening occlusion events. This is shown in Figure 5.6A. This simulation had the same input pattern as in Figure 5.3C. Now, rather than losing the continuity in the three input patterns, the self-sustaining peak effectively *glues* together the gaps in input. Note that the field does not do this perfectly—the field *expected* the input to the far left to reappear at a location to the left of where it actually reappeared. One can see a quick adjustment, however: the left peak (red line in Fig. 5.6B) dips down, but quickly blends with the new input and returns to a high activation value.

Although self-sustaining peaks do some things quite well, there are also some down sides to this type of cognitive stability. First, self-sustaining peaks are not quite as good at tracking spatial variations in input. Second, when there are large shifts in input, it takes self-sustaining peaks longer to shift from one input source to another. A third consequence is

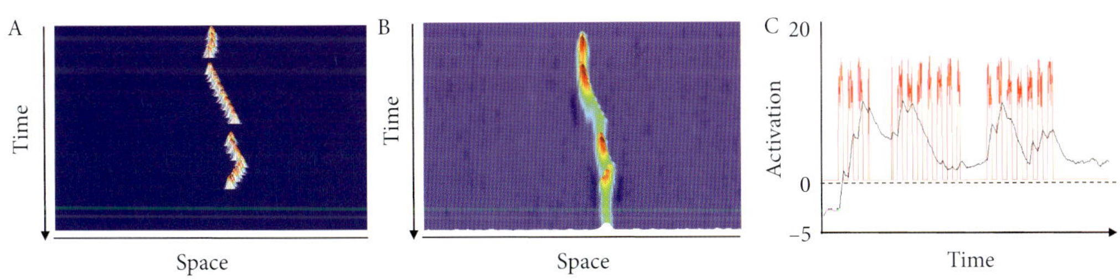


**Figure 5.5. Simulation depicting a self-sustaining peak**
This figure depicts a two-layer, dynamic neural field operating in the self-sustaining state (see panel B) when presented with variable input (see panel A; the same input used in Figure 5.2, panel G). A single peak is self-sustained in the field (see panel B and black trace/activation profile in panel C) despite the presence of noise and lengthy interruptions in the input (see red trace in panel C). The time course of simulations was 1 s (1000 ms).

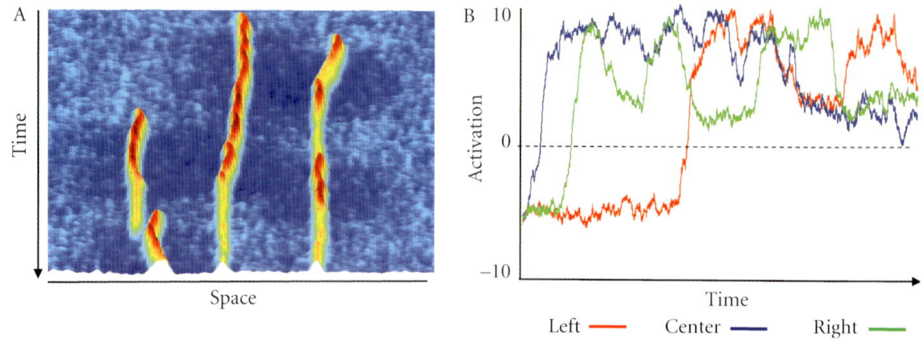


**Figure 5.6. Simulation depicting a self-sustaining peak**
This figure depicts a two-layer, dynamic neural field operating in the self-sustaining state (see panel A) tracking the same inputs used in Figure 5.3, panel C. Self-sustaining peaks are formed and effectively track the inputs across several intervening occlusion events (see activation profiles through time in panel B associated with each peak location in panel A; Left peak shown in red, Center peak shown in blue, Rigt peak shown in green). The time course of simulations was 10 s.

illustrated in Figure 5.7: fields with self-sustaining peaks have a lower capacity than fields with self-stabilized peaks. Panel A depicts a field with four simultaneously active self-sustaining peaks. In panel C, the input set size increases to six items. Rather than building six self-sustaining peaks, the field is able to sustain only four peaks once the input is removed. This occurs for the same reasons discussed previously: after the input is removed, the field settles into a state that balances the overall level of excitation and inhibition (compare the solid red and blue lines in panel B and panel D). More specifically, in the absence of the strong excitation provided by direct input, the field cannot maintain the level of excitation/inhibition required for six peaks. Thus, two peaks are *killed* and the excitatory/inhibitory balance returns to its maximal level. Note that we have added the overall excitation and inhibition levels for the self-stabilized simulation shown in Figure 5.4 for comparison (see dashed blue lines).

We conclude this discussion with one final note. The simulations in Figure 5.7 leave the impression that self-sustaining peaks will continue to exist forever under the right conditions. This is, of course, not the case in real neural systems; peaks can be de-stabilized by a variety of factors including neural noise, suppression by other inputs, and so on. Indeed, the likelihood that this will occur increases systematically with memory delays, leading to typical *forgetting curves.* Although this is the case, the forgetting mechanism in DFT—bifurcation into the resting state—predicts that when peaks drop out, they will drop out in an all-or-none fashion versus gradually degrading in resolution. Recent data from Zhang and Luck (2008) using a color recall task are consistent with this mechanism.

## Integrating Perception and Working Memory in Real Time: Multilayered Fields

The simple two-layer networks introduced thus far can serve either a perceptual or a working memory function. For instance, a two-layer network operating in the self-stabilized state, where peaks of activation are formed in response to input but die out when input is removed, could be used to capture the perceptual encoding of an array of items in the world. Additionally, a two-layer network operating in the self-sustaining state, where peaks of activation remain above threshold after input has been removed, can be used to implement a form of working memory. Critically, however, many real-world activities require the *integration of both functions.* For instance, to remember the locations of critical objects in the local surrounds, one needs to actively maintain information about these locations in working memory, but also keep those memories anchored to visible landmarks in the task space (Spencer et al., 2007). Similarly, to

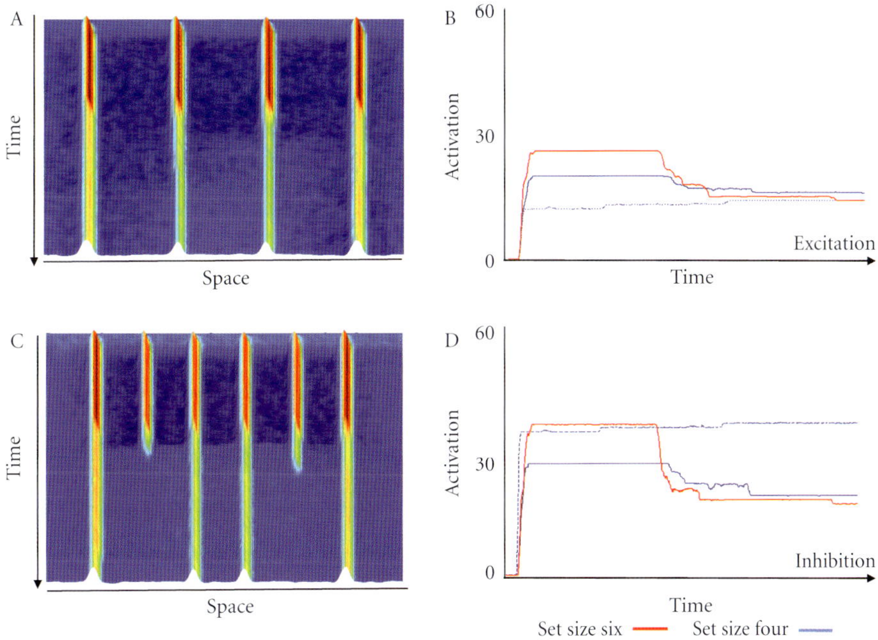


**Figure 5.7. Simulations depicting capacity limits of a field in the self-sustaining state**
This figure depicts a two-layer field operating in the self-sustaining state. The field successfully forms and maintains four self-sustained peaks of activation in the absence of input (A). However, when six inputs are presented, only four are selected and self-sustained throughout the 10-second simulation (C). Line plots show the sum of activation in excitatory (B) and inhibitory (D) layers for set-size 6 (red lines), set-size 4 (solid blue lines), and, for comparison, for set-size 4 with the field operating in the self-stabilized state, as in Figure 5.4 (dashed lines).

detect changes in the local surrounds, one must compare the currently perceived set of items to the set of items actively maintained in working memory. How are these types of functionality achieved within DFT?

To address the real-time integration of perception and working memory, we have developed a three-layer architecture consisting of two layers of excitatory neurons coupled to a single layer of inhibitory neurons (Johnson et al., in press; Simmering, Schutte, et al., 2008; Spencer et al., 2007). Figure 5.8 shows the architecture of this system. The model consists of an excitatory perceptual field (PF; panel A), an excitatory working memory field (WM; panel C), and a shared inhibitory layer (Inhib; panel B). Excitatory and inhibitory connections among the layers are indicated by green and red arrows, respectively. Note that this multilayered architecture was inspired, in part, by the cyto architecture of the visual cortex (Douglas & Martin, 1998). That said, it is possible to implement the same functionality we highlight below using two coupled two-layer fields where activation of the inhibitory layers in these networks inhibits *both* excitatory layers.

As its name suggests, PF is the primary target of perceptual input to the model; however, neurons in the WM field also receive weak, direct input. In addition, PF provides excitatory input to WM. Finally, both PF and WM pass input to the inhibitory layer and receive inhibition projected broadly back. These pairs of two-layer interactions are identical to the local excitatory and lateral inhibitory neural interactions evident in previous simulations.

What does this particular type of integration achieve? The three-layer architecture has been

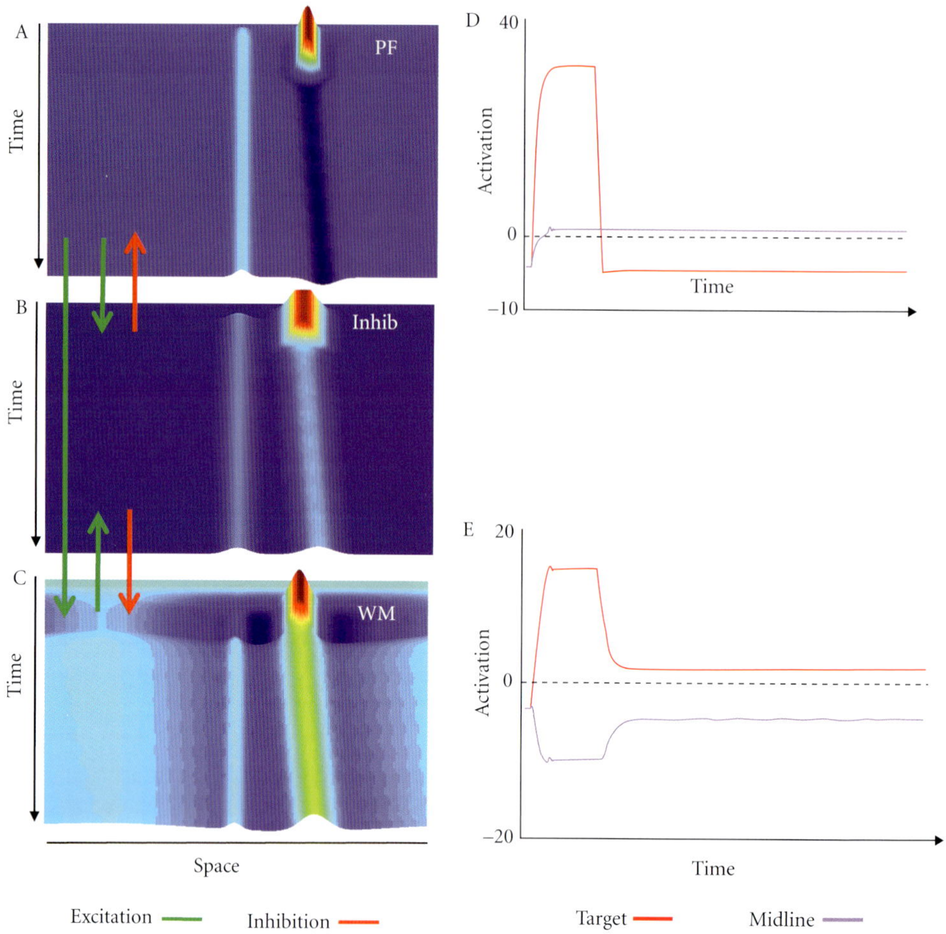


**Figure 5.8. Simulation illustrating a one-dimensional, three-layer dynamic neural field**
This figure depicts a simulation of a one-dimensional, three-layer dynamic neural field model depicting the anchoring of spatial working memory to the real-time perception of landmarks in a task space. The model consists of an excitatory perceptual field (PF; panel A), an excitatory working memory field (WM; panel C), and a shared inhibitory layer (Inhib; panel B). Excitatory and inhibitory connections among the layers are indicated by green and red arrows, respectively. Line plots (panels D and E) depict activation levels in PF and WM, respectively, in response to the target (red line) and midline (blue line) inputs. The time course of each simulation was 10 sec. See text for further details.

used to capture the anchoring of spatial working memory (SWM) to the real-time perception of landmarks in the task space (Schutte, Spencer, & Schöner, 2003; Spencer et al., 2007). An example can be seen in Figure 5.8, which shows a simulation of a three-layer model remembering a target location relative to a visible landmark during a delay. At the start of this simulation, two inputs are present: a landmark along the midline axis of the task space and a salient target input to the right of midline. In response to these inputs, PF (panel A) creates two self-stabilized peaks (see panel D), while WM (panel C) creates a single, self-sustaining peak associated with the stronger target input (see panel E).

Next, the target disappears from view. Consequently, the target peak returns to the resting level in PF, but remains in a self-sustaining state in WM. Thus, during the memory delay, PF is actively locked onto the visible landmark, while WM is actively maintaining a representation of the target. The online maintenance of a peak in PF and a peak in WM has both costs and benefits. One *cost* is readily apparent in panel C: the peak of activation in WM *drifts* rightward—away from midline—during the memory delay, consistent with empirical observations (see, for example, Spencer & Hund, 2002, 2003). This occurs because the peak of activation at midline in PF passes activation to the Inhib layer at midline (see panel B). This creates greater inhibition on the midline side of the self-sustaining peak in WM, effectively repelling this peak away from the landmark (for additional evidence of such effects, see Simmering & Spencer, 2008; Simmering, Spencer, & Schöner, 2006).

Although these interactions produce systematic bias in memory, Huttenlocher and colleagues (Huttenlocher et al., 1991; Huttenlocher, Hedges, & Vevea, 2000) pointed out that such biases can have adaptive consequences. For instance, this type of bias minimizes the likelihood of cases/errors that cross over to the wrong side of the landmark (that is, cases in which the target is shown to the right of midline but participants respond to the left). This bias also exaggerates the *left-ness* or *right-ness* of the target item, which could facilitate recall from long-term memory. Indeed, from a Bayesian perspective, such bias can help create an optimal spatial memory system, trading off systematic bias for an overall reduction in variance (see, Huttenlocher et al., 1991; Huttenlocher et al., 2000).

The simulation in Figure 5.8 shows how perceptual and working memory processes can be integrated in a real-time neural system using the concepts of DFT. Recent empirical and theoretical work has shown that the particular type of integration shown in Figure 5.8 has several emergent properties that can be linked to specific behavioral phenomena. For instance, we have recently demonstrated that the architecture shown in Figure 5.8 can capture children's (Simmering, Johnson, & Spencer, 2008, in preparation) and adults' (Johnson et al., in press) performance in a standard *change detection task* (see Luck & Vogel, 1997). In this task, subjects are shown an array of objects on a computer screen that vary in, say, location. The task is to remember these objects and report whether they are in the same or different locations relative to a second array of items presented 1 second later.

How does the three-layer model (as in Fig. 5.8) capture performance in this task? The initial presentation of the array creates multiple self-sustaining peaks of activation in working memory. During the delay, these peaks create *troughs* of inhibition in PF. An example is evident in panel A; during the memory delay, there is an inhibited region in PF associated with the self-sustaining peak in WM (panel C). What is the behavioral consequence of these troughs? Is there any behavioral evidence supporting them? If an item is presented in the same location as an item currently held in WM, this trough can suppress the build up of a self-stabilized peak in PF. By contrast, if an item is presented outside this trough, a self-stabilized peak can quickly form in PF. We use this difference—the absence of a peak in PF on *same* trials versus the presence of a peak in PF on *different* trials—to drive decision making in our model of change detection performance (Johnson, et al., in press). And this mechanism for generating same/different decisions has led to a host of novel behavioral predictions that have been successfully tested experimentally (Johnson et al., in press; Simmering & Spencer, 2008).

In summary, these examples highlight the adaptive consequences of integrating perceptual and working memory processes in a real-time neural system. But this provides only a beginning for thinking about embodied cognitive dynamics. In the next section, we add another novel ingredient—the integration of dynamics across multiple timescales.

### Integrating Multiple Timescales: Long-Term Memory Dynamics

As demonstrated in Figure 5.1, patterns of activation within DNFs can be influenced by the presence of sub-threshold inputs, that is, inputs that are too weak to drive neurons in a field from their negative resting level to above-zero activation values. Recall that in the absence of sub-threshold activation, the field shown in Figure 5.1 remains in the resting state. With a weak sub-threshold input, the field forms a self-stabilized peak. And with a stronger sub-threshold input, the field forms a self-sustaining peak.

One source of sub-threshold activation within DFT is input activation from a long-term memory (LTM) process. The formation of LTM traces in DFT relies on a very simple type of dynamics, which implements a form of Hebbian learning (see chapter 18). Specifically, whenever activation at a given field site climbs above threshold (i.e., above 0), activation traces begin to accumulate in an LTM field at that site. At subsequent points, such traces are fed back to the field of origin as an additional source of excitatory input. When there is no above-threshold activation in a field, no LTM traces are laid down, and previously accumulated traces gradually decay.

The accumulation and decay of LTM traces occur more slowly than activation evolves in the excitatory and inhibitory fields discussed thus far. This allows patterns of activation that reflect events unfolding over longer timescales to accumulate in LTM. For instance, over the course of an experimental session in the spatial paradigm we have been considering, LTM traces can be formed that reflect the probability of targets appearing at particular locations. Importantly, because LTM traces are only laid down when above-threshold peaks are formed, the memory here is not about the statistics of stimulation, but instead reflects the statistics of peaks, that is, the *decisions* made second-to-second by the dynamics within a neural field. We discuss this in greater detail below. For now, note that sometimes the history of a field's decisions in a given context diverges from the statistics of the input (see chapter 18 for discussion).

The series of simulations depicted in Figure 5.9 illustrates the build up of LTM traces every five trials across a series of thirty 10-second simulation trials, as well as the consequences of LTM input on patterns of activation in DNFs. For each simulated trial, a single input was presented at one of two possible mean locations (a left location and a right location). The inputs had the same types of variation used in previous simulations (i.e., amplitude, spatial location, presence/absence). As can be seen in Figure 5.9A, the field operated in a self-stabilized state in the first simulation, tracking the input on the left side, but also dipping into the resting state during two occlusions (see Fig. 5.9B).

Figures 5.9E–J depict how patterns of activation in LTM evolve over the course of 30 simulations, and how this can give rise to category formation and the emergence of *exemplars*. As can be seen in Figure 5.9E, after a few trials to each target (i.e., on trial 5), the activation in LTM is clustered around a central location. This occurs because, although inputs were presented at two different locations *across* trials, variations in the spatial positioning of the inputs *during* a given trial produce significant overlap and a consequent blending of LTM traces. Thus, early in learning the distribution of activation in LTM is unimodal, reflecting an average or *prototypical location*. Over many simulations, this centrally biased activation becomes stronger. By Figure 5.9H, however, LTM begins to differentiate, and by trial 30 (Fig. 5.9J), two distinct targets have emerged with humps centered at the actual mean target locations. As can be seen in Figure 5.9C (which depicts the field on trial 30) the LTM extracted these *targets* despite the continued presence of noise in the input. Figure 5.9C also illustrates an emergent effect that arises from the increased strength of activation from LTM into the field; with this added input, the field entered the self-sustaining state, surviving an occlusion event late in the trial (see Fig. 5.9D).

In summary, the simulations in Figure 5.9 illustrate how the LTM mechanism in DFT captures the emergence of a prototype-like representation early in learning; late in learning, this shifts to an exemplar-like representation, which correctly reflects the true mean input locations. This occurs in an unsupervised manner. Moreover, this form of learning has emergent consequences on the types of active decisions formed in real time during a trial. Thus, the two timescales of cognitive dynamics depicted

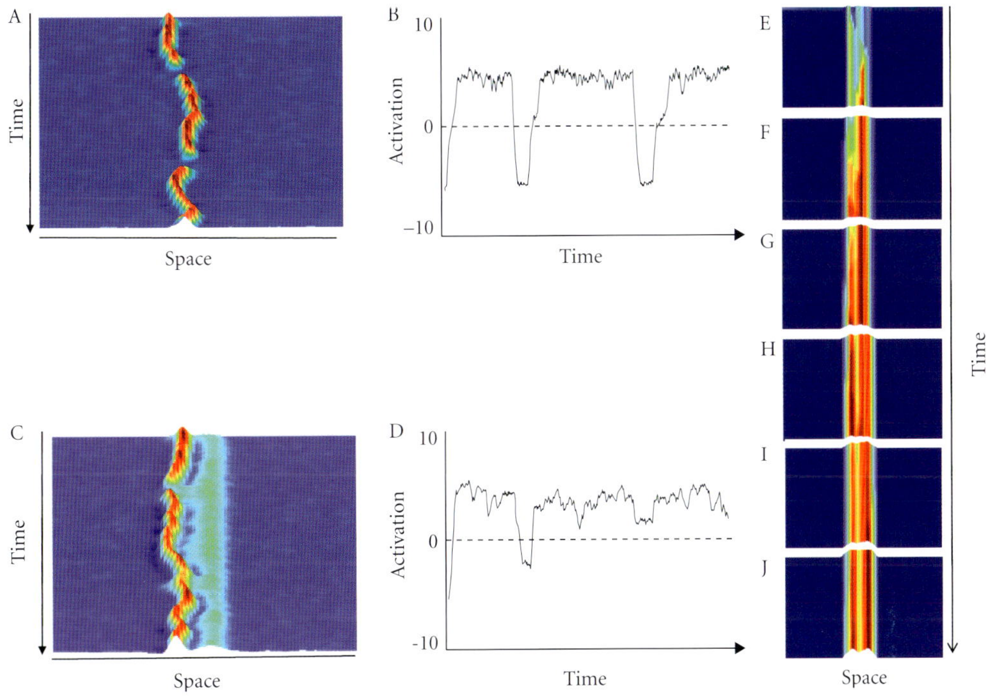


**Figure 5.9. Simulated buildup of LTM traces during a series of trials**
In this figure, each of the thirty 10-s trials consisted of the presentation of a single input at a mean left or right location. Inputs (not shown) were interrupted periodically and varied in amplitude and spatial location within a trial. At the beginning of the series, a self-stabilized peak tracks the input as long as it is present (panel A), but dies out when the input is interrupted (see activation profile in panel B). By the end of the series (panels C and D), the presence of robust LTM input to the field allows the peak to self-sustain despite numerous brief interruptions. Panels E–J depict activation in the LTM field during every fifth trial. Across the simulation runs, the distribution of activation in LTM shifts from a unimodal (prototypical) distribution (panels E, F, and G) to a bimodal (exemplar) distribution (panel J), which reflects the two target locations presented across the series of simulations.

here—the fast timescale where peaks form and the slower timescale of LTM formation—are truly interactive and lead to new phenomena over the trials.

In addition to contributing to a shift in the attractor states a field can enter, LTM can have other adaptive influences on the properties of peaks within an attractor state. For instance, Figure 5.10A depicts a simulation of a self-stabilized peak tracking an input amid distracting items. This simulation is identical to the simulation shown in Figure 5.3A with one exception—the field shown in Figure 5.10A is coupled to an LTM field. As can be seen by comparing the two figures, the simulations are quite similar initially. As the trial unfolds, however, the increased activation provided by input from LTM leads to a more robust self-stabilized peak. As a consequence, distractor suppression late in the trial is stronger and extends to distractors farther from the central peak (see arrows).

The simulations described thus far show that LTM input can be quite beneficial, making it more likely for peaks of activation to form in response to stimulation, increasing the stability of peaks in the face of variable and intermittent input, and making peaks more robust to distractors. In addition, the accumulation of traces in LTM can give rise to category formation and

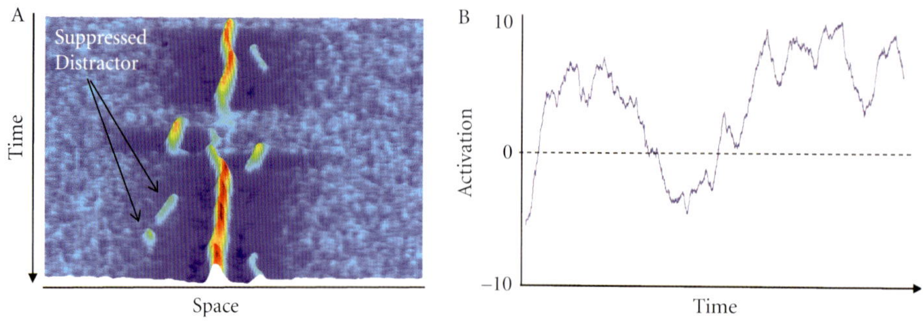


**Figure 5.10. Simulation demonstrating distractor suppression**
Replication of the simulation depicted in Figure 5.3, panel A, with the addition of input from a LTM field. LTM traces build up over the course of the 10-s simulation to produce a more robust self-stabilized peak (see panel A) that effectively suppresses far away distractors. (Compare suppressed distractors in Fig. 5.10, panel A, to activation values in Fig. 5.3, panel A.) Note that only yellow and red activation values are above zero in panel A.

can pull out exemplars when several inputs are tracked across multiple trials. Nevertheless, the stabilizing influence of LTM is sometimes achieved at the cost of flexibility in the face of changing inputs and task demands.

To appreciate this, consider the simulation shown in Figure 5.11, which captures a type of perseverative responding. At the beginning of the simulation, a single input is briefly presented and a stable, self-sustained peak forms at the location of the input (see Fig. 5.11A). Shortly thereafter, an activation trace begins to accumulate in the LTM field (Fig. 5.11B). The feedback from LTM serves to stabilize and strengthen the peak depicted in Figure 5.11A. The next event in the simulation is the presentation of a second, task-relevant input at a right location. Although there is now only a single stimulus present in the task space, the self-sustaining peak effectively treats this new input as a distractor and prevents a new peak from forming at the right location (see sub-threshold blue trace in Fig. 5.11C). As a consequence, the field perseverates—the ongoing state is determined by its past history rather than by the current input.

The simulations described in this section highlight a fundamental tension between stability and flexibility that arises in the nervous system. On one hand, it is highly advantageous for ongoing behavior to be open to the influence of past experiences. Such influences can have a stabilizing effect by, for instance, allowing the system to operate in a self-sustaining versus self-stabilized state. However, adaptive behavior is about more than stabilizing a particular behavior or mental state; adaptive behavior involves flexibly switching between stable states as the situation demands. This is the focus of the following section.

## Toward Behavioral Flexibility: Going into and out of (vs. near) Stable Attractors

Tensions among stability, instability, and flexibility have been dramatically illustrated in research on the Piagetian A-not-B error (Thelen, Schöner, Scheier, & Smith, 2001). In the A-not-B task, infants are shown a toy, the toy is hidden at location A, and the infants are allowed to search for the toy after a brief delay. This sequence is repeated several times. Next, the toy is hidden at nearby location B. After a 3–5-second delay, 8- to 10-month-old infants reliably search for the toy at the original location A. By contrast, 10- to 12-month-old infants search correctly at location B (Smith, Thelen, Titzer, & McLin, 1999).

The original DFT account of infants' performance in this task used a DNF coupled to an LTM or *pre-shape* field (Thelen et al., 2001). When the target was presented on each A-trial, the model formed a self-stabilized peak. Due to the short delay, the model was able to correctly *search* at this location. On the B-trials, however,

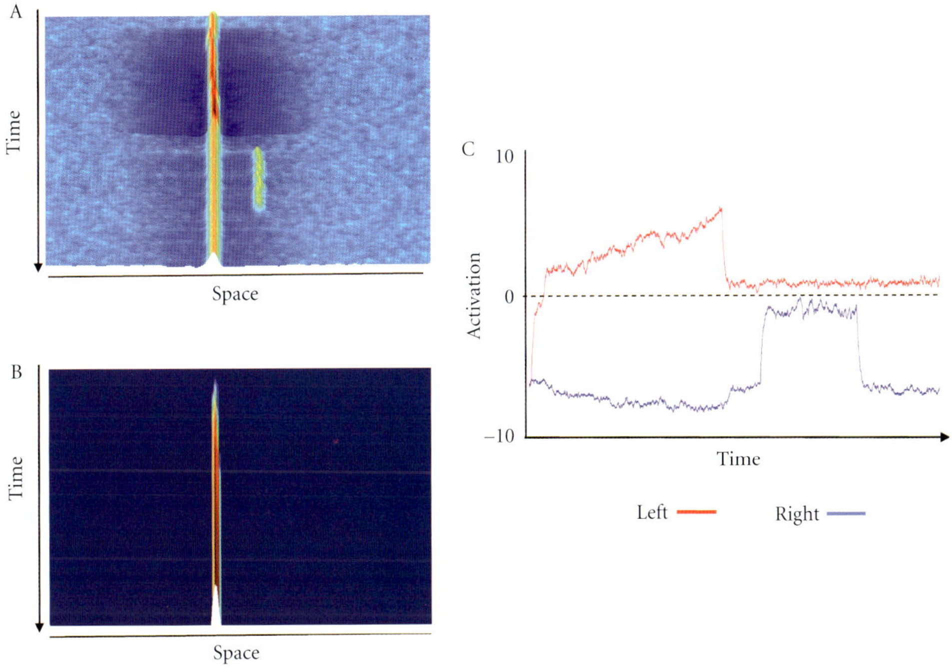


**Figure 5.11. Simulation highlighting tension between stability and flexibility in neural systems**
At the beginning of the simulation, a self-sustained peak is formed (see panel A and red trace in panel C) in response to a transient task-relevant input in the task space. As the 10-s simulation progresses, the peak is further stabilized by LTM traces building up at the left location (panel B). Thus, when a second task-relevant input appears to the right (see right activation hump in panel A), it is treated as a distractor and is suppressed by the robust peak at the left location—a form of perseveration. Line plots in panel C, show the activation profile through time at the left (red) and right (blue) locations.

the peak returned to the resting state during the longer memory delay. As a consequence, the model perseveratively searched at A, that is, the model's performance was dominated by the LTM of previous searches/reaches to A rather than a robust memory of the target at B. Thelen and colleagues proposed that the shift from perseverative responding to correct responding at 10–12 months of age reflected an increase in the strength of neural interactions. In terms of the dynamic field model, this represents a shift from the self-stabilized state to the self-sustaining state.

Although this dynamic field account captured key aspects of infants' performance in the A-not-B task and generated a host of novel predictions (e.g., Diedrich, Highlands, Spahr, Thelen, & Smith, 2001; Diedrich, Thelen, Smith, & Corbetta, 2000), a funny thing happened in the late 1990s. Esther Thelen was a strong proponent of embodied cognition; she also had a keen suspicion of computational models. Thus, she asked Gregor Schöner to demonstrate the embodiment of the dynamic field account of infant perseverative reaching by implementing the model on a robot with real sensors operating in real time. The result: the original A-not-B model failed this embodiment test! Rather than mimicking infants' performance, the robot spent trial after trial jiggling back and forth, unable to clearly respond to A or to B on any given trial.

What was the problem? The problem was that the original dynamic field model of the A-not-B error failed to generate a stable peak at the end of each trial. Rather, Thelen and colleagues used a basic modeling technique to *read off* the response—the model was said to *respond* to whichever location showed the strongest activation at the end of the simulation. The problem on the robot was that the field was in a resting state at the end of each trial and, as we showed in Figure 5.3E, fluctuations in amplitude in this state are a far cry from the robust neural signal a nervous system would need to drive a stable response. Thus, instead of reliably moving to A or to B, the robot jiggled back and forth as the location of maximal activation fluctuated from time step to time step.

To solve this problem, Dineva and colleagues (Dineva, Schöner, & Thelen, 2008, in preparation) used a novel feature of DFT; they raised the resting level of the field during the response phase of the A-not-B task (that is, when the box is moved to within an infant's reach). Raising the resting level of a neural field makes the field more excitable overall, allowing small humps of activation to rise above threshold and generate a peak. Simulations of a new model of infant perseverative reaching showed that this could capture not only the detailed distribution of infants' responses in the A-not-B task, but also the pattern of spontaneous errors infants show (i.e., reaches to B on A-trials) and critical trial-to-trial dependencies in infants' performance (e.g., how a single *spontaneous* reach to B on an A-trial effects the likelihood of reaching to A or B on subsequent trials).

Moreover, Dineva and colleagues (Dineva, Faubel, & Schöner, 2007) have recently demonstrated something possibly more impressive—that a DNF can, in fact, drive the behavior of an autonomous robot placed in an A-not-B task. This robot perceived the locations of colored targets in the task space via a camera. These inputs were fed into a DNF in real time. The field also received inputs from a continuously coupled LTM field which kept track of the robot's decisions from trial to trial. Finally, the DNF was coupled in real time to a motor system that turned the wheels of the robot. In particular, when the robot made a decision (i.e., formed a peak), this created an attractor for the motor system and moved the robot toward the represented location. Dineva and colleagues conducted a series of experiments with the robot and showed that this autonomous system reproduced the behavior of infants in the A-not-B situation—with real inputs and *live* decision making on the part of the robot. This provides a demonstration proof that DFT offers a coherent framework for thinking about embodied cognitive dynamics.

Dineva and colleagues' (Dineva et al., 2008, in preparation) model of the A-not-B situation demonstrates how a boost in the resting level of a field can move a system into a stable attractor state. This provides an initial view of a critical form of behavioral flexibility that is on firm neurophysiological ground. For instance, recent data suggest that one role of the prefrontal cortex is to selectively boost the gain on some neural populations over others (D'Esposito, 2007). This prefrontal boost could serve the same role as a resting-level boost in a neural field; it makes a neural field more excitable overall, allowing that neural population to make decisions (i.e., form peaks) even in the face of only partially certain information.

Interestingly, however, this is only part of the story. The simulation in Figure 5.12 shows the second part. This simulation is identical to the perseverative simulation in Figure 5.11 *except* that we modulated the resting level of the inhibitory field midway through the simulation (see Fig. 5.12C). By boosting the gain on neurons in the inhibitory layer, we can actively de-stabilize the peak at the left location. This allows the field to form a self-stabilized peak at the right location, effectively responding to this new input and overriding the strong LTM of the left location. Thus, a very simple mechanism—boosting the resting level of a population of neurons—can serve to move a neural field into a peak state or actively destabilize a peak.

We are just beginning to explore the types of behavioral flexibility this simple mechanism enables. Thus far, the picture that is emerging is quite powerful. For instance, Buss and Spencer (2008) used this type of mechanism to explain developmental changes in executive control in the dimensional change card sort task with 3- and 4-year-old children (see Morton & Munakata, 2002; Zelazo, 2006; Zelazo et al., 2003). And, critically, this mechanism makes several novel predictions, which we are currently testing.

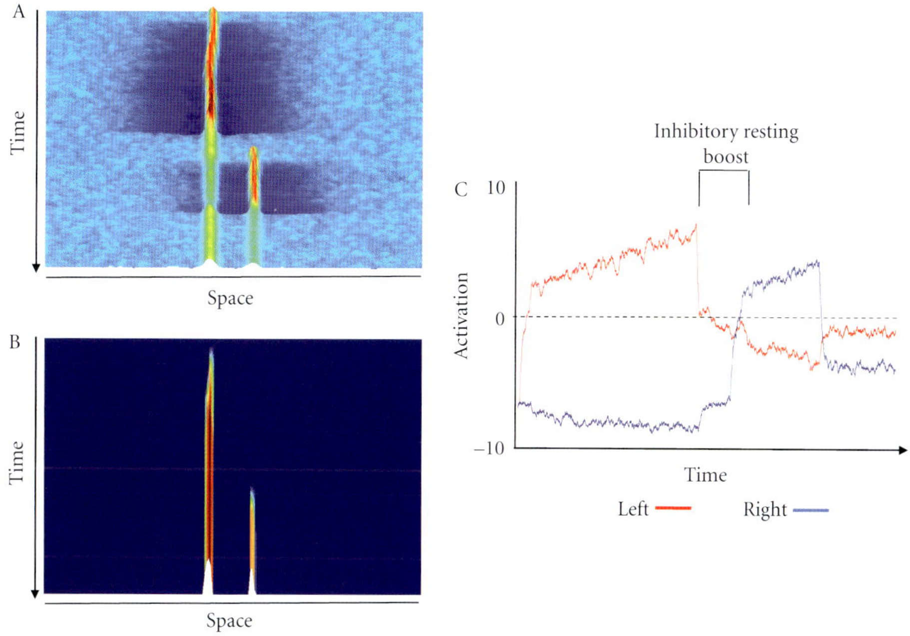


**Figure 5.12. A simple way to achieve stability and flexibility in dynamic neural fields**
This figure illustrates a simple way in which stability and flexibility can be achieved in dynamic neural fields. The simulation depicted is identical to the simulation depicted in Figure 5.11 *except* that the resting level of the inhibitory field was boosted slightly midway through the trial just before the second input appeared (see panel C). This served to destabilize the left peak (as seen in panel A), allowing a second peak to be built in response to the new task-relevant input to the right. As a result, LTM traces associated with the right target begin to build as the trial ends (see panel B). The time course of the simulation was 10 s.

Beyond specific models, however, we think there are some profound implications of this type of behavioral flexibility for dynamical systems theories more generally. Several recent dynamical systems accounts of cognition have correctly stressed that cognitive dynamics must be flexible to capture the impressive abilities of adult decision making (Kelso, 1995; Spivey, 2007). However, flexibility has been conceptualized as the ability to go near, rather than stably into, attractor states. Why? The reason these approaches have stressed this view of flexibility returns us to the simulation in Figure 5.11—if a system goes too deeply into a stable attractor state, it has difficulty getting *kicked* out of that state. The common image painted here is one of a ball rolling on a complex landscape of wells. If the ball rolls too deeply into a well, it takes a great deal of effort to get out of that well, and behavioral flexibility is limited. If, however, the ball rolls close enough to a well to support functional behavior, but then quickly moves on to the next behavioral response, the system—in theory—can show flexibility.

From our vantage point, this view of flexibility is missing something critical. Flexibility is not about going *near* attractors; it is about going *into* stable attractors and then having the ability to get back out to move on to the next stable attractor. We think this insight is critical, and our demonstration of embodiment with the robotics example backs up this claim. Simply put, a robot's behavior is poorly served by going near stable attractor states. What an autonomous system requires is stable decision making plus an active mechanism that can destabilize this state when the situation demands. We think peaks provide the stability such a system needs. And we think active changes in the resting level of a neural population provide the mechanism for getting into and out of such stable states.

This view of behavioral flexibility brings us to a final topic in our review of DFT concepts—development. In our view, a central issue in cognitive development is learning to regulate the dynamic balance among stability, instability, and flexibility (Schöner & Dineva, 2007; Spencer & Schöner, 2003). We discuss what this might mean in the section below.

## What about Development?

We have made promising advances in understanding development within the domain of spatial cognition using a relatively simple hypothesis—the *spatial precision hypothesis* (SPH). In the section below, we review this hypothesis and some of the novel, empirical findings it captures. We then address the issue of what might cause this particular type of developmental change.

The spatial precision hypothesis states that neural interactions become stronger, more precise, and more stable over development (Schutte et al., 2003; Spencer & Hund, 2003). This hypothesis is illustrated in Figure 5.13A. Each curve in this figure shows an example of a self-sustaining peak at some point in development with early development shown in darker lines and later development shown in lighter lines. As neural interactions strengthen from early to late development, the spatial precision of local excitatory interactions narrows and excitation increases. At the same time, lateral inhibitory interactions expand their spatial range and grow stronger. These changes result in relatively unstable peaks early in development, which are sensitive to input across a broad spatial range, and stable peaks later in development, which are only sensitive to input at narrow separations.

The developmental changes in neural interaction captured by the SPH can explain several behavioral results with children. For instance, in a sandbox task over development, children show a reduction in effects that are similar to the A-not-B error. In particular, when children are asked to find an object repeatedly at location A in a sandbox, their responses on a subsequent B-trial are biased in the direction of A. This supports the *drift* effect shown in Figure 5.8C, although drift in the sandbox case is toward a LTM of location A. More importantly for the present discussion, however, there is a

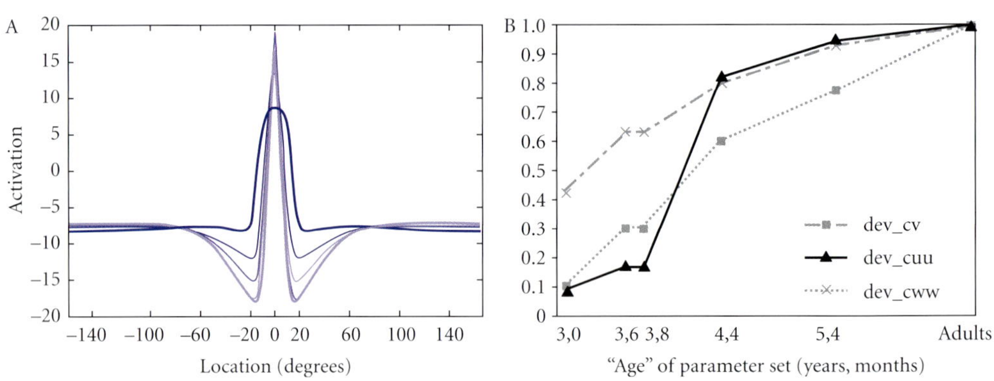


**Figure 5.13. Changes in strength of excitatory and inhibitory coupling over development**
This figure provides an illustration of how changes in the strength of excitatory and inhibitory coupling in a three-layer field model lead to increased stability and spatial precision of peaks across development. In panel A, darker activation profiles depict the structure of self-sustaining peaks for younger subject models, while lighter profiles depict self-sustaining peak profiles for older subject models. Note the narrowing and strengthening of peaks over development. Panel B depicts the strength changes in parameters used to produce the shift in peak structure over development shown in panel A. See text for further details.
*Source:* Adapted from simulations by Schutte and Spencer (in press).

change in the spatial range across which such effects are observed between 2 and 4 years of age; with enhanced precision over development, A and B must be spatially close for the A-not-B bias to be observed (for details, see Schutte et al., 2003). The enhanced stability of working memory peaks captured by the SPH also explains the reduction in mean and variable errors evident in a number of spatial recall tasks across a broad range of ages from 3 years to adulthood (Hund & Spencer, 2003; Plumert & Hund, 2001; Schutte & Spencer, 2002; Spencer & Hund, 2002, 2003).

Although the SPH captures a number of changes in performance across a range of ages, many of these developmental effects are quite intuitive (e.g., spatial memory improves over development). Does the SPH capture something less obvious? For instance, one of the more dramatic developmental changes in spatial cognition is the shift in spatial recall biases near a perceived reference frame: young children show biases toward symmetry axes, while older children show biases away from such axes (Huttenlocher, Newcombe, & Sandberg, 1994; Schutte & Spencer, 2002; Spencer & Hund, 2003). Can we capture this *qualitative* developmental transition using the same hypothesis?

A recent paper shows that the answer is yes. Schutte and Spencer (in press) showed that broader, weaker neural interactions in the three-layer model depicted in Figure 5.8 actually lead to attraction toward the landmark, rather than repulsion away from the landmark. This occurs because neural interactions are too weak to support the active maintenance of two peaks simultaneously. Using Figure 5.8 as an example, interactions in the model are too weak to support both the *reference* peak in PF and the *target* peak in WM. In the absence of a reference peak in PF, the peak in WM *drifts* toward the weak input associated with the landmark.

In addition to capturing the general direction of drift in early versus late development, Schutte and Spencer (in press) were able to capture the details of a complex developmental transition as children shifted month-by-month from an attraction pattern to a repulsion pattern. Critically, these details were captured via a *quantitative* shift in model parameters. In particular, to shift from the *child* parameters to *adult* parameters, we scaled the inputs and the strength of the projections among the layers of the three-layer model as specified by the SPH. The specific strength changes are depicted in Figure 5.13B. We increased the strength of local excitatory interactions in PF (see line dev_cuu), the strength of local excitatory interactions in WM (see line dev_cww), and the strength of the inhibitory projections from Inhib to both layers (see line dev_cv). Interestingly, these strength changes were sufficient to produce the *sharpening* of neural interaction evident in Figure 5.13A. That is, the enhancement in the precision of the peaks shown in Figure 5.13A was, in fact, completely emergent from changes in the strength of neural interactions.

Results from this study are consistent with other examples of nonlinear, dynamical systems that capture both quantitative and qualitative developmental change without positing new control processes. Rather, qualitative developmental changes can emerge via a reorganization of the system during the transition (Smith & Thelen, 1993; Thelen & Smith, 1994; van der Maas & Molenaar, 1992; van Geert, 1998). But this account goes beyond other dynamical systems account in that the SPH also has ties to the emerging literature on developmental cognitive neuroscience.

As an example, it is likely that development of the prefrontal cortex (PFC) is related to the changes in spatial recall discussed above. Data from several studies demonstrate that the prefrontal cortex plays a central role in SWM (Awh et al., 1999; Di Pellegrino & Wise, 1993; Kessels, Postma, Wijnalda, & De Haan, 2000; Nelson et al., 2000), and maturation of PFC is still occurring between 3 and 6 years of age. In fact, there is evidence that the development of the prefrontal cortex continues into the post-adolescent years (Gogtay et al., 2004; Sowell, Thompson, Tessner, & Toga, 2001). Synaptic density in the PFC reaches its peak at about 2 years of age in humans (Huttenlocher, 1979) and then declines until it finally reaches adult levels at about 16 years of age (Huttenlocher, 1990). Myelination is also still occurring in the frontal lobe between 2 and 6 years of age (Sampaio & Truwit, 2001). Thus, there is dramatic brain development happening during the period we see dramatic changes in spatial cognition. At this point, however, there is no strong neurophysiological support for the type of changes captured by the SPH. That said, to the extent that the changes noted above (e.g., myelination and pruning) lead to more efficient neural processing, they are, at least at a qualitative level, consistent with SPH.

Do these biological changes point toward a maturational picture? Perhaps, but they are equally consistent with an experience-dependent view of development. In particular, the types of change specified by the SPH could arise via simple Hebbian processes that strengthen neural connections among layers within cortical fields. We are currently examining whether the specific strength changes depicted in Figure 5.13B can arise via Hebbian learning, that is, in an experience-driven manner.

To summarize, the spatial precision hypothesis effectively captures a broad array of developmental changes in spatial cognition including changes in task performance (such as the A-not-B error) (Schutte et al., 2003), changes in the stability of SWM (Schutte, 2004, unpublished; Schutte & Spencer, in press; Spencer & Schöner, 2003), developmental differences in spatial recall versus spatial recognition performance (Spencer, Simmering, & Schutte, 2006), and the qualitative developmental shift in spatial recall biases discussed here (Schutte & Spencer, in press). Thus, this relatively simple, intuitive hypothesis about how neural interactions change systematically over the longer timescale of development has been quite generative. This is no accident; in our view, the richly structured real-time and learning-time dynamics of DFT can operate in concert with relatively simple developmental changes to yield dramatic, qualitative changes in behavior over long timescales. Importantly, we contend that the relatively modest view of development captured by the SPH brings an understanding of developmental mechanisms within reach.

Although the SPH has been an impressive workhorse thus far, we readily admit that this hypothesis captures only one aspect of developmental change. Another key issue is the emergence of behavioral flexibility over development (see discussions in Schöner & Dineva, 2007; Spencer & Schöner, 2003). We discussed a dynamic field approach to flexibility in the previous section. With regard to development, a central issue is how the neural system begins to gain active control over the mechanisms described above, that is, control over the resting level of neural populations such that cortical fields can be moved into and out of a peak state. Although we have clearly just scratched the surface on this topic, we contend that DFT has the potential to advance our understanding of this aspect of development.

## Conclusion

The goal of this chapter was to shed new light on the central concepts of dynamic field theory. Toward this end, we highlighted three neural attractor states that have been foundational in our approach, as well as many characteristics of activation peaks that *come for free* within our theoretical framework. Next, we used a multi-layered architecture to show how neural fields can integrate perceptual and working memory processes in real time, and highlighted some emergent features of the particular three-layer architecture we have developed.

We also focused on a different type of integration: integrating processes across multiple timescales using long-term memory dynamics. We showed how a type of Hebbian process in dynamic field theory can lead to category formation; however, this mechanism can also lead to less adaptive behaviors such as perseverative responding. This pointed toward another innovation of our framework—a novel account of behavioral flexibility. Here, we painted a picture of a neural system going stably into attractor states, but then quickly moving out of these states as the situation demands. We framed some of this discussion in the context of work on autonomous robotics, a field which has both demonstrated the embodied nature of dynamic field theory and highlighted the importance of stability for the generation of real actions in the context of real sensors.

In the final section, we highlighted aspects of our approach to development. Although this work is still in progress, our discussion focused on several novel contributions of our spatial precision hypothesis and pointed toward future work exploring the emergence of behavioral flexibility over development.

A central issue in the context of this chapter is whether the dynamic field theory basics we discussed solve the challenges we mentioned at the outset. For instance, one central challenge within cognitive science has been to develop a framework that can handle the real-time integration of perception, action, and cognition. Several examples from this chapter demonstrate that dynamic field theory can effectively tackle this problem head-on. The most direct example was work on autonomous robotics; however, the examples of our three-layer architecture also highlighted the integration of perceptual and working memory processes. A key insight in this work has been that stability in real time is necessary to realize a coherent interface between brain and behavior.

In addition to real-time integration, we also discussed the challenges inherent in dealing with lower level perception-action on one hand and higher level cognition on the other. Is dynamic field theory up to this challenge? More specifically, are the concepts of dynamic field theory relevant for higher level cognition? In our view, the answer is yes, and recent efforts to examine executive control within a dynamic field framework point in this direction (Buss & Spencer, 2008). In addition, Faubel and Schöner (2008) have recently demonstrated that dynamic field concepts can be used in an autonomous robotics context to support the fast learning of object labels (for related work, see chapter 4 and Samuelson et al., in press). This work uses coupled two-dimensional fields to show how higher dimensional categories of objects can emerge from a lower dimensional representation of features in cortical fields. We have used a similar approach to understand how the brain might solve the *binding problem* in vision (see Johnson et al., 2008).

We want to conclude by acknowledging that there are certainly many areas of overlap between our approach and connectionist approaches to development. Although we did not say much about connectionist approaches in the present chapter (for a more extensive discussion, see chapter 18), we concur with many of the authors of chapters in this volume that dynamic systems theory and connectionism offer complementary views of developmental processes. In our effort to clarify the concepts of dynamic field theory, we have presented material that will be central to a more substantive evaluation of the two approaches. We look forward to engaging in that evaluative process over the coming decade.

## Acknowledgments

Preparation of this manuscript was supported by NSF HSD 0527698 awarded to John P. Spencer, as well as support from the Obermann Center for Advanced Studies at the University of Iowa. We thank Gregor Schöner and Michael Thomas for helpful comments on an earlier version. Correspondence concerning this article should be addressed to John P. Spencer, Department of Psychology, University of Iowa, E11 Seashore Hall, Iowa City, Iowa 52242. E-mail: john-spencer@uiowa.edu.

## References

Amari, S.-I. (1977). Dynamics of pattern formation in lateral-inhibition type neural fields. *Biological Cybernetics, 27*(2), 77–87.

Amit, D. J., & Brunel, N. (1997). Model of global spontaneous activity and local structured activity during delay periods in the cerebral cortex. *Cerebral Cortex, 7*(3), 237–252.

Awh, E., Jonides, J., Smith, E. E., Buxton, R. B., Frank, L. R., Love, T., et al. (1999). Rehearsal in spatial working memory: Evidence from neuroimaging. *Psychological Science, 10*(5), 433–437.

Baddeley, A. D. (1986). *Working memory*. New York: Oxford University Press.

Barsalou, L. W., & Wiemer-Hastings, K. (2005). Situating abstract concepts. In D. Pecher & R. A. Zwaan (Eds.), *Grounding cognition: The role of perception and action in memory, language, and thinking* (pp. 129–163). New York: Cambridge University Press.

Bastian, A., Riehle, A., Erlhagen, W., & Schöner, G. (1998). Prior information preshapes the population representation of movement direction in motor cortex. *NeuroReport, 9*(2), 315–319.

Bastian, A., Schöner, G., & Riehle, A. (2003). Preshaping and continuous evolution of motor cortical representations during movement preparation. *European Journal of Neuroscience, 18*(7), 2047–2058.

Bicho, E., Mallet, P., & Schöner, G. (2000). Target representation on an autonomous vehicle with low-level sensors. *International Journal of Robotics Research, 19*(5), 424–447.

Braitenberg, V., & Schüz, A. (1991). *Anatomy of the cortex*. New York: Springer.

Braun, M. (1994). *Differential equations and their applications: An introduction to applied mathematics* (4 ed.). New York: Springer.

Bridgeman, B., Gemmer, A., Forsman, T., & Huemer, V. (2000). Processing spatial information in the sensorimotor branch of the visual system. *Vision Research, 40*(25), 3539–3552.

Brooks, R. A. (1991). New approaches to robotics. *Science, 253*(5025), 1227–1232.

Buss, A., & Spencer, J. P. (2008, July). *The emergence of rule-use: A dynamic neural field model of the DCCS*. Paper presented at the 30th Annual Conference of the Cognitive Science Society, Washington, DC. Retrieved October 17, 2008, from http://www.cogsci.rpi.edu/csjarchive/proceedings/2008/pdfs/p463.pdf

Camperi, M., & Wang, X.-J. (1998). A model of visuospatial working memory in prefrontal cortex: Recurrent network and cellular bistability. *Journal of Computational Neuroscience, 5*(4), 383–405.

Churchland, P. S., & Sejnowski, T. J. (1992). *The computational brain*. Cambridge, MA: MIT Press–Bradford Books.

Compte, A., Brunel, N., Goldman-Rakic, P. S., & Wang, X.-J. (2000). Synaptic mechanisms and network dynamics underlying spatial working memory in a cortical network model. *Cerebral Cortex, 10*(9), 910–923.

Constantinidis, C., & Steinmetz, M. A. (1996). Neuronal activity in posterior parietal area 7a during the delay periods of a spatial memory task. *Journal of Neurophysiology, 76*(2), 1352–1355.

Corbetta, D., & Thelen, E. (1996). The developmental origins of bimanual coordination: A dynamic perspective. *Journal of Experimental Psychology: Human Perception and Performance, 22*(2), 502–522.

Damasio, A. R., & Damasio, H. (1994). Cortical systems for retrieval of concrete knowledge: The convergence zone framework. In C. Koch & J. L. Davis (Eds.), *Large-scale neuronal theories of the brain* (pp. 61–74). Cambridge, MA: MIT Press–Bradford Books.

Dayan, P., & Abbott, L. F. (2001). *Theoretical neuroscience: Computational and mathematical modeling of neural systems*. Cambridge, MA: MIT Press.

D'Esposito, M. (2007). From cognitive to neural models of working memory. *Philosophical Transactions of the Royal Society B: Biological Sciences, 362*(1481), 761–772.

Diedrich, F. J., Highlands, T. M., Spahr, K. A., Thelen, E., & Smith, L. B. (2001). The role of target distinctiveness in infant perseverative reaching. *Journal of Experimental Child Psychology, 78*(3), 263–290.

Diedrich, F. J., Thelen, E., Smith, L. B., & Corbetta, D. (2000). Motor memory is a factor in infant perseverative errors. *Developmental Science, 3*(4), 479–494.

Dineva, E., Faubel, C., & Schöner, G. (2007, July). *Dynamic field theory and embodiment: Robotic demonstration of infant perseveration*. Paper presented at the 14th International Conference on Perception and Action, Yokohama, Japan.

Dineva, E., Schöner, G., & Thelen, E. (2008). Behavioral history matters: A dynamic field theory of spontaneous and perseverative errors in infant reaching. Manuscript in preparation.

Di Pellegrino, G., & Wise, S. P. (1993). Visuospatial versus visuomotor activity in the premotor and prefrontal cortex of a primate. *Journal of Neuroscience, 13*(3), 1227–1243.

Douglas, R., & Martin, K. (1998). Neocortex. In G. M. Shepherd (Ed.), *The synaptic organization of the brain* (4th ed., pp. 459–509). New York: Oxford University Press.

Durstewitz, D., Seamans, J. K., & Sejnowski, T. J. (2000). Neurocomputational models of working memory. *Nature Neuroscience, 3*(Suppl.), 1184–1191.

Edelman, G. M. (1987). *Neural Darwinism: The theory of neuronal group selection*. New York: Basic Books.

Erlhagen, W., & Schöner, G. (2002). Dynamic field theory of movement preparation. *Psychological Review, 109*(3), 545–572.

Faubel, C., & Schöner, G. (2008). Learning to recognize objects on the fly: A neurally based dynamic field approach. *Neural Networks, 21*(4), 562–576.

Finkel, L. H. (2000). Neuroengineering models of brain disease. *Annual Review Biomedical Engineering, 2*, 577–606.

Fuster, J. M. (1973). Unit activity in prefrontal cortex during delayed-response performance: Neuronal correlates of transient memory. *Journal of Neurophysiology, 36*(1), 61–78.

Georgopoulos, A. P., Kettner, R. E., & Schwartz, A. B. (1988). Primate motor cortex and free arm movements to visual targets in three-dimensional space. II. Coding of the direction of movement by a neuronal population. *Journal of Neuroscience, 8*(8), 2928–2937.

Georgopoulos, A. P., Lurito, J. T., Petrides, M., Schwartz, A. B., & Massey, J. T. (1989). Mental rotation of the neuronal population vector. *Science, 243*(4888), 234–236.

Georgopoulos, A. P., & Massey, J. T. (1988). Cognitive spatial-motor processes. 2. Information transmitted by the direction of two-dimensional arm movements and by neuronal populations in primate motor cortex and area 5. *Experimental Brain Research, 69*(21), 315–326.

Georgopoulos, A. P., Taira, M., & Lukashin, A. (1993). Cognitive neurophysiology of the motor cortex. *Science, 259*(5104), 47–52.

Glenberg, A. M. (1997). What memory is for: Creating meaning in the service of action. *Behavioral and Brain Sciences, 20*(1), 1–19, commentary 20–40, author's response 41–55.

Gogtay, N., Giedd, J. N., Lusk, L., Hayashi, K. M., Greenstein, D., Vaituzis, A. C., et al. (2004). Dynamic mapping of human cortical development during childhood through early adulthood. *Proceedings of the National Academy of Sciences, USA, 101*(21), 8174–8179.

Goldman-Rakic, P. S. (1995). Cellular bases of working memory. *Neuron, 14*(3), 477–485.

Grossberg, S. (1973). Contour enhancement, short-term memory, and constancies in reverberating neural networks. *Studies in Applied Mathematics, 52*, 217–257.

Hafting, T., Fyhn, M., Molden, S., Moser, M.-B., & Moser, E. I. (2005). Microstructure of a spatial map in the entorhinal cortex. *Nature, 436*(7052), 801–806.

Haskell, E., Nykamp, D. Q., & Tranchina, D. (2001). Population density methods for large-scale modelling of neuronal networks with realistic synaptic kinetics: Cutting the dimension down to size. *Network: Computation in Neural Systems, 12*(2), 141–174.

Hock, H. S., Schöner, G., & Giese, M. A. (2003). The dynamical foundations of motion pattern formation: Stability, selective adaptation, and perceptual continuity. *Perception and Psychophysics, 65*(3), 429–457.

Hund, A. M., & Spencer, J. P. (2003). Developmental changes in the relative weighting of geometric and experience-dependent location cues. *Journal of Cognition and Development, 4*(1), 3–38.

Huttenlocher, J., Hedges, L. V., Corrigan, B., & Crawford, L. E. (2004). Spatial categories and the estimation of location. *Cognition, 93*(2), 75–97.

Huttenlocher, J., Hedges, L. V., & Duncan, S. (1991). Categories and particulars: Prototype effects in estimating spatial location. *Psychological Review, 98*(3), 352–376.

Huttenlocher, J., Hedges, L. V., & Vevea, J. L. (2000). Why do categories affect stimulus judgment? *Journal of Experimental Psychology: General, 129*(2), 220–241.

Huttenlocher, J., Newcombe, N. S., & Sandberg, E. H. (1994). The coding of spatial location in young children. *Cognitive Psychology, 27*, 115–147.

Huttenlocher, P. R. (1979). Synaptic density in human frontal cortex: Developmental changes and effects of aging. *Brain Research, 163*(2), 195–205.

Huttenlocher, P. R. (1990). Morphometric study of human cerebral cortex development. *Neuropsychologia, 28*(6), 517–527.

Iossifidis, I., & Schöner, G. (2006, October). *Dynamical systems approach for the autonomous avoidance of obstacles and joint-limits for an redundant robot arm.* Paper presented at the IEEE International Conference on Intelligent Robots and Systems (IROS), Beijing, China.

Jackendoff, R. (1996). The architecture of the linguistic-spatial interface. In P. Bloom, M. A. Peterson, L. Nadel, & M. F. Garret (Eds.), *Language and space* (pp. 1–30). Cambridge, MA: MIT Press–Bradford Books.

Jancke, D., Erlhagen, W., Schöner, G., & Dinse, H. R. (2004). Shorter latencies for motion trajectories than for flashes in population responses of cat primary visual cortex. *Journal of Physiology (Paris), 556*(3), 971–982.

Johnson, J. S., Spencer, J. P., Luck, S. J., & Schöner, G. (in press). A dynamic neural field model of visual working memory and change detection. *Psychological Science.*

Johnson, J. S., Spencer, J. P., & Schöner, G. (2008). Moving to a higher ground: The dynamic field theory and the dynamics of visual cognition.

In F. Garzón, A. Laasko, & T. Gomila (Eds.), Dynamics and psychology [Special issue]. *New Ideas in Psychology, 26*(2), 227–251.

Kelso, J. A. S. (1995). *Dynamic patterns: The self-organization of brain and behavior.* Cambridge, MA: MIT Press–Bradford Books.

Kelso, J. A. S., & Tuller, B. (1984). A dynamical basis for action systems. In M. S. Gazzaniga (Ed.), *Handbook of cognitive neuroscience* (pp. 321–356). New York: Plenum Press.

Kessels, R. P. C., Postma, A., Wijnalda, E. M., & de Haan, E. H. F. (2000). Frontal-lobe involvement in spatial memory: Evidence from PET, fMRI, and lesion studies. *Neuropsychology Review, 10*(2), 101–113.

Kopecz, S., Engels, C., & Schöner, G. (1993). Dynamic field approach to target selection in gaze control. In S. Gielen & B. Kappen (Eds.), *Proceedings of the International Conference on Artificial Neural Networks, Amsterdam* (pp. 96–101). Berlin, Germany: Springer.

Lipinski, J., Spencer, J. P., & Samuelson, L. K. (in press). It's in the eye of the beholder: Spatial language and spatial memory use the same perceptual reference frames. In L. B. Smith, M. Gasser, & K. Mix (Eds.), *The spatial foundations of language.* New York: Oxford University Press.

Luck, S. J., & Vogel, E. K. (1997). The capacity of visual working memory for features and conjunctions. *Nature, 390*(6657), 279–284.

Mareschal, D., Johnson, M., Sirois, S., Spratling, M., Thomas, M. S. C., & Westermann, G. (2007). *Neuroconstructivism: Vol. 1. How the brain constructs cognition.* New York: Oxford University Press.

Morton, J. B., & Munakata, Y. (2002). Active versus latent representations: A neural network model of perseveration, dissociation, and decalage. *Developmental Psychobiology, 40*(3), 255–265.

Nelson, C. A., Monk, C. S., Lin, J., Carver, L. J., Thomas, K. M., & Truwit, C. L. (2000). Functional neuroanatomy of spatial working memory in children. *Developmental Psychology, 36*(1), 109–116.

Newell, A., & Simon, H. A. (1976). Computer science as empirical enquiry: Symbols and search. *Communications of the ACM, 19*(3), 113–126.

Palmer, S. E., & Kimchi, R. (1986). The information processing approach to cognition. In T. J. Knapp & L. C. Robertson (Eds.), *Approaches to cognition: Contrasts and controversies* (pp. 37–77). Hillsdale, NJ: Erlbaum.

Plumert, J. M., & Hund, A. M. (2001). The development of memory for location: What role do spatial prototypes play? *Child Development, 72*(2), 370–384.

Port, R. F., & van Gelder, T. (Eds.). (1995). *Mind as motion: Explorations in the dynamics of cognition.* Cambridge, MA: MIT Press–Bradford Books.

Rao, S. C., Rainer, G., & Miller, E. K. (1997). Integration of what and where in the primate prefrontal cortex. *Science, 276*(5313), 821–824.

Rumelhart, D. E., McClelland, J. L., & the PDR Research Group. (1986). *Parallel distributed processing: Explorations in the microstructure of cognition: Vol. 1. Foundations.* Cambridge, MA: MIT Press–Bradford Books.

Salinas, E. (2003). Self-sustaining activity in networks of gain-modulated neurons. *Neurocomputing, 52–54*, 913–918.

Sampaio, R. C., & Truwit, C. L. (2001). Myelination in the developing human brain. In C. A. Nelson & M. Luciana (Eds.), *Handbook of developmental cognitive neuroscience* (pp. 35–44). Cambridge, MA: MIT Press–Bradford Books.

Samuelson, L. K., Schutte, A. R., & Horst, J. S. (in press). The dynamic nature of knowledge: Insights from a dynamic field model of children's novel noun generalizations. *Cognition.*

Schöner, G. (1995). Recent developments and problems in human movement science and their conceptual implications. *Ecological Psychology, 7*(4), 291–314.

Schöner, G., & Dineva, E. (2007). Dynamic instabilities as mechanisms for emergence. *Developmental Science, 10*(1), 69–74.

Schöner, G., & Thelen, E. (2006). Using dynamic field theory to rethink infant habituation. *Psychology Review, 113*(2), 273–299.

Schutte, A. R. (2004). A developmental transition in spatial working memory. Unpublished doctoral dissertation, University of Iowa, Iowa City.

Schutte, A. R., & Spencer, J. P. (2002). Generalizing the dynamic field theory of the A-not-B error beyond infancy: Three-year-olds' delay- and experience-dependent location memory biases. *Child Development, 73*(2), 377–404.

Schutte, A. R., & Spencer, J. P. (in press). Tests of the dynamic field theory and the spatial precision hypothesis: Capturing a qualitative developmental transition in spatial working memory. *Journal of Experimental Psychology: Human Perception and Performance.*

Schutte, A. R., Spencer, J. P., & Schöner, G. (2003). Testing the dynamic field theory: Working memory for locations becomes more spatially precise over development. *Child Development, 74*(5), 1393–1417.

Simmering, V. R., Johnson, J. S., & Spencer, J. P. (2008). Does change detection underestimate capacity? Insights from a dynamic field theory of visual working memory. Manuscript in preparation.

Simmering, V. R., Schutte, A. R., & Spencer, J. P. (2008). Generalizing the dynamic field theory of spatial cognition across real and developmental time scales. In S. Becker (Ed.), Computational cognitive neuroscience [Special issue]. *Brain Research, 1202*(C), 68–86.

Simmering, V. R., & Spencer, J. P. (2008). Generality with specificity: The dynamic field theory generalizes across tasks and time scales. *Developmental Science, 11*(4), 541–555.

Simmering, V. R., Spencer, J. P., & Schöner, G. (2006). Reference-related inhibition produces enhanced position discrimination and fast repulsion near axes of symmetry. *Perception and Psychophysics, 68*(6), 1027–1046.

Skarda, C. A., & Freeman, W. J. (1987). How brains make chaos in order to make sense of the world. *Behavioral and Brain Sciences, 10*(2), 161–173, commentary 174–183, authors' response 183–192.

Smith, L. B., & Thelen, E. (Eds.). (1993). *A dynamic systems approach to development: Applications.* Cambridge, MA: MIT Press–Bradford Books.

Smith, L. B., Thelen, E., Titzer, R., & McLin, D. (1999). Knowing in the context of acting: The task dynamics of the A-not-B error. *Psychological Review, 106*(2), 235–260.

Sowell, E. R., Thompson, P. M., Tessner, K. D., & Toga, A. W. (2001). Mapping continued brain growth and gray matter density reduction in dorsal frontal cortex: Inverse relationships during postadolescent brain maturation. *Journal of Neuroscience, 21*(22), 8819–8829.

Spencer, J. P., & Hund, A. M. (2002). Prototypes and particulars: Geometric and experience-dependent spatial categories. *Journal of Experimental Psychology: General, 131*(1), 16–37.

Spencer, J. P., & Hund, A. M. (2003). Developmental continuity in the processes that underlie spatial recall. *Cognitive Psychology, 47*(4), 432–480.

Spencer, J. P., & Perone, S. (2008). Defending qualitative change: The view from dynamical systems theory. *Child Development, 79,* 1639–1647.

Spencer, J. P., & Schöner, G. (2003). Bridging the representational gap in the dynamical systems approach to development [Special section/issue]. *Developmental Science, 6*(4), 392–412.

Spencer, J. P., Simmering, V. R., & Schutte, A. R. (2006). Toward a formal theory of flexible spatial behavior: Geometric category biases generalize across pointing and verbal response types. *Journal of Experimental Psychology: Human Perception and Performance, 32*(2), 473–490.

Spencer, J. P., Simmering, V. R., Schutte, A. R., & Schöner, G. (2007). What does theoretical neuroscience have to offer the study of behavioral development? Insights from a dynamic field theory of spatial cognition. In J. M. Plumert & J. P. Spencer (Eds.), *The emerging spatial mind* (pp. 320–361). New York: Oxford University Press.

Spivey, M. J. (2007). *The continuity of mind.* New York: Oxford University Press.

Steinhage, A., & Schöner, G. (1997). Self-calibration based on invariant view recognition: Dynamic approach to navigation. *Robotics and Autonomous Systems, 20*(2–4), 133–156.

Steinhage, A., & Schöner, G. (1998). Dynamical systems for the behavioral organization of autonomous robot navigation. In P. S. Schenker & G. T. McKee (Eds.), *Proceedings of SPIE: Vol. 3523. Sensor fusion and decentralized control in robot systems* (pp. 160–180). Bellingham, WA: SPIE Publishing.

Thelen, E. (1995). Motor development: A new synthesis. *American Psychologist, 50*(2), 79–95.

Thelen, E., Schöner, G., Scheier, C., & Smith, L. B. (2001). The dynamics of embodiment: A field theory of infant perseverative reaching. *Behavioral and Brain Sciences, 24*(1), 1–34, commentary 34–70, authors' response 70–80.

Thelen, E., & Smith, L. B. (1994). *A dynamic systems approach to the development of cognition and action.* Cambridge, MA: MIT Press–Bradford Books.

Thelen, E., & Ulrich, B. D. (1991). Hidden skills: A dynamic systems analysis of treadmill stepping during the first year. *Monographs of the Society for Research in Child Development, No. 223, 56*(1), 1–98, discussion 99.

Townsend, J. T., & Busemeyer, J. (1995). Dynamic representation of decision making. In R. F. Port & T. van Gelder (Eds.), *Mind in motion: Explorations in the dynamics of cognition.* Cambridge, MA: MIT Press.

van der Maas, H. L. J., & Molenaar, P. C. M. (1992). Stagewise cognitive development: An application of catastrophe theory. *Psychological Review, 99*(3), 395–417.

van Geert, P. (1998). A dynamic systems model of basic developmental mechanisms: Piaget, Vygotsky, and beyond. *Psychological Review, 105*(4), 634–677.

van Gelder, T., Bechtel, W., Beer, R. D., Braisby, N., Cooper, R., Franks, B., et al. (1998). The dynamical hypothesis in cognitive science. *Behavioral and Brain Sciences, 21*(5), 615–628, commentary 629–654, author's response 654–661.

Wilson, H. R., & Cowan, J. D. (1972). Excitatory and inhibitory interactions in localized populations of model neurons. *Biophysical Journal, 12*(1), 1–24.

Zelazo, P. D. (2006). The dimensional change card sort (DCCS): A method of assessing executive function in children. *Nature Protocols, 1*, 297–301.

Zelazo, P. D., Müller, U., Frye, D., Marcovitch, S., Argitis, G., Boseovski, J., et al. (2003). The development of executive function in early childhood. *Monographs of the Society for Research in Child Development, 68*(3), 1–137.

Zhang, W., & Luck, S. J. (2008). Discrete fixed-resolution representations in visual working memory. *Nature, 453*(7192), 233–235.

# CHAPTER 6

## Timescales of Change in Connectionist and Dynamical Systems Approaches to Learning and Development

*Karl M. Newell, Yeou-Teh Liu, and Gottfried Mayer-Kress*

Both connectionist and dynamical systems frameworks have invoked the construct of timescales in theorizing about the central problem of the laws of change over time in learning and development. Indeed, the construct of timescales is fundamental in a number of contemporary connectionist and dynamical theories and models (e.g., Elman et al., 1996; Smith & Thelen, 1993; Thelen & Smith, 1994). Although the phrase *multiple timescales of change* has been used with increasing propensity in contemporary accounts of learning and development, there is actually a paucity of research (theoretical or empirical) structured by either the connectionist or dynamical framework that investigates the core construct of timescales.

In this chapter, we examine the origins of the construct of timescales as expressed in Waddington's (1957) metaphor for development of the epigenetic landscape, and its subsequent use as a metaphor for change in learning/development in theoretical frameworks, both connectionist (e.g., Elman et al., 1996) and dynamical systems (Kugler, 1986; Muchisky, Gershkoff-Stowe, Cole, & Thelen, 1996; Smith & Thelen, 1993; Thelen & Smith, 1994; van Geert, 1994). From this background, we outline the dynamical origins of timescales in behavioral change through a consideration of the timescales arising from growth/decay and oscillatory processes in living and nonliving systems. A related theme is how the constraints to action (evolutionary, ontogenetic, physiological, and task goals) channel the dynamics of behavior and the rates of change of learning and development in systems at all levels of the organism.

The theoretical framework for the ideas on timescales presented here is our epigenetic landscape model for motor learning and development, which is constructed as a low-dimensional approximation of a high-dimensional (or infinitely dimensional) complex dynamical system that possesses a score, error, or fitness function (Mayer-Kress, Liu, & Newell, 2006; Newell, Liu, & Mayer-Kress, 2001, 2003). This approach was driven by the overarching principles of self-organizing systems (Haken, 1983; Iberall, 1972; Prigogine & Nicolis, 1971) together with the operational developments of nonlinear dynamical systems in physical nonliving systems (Boccara, 2003). In this context, we note that the fields of motor control (Haken, Kelso, & Bunz, 1985; Kelso, 1995; Kugler & Turvey, 1987), motor learning (Newell, Kugler, Van Emmerik, & McDonald, 1989; Schöner, 1989), and motor development (Thelen, Fogel, & Kelso, 1987) were influenced by the self-organization metaphor and the principles and methods of nonlinear dynamics earlier than many areas of psychology. Interestingly, and in contrast, the movement domain has been relatively uninfluenced by the ideas of connectionism (although, see Simms, 1994).

In our dynamical framework for the evolution of behavioral change in motor learning and development, we share the views of connectionist approaches (Elman et al., 1996) in regard to the important role of the task goal (i.e., target) in organizing the landscape. Here, we contrast the role of stability (as emphasized in some dynamical systems approaches) and the role of task goals (or *zero error* in the connectionist frameworks) to the determination of the timescales of change. We conclude by illustrating briefly three case studies of the timescales of change in our epigenetic landscape framework to motor learning and development.

## What Are Timescales of Change?

The construct of timescales is being increasingly used in neuroscience, psychology, and related sub-domains of the study of human behavior. And as noted above, the phrase *multiple timescales* is now common in dynamical systems and connectionist approaches to behavior and physiology. The concept of multiple timescales in the extreme situation of infinitely many timescales in physiology and behavior is most evident in phenomena that scale as a 1/*f*-like fractal process (Bassingthwaighte, Liebovitch, & West, 1994; Ward, 2001). Nevertheless, the origin and meaning of the term *timescale* is rarely articulated—even in the life science fields of human behavior where it is used—perhaps because its long historical use in the development of the concept of time is seen to make such an elaboration unnecessary.

The concept of time has been refined throughout history, in part due to the development of new timescales. The circadian rhythm of earth rotation is the most obvious indicator of the passage of time, and the timescale of the day is the fundamental element of all calendars. Not all frameworks of time, however, are based on the rotation of the earth. Indeed, the forms of time known as dynamical, atomic, and coordinate time are not based on the idea of the day in the standard sense (Haliday, Resnick, & Walker, 2005), but they do relate to properties of dynamical systems.

In the traditions of dynamical systems there are two types of idealized motions that naturally introduce the construct of timescales (Haken, 1983; Kaplan & Glass, 1995). These motions have either periodic oscillations (such as the rotation of the earth) or *growth/decay* at a constant rate (such as in radioactive decay used, for instance, in carbon dating of fossils). In oscillatory systems, the period (the inverse of the frequency) determines the intrinsic timescale of the system. In growth/decay systems, the intrinsic timescale is the inverse of the growth/decay rate sometimes also expressed as *doubling time* or *half-life*.

For linear dynamical systems, these two classes of behavior/motion and combinations thereof are the only forms of movement observed, apart from the trivial dynamics of a nonmoving fixed point. These forms of movement are also the basis of almost all known clocks; rotation of the earth, swinging of a pendulum, vibrations of a quartz crystal, and frequencies of cesium atoms, all represent clocks based on periodic movements. Water clocks, carbon dating, and the red shift of distant galaxies are examples of methods based on constant rates of change that are used to measure time at a vast range of scales. For both types of exponential trajectory, a variable x($t$) at time $t$ can be expressed as a complex exponential function:

$$x(t) = A\, e^{(\gamma + i\omega)t} \qquad (1)$$

Here A is the position at time $t = 0$, the real part $\gamma$ of the exponent is the growth/decay rate, and the imaginary part $\omega$ is the frequency of the oscillation. The reciprocal of the growth or decay rate is the intrinsic or characteristic timescale.

The two types of idealized motion, oscillatory and growth/decay, are fundamental to describing behavior close to a fixed point, although nonlinear dynamical systems have a much richer repertoire of trajectories. The fixed point is the mathematical concept that is associated with the equilibrium state of the system (Kaplan & Glass, 1995; Strogatz, 1994). Fixed points correspond to the absence of motion as in a pendulum at rest. Furthermore, close to fixed points, a nonlinear system can be approximated to be linear, so that the motion of the trajectory can be characterized by the exponential function of equation (1). Thus, the timescale within the growth or decay process to a fixed point is characterized by the time it takes for the dynamics to double/half the distance to the fixed point. In this sense then, a timescale is not merely the time of an event as it is usually interpreted, but a particular duration (time) of a *dynamical* growth/decay process. (Note that reduction of error or distance to the goal is formally a decay process.)

This set of basic assumptions can be developed so that what is called *the multiple timescales of change* in learning and development can be interpreted as originating from the system's trajectory on an *evolving* attractor landscape. Different attractor landscapes (and their unique evolution along the pathway of change over time in terms of bifurcations and transient phenomena) can lead to different patterns of change at the behavioral level, including multiple timescales as expressed in learning and development. For instance, the laws of learning in exponential, power law, and S-shaped change can be mapped to the landscape interpretations of attractor dynamics (Newell et al., 2001) as can the development of the fundamental movement patterns in infancy (Newell et al., 2003).

Thus, the different functions of change observed at the behavioral level in learning and development map to different dynamical phenomena and hold a particular meaning in the determination of the timescales of change. For example, the power law, when interpreted, formally has an infinite number of timescales within the range established for the phenomena at hand (Schroeder, 1991). The exponential has a single timescale given its single proportional rate of change. Furthermore, the function for the change in behavior over time might reflect a small number of characteristic exponential timescales but not actually be a power law. Thus, the phrase *multiple timescales of change* has several potential instantiations that are not well understood in learning and development.

The traditions of curve fitting and the evaluation of the nature of change through percent of variance criteria are prone to false interpretations of the timescales of change (Liu, Mayer-Kress, & Newell, 2003, 2004; Newell et al., 2001; Pitt & Myung, 2002). First, the universality of the power law in learning (Newell & Rosenbloom, 1981) is not as apparent as has long been assumed (see Gallistel, Fairhurst, & Balsam, 2004; Heathcote, Brown, & Mewhort, 2000; Newell et al., 2001). Second, the traditional practice of averaging data over subjects and trials can also lead to a false interpretation of the dynamical model of change (Newell et al., 2001). Third, even the information criterion of Akaike (Akaike, 1974; Burnham & Anderson, 2002) treats the timescales of change in a neutral fashion in establishing criteria to evaluate functions and model selection with a different number of parameters. We claim that timescales play a more significant role in characterizing a system than other system parameters, such as the initial amplitude of a signal. Therefore, it is not appropriate to treat all statistical fit parameters in an indiscriminate way and thus view the data analysis tool as a black box. In summary, it appears that there is not simply a single function of learning, and it appears that the veridical estimate of persistent change over time has been biased by the theoretical shallowness of following a curve-fitting strategy and taking at face value the criterion of a percent of variance accounted for from fitting packaged programs to data sets.

It is our position that the timescales of change in learning and development need to be established from a theoretical background *prior to* curve-fitting procedures, rather than on a post hoc determination of model function based on measures to account for percent of variance (Liu et al., 2003, 2004; Newell et al., 2001). This general theoretical orientation should be applied to properties of the change over time in individual behavior if a veridical estimate of the timescale(s) of change is to be made. Our approach to this is an a priori dynamical landscape model that seriously considers the roles of timescales in determining the learning model (Newell et al., 2001). From this approach, the modeling of individual learning data is emphasized, but averaging of aspects of the data can be done judiciously from properties of the individual timescale of change.

## Constraints and Timescales of Change in Learning and Development

There have been a number of approaches to categorizing the primary constraints to action and their respective timescales. The general view, although expressed in somewhat different labels and terms in each approach, is that there are four classes of constraint to action that reflect orders of magnitude differences in their timescales of influence. These categories of constraint are evolutionary, ontogenetic (developmental), physiological (epigenetic biological), and task (Eckhardt, 2000; Mittenthal & Baskin, 1992; Waddington, 1957).

The timescale of evolutionary constraints is over generations of the species and, although recognized as a fundamental category of influence to the learning and performance of individuals, it is not typically pursued theoretically or experimentally in studies of learning and development in context (although see Johnston & Lickliter, chapter 14; Nolfi, Elman, & Parisi, 1994). Indeed, most emphasis is given to either the timescales of ontogeny or, more broadly, those arising from the local dynamics of physiology and biology (epigenetic dynamics). Moreover, in learning and developmental theorizing, the role of the task in organizing/influencing the timescale is typically minimized or ignored.

The timescales of the dynamics of biological systems are most usually approached (directly or indirectly) by a consideration of the rhythms of the system output. It is well established that the timescales of change in the rhythms of ontogeny can occur over a wide range of scales from years to months, weeks, days, hours, minutes, and even fractions of a second (Foster & Kreitzman, 2004; Luce, 1971). The timescales are usually shorter the closer to micro the level of analysis of the system, but there are counter examples to this general rule of thumb. For instance, the lengths of telomeres, the end regions of DNA which shorten with each cell division, constitute a molecular clock, measuring the biological age of an organism. Depending on the species, the characteristic timescales of this clock can, therefore, go up to tens of years.

The rhythms of biological systems provide an intrinsic metric for the timescales of many basic biological phenomena across a range of levels of analysis of the system. The premise of our attractor landscape framework to learning and development is that the link between timescales and oscillatory or growth/decay phenomena can be elaborated to essentially all aspects of human behavior. This includes the range of learned behaviors that are evident throughout the life span and that have been the focus of different sub-domains of learning and development.

In our framework, as noted above, the term *timescale* refers to the particular time duration of a process that is supported by the dynamical principles arising from oscillatory and growth/decay changes in the system. However, in learning and development, the idea of timescales has also been used less restrictively—as in the context of just referring to the so-called real time of an event—without any link to oscillatory or growth/decay processes. Given the context of the intrinsic dynamics of oscillatory and growth/decay processes, it is intuitive that the real time of a behavioral action episode can be viewed as an emergent property rather than a specified feature or property of an action.

The phrase *multiple timescales of action* has been introduced principally through dynamical accounts of learning and development (Smith & Thelen, 1993; Thelen & Smith, 1994), and that will be our emphasis here. However, the specific development of the construct of timescales to the problems of the persistent and transitory change in learning and development has been less forthcoming. The most prevalent orientation is to link, in a metaphorical way, the notion of timescales to the epigenetic landscape of Waddington (1957). We will develop the fundamental assumptions of this approach in the next section of this chapter; but before doing that, we outline the essence of two approaches to timescales in learning and development to provide some practical implications of the possibilities.

van Geert (1994) presented the case that the major transitions in development reflect a fractal process. The fundamental rule that he examines is the fractal relation in which the frequency of behavioral change is a power function of its magnitude. In this view, the developmental changes that have the greatest magnitude are those identified as stages; these are also small in frequency of occurrence (that is, very few). van Geert outlined the basis of timescales for the major transitions or stages of development and used the framework of Fischer and Rose (1993) to provide developmental examples of these changes in timescale. In van Geert's view (1994), the underlying principle for the fractal pattern of transitions is the logistic (S-shaped) function.

In more of a life span perspective, Lemke (2002) examined multiple timescales over many orders of magnitude in the social organization of learning. A key feature of this framework is the attempt to explain why some learning experiences contribute to long-term patterns of action and personal development, while other experiences have only short-term effects and transitory influences. The center piece of this analysis is a table of timescales (in seconds) for human activity; the table goes from very short timescales, as in chemical synthesis ($10^{-5}$ s), membrane processes ($10^{-5}$ s), neural firings ($10^{-5}$ s), and neuronal patterns ($10^{-5}$ s) to the very long timescales of world system change, as in new cultures and languages ($10^{11}$ s), ecosystem climate change ($10^{12}$ s), and evolutionary changes (the $10^{14}$ s scale of human evolution). Embedded in the middle range of these extreme fast and slow timescales are human activities of vocal articulation ($10^{-5}$ s), utterance (1–10 s), exchange ($2$–$10^2$ s), episode ($10^3$ s, 15 min), lesson ($10^3$–$10^4$ s, 1 hour), lesson sequence ($10^4$ s, 2.75 hours), school day ($10^5$ s, 1 day), unit ($10^6$ s, 11.5 days), semester/year curriculum ($10^7$ s, 4 months), multiyear curriculum ($10^8$ s, 3.2 years), and life span educational development ($10^9$ s, 32 years). Lemke observed that the learning in social organization systems that has longer timescales tends also to operate in more spatially embedded systems. There does not appear to be any experimental evidence for the intrinsic dynamics of the timescales arising from these levels of social organization.

Thus, and in summary, we see that the term *timescale* refers to a period of time, but it has been used in two different ways in learning and development. In the most generic and loose way, it refers to the duration of a process or event. In the more formal interpretation from dynamical systems, it relates to time periods that emerge from the intrinsic oscillatory and growth/decay processes of the system. In either approach the timescales are assessed as a first pass in regard to their order of magnitude difference for various processes, such as features of learning and development.

## Evolving Landscapes as Metaphor in Motor Learning and Development

The origins of the dynamical interpretation of change in learning and development can be traced to the idea of an evolving epigenetic landscape (Waddington, 1957). Waddington was interested in embryogenesis, and the landscape model was a metaphor for considering the dynamics of developmental growth and change. The idea of the epigenetic landscape has since been used to exemplify the emergence of a range of developmental phenomena (e.g., Connolly, 1986; Meinhardt, 1982) and also changes in gene expression triggered by behavioral changes such as exercise and practice in learning a task (Kaczmarek, 2000).

The landscape in Figure 6.1 illustrates the pathways of developmental change and the

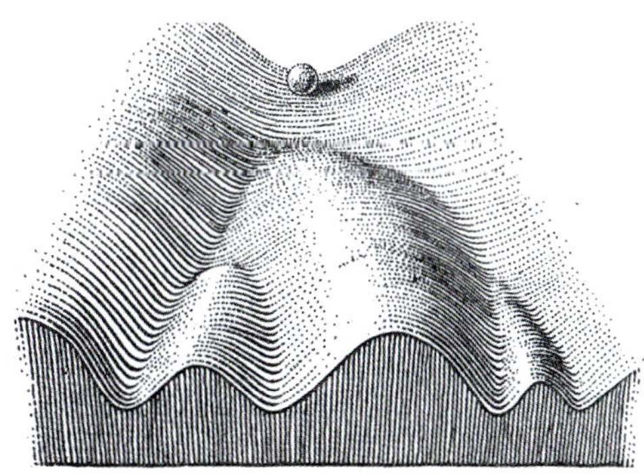

**Figure 6.1. Waddington's schematic of the epigenetic landscape**
This figure depicts Waddington's proposed epigenetic landscape. The dimensions are as follow: fore to aft distance represents time or developmental age, horizontal distance represents the emergence/dissolution of activities, and the slope of the surface captures the rate of developmental change. The presumed motion of the ball on the landscape would reflect a developing phenotype. See text for additional interpretation.
*Source*: Adapted from Waddington (1957).

presumed motion of the ball on the landscape would reflect the developing phenotype. The fore to aft dimension is time or developmental age, the horizontal axis instantiates the emergence and dissolution of particular activities that hold dynamic equilibria, and the slope of the surface captures the rate of developmental change. The stability of the system is inferred from the depth of the landscape wells. This 2-D landscape in a 3-D space (two geographical dimensions plus one elevation dimension) is a low-dimensional schematic of the basic dynamic processes of development that are derived from a high-dimensional biological system. Waddington's theoretical approach had the properties of the landscape emerging from genetic control, but the landscape metaphor for the dynamic processes can be elaborated to include the confluence of constraints from the organism, environment, and task (Connolly, 1986).

The general features of the landscape have been used to capture the dynamic processes of change in motor development (Connolly, 1986; Kugler, 1986; Muchisky et al., 1996). Muchisky and colleagues (Muchisky et al., 1996) elaborated the Waddington schematic to that shown in Figure 6.2 to reflect a broader range of the dynamic changes evident in the development of the fundamental movement sequence in infants. This figure in essence shows a longer period of developmental time than implied in the original Waddington schematic and the potential for the evolution and dissolution of a broader range of stable states with the passage of time. Indeed, the image here could reflect the learning of actions and tasks throughout the lifespan and

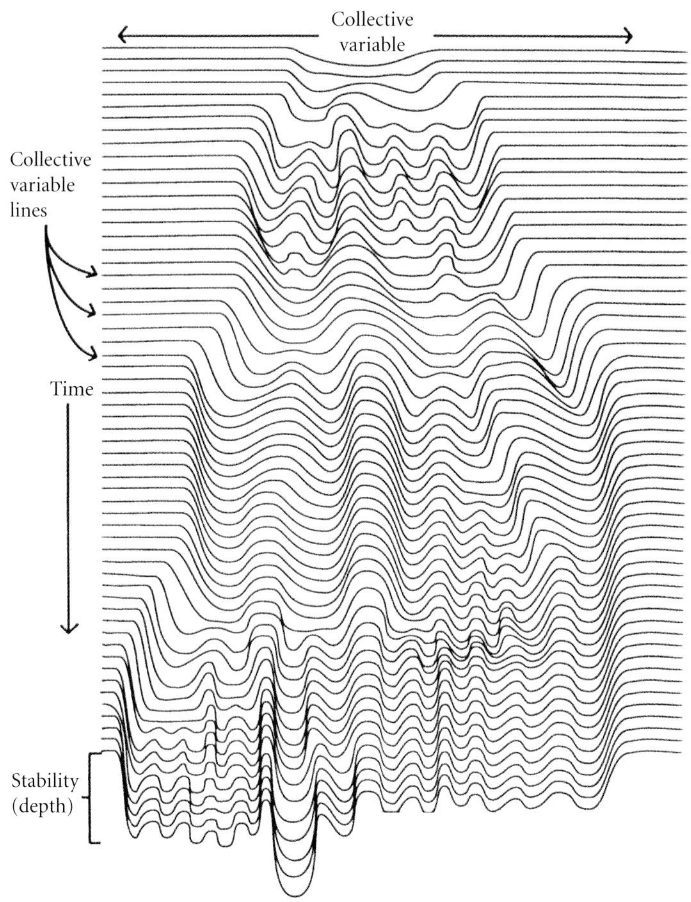


**Figure 6.2. Multiple timescales of development**
This figure depicts an elaborated form of Waddington's landscape (see Fig. 6.1). See text for interpretation.
*Source*: Adapted from Muchisky, Gershkoff-Stowe, Cole, and Thelen (1996).

the spiral of stability and instability that reflects the evolving nature of the dynamics of learning and development.

Figure 6.3 shows a hypothetical error surface from a neural network (Elman et al., 1996). These authors pointed up the similarity in the features of the artificial network landscape to that of the original Waddington epigenetic landscape (contrast to Fig. 6.1). In the artificial network image of Figure 6.3, the phenotype is the evolving set of connection strengths among artificial synapses. As in the Waddington landscape, the network seeks the *low ground* of error in its weight space. There are often many pathways to equivalent solutions—an instance of equifinality (Bertalanffy, 1975). Elman and colleagues (Elman et al.,1996) claimed that the similarity between the natural and artificial landscapes is more than a coincidence in that they reflect fundamentally the same process. More importantly, for the purposes of this chapter, it follows that there is common ground between the dynamical systems and connectionist frameworks in considering the role of timescales in learning and development.

The images of various versions of the epigenetic landscape give a common view to the metaphor of dynamical processes in learning and development. But, like many metaphors, when one delves beyond the surface interpretations and analogies, noteworthy difficulties and problems begin to emerge. Indeed, a more formal examination of dynamical processes is important if we are to go beyond the metaphor (van der Maas, 1996) and develop formal models of the processes of learning and development.

We have previously discussed several limitations to the theoretical and experimental messages inherent in the landscape images of the above examples (Newell et al., 2003). Here, we reemphasize briefly some of these limitations to the extant landscape metaphors with an emphasis on examples of motor learning and development, although we would propose that the points are general to landscape modeling of change in all behavioral contexts. This section also provides some background to the subsequent consideration of the contrasting roles of distance from the goal (i.e., *error* in neural networks) and stability (dynamical systems) in landscape models. A listing of limitations follows:

1. The standard landscape image has the most stable state at the beginning of phenotypical development dissolving with the passage of time. Thus, in motor development, for example, it implies that lying down dissolves as a stable activity. Yet, the postural state of lying down remains stable through the life span.
2. The horizontal dimensions capture the qualitative stable states as reflected in posture,

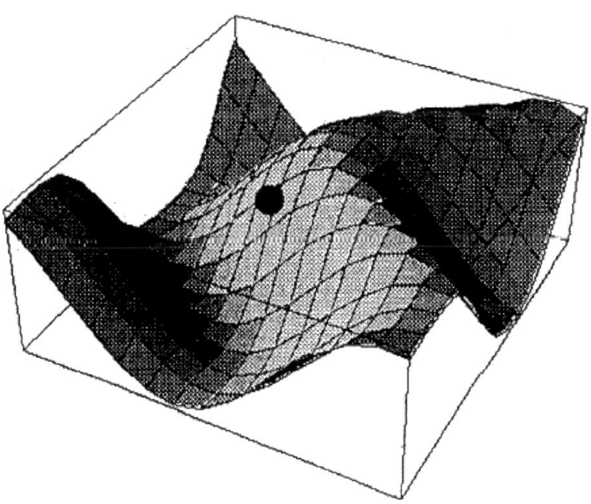

**Figure 6.3. Hypothetical error surface from a neural network**
This figure depicts a hypothetical error surface from a neural network. Here, the phenotype is the evolving set of connection strengths among artificial synapses. See text for interpretation. *Source*: Adapted with permission from Elman, Bates, Johnson, Karmiloff-Smith, Parisi, and Plunkett (1996, Figure 1.2.)

locomotion, and manipulation. However, the horizontal or activity dimension extension of the manifold in general is multidimensional. Without considering the stability of the activities in a multidimensional framework, it is difficult to assess the dynamical relations among the stable states and how these relations change over developmental time.

3. The landscape models do not reflect that the dynamical landscape supporting physical activities can change due to the shorter timescales at which goals and intentions can emerge and change. Thus, the standard landscape metaphor does not capture sufficiently the wide range of timescales that influence behavioral change. For instance, at one developmental state, a certain behavioral attractor (e.g., playing with a certain age-appropriate toy) could be reached (perhaps after a period of practice) when, for instance, constraints of arm length can be overcome.
4. The landscape metaphor shows the collection of states that are possible at a particular age, but it does not capture the relative dynamic properties of these states at any moment in time. Another way of saying this is that the landscape depicts the potential scope of stable states at a point in development, but does not depict the actual nature of the landscape as it evolves in real time in the implementation of a given activity, in spite of the apparent concerns and claims for this. For instance, at any developmental stage, multiple options for stable behavior coexist. The intention of selecting one of those as an actual behavioral task can be interpreted that the landscape now changes on a short timescale, destabilizing all but the target attractor, so that the performed movement will follow that landscape on the short timescale.

Thus, the landscape metaphor is just that—a metaphor. It has limitations for the development of formal models be they connectionist or dynamical. These limitations also provide a context for highlighting the important differences between the two frameworks with respect to the role of distance-from-goal and stability in landscape models.

## Distance to Goal and Stability in Landscape Models

The original use of landscape models in physics described how the potential or *free* energy of a system changes with the state of the system. In the simple case of a ball in a geographical landscape, the state of the system is given by the position of the ball. In our context, we assume the surface of the landscape is slippery but with high friction (what has been described as covered with molasses) so that the speed of the ball is always determined by the local slope of the landscape; therefore, momentum of the ball need not be treated as an independent degree of freedom. This implies that there is no overshoot; that is, if the ball reaches a minimum, it will come asymptotically to a stop at the minimum.

A more realistic ball in a smooth landscape would have finite speed and kinetic energy as it reaches the minimum. Therefore, it would continue moving uphill until friction losses and the build up of potential energy equals the kinetic energy. At that point, the ball comes to a stop before it returns back toward the lowest point. In this case, the approach to the minimum would be described by a damped oscillation (*bracketing*) instead of the *creeping* approach in the landscape models used in the context of epigenetic landscapes. At any point in the landscape we can calculate the free energy of the system, namely the maximum mechanical work that the system can do (without friction, that is, at infinitesimally small speed) while sliding down the landscape to its lowest point. It is proportional to the elevation of that point in the landscape; points at higher elevations can do more work on the way down than points close to the bottom. In an anthropomorphic sense, one can also say that the system *wants to go to the lowest point.*

In the context of learning, we therefore identify the lowest point in the landscape under consideration as the *goal* of the learning task. This position in the landscape represents a participant's (potential) coordinated movement pattern, whereas the elevation corresponds to a score function that becomes zero when the goal is reached. Therefore, we refer to the elevation in the landscape as *distance to the goal* (*d2g*) in the sense of score difference. The distance in the behavioral space (corresponding to the geographic distance along the horizontal dimension in the landscape) describes how much the behavior must be changed in order to reach the goal performance that is the bottom of the landscape. Note that we call this lowest point the *attractive* fixed point since the slope at this point is zero, which implies no force acting on the system and, therefore, no change in behavior.

As is clear from our experience in geographical landscapes, this geographical or horizontal distance depends on the slope of the landscape: If the slope is very steep close to the goal point, small changes in behavior (changes in horizontal position) yield big improvements in performance score (changes in vertical elevation). In physical terms, a large slope, therefore, corresponds to a strong attractive force of a goal point. In the context of movement, a shallow landscape makes it more difficult to recognize which (seemingly random) change of a movement pattern leads closer to the goal; this is because the change in the score value will be very small.

Thus, the stability of a goal point (a minimum in a landscape) is related to the slope of the neighboring part of the landscape. Let us take snapshots of a random collection of balls close to the goal, or let us look at the trajectory of a ball close to the minimum under the influence of a random force. We would observe that in a minimum with steep surrounds, the spread of the distribution in the horizontal dimensions of probable locations of the system is smaller than in the case of a shallower minimum. Thus, we can characterize the (linear) stability of a minimum by the width of a random distribution of points under the influence of fluctuations.

In some discussion of landscapes, we see a different notion of stability, namely the depth of the potential well: *The deeper the well the more stable the system*. This notion is related to the amount of work it would take to move the system out of the well. We call this state *integrated stability*; in the context of chemical reaction, it is also known as *activation energy*. In the landscape model, linear stability (or the steepness of the slope) indicates how much maximum force is necessary to leave the well. One can imagine the difference between a *shallow well* (one with very steep walls but not very deep) with a *dried out lake* (the slope of the walls is not very steep, but the bottom of the lake might be very deep). To get out of the well, one would need to overcome the full gravitational force of one's weight, but the work would be relatively small if the well is not very deep. In contrast, to get out of the lake, one needs to climb what is a potentially very shallow slope but must cover a much higher vertical distance. Thus, the maximum force could be much smaller than in the well example; while at the same time, the overall work required for the escape could be much larger.

We hope that the above conceptual example helps the reader to understand that the stability concepts in dynamical systems applications refer to the steepness properties (with steep slopes corresponding to narrow distributions with small variances); whereas the d2g approach in our models (or the error parameters in connectionist models) corresponds to the second case that measures the work that can be gained (modeled as the total vertical distance covered) by moving from the current location to the goal fixed point.

When close to a fixed point, *linear stability* is calculated by how much the restoring force increases with the distance from the fixed point. In mathematical terms, in the context of a potential landscape, linear stability is equivalent to the eigenvalues of the Jacobian matrix of the force field $\mathbf{F}(x^*)$ for a fixed point $x^*$. In a two-dimensional landscape, if the eigenvalues are both negative, then the fixed point is stable and the direction corresponding to the eigenvalues with the larger magnitude has a steeper increase in slope. Figure 6.4 shows a special example of a Gaussian peak superimposed on a Gaussian valley with smaller range. The result resembles a volcano with an asymmetrical crater.

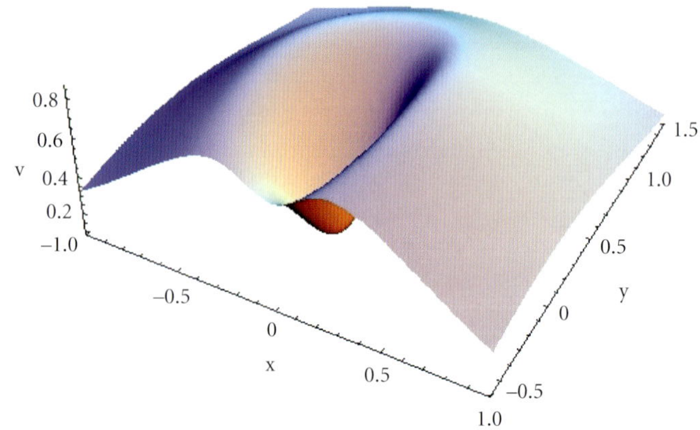


**Figure 6.4. Example of a metastable fixed point**
This figure depicts a metastable fixed point at $(x^*, y^*) = (0,0)$. The eigenvalue related to the x direction is –43.0; the one associated with the y direction is –6.3, indicating that the valley is steeper in x direction than in y direction. That implies that here, x would correspond to the fast timescale and y to the slow timescale. See text for interpretation.

The explicit equation for the potential landscape shown in Figure 6.4 is given by:

$$V\,(x, y) = e^{-\left(\frac{x^2}{\sigma x^2} + \frac{y^2}{\sigma y^2}\right)} - re^{-\left(\frac{x^2}{\tilde{\sigma} x^2} + \frac{y^2}{\tilde{\sigma} y^2}\right)} \qquad (2)$$

Here, $r = 0.9$, $\sigma_x = 1$, $\sigma_y = 1.5$, $\tilde{\sigma}_x = 0.2$, $\tilde{\sigma}_y = 0.5$. From equation (2), *V(x,y)* is the height of the landscape at a given point *(x,y)*. Both the local linear and integrated stability of a fixed point contribute to the attractiveness of the goal.

Note, however, that both notions of stability are a function of and can be calculated from the elevation function. Knowing the *V* or *d2g* for every point in the landscape, therefore, provides us with the stability information of a goal state but many different landscapes can have local minima (goals) that have identical stability indicators both for local linear and integrated stability. This is one manifestation of the concept of degeneracy in landscapes reflecting the many-to-one nature of mappings from locations to elevations. In cartographic terms, *degeneracy* is the idea that any regular landscape has smooth contour lines for which elevation does not change. For each one of those points on a contour line, the performance measure, therefore, would be identical.

## Case Studies of Landscapes and Timescales of Change in Learning and Development

In this last section of the chapter, we present three case studies of epigenetic landscapes from our work in motor learning and development. The examples focus on the acquisition of different motor skills at different ontogenetic stages. The emphasis is on timescales and their role in determining the nature of the dynamical processes that capture the persistent change in behavior over time (Newell, 1986). Each example captures one of the principles about the role of timescales in learning and development outlined previously.

### Case Study 1: Development of Prone Progression in Infancy

The emergence of the infant motor development sequence represents one of the most fundamental and pronounced sequences of changes in the dynamics of behavior over the lifespan. The motor development sequence was the focus of the early classic studies of development (Gesell, 1929; McGraw, 1943; Shirley, 1931), and interest in this developmental process has been revived by dynamical approaches to development (Kugler, 1986; Kugler, Kelso, & Turvey, 1982; Smith & Thelen, 1993; Thelen & Smith, 1994). Developmentalists were attracted by the intuitiveness of the dynamical constructs of stability and change; the face validity of dynamical constructs seemed relevant to the study of the emergence of new movement forms.

Here, we consider the landscape for the development of prone progression in the postural states of the infant (Gesell, 1929, 1946). This prone progression includes the initial posture of lying down and the subsequent emergence of sticking the chin up, sitting, creeping, crawling, and standing. Gesell showed that this developmental progression is nonlinear. We have constructed a developmental landscape of attractor dynamics for this postural progression (Newell et al., 2003). Isolating the development of postural activities, we do not imply that there is no relation to other activity classes of the motor development sequence. Indeed, we assume that our strategy has some general implications to the consideration of all action subcategories in principle. The phenomenological model is developed for a land-based surface of support (such as the room of a house) and is built on the empirical findings of Gesell (1946).

In the landscape model, head height and locomotion speed are the *order parameters* for the emergence of the prone progression. We use the term *order parameter* more loosely than in the synergetic tradition; we acknowledge that the nature of infants' multidimensional systems presents challenges to modeling that have not been faced to date in developments of other well-known dynamical models (e.g., the Haken-Kelso-Bunz model; Haken et al., 1985). We also recognize the redundancy that allows for many micro level states to realize an identical configuration in the order parameter space. (For example, many different combinations of muscle fibers can produce the same observed movement pattern.) The contrasting interpretation to be drawn from the double (Fig. 6.5) and single (Fig. 6.6) order parameter confirms the point made earlier of the importance of considering the multidimensional nature of the activity dimensions.

The dynamical landscape in Figure 6.5 shows several basic features about the emergence of the prone progression. First, the landscape is similar to that of Waddington (1957) because it represents a prototypical landscape of the potential stable states that exist over developmental time. It does not capture the changing landscape that relates to intention and the implementation of action. Second, the shape of the potential well for each activity is consistent with the behavioral properties shown by Gesell (1946) with regard to the emergence and preservation of the activity over time. Third, the landscape reveals that there is little equivalence for the infant to reach a certain head height and locomotion speed. In most cases, the infant at the given developmental time has only one stable state to achieve at a given point in the coordinate space.

The single order parameter of head height organizes the same behavioral data for the landscape in Figure 6.6. The time axis corresponds to developmental time. It is clear that the single order parameter is insufficient to capture the developmental pathway of change as the transition from standing to lying down would have to pass through the other activities of sitting, creeping, and chin up in the single dimension case. Thus, the difference in interpretation of Figures 6.5 and 6.6 confirms the need to consider the multiple dimensions of activities in the organization of the timescales of the landscape.

This example shows the importance of a principled basis for choosing the dimensions in which to organize the data. In particular, we have shown the potential use of more than a single order parameter in the consideration of whole-body movements with the involvement of many biomechanical degrees of freedom. There needs to be a concerted effort to characterize the order parameters of a broader range of actions and tasks.

### Case Study 2: Landscape of Learning with Warm-Up Decrement

The simplest landscape can be associated with a single goal that is the only movement pattern that is attractive (the local minimum in the landscape); this goal can be reached without any transitions to new movement patterns that involve changes in the landscape itself or external forces (e.g., teacher or instructions). Let us assume for the moment that two order parameters contribute to learning and the resulting performance improvement. The general construct of order parameters would lead to the prediction that these dimensions are not degenerate and lead to two different timescales. The situation is, therefore, locally that of Figure 6.7, but now the fixed point would possess unlimited integrated stability or activation energy; thus, the rim of the crater in Figure 6.7 would have infinite elevation.

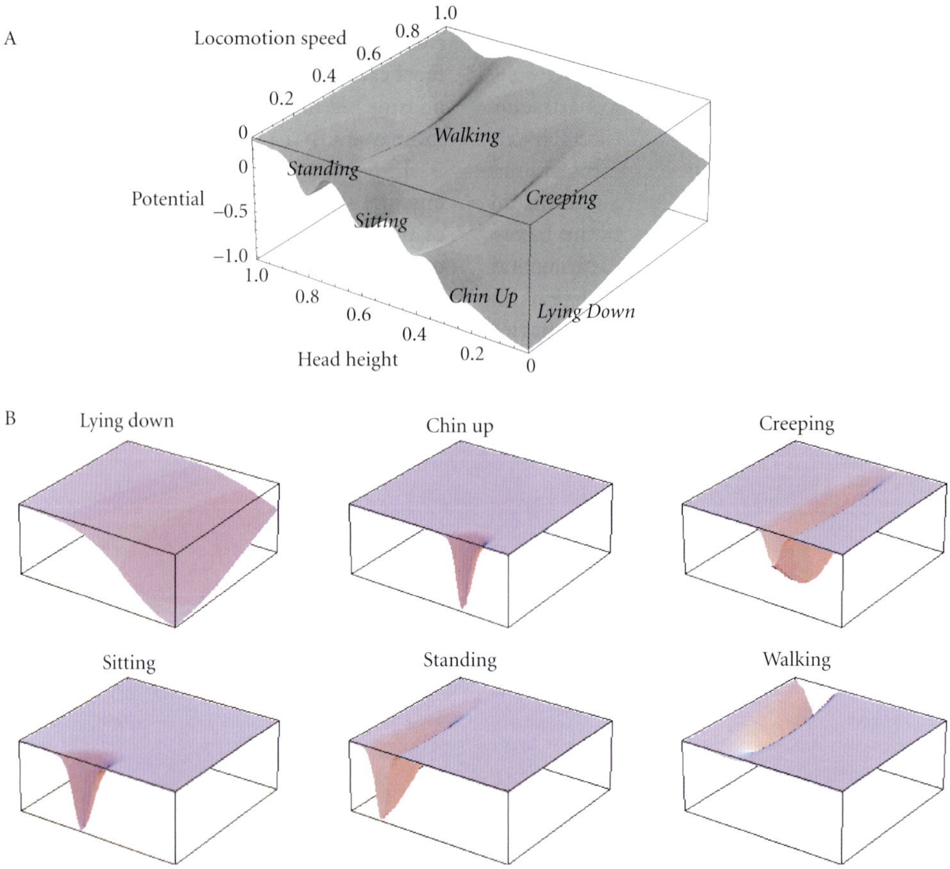


**Figure 6.5. Attractor landscape dynamics in the emergence of the prone progression**
(Top panel) The upper panel depicts the attractor landscape dynamics of the development of the infant prone progression. See text for details of simulation. (Bottom panel) The lower panel depicts the individual contributions of the attractors for each activity corresponding to the different behavioral attractors. If the contributions of these wells are superimposed, the landscape in the upper figure is the result.
*Source*: Adapted with permission from Newell, Liu, and Mayer-Kress (2003).

The theoretical movement trajectory, therefore, can be easily predicted. With the high descent rate associated with the fast timescale the system would quickly approach the bottom of the valley (the slow stable manifold of the fixed point) and then continue the approach to the fixed point at a constant modest rate given by the slow timescale. This pattern of fast → slow transitions in the learning curve seems to be universal and also seems to apply to higher dimensional landscapes where it could be mistaken by power-law curves that show a similar fast → slow ordering of (infinitely many) timescales.

Figure 6.8 simulates the trajectory point on a 3-D landscape, which means the landscape has three distinct timescales. Here, the timescales range from the slowest one in $x$-direction, one that is three times faster in the $y$-direction, and still three times faster in the $z$-direction. We start the system off at a position of the landscape with equal distance from the goal in each of the dimensions, for instance, initial point $(x_0, y_0, z_0) = (1, 1, 1)$. As described in words

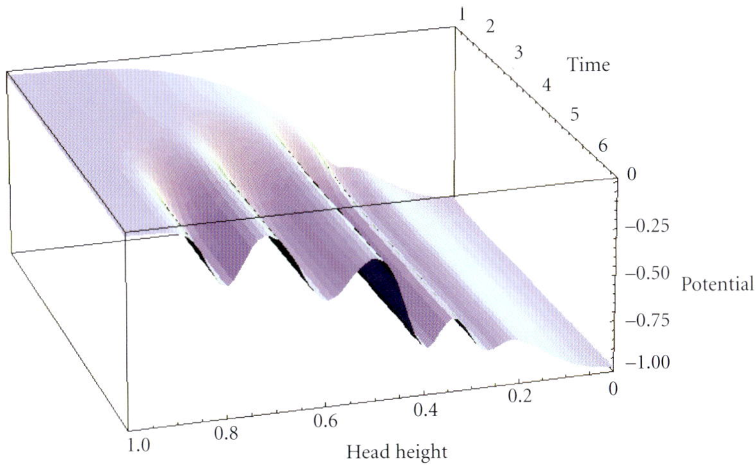


**Figure 6.6. Landscape dynamics in the emergence of the prone progression**
This figure depicts a one-dimensional version of the development of the prone progression. Onset times are 1, 2, . . . 6 units.
*Source*: Adapted with permission from Newell, Liu, and Mayer-Kress (2003).

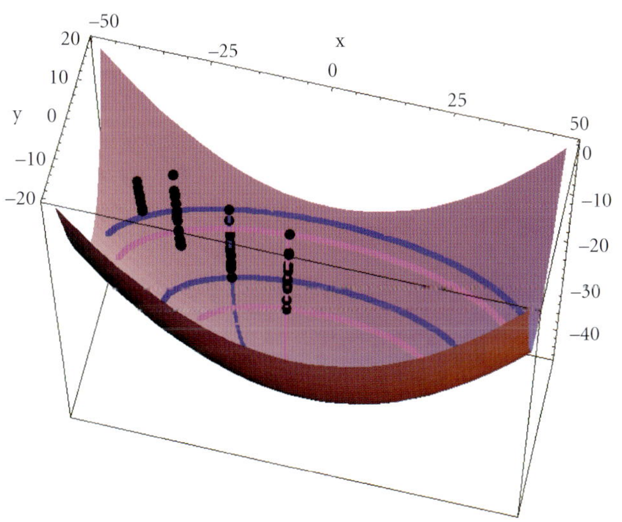


**Figure 6.7. Landscape dynamics with two timescales**
This figure depicts a simple landscape as described by Snoddy (1926) with two timescales associated with the x (slow) and y (fast) dimensions of a learning process. The elliptical curves in the landscape correspond to contour lines with constant elevation. The black dots correspond to empirical data that appear to follow the gradient lines in the landscape; two gradient lines have been extrapolated from the data.
*Source*: Reproduced from Newell, Liu, and Mayer-Kress (2005).

above, we observe that the system quickly approaches the $(x, y)$ plane, while at the same time also progressing along those dimensions closer to the goal. As $z$ becomes smaller than about 0.2 the movement in the y direction becomes dominant, until finally the only significant distance from the goal is along the slow x direction. We stop the simulation at twice the slowest characteristic timescale when the system has approached the goal to about 90% and all the remaining approach is dominated by the slowest timescale alone.

In Figure 6.9 we show directly the distance of the system from the goal in a log linear representation (solid line). A single timescale would be represented by a straight line with a single slope. The universal fast-slow pattern is clearly visible in that the initial slope is significantly

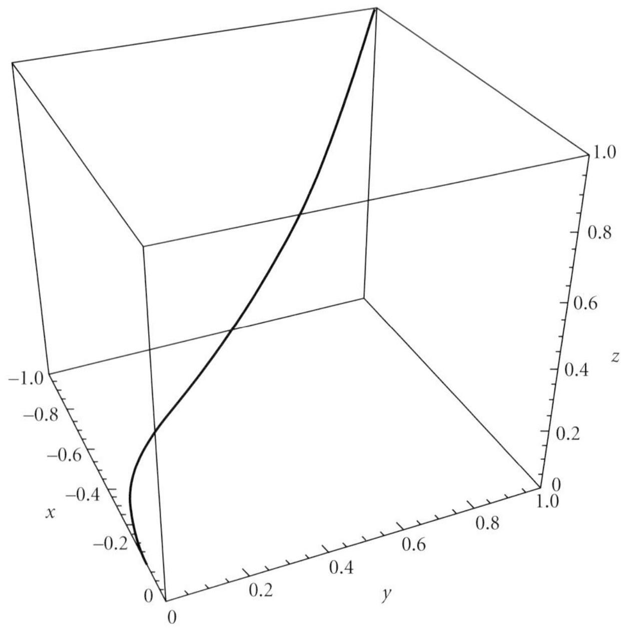


**Figure 6.8. Three-dimensional landscape dynamics**
This figure depicts a 3-D landscape. Note, that the simulated movement trajectory follows the steepest descent path toward the minimum, fixed point at zero. Elevation is now in the fourth dimension.

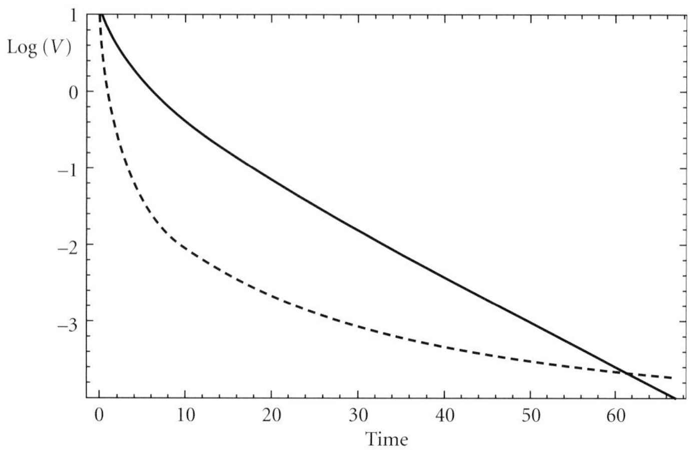


**Figure 6.9. Temporal dynamics for exponential model.**
Distance to the goal (logarithmic scale) for the three timescale exponential model of Fig. 6.8 (solid cure). Note the pronounced fast-slow pattern in the first 20 time units. For comparison a power law function (dashed line) is shown in the same log linear scale.

steeper that the slope of the later part of the curve. To illustrate the similarity of the fast-slow patterns in the power-law function (dashed line in Fig. 6.9), we provide an example with similar variation over the same time range (two characteristic timescales). We can observe that the infinitely many timescales of the power law function lead to much more extreme fast-slow patterns than in our exponential model with three timescales.

This theoretical case becomes slightly more complicated when we consider actual human behavior. As can be easily seen, a constant rate of approach for any timescale will only bring one exponentially close to but never reach the fixed point. Realistic timescales also make it obvious that there are very few, if any, coordination skills where a participant starts as a novice and reaches expert performance level within a single practice session.

This raises the issue of practice schedules, an issue that has been discussed in the motor learning literature for a long time. In contrast to a mechanical system where we can reasonably expect that we can continue a measurement curve more or less in the same condition if we start a new session, this is obviously not the case with human motor learning or even learning in general. Activities between sessions will influence the state of the participant upon resumption of practice. Moreover, practice activity itself will initiate epigenetic physiological processes that will continue long after the end of the practice session.

Thus, it is not surprising that if we reconstruct a trajectory in a learning landscape the end point of a practice segment of one session will be significantly different from the starting point of the next segment of practice trajectory. Observations—going back all the way to Snoddy's (1926) ground-breaking work—indicate that there is a class of learning situations where the performance at the beginning of a practice session is considerably depressed (*warm-up decrement*), but then recovers to a level that would be expected from a second, slower learning rate that reflects the persistent changes due to practice. We were able to implement a quantitative reconstruction of a four segment learning trajectory from Snoddy's original data (Newell, Liu, & Mayer-Kress, 2005; Snoddy, 1926). We showed that warm-up decrement is primarily caused by a change in initial conditions away from the fixed point in the fast dimension. Surprisingly the reconstruction also showed that during the rest period, the state of the system moved closer to the fixed point in the direction of the slow dimension. This phenomenon strongly suggests the presence of (at least) two independent mechanisms of change in the performance variable, acting at different characteristic timescales. We were able to associate those mechanisms with adaptation (Liu, Mayer-Kress, Lee, & Newell, 2006).

These patterns and quantitative results would be difficult to obtain without the theoretical framework of a landscape model associated with the different timescales. This perspective now leads to new questions and predictions about changes in learning and development that will be tested in future research. This line of research promises *new* ways to look at *old* problems of the performance characteristics that define learning curves and stages of development.

### Case Study 3: Landscape Model of Learning a Coordination Task

In the previous case we saw that an $n$-dimensional landscape generically (i.e., without degeneracy of the fixed point) will have $n$ timescales given by the different eigenvalues of the fixed point associated with the attracting minimum in the landscape. This will describe the complete dynamics of the system under the condition that the parameters describing the landscape (including the eigenvalues of the fixed points) are constant. This assumption is only approximately true to a degree that depends on the context of the specific application. In some situations, learning might be modeled to include a new timescale (typically much longer) that is given by the dynamics of the landscape itself. This phenomenon has a true analogue in geographic landscapes where, for instance, the timescales associated with movement of water in streams is much shorter than the geological landscapes that can alter the location of riverbeds. In a psychological context, the performance of a habitual behavior is usually occurring at much slower timescales than required to change the habit (i.e., getting out of one's rut).

In the following, we discuss an example that describes the learning of a specific motor task related to the celebrated Haken-Kelso-Bunz (HKB) model (Haken et al., 1985) of synchronized finger oscillations. The order parameter is given here by the relative phase $\Delta\varphi$ of the finger oscillators. We know that for high oscillation frequencies, only the *in-phase* ($\Delta\varphi = 0$) oscillator is stable. At lower frequencies, a second attractor with *anti-phase* oscillations ($\Delta\varphi = \pi$) emerges. That means that depending on the initial condition (i.e., how we started the oscillation), two distinct movement patterns can be observed. If one finger is just a little out of phase regarding the other finger, the difference will go to zero and both finger move in sync. On the other hand, if we start off close to an *opposite* condition of one finger pointing up while the other is pointing down, then we expect both fingers to be attracted to an *anti-phase* movement pattern. With special practice, one can learn other phase relationships, such as $\Delta\varphi = \pi/2$ (Zanone & Kelso, 1992). As we increase the frequency of the oscillation, the chance goes to zero that a random initial phase will be attracted to any stable pattern other than the in-phase pattern. It is a common phenomenon in nonlinear systems that when approaching transitions to new patterns of behavior the dynamics of the system will depend on the influence of other dimensions such as the history of the system or previous states (i.e., *hysteresis*).

In our approach, we model this situation by embedding the 1-D HKB landscape (*x* coordinate) into a landscape $V_{x0}(x,y)$ with an extra dimension (*y* coordinate) reflecting this influence. A new bifurcation parameter *c* measures how well a participant has learned the new task of finger oscillations at a 90° phase difference ($\Delta\varphi = \pi/2$). Note that for $c = 0$, our model contains the HKB model as a special case. The parameter $\sigma$ is related to the range of attraction of the new fixed point. The dynamics in the new *y* direction is chosen such that the system is slowly damped and approaches the $y = 0$ hyperplane. The *x* variable is related to the phase difference $\Delta\varphi$ such that a phase difference $\Delta\varphi = \pi/2$ corresponds to a target of $x_0 = 1/4$. The explicit form of our generalized HKB system for 90° phase learning is given by:

$$V_{x_0}(x,y) = y^2 - \cos(2\pi x) - \frac{1}{4}\cos(4\pi x) - c\left(1 - \frac{y}{4}\right)e^{-\frac{(x-x_0)^2}{\sigma^2}} \quad (3)$$

In this interpretation of the landscape, which is different from the interpretation previously used (Newell et al., 2001), we only have the in-phase attractor stable at the beginning of practice. Independent of the initial condition, the system will converge to an in-phase oscillation (Fig. 6.10, top left panel). After a certain amount of practice, the landscape has deformed and a stable fixed point emerges at the desired $x_0 = 90°$ attractor. Note that still the majority of initial conditions will converge to the in-phase attractor, and special preparations are required to reach the new fixed point (Fig. 6.10, top right panel). Finally, after further practice, a large area of initial conditions will be attracted to the new fixed point, and the chance of performing the task is significantly increased (Fig. 6.10, bottom panel). This landscape model would predict that initial conditions close to $x = 270°$ would be attracted to the in-phase fixed point.

So far we have treated the change in the landscape as a nonautonomous, external modification of a control parameter. It is straightforward, however, to extend the dimension of the autonomous state space by assigning a new dynamical equation to the evolution of the bifurcation parameter *c*. The complexity of this case, however, is also enhanced compared to the previous cases because it includes a bifurcation or nonequilibrium phase-transition which is known to exhibit the phenomenon of *critical slowing down* (that is, the divergence of timescales at the critical point of the transition).

In this example, as the system approaches the transition point, it takes increasingly longer to approach the attractor state. For instance, if one learns to juggle, one will typically experience a critical phase where it will take increasingly longer to drop the ball (i.e., one approaches the *fail state*) or to get into a stable movement pattern that can be sustained indefinitely (i.e., one enters a *success state*). Therefore, it might be difficult to extract quantitative estimates of the timescales of change from limited experimental data. Nevertheless, this example demonstrates two timescales in operation, the drift of the system to a stable state, and the change in stable states that are available.

## Conclusion

A central premise arising in our epigenetic landscape models of learning and development is that the construct of timescales provides a fundamental way to assess the core processes underlying the persistent change in behavior over time. In that sense, landscapes both reveal

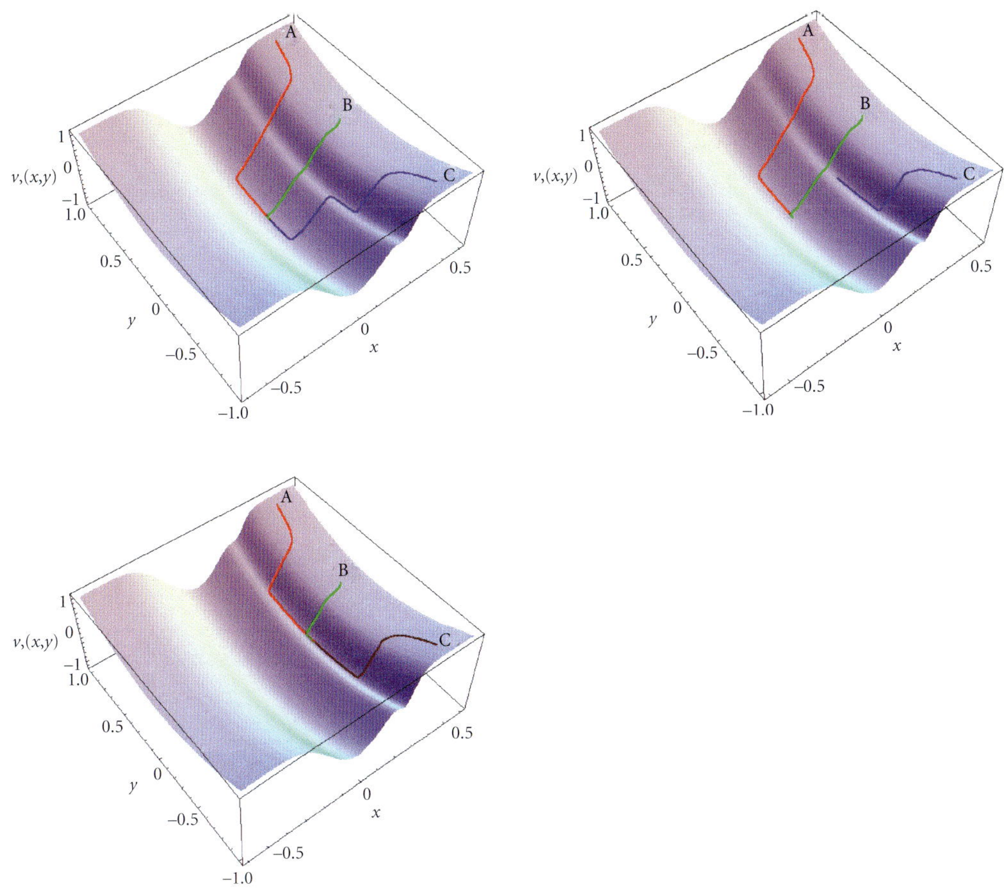


**Figure 6.10. Landscape of learning the 90 degree phase task of the Haken-Kelso-Bunz model**
Top left panel shows three trajectories at the beginning of practice (c = 0.4). Only temporary stabilization of the target phase $x_0 = 0.25$ can be achieved when starting from special initial conditions indicated by trajectory C. Top right panel shows performance right at the transition (c = 0.425). Here, the target phase $x_0 = 0.25$ shows one-sided stability: Initial conditions close to C will be attracted to the new attractor. Note that in this situation the system is very sensitive to noise perturbations. Bottom panel shows performance after sufficient practice (c = 0.525). Now, all initial conditions close to the target attractor $x_0 = 0.25$ will converge to the fixed point. Source: Reproduced from Newell, Liu, & Mayer-Kress (2001).

mechanisms of change and offer tools to better describe how the system is changing. The timescales provide a window into the dimensions of behavioral change and how these evolve over time. A consideration of timescales has allowed us to distinguish dimensions that are often confounded in other modeling approaches.

A point of distinction or at least emphasis between the dynamical and connectionist approaches to learning and development is in their respective use of the term *timescale.* Dynamical approaches give importance to the intrinsic dynamics of the system in determining timescales, but this formality is less evident in the use of timescales in connectionist approaches. This distinction also gives rise to the stronger emphasis on stability in dynamical systems modeling than in connectionist modeling. However, these theoretical frameworks may just be on different timescales of development; perhaps it will not be long before it is difficult to tell one framework from the other (Spencer & Schöner, 2003).

In the connectionist networks literature, little attention seems to be given to the mathematical shape of the learning/error curve. We claim that the landscape modeling and associated structure of the learning curve, especially the presence of multiple timescales, is a universal feature of natural and artificial learning systems. We would, therefore, predict that in carefully designed learning simulations, it should be possible to extract different timescales and use them to quantitatively reconstruct a connectionist landscape that goes well beyond the metaphor. At least issues of stationarity and limitations in data collection time should be easier to control in connectionist simulations than in the wetware of biological learners. In future publications, we want to address the issue of epigenetic landscapes that describe the change in the learning system between practice sessions. Here issues of unlearning and memory consolidation appear to be of central importance in driving the timescales of learning and development. Again, a parallel investigation in connectionist simulations could prove most fruitful.

## Acknowledgments

This work was supported in part by NSF #0114568.

## References

Akaike, H. (1974). A new look at the statistical model identification. *IEEE Transactions on Automatic Control, 19*(6), 716–723.

Bassingthwaighte, J. B., Liebovitch, L. S. West, B. J. (1994). *Fractal physiology.* New York: Oxford University Press.

Bertalanffy, L. von (1975). *Perspectives on general system theory: Scientific-philosophical studies.* New York: G. Braziller.

Boccara, N. (2003). *Modeling complex systems* [Series: Graduate texts in contemporary physics]. New York: Springer.

Burnham, K. P., & Anderson, D. R. (2002). *Model selection and multi-modal inference: A practical information-theoretic approach* (2nd ed.). New York: Springer.

Connolly, K. J. (1986). A perspective on motor control. In M. G. Wade & H. T. A. Whiting (Eds.), *Motor development in children: Aspects of coordination and control* [Series: NATO Advanced Science Institutes: D. Behavioural and social sciences] (pp. 3–22). Dordrecht, The Netherlands: Martinus Nijhoff.

Eckhardt, R. B. (2000). *Human paleobiology.* New York: Cambridge University Press.

Elman, J. L., Bates, E. A., Johnson, M. H., Karmiloff-Smith, A., Parisi, D., & Plunkett, K. (1996). *Rethinking innateness: A connectionist perspective on development.* Cambridge, MA: MIT Press–Bradford Books.

Fischer, K. W., & Rose, S. P. (1993). Development of coordination of components in brain and behavior: A framework for theory and research. In G. Dawson and K. W. Fischer (Eds.), *Human behavior and the developing brain.* New York, NY: Guilford.

Foster, R. G., & Kreitzman, L. (2004). *Rhythms of life: The biological clocks that control the daily lives of every living thing.* New Haven, CT: Yale University Press.

Gallistel, C. R., Fairhurst, S., & Balsam, P. (2004). The learning curve: Implications of a quantitative analysis. *Proceedings of the National Academy of Science, USA, 101*(36), 13124–13131.

Gesell, A. (1929). Maturation and infant behavior pattern. *Psychological Review, 36*(4), 307–319.

Gesell, A. (1946). The ontogenesis of infant behavior. In L. Carmichael (Ed.), *Manual of child psychology* (pp. 295–331). New York: Wiley.

Haken, H. (1983). *Synergetics—an introduction: Nonequilibrium phase transitions and self-organization in physics, chemistry, and biology* (3rd ed.). New York: Springer.

Haken, H., Kelso, J. A. S., & Bunz, H. (1985). A theoretical model of phase transitions in human hand movements. *Biological Cybernetics, 51*(5), 347–356.

Haliday, D., Resnick, R., & Walker, J. (2005). *Fundamentals of physics* (7th ed.). New York: Wiley.

Heathcote, A., Brown, S., & Mewhort, D. J. K. (2000). The power law repealed: The case for an exponential law of practice. *Psychonomic Bulletin and Review, 7*(2), 185–207.

Iberall, A. S. (1972). *Toward a general science of viable systems.* New York: McGraw-Hill.

Kaczmarek, L. (2000). Gene expression in learning processes. *Acta neurobiologiae experimentalis,* 60: 419–424.

Kaplan, D., & Glass, L. (1995). *Understanding nonlinear dynamics* [Series: Texts in applied mathematics, Vol. 19]. New York: Springer.

Kelso, J. A. S. (1995). *Dynamic patterns: The self-organization of brain and behavior.* Cambridge, MA: MIT Press–Bradford Books.

Kugler, P. N. (1986). A morphological perspective on the origin and evolution of movement patterns. In M. G. Wade & H. T. A. Whiting (Eds.), *Motor development in children: Aspects of coordination and control* [Series: NATO Advanced Science Institutes: D. Behavioural and social sciences], (pp. 459–525). Dordrecht, The Netherlands: Martinus Nijhoff.

Kugler, P. N., & Turvey, M. T. (1987). *Information, natural law, and the self-assembly of rhythmic movement: Theoretical and experimental investigations.* Hillsdale, NJ: Erlbaum.

Kugler, P. N., Kelso, J. A. S., & Turvey, M. T. (1982). On the control and coordination of naturally developing systems. In J. A. S. Kelso & J. E. Clark (Eds.), *The development of movement control and co-ordination* (pp. 5–78). New York: Wiley.

Lemke, J. L. (2002, October). Keeping learning alive: Multiple timescales in the social organization of learning. Paper presented at the *Fifth International Conference of the Learning Sciences*, Seattle, WA.

Liu, Y.-T., Mayer-Kress, G., & Newell, K. M. (2003). Beyond curve fitting: A dynamical systems account of exponential learning in a discrete timing task. *Journal of Motor Behavior, 35*(2), 197–207.

Liu, Y.-T., Mayer-Kress, G., & Newell, K. M. (2004). Beyond curve fitting to inferences about learning. *Journal of Motor Behavior, 36*(2), 233–238.

Liu, Y.-T., Mayer-Kress, G., & Newell, K. M. (2006). Qualitative and quantitative change in the dynamics of motor learning. *Journal of Experimental Psychology: Human Perception and Performance, 32*(2), pp. 380–393.

Luce, G. G. (1971). *Biological rhythms in human and animal physiology.* New York: Dover.

Mayer-Kress, G., Liu, Y.-T., & Newell, K. M. (2006). Complex systems and human movement. *Complexity, 12*(2), 40–51.

McGraw, M. B. (1943). *The neuromuscular maturation of the human infant.* New York: Columbia University Press.

Meinhardt, H. (1982). *Models of biological pattern formation.* New York: Academic Press.

Mittenthal, J. E., & Baskin, A. B. (Eds.). (1992). *Principles of organization in organisms: Proceedings of the Workshop on Principles of Organization in Organisms, held June, 1990, in Santa Fe, New Mexico.* Reading, MA: Addison-Wesley.

Muchisky, M., Gershkoff-Stowe, L., Cole, E., & Thelen, E. (1996). The epigenetic landscape revisited: A dynamic interpretation. In B. Hopkins, S. Johnson (Series Eds.), C. Rovee-Collier, & L. P. Lipsitt (Eds.), *Advances in Infancy Research* (Vol. 10, pp. 121–159). Norwood. NJ: Ablex.

Newell, A., & Rosenbloom, P. S. (1981). Mechanisms of skill acquisition and the law of practice. In J. R. Anderson (Vol. Ed.), *Carnegie Mellon symposia on cognition series: Vol. 16. Cognitive skills and their acquisition* (pp. 1–55). Hillsdale, NJ: Erlbaum.

Newell, K. M. (1986). Constraints on the development of coordination. In M. G. Wade & H. T. A. Whiting (Eds.), *Motor development in children: Aspects of coordination and control* (pp. 341–360). Leiden, The Netherlands: Martinus Nijhoff.

Newell, K. M., Liu, Y.-T., & Mayer-Kress, G. (2001). Timescales in motor learning and development. *Psychological Review, 108*(1), 57–82.

Newell, K. M., Liu, Y.-T., & Mayer-Kress, G. (2003). A dynamical systems interpretation of epigenetic landscapes for infant motor development. *Infant Development and Behavior, 26*(4), 449–472.

Newell, K. M., Liu, Y.-T., & Mayer-Kress, G. (2005). Learning in the brain-computer interface: Insights about degrees of freedom and degeneracy in a landscape model of motor learning. *Cognitive Processing, 6*(1), 37–47.

Newell, K. M., Kugler, P. N., & van Emmerik, R. E. A., & McDonald, P. V. (1989). Search strategies and the acquisition of coordination. In S. A. Wallace (Ed.), *Perspectives on the coordination of movement* (pp. 85–122). Amsterdam: North-Holland.

Nolfi, S., Elman, J. L., & Parisi, D. (1994). Learning and evolution in neural networks. *Adaptive Behavior, 3*(1), 5–28.

Pitt, M. A., & Myung, I. J. (2002). When a good fit can be bad. *Trends in Cognitive Sciences, 6*(10), 421–425.

Prigogine, I., & Nicolis, G. (1971). Biological order, structure and instabilities. *Quarterly Reviews of Biophysics, 4*(2), 107–148.

Schöner, G. (1989). Learning and recall in a dynamic theory of coordination patterns. *Biological Cybernetics, 62*(1), 39–54.

Schroeder, M. R. (1991). *Fractals, chaos, power laws: Minutes from an infinite paradise.* New York: W. H. Freeman.

Shirley, M. M. (1931). *The first two years: A study of twenty-five babies: Vol. I. Postural and locomotor development.* Minneapolis: University of Minnesota Press.

Simms, K. (1994). Evolving 3D morphology and behavior by competition. In R. Brooks & P. Maes (Eds.), *Artificial Life IV: Proceedings of the Fourth International Workshop on the Synthesis and Simulation of Living Systems* (pp. 28–39). Cambridge, MA: MIT Press. Retrieved on October 26, 2008, from http://www.cs.bham.ac.uk/~wbl/biblio/cache/http___web.genarts.com_karl_papers_alife94.pdf

Smith, L. B., & Thelen, E. (Eds.). (1993). *A dynamic systems approach to development: Applications.* Cambridge, MA: MIT Press–Bradford Books.

Snoddy, G. S. (1926). Learning and stability: A psychophysiological analysis of a case of motor learning with clinical applications. *Journal of Applied Psychology, 10*(1), 1–36.

Spencer, J. P., & Schöner, G. (2003). Bridging the representational gap in the dynamic systems approach to development. *Developmental Science, 6*(4), 392–412.

Strogatz, S. H. (1994). *Nonlinear dynamics and chaos: With applications to physics, biology, chemistry, and engineering* [Series: Studies in nonlinearity]. Reading, MA: Addison-Wesley.

Thelen, E., & Smith, L. B. (1994). *A dynamic systems approach to the development of cognition and action.* Cambridge, Mass: MIT Press–Bradford Books.

Thelen, E., Kelso, J. A. S., & Fogel, A. (1987). Self-organizing systems and infant motor development. *Developmental Review, 7*(1), 39–65.

van der Maas, H. L. J. (1996). Beyond the metaphor? Reviews of *A dynamic systems approach to development: Applications* by Linda B. Smith and Thelen Esther (1993) and *A dynamic systems approach to the development of cognition and action* by Esther Thelen and Linda B. Smith (1994) [Book review]. *Cognitive Development, 10*(4), 621–642.

Van Geert, P. (1994). *Dynamic systems of development: Change between complexity and chaos.* New York: Harvester Wheatsheaf.

Waddington, C. H. (1957). *The strategy of the genes: A discussion of some aspects of theoretical biology.* London: Allen & Unwin.

Ward, L. M. (2001). *Dynamical cognitive science.* Cambridge, MA: MIT Press–Bradford Books.

Zanone, P. G., & Kelso, J. A. S. (1992). Evolution of behavioral attractors with learning: Nonequilibrium phase transitions. *Journal of Experimental Psychology: Human Perception and Performance, 18*(2), 403–421.

# PART III

# Connectionism: Case Studies

# CHAPTER 7

## Connectionist Approaches to Perseveration: Understanding Universal and Task-Specific Aspects of Children's Behavior

*J. Bruce Morton and Yuko Munakata*

Everyone perseverates at one time or another, repeating behaviors that were adaptive in an earlier context but that are no longer appropriate. After reorganizing the objects on our desk, for example, we might reach for the phone in its old rather than its new location. Infants and children are particularly robust perseverators. After repeatedly watching as an object is hidden in one location (location A) and searching for it there, infants perseverate by reaching back to that location after watching as the object is hidden in a new location (location B) (Diamond, 1985; Piaget, 1954). Infants will even make such perseverative A-not-B errors when the visible environment clearly specifies the appropriate behavior. For example, after repeatedly pulling a towel in one location (A) to retrieve a distant toy on the towel, infants perseverate by pulling the towel in that location when the toy is visibly no longer on it, instead of pulling a towel in a new location (B) that has a toy on it (Aguiar & Baillargeon, 2000). Similarly, 3-year-old children often perseverate in simple card-sorting tasks by continuing to sort cards one way (e.g., by color) when asked to switch and sort them in a new way (e.g., by shape) (Zelazo, Frye, & Rapus, 1996). Finally, 6-year-old children often persist in judging a speaker's feelings from what she says when asked to switch and judge her feelings from the sound of her voice (Morton, Trehub, & Zelazo, 2003).

These kinds of perseverative errors and children's eventual success on such tasks, have been studied extensively, leading to rich characterizations of children's performance and the many factors that influence it (e.g., Brace, Morton, & Munakata, 2006; Brooks, Hanauer, Padowska, & Rosman, 2003; Diamond, 1985; Jacques, Zelazo, Kirkham, & Semcesen, 1999; Kirkham, Cruess, & Diamond, 2003; Marcovitch & Zelazo, 1999; Morton & Munakata, 2002b; Perner & Lang, 2002; Smith, Thelen, Titzer, & McLin, 1999; Towse, Redbond, Houston-Price, & Cook, 2000; Wellman, Cross, & Bartsch, 1986; Zelazo et al., 2003). As elaborated below, this work has established both commonalities in perseverative behaviors across task paradigms, as well as unique aspects of perseveration that are tied to specific details within a given task.

In terms of commonalities across task paradigms, children show similar patterns of perseveration across tasks that vary greatly in their response requirements (e.g., reaching vs. answering verbal queries; card-sorting vs. error detection; Jacques et al., 1999; Munakata & Yerys, 2001; Piaget, 1954; Zelazo et al., 1996). In addition, children's performance improves with age across task paradigms (Diamond, 1985; Jacques & Zelazo, 2001; Zelazo et al., 1996). Children show similar dissociations across task paradigms, demonstrating some knowledge of appropriate behaviors despite responding perseveratively. For example, in the A-not-B task, infants will occasionally look at the correct hiding location for a toy (Diamond, 1985; Hofstadter & Reznick, 1996) or appear to recognize where the hidden toy should be (Ahmed & Ruffman, 1998), while nonetheless reaching perseveratively to a previous hiding location.[1] Similarly, in the card-sorting task, children will reliably answer questions correctly about what they are supposed to be doing (e.g., "Where do trucks go in the shape game?"), while sorting cards perseveratively according to a previously relevant rule (Zelazo et al., 1996).

In addition to common patterns across tasks, specific details of individual tasks strongly influence whether children perseverate. For example, the number and distinctiveness of hiding locations affects whether infants perseverate in the A-not-B task (Bremner, 1987; Cummings & Bjork, 1983a, 1983b; Wellman et al., 1986). Whether a toy is placed on a towel versus physically attached to the towel affects whether infants perseverate in the towel-pulling task (Aguiar & Baillargeon, 2000). The types of cards used and how they are labeled affects whether children perseverate in the card-sorting task (Kirkham et al., 2003; Perner & Lang, 2002; Yerys & Munakata, 2006; Zelazo et al. 2003). And, children show *décalages* in their behavior, to use the Piagetian term, succeeding on one task (e.g., card-sorting) at a given age while perseverating on another task (e.g., speech-interpretation) that is similar in structure but different in its details (Morton et al., 2003).

Both connectionist and dynamic systems models have played an important role in helping understand this diversity of findings. In this chapter, we focus on case studies of connectionist models of perseveration for understanding infants' reaching behavior (Munakata, 1998; Stedron, Sahni, & Munakata, 2005) and children's rule-guided behavior (Morton & Munakata, 2002a, 2002b). These models demonstrate how general connectionist principles (1) provide a unified framework for understanding perseveration across ages, task paradigms, and response modalities, (2) capture important details of children's performance, and (3) lead to novel and testable predictions.

We compare and contrast these contributions from connectionist models with those from dynamic systems models of the A-not-B task, which place greater emphasis on the details of visually guided reaching. Such details may be essential for understanding nuances of infant performance on the A-not-B task, but the more general principles may have greater utility for understanding perseverative behavior across the life span.

The connectionist models we discuss focus on the nature of the representations supporting behavior, specifically the importance of graded and distinct types of representations. First, these models capture the assumption that knowledge is graded in nature rather than all-or-nothing, and can be weaker or stronger depending on a variety of factors, including an individual's state of development, experience with a task, and the amount of support from the environment (McClelland, 1989; Munakata, 2001; Sitton, Mozer, & Farah, 2000). Second, these models capture the assumption that knowledge can take multiple forms rather than being monolithic. One important distinction is between active versus latent representations (Cohen, Dunbar, & McClelland, 1990; Munakata, Morton, & Stedron, 2003). Active representations take the form of sustained neuronal firing, which can be communicated to other processing regions. Active representations are thought to tap prefrontal cortical regions and develop relatively late. Latent representations take the form of changes in connections between neurons; these changes influence subsequent processing of stimuli by those neurons without otherwise being accessible to other processing regions. Latent representations are thought to tap posterior cortical and subcortical areas and develop relatively early. This active-latent distinction is captured naturally in connectionist models. Active representations take the form of sustained activation of processing units, and latent representations take the form of changes in the strength of connection weights between

[1] Although some researchers have challenged the reliability and interpretation of such dissociations because, for example, looking and reaching behaviors are often closely coupled (e.g., Smith & Scheier, 1998; Spencer & Schutte, 2004), looking and reaching measures nonetheless reveal differential sensitivity when compared systematically in the A-not-B task.

processing units. Both kinds of representations can vary continuously in strength, capturing the notion of gradedness. In what follows, we discuss models that demonstrate how such graded, active and latent representations provide a unified framework for understanding both general and specific aspects of children's perseveration.

## Case Studies: Connectionist Models of Infant Perseveration in A-not-B Tasks

We focus first on a model of the Piagetian A-not-B task with hidden objects (Munakata, 1998) and then briefly describe extensions of this A-not-B model to other variants of the A-not-B task (Stedron et al., 2005).

### Architecture, Stimuli, and Learning Algorithm

The model includes three input layers, a hidden layer (also referred to as an internal representation layer), and two output layers (Fig. 7.1). Input layers allow the network to encode information about task-related events: (1) location of events, with a separate unit encoding each of three locations (A, B, and C); (2) toys presented in the events, with one unit encoding a default toy (T1) and a second unit encoding an alternate (T2); and (3) covers that conceal the toys, with one unit encoding the default cover (C1) and a second unit encoding an alternate (C2). Hidden and output layers each contain three units that correspond with the three possible hiding locations. The two output layers allow the network to indicate separate gazing and reaching responses and are essentially the same except that units in the gazing layer respond to every event, whereas units in the reaching layer respond only to events that are within reaching distance. This difference captures the fact that during the actual A-not-B task, infants are able to gaze to particular locations throughout the entire procedure but can reach only at select times.

Recurrent and feed-forward connections link the units of the model and allow the network to form active and latent representations of its experience. Feed-forward connectivity includes connections between each of the three input layers and the hidden layer, as well as connections between the hidden layer and the two output layers.

Consistent with infant behavior, this initial feed-forward connectivity includes a bias to respond appropriately to location information such that events that occur at location A lead the network to represent A and gaze to location A. This bias is instantiated as a set of connection weights such that connections between corresponding units (e.g., A input unit to A hidden unit) have larger weights than connections between noncorresponding units (e.g., A input

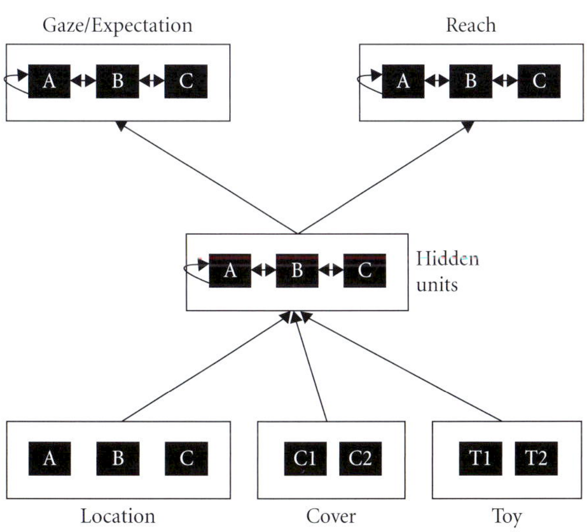


**Figure 7.1. An AB network**

unit to B or C hidden units). These feed-forward connection weights change with experience according to a Hebbian learning rule such that units that fire together (i.e., show correlated activity) wire together (i.e., increase in connection weights). For example, input at location A leads to simultaneous activity in the A units of the internal representation and gaze layers. Consequently, with repeated experience representing and gazing at a target (e.g., a toy) presented at location A, connections between A input, hidden, and output units are further strengthened and establish a latent representation of the target at A. This latent representation in turn causes the network to gaze preferentially to location A on subsequent trials.

In addition to feed-forward connections, hidden and output units also make connections to themselves. These so-called recurrent connections allow the model to form active representations of its experience—essentially self-sustaining activities of select units of the model that can be maintained in the absence of continued input from the input layers. In the present model, the strength of recurrent connections is varied by hand to simulate changes in active memory hypothesized to occur with development. Other connectionist models have shown that changes in the strength of recurrent connections and active memory can occur through experience with objects (Munakata, McClelland, Johnson, & Siegler, 1997).

The task is administered as a series of A-trials followed by a B-trial. Each trial consists of four segments that are presented as patterns of activity over the input units and that correspond directly with the procedure administered to infants. A segment includes the following.

*Start*: Covers sit in place on the apparatus until the experimenter draws the infant's attention to a particular location.

*Presentation*: Experimenter presents a toy at a particular location.

*Delay*: The toy has been hidden, and the apparatus sits with the lids in place.

*Choice*: Experimenter pushes the apparatus toward the infant until the locations are within reach. Reaching is permitted only in this segment of the trial.

To illustrate, Figure 7.2 displays patterns of input corresponding to an A-trial. Inputs in each segment correspond with the visible aspects of the display, with more salient aspects of the display producing more activity. At Start, there is only a small amount of activation for each location and cover, reflecting the fact that at this point, the experimenter has not drawn the infant's attention to the apparatus. At Presentation, the input units for toy T1 and location A are highly active, reflecting presentation of the target at location A. During Delay, the cover and location units are moderately active, as these aspects of the display remain visible to the infant, but the toy unit is not active, as the toy is hidden. Finally, at Choice (when the apparatus is pushed within reach of the infant), activity in all location units increases to reflect the increased salience of the three locations. Following Choice, activations are reset to zero, and the next trial is presented. B-trials (not shown) are identical, except that the target is presented at location B (i.e., the input unit for location B is activated).

### Performance in the A-not-B Task

The model provides a comprehensive framework for understanding general aspects of infant performance in the A-not-B task, including perseveration, age-related change, and dissociation. The model also provides a framework for understanding more detailed aspects of children's performance and generates counterintuitive predictions that have been supported in subsequent studies.

#### General Aspects of Infant Performance

*Perseveration* Figure 7.3 depicts the network's gazing and reaching performance on A-trials and B-trials as a function of age or recurrence strength. The model gazes and reaches correctly on A-trials independent of age (i.e., recurrence strength). By contrast, when recurrence is low, the model perseverates strongly, gazing and reaching correctly on only 15% of trials. Why does the model search correctly on A-trials but perseverate on B-trials?

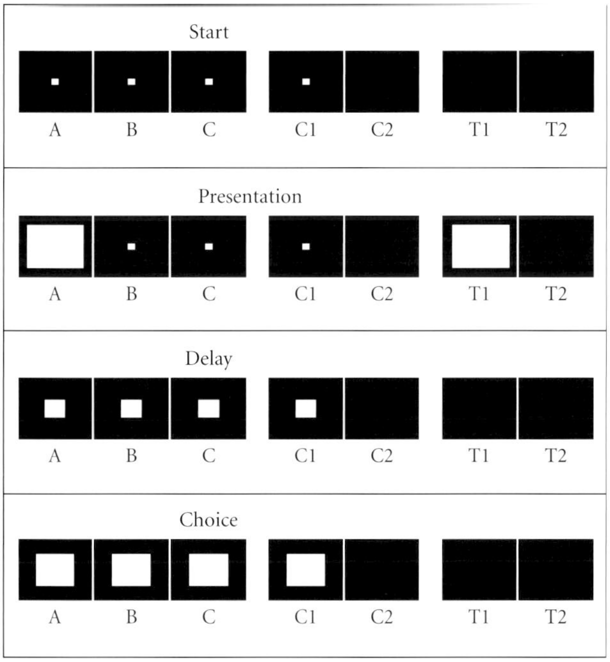


**Figure 7.2. Input pattern for an A-trial segment**
This figure illustrates the input patterns for the four segments of an A-trial (Start, Present [the presentation segment], Delay, and Choice) used in the standard two-location A-not-B task. In this schematic, the Present segment includes the so-called cover presentation segment used by other investigators. The amount of activity on a given input unit is represented by the size of the square in that unit. The three sets of units, from left to right, represent location (A, B, and C), cover type (C1 and C2), and toy type (T1 and T2).

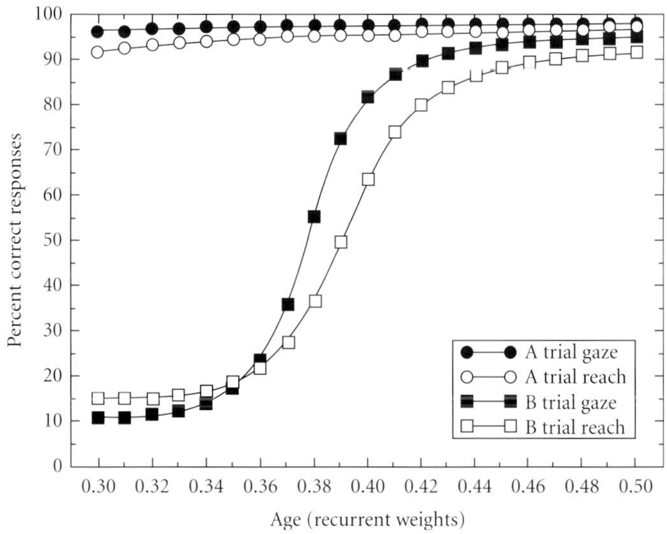


**Figure 7.3. A-not-B task: Network performance in gaze and reach as a function of simulated age**
On A-trials, the network is accurate across all levels of recurrence shown because any ability to maintain activity for location A is supported by latent weight changes (i.e., is built up across previous trials) that favor A over B. On B-trials, the network responds non-perseveratively only as the recurrent weights become stronger. The gaze/expectation system shows non-perseverative responding earlier than the reach system.

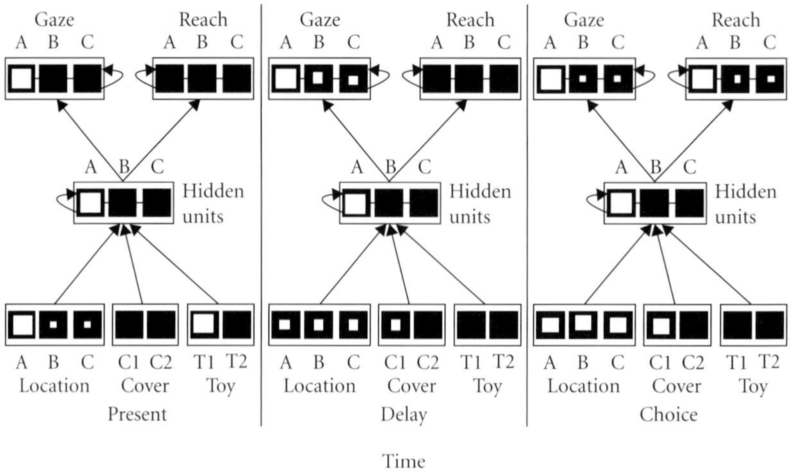


**Figure 7.4. The A-not-B model: Patterns of network activation for A-trial when recurrence is low** This figure depicts patterns of network activation during the Present, Delay, and Choice segments of an A-trial; recurrence is low in this figure. The network correctly represents A, and gazes and reaches to location A.

When recurrence is low, the model is unable to actively maintain information about prior events at location A, but gazes and reaches/searches correctly at location A by virtue of changes in latent weights that favor location A. Figure 7.4, which shows an A-trial of a young model with relatively weak recurrent connections, illustrates this concept. Because the model receives several practice trials at location A, the strength of connections between A units changes due to Hebbian learning and leads to a relatively strong latent bias to respond to location A. Therefore, in the presentation segment, when a target (a toy) is presented at location A, the model represents that something is happening at location A and gazes at location A.[2] Even during the delay segment, when the model is given identical input to location A, B, and C units, the model has no difficulty representing A and gazing to location A, even though it cannot actively maintain an internal representation of A. This is because feed-forward connections are strongly biased in favor of location A. Similarly, in the choice segment, even though the model receives equal input from the location A, B, and C units, it still actively represents A and gazes and reaches to location A. This would correspond to the following account of very young infants: while they are not terribly good at actively remembering that a toy was hidden at location A, they have built up strong biases to represent A and look and reach to location A; and they are inclined to reach in that direction.

The model's ability to maintain activity is, however, critical for B-trial performance, because the model needs to maintain a representation of B in order to overcome a latent bias to respond to location A. This is illustrated in Figure 7.5, which shows the same model performing a B-trial. At the beginning of the trial, when the target is presented at location B, the model represents B and gazes toward location B, just as infants gaze correctly when the toy is being hidden. However, in the delay segment, input to location A, B, and C units are equally active, and the model has a strong bias to location A and a weak ability to actively maintain new information. Consequently, the model's internal representation of B begins to fade. By the time the choice segment comes, the active representation of B has faded almost entirely; the model is representing A and perseveratively gazing and reaching to location A. Thus, in the face of strong bias to location A, the model's

[2] The model does not reach to A because the reach output layer is not allowed to update until the choice segment.

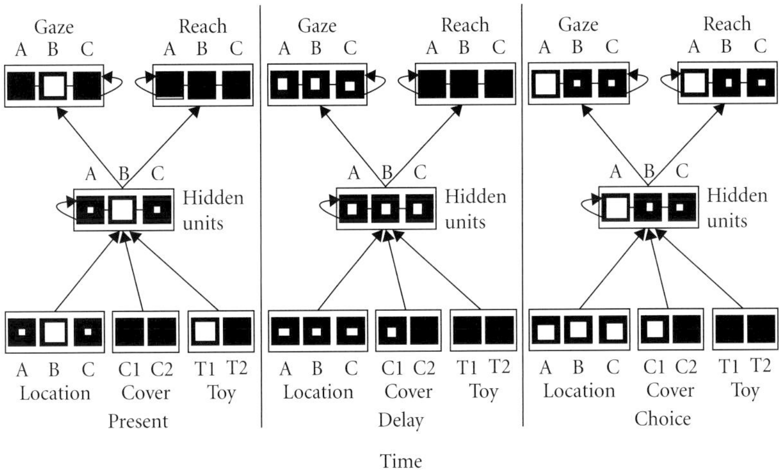


**Figure 7.5. The A-not-B model: Patterns of network activation for B-trial when recurrence is low**
This figure depicts patterns of network activation during the Present, Delay, and Choice segments of a B-trial; recurrence is low in this figure. During the Present segment, the network represents B. However, because recurrence is low, this representation fades during the Delay segment. By the Choice segment, the model more strongly represents A and gazes and reaches perseveratively to location A.

weak ability to actively maintain information about recent events at location B is inadequate for representing B and gazing and reaching to location B; the model perseverates by gazing and reaching to location A.

*Effects of Age* Performance in the A-not-B task is related to age, with older and younger infants (e.g., 12- and 10-month-old infants) performing comparably on A-trials but older infants performing better than younger infants on B-trials given a particular delay (Marcovitch & Zelazo, 1999; Wellman et al., 1986). The model's performance on B-trials improves incrementally with age, simulated as an increase in the strength of recurrent weights. This is illustrated in Figure 7.6, which shows an older model performing a B-trial. The critical difference in performance begins to emerge in the delay segment of the trial when the network is receiving equal input to the location A, B, and C units. In previous trials when recurrent weights were low, the model's internal representation of B began to fade; now the older model's internal representation of B is strongly maintained, despite a strong bias to represent A. The model's strong active representation of B carries through to the choice segment, and therefore, even in the face of a strong bias to location A, the model correctly represents B and gazes and reaches to location B.

*Dissociation* Although infants' memory for hidden toys seems poor when tested by means of reaching—most infants perseverate by reaching to location A when a toy is hidden at location B—their memory seems considerably better when tested by means of gazing. For example, infants that reach perseveratively to location A in the A-not-B task nevertheless look longer when a toy hidden at location B is revealed at location A than when it is revealed at location B (Ahmed & Ruffman, 1998). Similarly, infants are more likely to gaze correctly than reach correctly to location B on B-trials (Hofstadter & Reznick, 1996). Thus, infants exhibit dissociations in their behavior, appearing more knowledgeable when tested one way than another. The model exhibits the same dissociation in its performance by showing strong perseveration in its reaching system but more accurate performance in its gaze system. Evidence of this dissociation can be observed in Figure 7.3 which shows the gazing and reaching performance of the model during A-trials and B-trials at various levels of recurrence. A moderately old model with recurrence of 0.4, for example, reaches correctly on

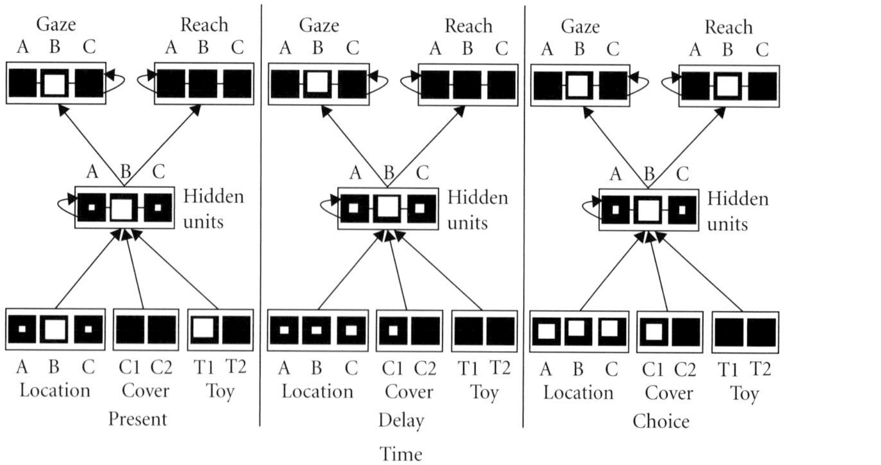


**Figure 7.6. The A-not-B model: Patterns of network activation for B-trial when recurrence is high**
This figure depicts patterns of network activation during the Present, Delay, and Choice segments of a B-trial; recurrence is high in this figure. The model accurately represents B, and gazes and reaches to location B, despite a strong latent bias to represent A and gaze and reach to location A.

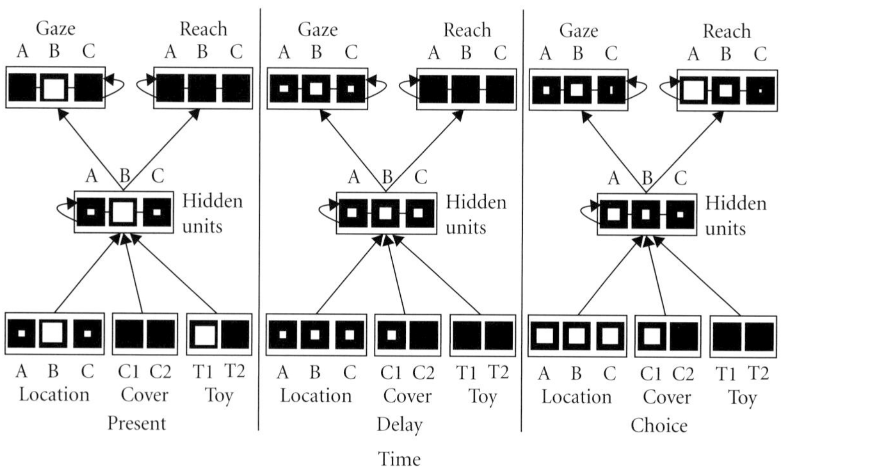


**Figure 7.7. The A-not-B model: Patterns of network activation for B-trial when recurrence is moderate**
This figure depicts patterns of network activation during the Present, Delay, and Choice segments of a B-trial; recurrence is moderate in this figure. A weak representation of B at the Choice segment is adequate to drive correct gaze but inadequate to drive correct reach.

approximately 60% of B-trials but gazes correctly on over 80% of B-trials. Why does the model show this dissociation?

The discrepancy arises because the underlying representation of B is adequately strong to drive correct gazing but not correct reaching. This effect is illustrated in Figure 7.7, which shows activity in a moderately old model (recurrence = 0.4) during the presentation, delay, and choice segments of a B-trial. In the presentation segment, a target toy (T1) is presented at location B, and the model represents B and gazes to location B. During the delay segment, when location A, B, and C input units are equally active, recurrent connections are moderately strong allowing the model to actively maintain a representation of B and correctly gaze to location B, but are not sufficiently strong to completely override the model's strong latent representation of A. Thus, the internal representation is gradually pulled back toward A. When the choice segment begins and the reach system is able to update, A is now more strongly represented than B due to the influence of the model's latent bias to A, and the model reaches incorrectly to location A. This weak representation of B is, however, sufficient to drive a correct gaze to location B. This is because the gazing system has updated throughout the delay segment and allowed activation of the B gaze unit. Therefore, even though inputs to the gaze system in the choice segment favor A, the gaze system responds correctly. This parallels what is observed with infants, who may show earlier success in gaze/expectation variants of the A-not-B task because they can constantly update their gazing and their expectations about B during the presentation, delay, and choice segments of B-trials which would allow them to more effectively counter perseverative tendencies in their gaze responses. In contrast, infants can only reach at the choice point, by which time their representations have become more susceptible to perseverative biases. Therefore, in the model, even though inputs to the gaze system in the choice segment favor A, the gaze system responds correctly. Dissociations of this kind illustrate a central idea of graded representation accounts, namely that a single shared representation (in this case a weak representation of B) can be sufficient to drive correct responses in one system (i.e., the gaze system) and yet insufficient to drive correct responses in a different system (i.e., the reaching system).

### Detailed Aspects of Infant Performance

In addition to capturing general aspects of performance like perseveration, age-related change, and dissociation, the model can also account for a number of detailed aspects of infant performance in the A-not-B task. These include the effects of the number of hiding locations, the distinctiveness of the covers, the use of covers alone versus covers and toys, and merely observing toys being hidden at location A rather than reaching for them. The following discussion focuses on two of these effects, the effects of covers alone and observation alone.

One finding that seems to challenge the idea that perseveration in the A-not-B task relates to memory of hidden objects is that infants perseverate in a covers-only version of the task that has no hidden objects. In this cover-cover version, infants watch and then reach for a cover that is lifted and waved at location A and then perseverate by reaching back to location A when a cover is subsequently lifted and waved at location B (Munakata, 1997; Smith et al., 1999). These findings suggest that motor history rather than memory for hidden objects is critical for eliciting perseverative search (Smith et al., 1999).

However, although the covers remain visible throughout cover-only versions of the task, memory is still required to determine which of the two covers was waved. Moreover, hidden toys do in fact influence infant performance in variants of the covers-only paradigm. For example, infants perseverate less in a cover-toy version of the task (in which they reach for a cover waved at location A and then watch a toy hidden and covered at location B) than they do in the cover-cover version (Munakata, 1997).

The model captures all of these effects in terms of levels of input activity (i.e., higher input activity for toy and cover than for cover alone) and resulting changes in latent representations. As seen in Figure 7.8, which shows performance on two types of A-trials (cover-only [*cover*] and cover + toy [*toy*]) and the four types of B-trials (i.e., two types of B-trials that each follow two types of A-trials). The model performs accurately on both the *cover* and *toy* A-trials at all levels of recurrence. The network has a bias to respond to A based on weight changes that occur in pretraining and A-trials. Consequently, differences in the level of input activity for toy and cover events and the ability to maintain information have little effect on performance. The model perseverates strongly and comparably on B-trials of *toy-toy* and *cover-cover* versions, as strong activation of B in the hidden layer competes against a strong latent representation of A in the former; weak activation of B competes against a weak latent representation of A in the latter. The model performs better in the *cover-toy* version because in this condition, strong activation of B competes against a weak latent representation of A. Thus, for these three conditions, the model's B-trial performance reflects the relative strength of active and latent representations.

Somewhat counterintuitively, performance on B-trials in the *toy-cover* version is comparable instead of worse than performance in the *toy-toy* and *cover-cover* versions; this is true for both the network (Munakata, 1998) and infants (Munakata, 1997). The reason is that A-trials and B-trials are more similar to one another in *toy-toy* and *cover-cover* versions than in the *toy-cover* version. As a result, B-trials *remind* the network of A-trials more in *toy-toy* and *cover-cover* versions than in the *toy-cover* version. For example, on B-trials, the toy in the *toy-toy* version provides strong activation to B *and* A, since the toy is presented at B but has also been previously associated with A. In effect, models and infants are more strongly reminded of A in the *toy-toy* version than in the *toy-cover* version, leading to comparable levels of preservation in both versions.

Further evidence for the importance of factors other than motor history in the A-not-B task is that infants perseverate after merely observing events at location A (Butterworth, 1974; Diamond, 1983; Evans, 1973; Spencer & Schutte, 2004). The model showed the same behavior, due to latent biases to attend to A that were formed after merely observing events at location A (Munakata, 1998). To simulate the absence of reaching during A-trials, inputs corresponding to regular A-trials were administered to a standard network, but the learning rate for connections between hidden and reaching units were set to 0. Consequently, these connection weights did not change across A-trials, which effectively simulated an absence of experience reaching to location A. The model accurately simulated infant performance by perseverating strongly on B-trials. The model perseverated in this way because although it had never reached to location A, it nevertheless learned to attend to A. Repeated experience watching a toy hidden at location A strengthened input-to-hidden unit weights, causing the model to preferentially represent A on its hidden units. Although hidden-to-reaching unit weights did not change, preferential attention to A was sufficient to drive perseveration on B-trials. The model therefore provides a comprehensive account of specific effects involving

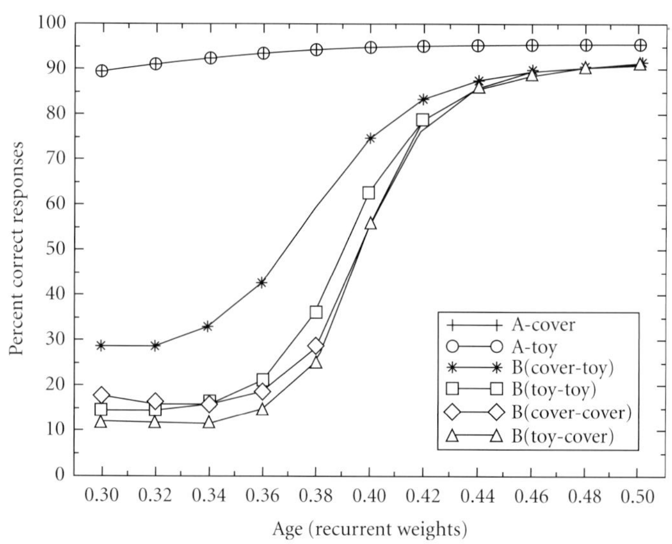


**Figure 7.8. Modified A-not-B task: Network performance by simulated age and toy presence** On A-trials, the networks are accurate with both *cover* and *toy* trials across all levels of recurrence shown, because any activity for A is supported by latent weight changes that favor A over B. On B-trials, the networks perseverate on all task versions at some point in development, but show non-perseverative responding earliest on the A-cover/B-toy (*cover-toy*) version.

various combinations of hidden toys and reaching experience, and clarifies how motor history is just one of many factors influencing infant perseveration in the A-not-B task.

### Predictions

The model generates a number of novel and testable predictions about infant performance in the A-not-B task. One counterintuitive prediction of particular relevance for current purposes is that infants should show a developmental U-shaped curve in A-not-B performance. This is illustrated in Figure 7.9, which shows B-trial performance at various levels of delay and recurrence. For delays greater than 1, as recurrence strength increases from the lowest to more moderate values, B-trial performance actually gets worse before getting better. This pattern of performance is related to the fact that during A-trials, the model's latent representation of A gets stronger partly as a function of how active A hidden units are, which is in turn a function of recurrence strength. The parallel in infants would be that very young infants do not think very much at all about A after the toy is hidden; they therefore build up less of a bias in that direction compared with older infants, who continue to maintain information about A across the delay. Thus, the model generates the novel prediction that young infants would actually do better on B-trials than somewhat older infants since the younger infants do not maintain information about A across the delay segment. This prediction has been supported using a *lids-only* variant of the A-not-B task administered to 5-, 6-, 7-, and 8-month-old infants (Clearfield, Diedrich, Smith, & Thelen, 2006). Consistent with the predictions of the model, B-trial performance actually declines between 5-months and 8-months; B-trial performance improves thereafter.

### Performance in Modified A-not-B Tasks: Perseveration in the Absence of Apparent Memory Demands

One criticism of the idea that perseveration is linked to underdeveloped active memory mechanisms is that perseveration can occur in instances in which all of the information required to solve a problem is immediately available. After reorganizing the office, for example, it is not uncommon to reach for the telephone in its old location even though its new location is in clear view. Young infants show a similar form of perseveration in a simple task depicted in Figure 7.10. An attractive toy is placed on the A-towel, while a similar toy is placed *behind* the B-towel. Although initially infants successfully retrieve the toy by pulling the A-towel, when the position of the toys is reversed, infants

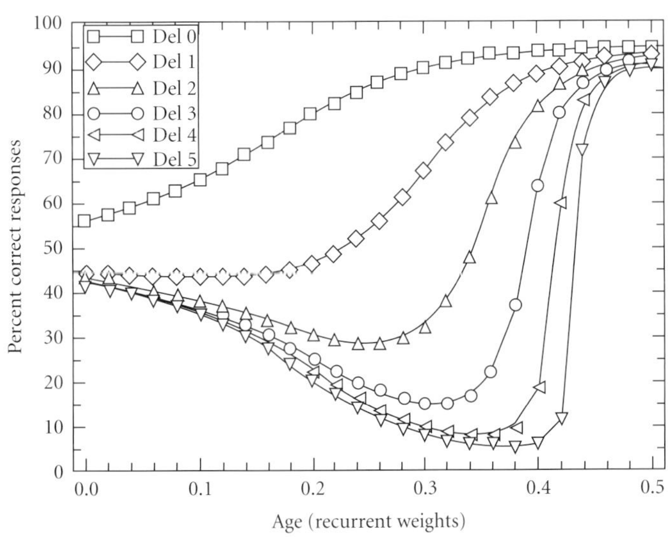


**Figure 7.9. The A-not-B task: Network performance as a function of simulated age and delay**
This figure presents the percentage of (correct) responses by the networks to B on B-trials as a function of age and delay (0–5 delay units). Data show a U-shaped developmental curve, such that when delay > 1, B-trial accuracy initially declines with increasing recurrence before increasing at higher levels of recurrence.

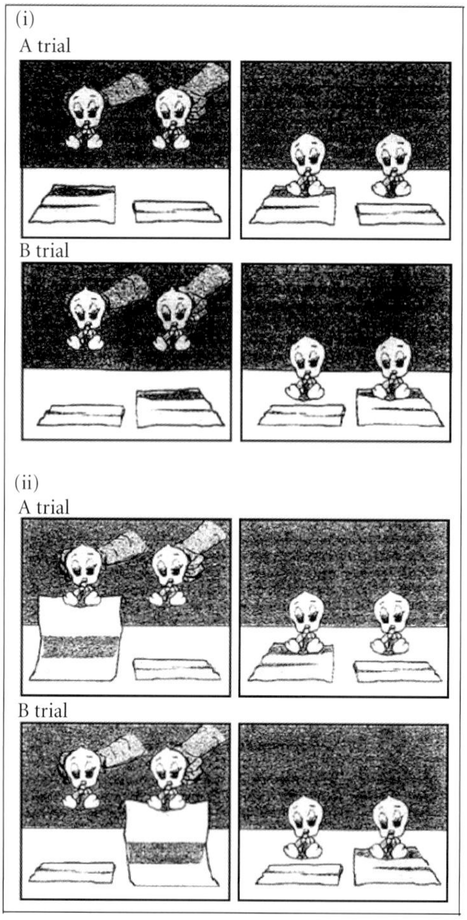


**Figure 7.10. Modified A-not-B task: Cloth-pulling version, toy-unattached or toy-attached** This figure pictures A- and B-trials in toy-unattached (i) and toy-attached (ii) adaptations of the cloth-pulling version of the A-not-B task (Aguiar & Baillargeon, 2000). On A-trials, infants pull the towel with the toy on it. On B-trials, infants reach perseveratively to the prior location, although the correct towel is specified by the visible environment.

perseverate by continuing to pull the A-towel (Aguiar & Baillargeon, 2000). Strikingly, infants show very different performance across similar versions of the task. Infants at 9 months of age, for example, succeed in a *toy-attached* version in which the toy is attached to one of the towels. However, not until 11 months of age do infants succeed in a very similar *toy-unattached* version in which the toy is placed on but is not attached to one of the towels (Aguiar & Baillargeon, 2000). Why do infants perseverate when the solution to the problem is in full view? And why does infant performance differ across very similar versions of the same task?

Although these findings appear to challenge the idea that infant perseveration is linked to active memory, it is possible that active maintenance mechanisms support not only memory but also attention to task-relevant information in the environment (Chafee & Goldman-Rakic, 2000; De Fockert, Rees, Frith, & Lavie, 2001; Desimone & Duncan, 1995; Fuster, Bauer, & Jervey, 1985). On this account, selection of information from a complex visual scene (i.e., selective attention) is an emergent process involving competition among objects for representation and analysis in posterior cortical regions. Active maintenance mechanisms, mediated by the prefrontal cortex (PFC), support the selection of task-relevant information by biasing competition for representation in the posterior cortex (Desimone & Duncan, 1995; Miller & Cohen, 2001). When active maintenance is robust, competition for representation can be efficiently biased in favor of task-relevant information. However, when active maintenance is weak, this competition favors previously relevant information.

This hypothesis was tested using a modified version of the A-not-B model in which PFC-mediated active maintenance mechanisms and competition for representation were explicitly implemented (Stedron et al., 2005). The model included distinct PFC units whose function was to maintain information about the current location of the toy and bias competition for representation in the hidden layer. As well, The input layer was modified so that inputs included information about whether toys were placed on cloths or were attached to cloths (Fig. 7.10).

As shown in Figure 7.11, the model simulated general aspects of infant performance, including accurate performance on A-trials, perseveration on B-trials, and age-related performance improvements. More detailed aspects of performance that are related to the earlier mastery of the toy-attached version of the task were

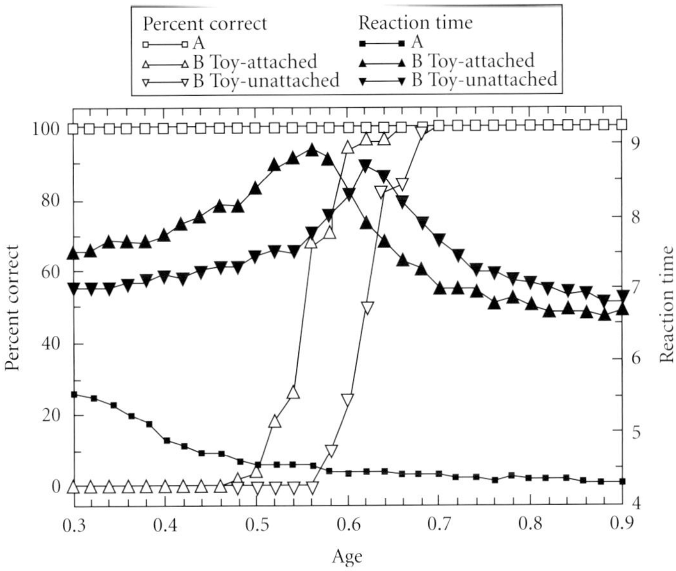


**Figure 7.11. Modified A-not-B task: Network performance as a function of simulated age and toy presentation**

This figure presents percent correct and reaction time results from networks at different ages (simulated by varying strength of recurrent connections supporting sustained activations) on A-trials (shown for just the *toy-attached* condition; similar results were obtained in the *toy-unattached* condition) and B-trials. Networks showed the developmental progression previously observed in infants. Changes in recurrent connections alone allowed networks to progress from perseverating to succeeding on B-trials, first in the toy-attached and then in the toy-unattached condition. Thus, performance improved through the same changes that increased working memory abilities in prior models, despite the lack of apparent memory demands in these tasks. Reaction times were slowest around the transitions from perseverating to succeeding.

evidenced, as well. Both general and detailed aspects of performance are discussed below.

### General Aspects of Infant Performance

*Accurate A-Trial Performance across Ages* Like infants, the model performs accurately on A-trials across all ages (i.e., all levels of recurrence) because of the greater input activation associated with the correct side. This greater activation to the A side propagates through feed-forward connections and allows A units in the hidden, PFC, and output layers to win the competition for representation over the B units. Successful reaches to A subsequently strengthen these feed-forward connections, thereby establishing a latent bias to attend and reach to A.

*Perseveration* On B-trials, at low levels of recurrence, the model perseverates strongly, inappropriately pulling the A towel. On B-trials, greater input activation is again associated with the correct side, but this activation has to compete against a latent bias to attend and respond to A. When recurrence strength is low, the representation of the B location in PFC units is insufficiently strong to override this bias, and the model perseverates by reaching to A, even though all the information required to make the correct response is available to the network on the input layers.

*Age-Related Performance Improvements* Hypothesized age-related changes in active maintenance mechanisms were simulated by increasing the strength of recurrent connections PFC units made to themselves. Stronger recurrent connections amplified the representation of B in the PFC layer, leading to greater top-down support for the B unit in the hidden layer and improved B-trial performance. Thus, these models offer an existence proof that active maintenance can provide a common mechanism for improving memory for hidden objects as well as attention to relevant information in the environment.

### Detailed Aspects of Infant Performance

Like infants (Aguiar & Baillargeon, 2000), the model showed earlier mastery of the toy-attached than toy-unattached version of the task (see Fig. 7.11). The model showed this specific effect because the difference in input activation between the correct and incorrect sides was greater in the toy-attached version (4 units of input activation for the correct side versus 2 units for the incorrect side) than in the toy-unattached version (3 units versus 2 units). Consequently, the correct location *stands out* more in the toy-attached than in the toy-unattached version, and thus is mastered earlier (i.e., at lower recurrence). The simulation results suggest that the toy-attached version is easier because additional features of the task make the correct location more salient to infants and help support attention to that location despite underdeveloped active maintenance mechanisms. When fewer features highlight the correct location, as is the case in the toy-unattached version, infants have to rely more on active maintenance mechanisms to overcome bias and attend to the relevant information in the environment.

### Predictions

The model leads to the novel prediction that infants should show a developmental U-shaped curve in their response times on B-trials (Fig. 7.11). In particular, the model predicts that young infants who perseverate strongly to A and older infants who reach correctly to B should show shorter response times than infants of an intermediate age. These predictions, currently untested, are based on an analysis of networks' response time at various levels of recurrence, where response time is measured as the number of processing cycles a model requires to settle on a particular trial. In very young and very old models, response times are very short because in both cases there is a large imbalance between the PFC unit's representation of B (i.e., very weak in young networks and very strong in old networks) and the latent bias to A. By contrast, models of intermediate age show substantially longer response times. This behavior occurs because the active representation of B and the latent representation of A are more equal in strength, making it difficult for the model to settle on a particular response; such responses are characteristic of developing systems that are in transition (e.g., Van der Maas & Molenaar, 1992; Siegler, 1995).

## Case Studies: Connectionist Models of Perseveration in Children's Rule-Guided Behavior

The general principles of graded, active, and latent representations provide a framework for understanding general and specific aspects of perseveration in children's rule-guided behavior. In card-sorting tasks, for example children are presented cards that can be sorted either by shape or color (e.g., red trucks) and sort them using one set of rules (e.g., based on shape) before being asked to switch and use new rules ( i.e., based on color). Similarly, in a speech interpretation task (SIT; Morton et al., 2003), children are presented with utterances containing conflicting emotion cues (e.g., "I won a prize" spoken sadly) and are asked to judge the speakers feelings from what she says and then switch and judge her feelings from her tone of voice. In both tasks, children persist in using old rules when instructed to switch to the new rules even though they accurately answer questions about the new rules. Although both tasks involve flexible rule-use, whose development is thought to be related to age-related changes in PFC functioning [Kirkham et al., 2002]), they are mastered at very different ages (Morton et al., 2003; Zelazo et al., 1996). We briefly describe a model that captures general aspects of children's card-sort performance and specific differences between card-sort and SIT performance, and which leads to novel and testable predictions.

### Architecture and Task Administration

The card-sort model, shown in Figure 7.12, features separate pathways, that connect visual features of the test cards (i.e., color and shape) with potential responses. Additional input layers (i.e., verbal features and rules) allow the model to encode sorting rules, descriptions of the cards, and verbal queries. As in previous models, latent representations take the form of changes in feed-forward connection weights, whereas active representations take the form of self-sustaining activity in PFC units. This activity is made possible by recurrent connections and supports rule switching by biasing competition among hidden units. The task is administered as patterns of input delivered to the input units and includes preswitch trials in which the model sorts cards by one rule (e.g., color; see Fig. 7.12) followed by postswitch trials in which the sorting rule changes. The model is then asked questions about the postswitch rules.

### Perseveration and Age-Related Change

Figure 7.13 depicts the performance of the model across preswitch and postswitch performance as a function of age (i.e., recurrence). At all ages, models sort accurately in preswitch trials and establish a strong latent representation of the preswitch rule, but young models perseverate in postswitch trials. When recurrence strength is low, PFC units only maintain a weak representation of the postswitch rule and can only minimally bias competition among units in the internal representation layer. As a result, they fail to override a strong latent bias to sort by the preswitch rule. Older models with stronger recurrent connections robustly maintain an active representation of the postswitch rule

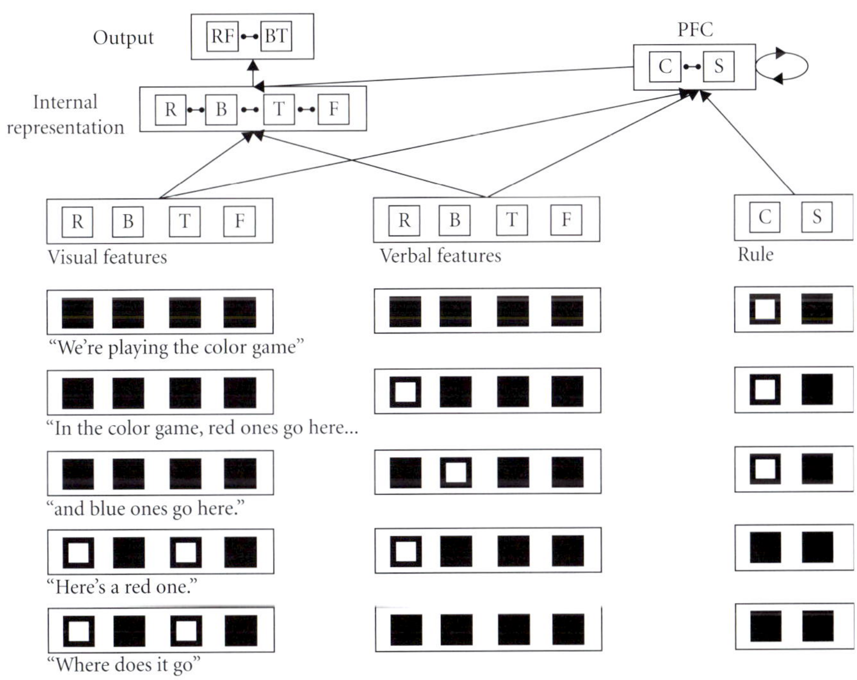


**Figure 7.12. Card-sort model: Simplified version of the network**
This figure depicts a simplified version of the card-sort network; five inputs for a trial are shown in the five rows below. Each trial included a presentation of the rule (first row), a description of the rule (second and third rows), a description of a test card (fourth row), and a presentation of a test card (fifth row). The description of the rules also included activation of the relevant output units (not shown). (R = red, B = blue, T = truck, F = flower; C = color, S = shape; PFC = prefrontal cortex units)

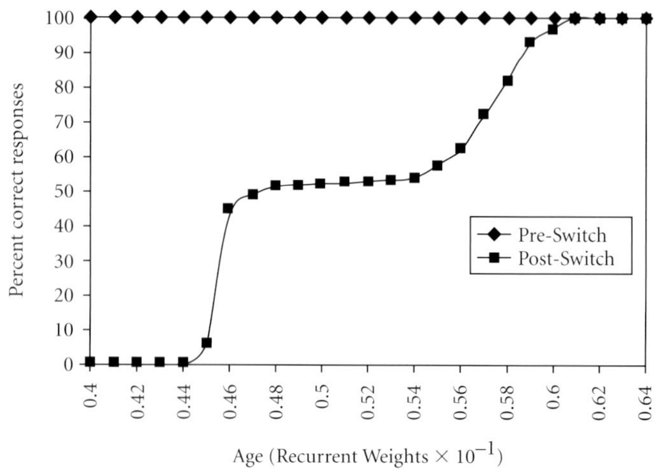


**Figure 7.13. Card-sort model: Network performance as a function of simulated age**
This figure depicts percent correct responses by the networks in the card-sort task as a function of age (as simulated by varying recurrent weights). The networks performed perfectly on preswitch trials but perseverated on postswitch trials when recurrence was low.

and more effectively override latent biases. As a result, older models are more likely to switch appropriately in postswitch trials. The model therefore simulates critical aspects of children's card-sorting performance, including the accurate preswitch performance across age and the incremental improvement in postswitch performance with increasing age.

## Dissociations

Like 3-year-old children, young models also show apparent knowledge–action dissociations by perseverating strongly on postswitch trials while responding correctly to simple questions about the postswitch rules (Zelazo et al., 1996). This apparent dissociation can be understood in terms of differences in the amount of conflict contained in knowledge and action measures (Morton & Munakata, 2002a, 2002b; Munakata & Yerys, 2001). Standard knowledge questions pertain to only a single feature of the cards (e.g., "When sorting by shape, where do trucks go?") and therefore contain no conflict. Thus, young networks that form weak active representations of the sorting rules can nevertheless answer these questions correctly. By contrast, the sorting task (i.e., the action measure) presents conflicting shape and color information (i.e., cards that would be sorted one way if sorted by shape but a different way if sorted by color). Thus, only older networks that can form strong active representations of the new shape rule can resolve the conflict between shape and color in postswitch trials. Thus, knowledge–action dissociations in the card-sort task may be more apparent than real, in that they reflect differences in the amount of conflict contained in measures of knowledge and action.

Consistent with this account, apparent dissociations virtually disappear when conflicts across measures of knowledge and action are controlled (Morton & Munakata, 2002a; Munakata & Yerys, 2001). In one study, 3-year-old children were presented with the standard card-sort task and were asked questions about the postswitch rules. Simple questions contained no conflict (e.g., "In the shape game, where do the trucks go?"); conflict questions contained conflicting shape and color information (e.g., "In the shape game, where do red trucks go?"). Children's knowledge of the postswitch rules appeared to outstrip their sorting performance when knowledge was assessed by simple questions—there was evidence of dissociation between knowledge and sorting/action. This apparent dissociation was not evidenced when knowledge was assessed by means of the conflict questions (Munakata & Yerys, 2001). Processes underlying apparent dissociations may be general across the lifespan (Morton & Munakata, 2008, in preparation). Adults show an apparent dissociation in their reaction times on card-sorting tasks by responding relatively slowly when switching under conditions of conflict but responding relatively quickly when answering nonconflict questions about the new rule. However, consistent with the graded representations account, these apparent dissociations disappeared when knowledge and action measures contained the same amount of conflict.

### Detailed Aspects of Performance

Although the card-sort and the SIT both involve switching, children master the card-sorting task long before the SIT. One potentially important difference is that children exhibit a strong prepotent bias to judge a speaker's feelings from what she says rather than how she speaks (Friend, 2000; Morton & Trehub, 2001) but no strong prepotent bias to sort by color or shape (Zelazo et al., 1996). Consequently, the degree of conflict operative in postswitch SIT trials is greater than in the card-sort task; thus, postswitch performance in SIT trials may demand a more robust active representation of postswitch rules. Consistent with this hypothesis, a card-sort and a SIT model that were identical except for the definitions of the constituent units and the presence/absence of a preexisting latent bias for one sorting dimension showed the same general performance patterns (i.e., perseveration, age-related change, dissociation), *except* that the SIT model required stronger recurrence to achieve post-switch performance comparable to that of the card-sort model (Morton & Munakata, 2002). Thus, the active–latent distinction provides a parsimonious framework for understanding differences in when children master formally similar tasks.

### Predictions

The model makes the novel and counterintuitive prediction that guided practice may be more effective in eliciting flexible behavior in preschoolers than direct instruction. The prediction stems from the fact that young models have difficulty actively maintaining a strong representation of a new sorting rule (e.g., color) due to weak levels of recurrence in PFC units. Consequently, instructing young networks to use a new sorting rule (in this case, by activating the color rule input unit) is relatively ineffective in eliciting a switch in sorting criteria. By contrast, guided practice in the use of the new rule, implemented as a series of trials sorting cards that contained only color, was highly effective in eliciting a switch to color. In the absence of competition from the previously relevant shape features, the models correctly sorts by color during the guided practice trials, which increases the strength of the models latent representation of color. When shape features were then reintroduced, the models continued to sort by color. This prediction has been confirmed (Brace et al., 2006). In a modified version of the card-sort task, 3-year-old children sorted cards by one rule and then received either guided practice in the use of a new rule (like that administered to the networks) or instructions to switch to a new rule. Consistent with the predictions of the model, significantly more children switched to the new rule given guided practice alone than given instructions alone.

## Connectionism and Dynamic Systems Theory: Discussion of Crosscutting Conceptual Issues

The connectionist case studies described in this chapter provide a unified framework for understanding perseveration and flexibility—general aspects observed across ages, task paradigms, and response modalities as well as important details specific to particular tasks. In this section, we consider the implications of these case studies for broader issues about core concepts in cognition, claims about what develops, timescales of change, levels of explanation, and data constraints. We indicate where answers from these case studies may not be representative of connectionist approaches in a general sense and draw comparisons and contrasts with dynamic systems approaches throughout (see our related discussions in Munakata, 2006; Munakata & McClelland, 2003; Munakata, Sahni, & Yervys, 2001).

### Core Concepts

Our connectionist case studies capture many assumptions about core concepts of cognition that are representative of connectionist approaches more generally and highly overlapping with dynamic systems approaches. However, the focus of these case studies on the distinction between active and latent representations may be unique.

Common aspects of connectionist and dynamic systems approaches include a dynamic, highly interactive view of behavior as reflecting a confluence of multiple, graded factors that interact in nonlinear ways (Elman et al., 1996; Kelso, 1995; McClelland, Rumelhart, & PDP Research Group, 1986; Smith & Thelen, 2003). One example from our A-not-B case study (Munakata, 1998) concerned the effects of changing the strength of recurrent connections. Very young models (with very low levels of recurrence) could actively maintain information about a recent hiding location for only a brief period of time, but they actually performed better on B-trials than slightly older models. As discussed earlier, this counterintuitive effect occurred because recurrence strength affected not only how well the networks could override latent biases on B-trials, but also how strongly latent biases to A were established during A-trials. Thus, changes in recurrence could not be understood in isolation at any given point, but rather, needed to be considered in terms of their influence on the dynamics of the entire model across time.

Multicausality and interactivity are also at the heart of Dynamic systems accounts, such as a dynamic field model of infant A-not-B performance(Spencer & Schöner, 2003; Thelen, Schöner, Scheier, & Smith, 2001). Central to such an approach is the concept of stability, which refers to a model's ability to maintain a pattern of activation corresponding to, for example, a previously viewed object. The ability to enter into self-sustaining, stable states has important implications for A-not-B performance because it allows models to establish *representational states* (Spencer & Schöner, 2003) and carry forward activation related to prior events that are no longer perceptually available so that these events can influence subsequent processing. This capacity is not determined by one factor but arises through the confluence of multiple factors, including the nature of inputs and the strength of excitatory and inhibitory intra-field interactions.

Another parallel in connectionist and dynamic systems approaches to cognition is that no clear distinction is made between processing and knowledge in these approaches. That is, knowledge is not treated as separable from processes that operate on knowledge; instead, knowledge and processing are intrinsically related. One example from the A-not-B and card-sorting case studies (Morton & Munakata, 2002a; Munakata, 1998) concerned behavioral dissociations, where infants look to the correct hiding location for a toy but reach perseveratively, or children answer questions correctly about a new card-sorting rule but sort perseveratively. Standard accounts explained these dissociations in terms of a disconnect between children's knowledge (e.g., knowledge is viewed as intact) and their processing (e.g., inhibitory processing is viewed as underdeveloped) (Ahmed & Ruffman, 1997; Diamond, 1985; Hofstadter & Reznick, 1996). In contrast, the connectionist case study models demonstrated how such dissociations could arise through highly interactive systems, as a result of the strength of representations formed during processing. Similarly, dynamic systems accounts do not draw a strong distinction between knowledge and processing. For example, dynamic systems accounts emphasize that the A-not-B error is not about what infants have or don't have as enduring concepts or traits, but about what infants do and what they have done (Thelen et al., 2001). Thus, knowledge emerges through the course of infant processing, rather than taking the form of reified entities in the infant's head.

Our case studies point to one important potential difference between connectionist and dynamic systems approaches to cognition: the focus on distinct types of representations. The distinction between active and latent representations is central to connectionist accounts of perseveration and flexibility. This distinction is motivated by behavioral, physiological, and computational data (reviewed in Ashby & Todd, 2005; Bunge, Kahn, Wallis, Miller, & Wagner, 2003; Casey, Durston, & Fossella, 2001; Frank, 2005; Jog, Kubota, Connolly, & Graybiel, 1999; Kharitonova, Chien, Colunga,

& Munakata, 2007, 2008—in preparation; Miller & Desimone, 1994; Miller, Erickson, & Desimone, 1996; Patalano, Smith, Jonides, & Koepee, 2001; Rougier, Noelle, Braver, Cohen, & O'Reilly, 2005; Wallis, Anderson, & Miller, 2001). The active–latent distinction has also led to novel, counterintuitive predictions that have since been confirmed. In contrast, dynamic systems approaches have not emphasized such representational distinctions, focusing instead on the integration of various factors such as environmental stimuli and memories for prior events within a dynamic field (Spencer & Schöner, 2003; Thelen et al., 2001).

### Development: What changes?

Both connectionist and dynamic systems approaches have emphasized the importance of multiple interacting factors and nonlinear, emergent effects in development (e.g., Elman et al., 1996; Smith & Thelen, 2003). For example, connectionist models have been used to investigate how small changes in the starting state of a system (e.g., in the firing rates of processing units) could interact with other processes to lead to specializations of different brain regions (O'Reilly & Johnson, 2002) or to developmental disorders that appear to be highly modular (Joanisse & Seidenberg, 2003; Thomas & Karmiloff-Smith, 2003). Similarly, dynamic systems models have been used to investigate how small changes in the visual environment could lead infants to very different trajectories in perceptual state space in ways that might explain their patterns of looking in habituation studies (Smith & Thelen, 1993; Thelen & Smith, 1994).

There are some contrasting views in the two approaches. Connectionist approaches have typically emphasized the role of learning in developmental change by implementing neurobiologically plausible mechanisms or algorithms to guide such learning. Dynamic systems approaches have instead focused on the importance of control variables that change with age without explanations or mechanisms of implementation that lead to such changes. For example, all of our case studies focused on Hebbian learning mechanisms for explaining how habitual behaviors could build across development. These case studies also linked the reduction of perseveration across development with age-related advances in PFC-mediated active maintenance mechanisms. Although these advances were implemented by hand in these case studies (through increases to recurrent connections), other connectionist models have focused on understanding how such PFC-mediated, active maintenance abilities can develop through the course of learning (e.g., Munakata et al., 1997; Rougier et al., 2005).

In contrast, in dynamic systems models of the A-not-B task, development is simulated in terms of changes to parameters without explanation of the mechanisms driving such changes. For example, in one model development was simulated through changes to a parameter, $h$, the resting level of a dynamic field (Thelen et al., 2001). In another model, development was simulated through changes in a local excitation–lateral inhibition function (Spencer & Schöner, 2003), which determined the extent to which an activated site in the field increased activation in neighboring sites and suppressed activity in distant sites. Fields that were weakly interactive had difficulty establishing a self-sustaining state and consequently were dominated by current inputs. By contrast, fields that were strongly interactive could establish self-sustaining states or attractors and could, therefore, maintain a representation of prior input, even when the input was no longer delivered to the field. This concept of stability is very similar to active maintenance in our case studies, but the plausibility of changes in the resting level or interactivity of a dynamic field as a mechanism of developmental change is less clear.

### Timescales of Change and Levels of Explanation

Both connectionist and dynamic systems approaches have the potential to address change over very broad timescales, from milliseconds to months. For example, our case study A-not-B model led to predictions about U-shaped developmental curves on the order of months. These predictions have since been confirmed through studies motivated by a dynamic systems approach (Clearfield et al., 2006). Similarly, both our case study A-not-B model and dynamic systems approaches can explain effects on a much smaller timescale of change, for example, trial-by-trial effects of toys versus covers on A-not-B performance. Although dynamic systems discussions of the A-not-B task have focused on the precise dynamics of infant reaching (e.g., Diedrich, Thelen, Smith, & Corbetta, 2000) more than connectionist approaches, dynamic systems models have not always addressed such real-time reaching dynamics (Spencer & Schöner, 2003; Thelen et al., 2001). Moreover, connectionist models have been used to explore real-time dynamics at the millisecond level in other domains (e.g., Huber & O'Reilly, 2003; Usher & McClelland, 2001). Thus, we do not see inherent differences between connectionist and dynamic systems approaches in their potential to address change over multiple timescales in an integrated framework.

Similarly, both connectionist and dynamic systems approaches have the potential to address multiple levels of explanation. For example, as we have emphasized throughout this chapter, our case study connectionist models explain both very general aspects of perseveration observed across ages, paradigms, and response modalities, as well as more detailed aspects of perseveration that are specific to particular tasks. In addition, abstract connectionist models have been used to investigate a range of behaviors (e.g., Arbib, 2002; Elman et al., 1996; McClelland, Rumelhart, & PDP Research Group, 1986), while more detailed connectionist models have been used to investigate neurobiological mechanisms (e.g., Arbib, 2002; O'Reilly & Munakata, 2000). Again, although dynamic systems discussions of the A-not-B task have focused on the importance of embodied cognition and the details of perceptual-motor systems (e.g., Diedrich et al., 2000; Thelen et al., 2001), dynamic systems models have not always incorporated details at such levels, for example, details regarding the kinematics of reaching (Spencer & Schöner, 2003; Thelen et al., 2001). However, connectionist models have been used to investigate the detailed kinematics of learning to reach (e.g., Berthier, Rosenstein, & Barto, 2005). Thus, again, we do not see inherent differences between connectionist and dynamic systems approaches in their potential to address multiple levels of explanation, each of which is important. Higher level explanations and general principles may be critical for understanding general aspects of perseveration, such as the fact that perseveration is observed across response modalities and even after infants have merely observed hidings at location A in the A-not-B task (Butterworth, 1974; Diamond, 1983; Evans, 1973; Spencer & Schutte, 2004). More detailed models of visually guided reaching may be needed to explain effects of body orientation and arm weights on infant performance on the A-not-B task (Smith & Thelen, 2003).

### Constraints from Empirical Data

Both neural network and dynamic systems models are constrained by various types of empirical data: behavioral, neurobiological, qualitative, quantitative, and so on. Connectionist models may have a relative strength in incorporating constraints from neurobiology. Dynamic field theory models establish stabilities or attractor states through local excitatory and lateral inhibitory intra-field connections, a well-known brain-like interactive mechanism. However, connectionist models also capture these more abstract aspects of neurobiological processing, while providing a more transparent mapping between the computational elements of units and weights and the corresponding biological elements of neurons and synapses—between network architectures and learning algorithms and the corresponding neural systems and processes (e.g., Frank, 2005; O'Reilly & Rudy, 2001).

## Conclusion

As many others have argued (e.g., Elman, 2003; Spencer & Schöner, 2003; Thelen & Bates, 2003), connectionist and dynamic systems approaches to development show both important similarities and important differences. We believe that the similarities highlight essential characteristics of development, such as nonlinear, emergent effects that arise in highly interactive systems in which knowledge and processing are intertwined. These characteristics are ignored or even challenged by alternative approaches to development, such as some modular or nativist characterizations (e.g., Baron-Cohen, 1998; Fodor, 1983; Leslie, 1992). We believe that many of the differences between connectionist and dynamic systems approaches (for example, in timescales of change and levels of explanation) reflect the foci of particular models/modelers rather than inherent differences between these frameworks. Each approach alone may serve to provide a unified framework for understanding behavior and development, as demonstrated by the case studies of connectionist models of perseveration discussed here. However, the most complete accounts and understanding may emerge through a synergy of the complementary strengths of connectionist and dynamic systems approaches.

## Acknowledgments

Preparation for this chapter was supported by a grant from the National Science and Engineering Research Council to J. Bruce Morton and a grant from the National Institute of Child Health and Human Development to Yuko Munakata.

## References

Aguiar, A., & Baillargeon, R. (2000). Perseveration and problem solving in infancy. In H. W. Reese (Ed.), *Advances in Child Development and Behavior* (Vol. 27, pp. 135–180). New York: Academic Press.

Ahmed, A., & Ruffman, T. (1998). Why do infants make A not B errors in a search task, yet show memory for the location of hidden objects in a nonsearch task? *Developmental Psychology, 34*(3), 441–453.

Arbib, M. A. (Ed.). (2002). *The handbook of brain theory and neural networks* (2nd ed.). Cambridge, MA: MIT Press–Bradford Books.

Ashby, F. G., & Todd, M. W. (2005). Human category learning. *Annual Review of Psychology, 56*, 149–178.

Baron-Cohen, S. (1998). Modularity in developmental cognitive neuropsychology: Evidence from autism and Gilles de la Tourette syndrome. In J. A. Burack, R. M. Hodapp, & E. Zigler (Eds.), *Handbook of mental retardation and development* (pp. 334–348). New York: Cambridge University Press.

Berthier, N. E., Rosenstein, M. T., & Barto, A. G. (2005). Approximate optimal control as a model for motor learning. *Psychological Review, 112*(2), 329–346.

Brace, J. J., Morton, J. B., & Munakata, Y. (2006). When actions speak louder than words: Improving children's flexibility in a card-sorting task. *Psychological Science, 17*(8), 665–669.

Bremner, J. G. (1978). Spatial errors made by infants: Inadequate spatial cues or evidence of egocentrism? *British Journal of Psychology, 69*(1), 77–84.

Brooks, P. J., Hanauer, J. B., Padowska, B., & Rosman, H. (2003). The role of selective attention in preschoolers' rule use in a novel dimensional card sort. *Cognitive Development, 18*(2), 195–215.

Bunge, S. A., Kahn, I., Wallis, J. D., Miller, E. K., & Wagner, A. D. (2003). Neural circuits subserving the retrieval and maintenance of abstract rules. *Journal of Neurophysiology, 90*(5), 3419–3428.

Butterworth, G. E. (1974). *The development of the object concept in human infants.* Unpublished doctoral dissertation, University of Oxford.

Casey, B. J., Durston S., & Fossella J. A. (2001). Evidence for a mechanistic model of cognitive control. *Clinical Neuroscience Research, 1*(4), 267–282.

Chafee, M. V., & Goldman-Rakic, P. S. (2000). Inactivation of parietal and prefrontal cortex reveals interdependence of neural activity during memory-guided saccades. *Journal of Neurophysiology, 83*(3), 1550–1566.

Clearfield, M. W., Diedrich, F. J., Smith, L. B., & Thelen, E. (2006). Young infants reach correctly in A-not-B task: On the development of stability and perseveration. *Infant Behavior and Development, 29*(3), 435–444.

Cohen, J. D., Dunbar, K., & McClelland, J. L. (1990). On the control of automatic processes: A parallel distributed processing account of the Stroop effect. *Psychological Review, 97*(3), 332–361.

Cummings, E. M., & Bjork, E. L. (1983a). Perseveration and search on a five-choice visible displacement hiding task. *Journal of Genetic Psychology, 142*, 283-291.

Cummings, E. M., & Bjork, E. L. (1983b). Search and behavior on multi-choice hiding tasks: Evidence for an objective conception of space in infancy. *International Journal of Behavioral Development, 6*(1), 71–87.

De Fockert, J. W., Rees, G., Frith, C. D., & Lavie, N. (2001). The role of working memory in visual selective attention. *Science, 291*(5509), 1803–1806.

Desimone, R., & Duncan, J. (1995). Neural mechanisms of selective visual attention. *Annual Review of Neuroscience, 18*, 193–222.

Diamond, A. (1983). Behavior changes between 6–12 months of age: What can they tell us about how the mind of the infant is changing? *Dissertation Abstracts International, 44*(01B), 337.

Diamond, A. (1985). Development of the ability to use recall to guide action, as indicated by infants' performance on AB. *Child Development, 56*(4), 868–883.

Diedrich, F. J., Thelen, E., Smith, L. B., & Corbetta, D. (2000). Motor memory is a factor in infant perseverative errors. *Developmental Science, 3*(4), 479–494.

Elman, J. L. (2003). Development: It's about time. *Developmental Science, 6*(4), 430–433.

Elman, J. L., Bates, E. A., Johnson, M. H., Karmiloff-Smith, A., Parisi, D., & Plunkett, K. (1996). *Rethinking innateness: A connectionist perspective on development*. Cambridge, MA: MIT Press–Bradford Books.

Evans, W. F. (1973). The stage-IV error in Piaget's theory of object concept development: An investigation of the role of activity. *Dissertation Abstracts International, 34*(11B), 5651.

Fodor, J. A. (1983). *The modularity of mind*. Cambridge, MA: MIT Press–Bradford Books.

Frank, M. J. (2005). Dynamic dopamine modulation in the basal ganglia: A neurocomputational account of cognitive deficits in medicated and nonmedicated Parkinsonism. *Journal of Cognitive Neuroscience, 17*(1), 51–72.

Friend, M. (2000). Developmental changes in sensitivity to vocal paralanguage. *Developmental Science, 3*(2), 148–162.

Fuster, J. M., Bauer, R. H., & Jervey, J. P. (1985). Functional interactions between inferotemporal and prefrontal cortex in a cognitive task. *Brain Research, 330*(2), 299–307.

Hofstadter, M. C., & Reznick, J. S. (1996). Response modality affects human infant delayed-response performance. *Child Development, 67*(2), 646–658.

Huber, D.E., & O'Reilly, R.C. (2003). Persistence and accommodation in short-term priming and other perceptual paradigms: Temporal segregation through synaptic depression. *Cognitive Science, 27*(3), 403–430.

Jacques, S., & Zelazo, P. D. (2001). The flexible item selection task (FIST): A measure of executive function in preschoolers. *Developmental Neuropsychology, 20*(3), 573–591.

Jacques, S., Zelazo, P. D., Kirkham, N. Z., & Semcesen, T. K. (1999). Rule selection versus rule execution in preschoolers: An error-detection approach. *Developmental Psychology, 35*(3), 770–780.

Joanisse, M. F., & Seidenberg, M. S. (2003). Phonology and syntax in specific language impairments: Evidence from a connectionist model. *Brain and Language, 86*(1), 40–56.

Jog, M. S., Kubota, Y., Connolly, C. I., & Graybiel, A. M. (1999). Building neural representations of habits. *Science, 286*(5445), 1745–1749.

Kelso, J. A. S. (1995). *Dynamic patterns: The self-organization of brain and behavior*. Cambridge, MA: MIT Press–Bradford Books.

Kharitonova, M., Chien, S., Colunga, E., & Munakata, Y. (2007, March). The emergence of human intelligence: An early synergy of flexibility and generalization. Paper presented at the biennial meeting of the Society for Research on Child Development, Boston, MA.

Kharitonova, M., Chien, S., Colunga, E., & Munakata, Y. (in press). More than a matter of getting "unstuck": Flexible thinkers use more abstract representations than perseverators. Developmental Science.

Kirkham, N. Z., Cruess, L., & Diamond, A. (2003). Helping children apply their knowledge to their behavior on a dimension-switching task. *Developmental Science, 6*(5), 449–476.

Leslie, A. M. (1992). Pretence, autism, and the theory-of-mind-module. *Current Directions in Psychological Science, 1*(1), 18–21.

Marcovitch, S., & Zelazo, P. D. (1999). The A-not-B error: Results from a logistic meta-analysis. *Child Development, 70*(6), 1297–1313.

McClelland, J. L. (1989). Parallel distributed processing: Implications for cognition and development. In R. G. M. Morris (Ed.), *Parallel distributed processing: Implications for psychology and neurobiology* (pp. 8–45). New York: Oxford University Press.

McClelland, J. L., Rumelhart, D. E., & PDP Research Group. (1986). *Parallel distributed processing: Explorations in the microstructure of cognition: Vol. 2. Psychological and biological models*. Cambridge, MA: MIT Press.

Miller, E. K., & Cohen, J. D. (2001). An integrative theory of prefrontal cortex function. *Annual Review of Neuroscience, 24,* 167–202.

Miller, E. K., & Desimone, R. (1994). Parallel neuronal mechanisms for short-term memory. *Science, 263*(5146), 520–523.

Miller, E. K., Erickson C. A., & Desimone, R. (1996). Neural mechanisms of visual working memory in the prefrontal cortex of the macaque. *Journal of Neuroscience, 16*(16), 5154–5167.

Morton, J. B., & Munakata, Y. (2002a). Active versus latent representations: A neural network model of perseveration, dissociation, and decalage. *Developmental Psychobiology, 40*(3), 255–265.

Morton, J. B., & Munakata, Y. (2002b). Are you listening? Exploring a developmental knowledge-action dissociation in a speech interpretation task. *Developmental Science, 5*(4), 435–440.

Morton, J. B., & Munakata, Y. (2008). The appearance (and disappearance) of knowledge-action dissociations in adults. Manuscript in preparation.

Morton, J. B., & Trehub, S. E. (2001). Children's understanding of emotion in speech. *Child Development, 72*(3), 834–843.

Morton, J. B., Trehub, S. E., & Zelazo, P. D. (2003). Sources of inflexibility in 6-year-olds' understanding of emotion in speech. *Child Development, 74*(6), 1857–1868.

Munakata, Y. (1997). Perseverative reaching in infancy: The roles of hidden toys and motor history in the A-not-B task. *Infant Behavior and Development, 20*(3), 405–416.

Munakata, Y. (1998). Infant perseveration and implications for object permanence theories: A PDP model of the AB task. *Developmental Science, 1*(2), 161–184.

Munakata, Y. (2001). Graded representations in behavioral dissociations. *Trends in Cognitive Sciences,*(7), 309–315. *5*

Munakata, Y. (2006). Information processing approaches to development. In W. Damon, R. Lerner (Series Eds.), D. Kuhn, & R. S. Siegler (Vol. Eds.), *Handbook of child psychology: Vol. 2. Cognition, perception, and language* (6th ed., pp. 426–463). New York: Wiley.

Munakata, Y., & McClelland, J. L. (2003). Connectionist models of development. *Developmental Science, 6*(4), 413–429.

Munakata, Y., McClelland, J. L., Johnson, M. H., & Siegler, R. S. (1997). Rethinking infant knowledge: Toward an adaptive process account of successes and failures in object permanence tasks. *Psychological Review, 104*(4), 686–713.

Munakata, Y., Morton, J. B., & Stedron, J. M. (2003). The role of prefrontal cortex in perseveration: Developmental and computational explorations. In P. T. Quinlan (Ed.), *Studies in Developmental Psychology* [series]. *Connectionist models of development: Developmental processes in real and artificial neural networks* (pp. 83–114). London: Taylor & Francis–Psychology Press.

Munakata, Y., Sahni, S. D. & Yervys, B. E. (2001). An embodied theory in search of a body: Challenges for a dynamic systems model of infant perseveration [Commentary]. *Behavioral and Brain Sciences, 24*(1), 56–57.

Munakata, Y., & Yerys, B. E. (2001). All together now: When dissociations between knowledge and action disappear. *Psychological Science, 12*(4), 335–337.

O'Reilly, R. C., & Johnson, M. H. (2002). Object recognition and sensitive periods: A computational analysis of visual imprinting. In M. H. Johnson, Y. Munakata, & R. O. Gilmore (Eds.), *Brain development and cognition: A reader* (2nd ed., pp. 392–413). Oxford, United Kingdom: Blackwell.

O'Reilly, R. C., & Munakata, Y. (2000). *Computational explorations in cognitive neuroscience: Understanding the mind by simulating the brain.* Cambridge, MA: MIT Press–Bradford Books.

O'Reilly, R. C., & Rudy, J. W. (2001). Conjunctive representations in learning and memory: Principles of cortical and hippocampal function. *Psychological Review, 108*(2), 311–345.

Patalano, A. L., Smith, E. E., Jonides, J., & Koeppe, R. A. (2001). PET evidence for multiple strategies of categorization. *Cognitive, Affective and Behavioral Neuroscience, 1*(4), 360–370.

Perner, J., & Lang, B. (2002). What causes 3-year-olds' difficulty on the dimensional change card sorting task? *Infant and Child Development 11*(2), 93–105.

Piaget, J. (1954). *The construction of reality in the child* (M. Cook, Trans.). New York: Basic Books.

Rougier, N. P., Noelle, D. C., Braver, T. S., Cohen, J. D., & O'Reilly, R. C. (2005). Prefrontal cortex and flexibility cognitive control: Rules without symbols. *Proceedings of the National Academy of Sciences, USA, 102*(20), 7338–7343.

Siegler, R. S. (1995). How does change occur: A microgenetic study of number conservation. *Cognitive Psychology, 28*(3), 225–273.

Sitton, M., Mozer, M. C., & Farah, M. J. (2000). Superadditive effects of multiple lesions in a connectionist architecture: Implications for the neuropsychology of optic aphasia. *Psychological Review, 107*(4), 709–734.

Smith, L. B., & Scheier, C. (1998). Babies have bodies: Why Munakata's net fails to meet its own goals. *Developmental Science, 1*(2), 202–204.

Smith, L. B., & Thelen, E. (Eds.). (1993). *A dynamic systems approach to development: Applications.* Cambridge, MA: MIT Press–Bradford Books.

Smith, L. B., & Thelen, E. (2003). Development as a dynamic system. *Trends in Cognitive Sciences, 7*(8), 343–348.

Smith, L. B., Thelen, E., Titzer, R., & McLin, D. (1999). Knowing in the context of acting: The task dynamics of the A-not-B error. *Psychological Review, 106*(2), 235–260.

Spencer, J. P., & Schöner, G. (2003). Bridging the representational gap in the dynamic systems approach to development. *Developmental Science, 6*(4), 392–412.

Spencer, J. P., & Schutte, A. R. (2004). Unifying representations and responses: Perseverative biases arise from a single behavioral system. *Psychological Science, 15*(3), 187–193.

Stedron, J. M., Sahni, S. D., & Munakata, Y. (2005). Common mechanisms for working memory and attention: The case of perseveration with visible solutions. *Journal of Cognitive Neuroscience, 17*(4), 623–631.

Thelen, E., & Bates, E. (2003). Connectionism and dynamic systems: Are they really different? [Special section/issue]. *Developmental Science, 6*(4), 378–391.

Thelen, E., Schöner, G., Scheier, C., & Smith, L. B. (2001). The dynamics of embodiment: A field theory of infant perseverative reaching. *Behavioral and Brain Sciences, 24*(1), 1–34, commentary 34–70, authors' response 70–80.

Thelen, E., & Smith, L. B. (1994). *A dynamic systems approach to the development of cognition and action.* Cambridge, MA: MIT Press–Bradford Books.

Thomas, M. S. C., & Karmiloff-Smith, A. (2003). Connectionist models of development, developmental disorders, and individual differences. In R. J. Sternberg, J. Lautrey, and T. I. Lubart (Eds.), *Models of intelligence: International perspectives* [Series: Decade of behavior] (pp. 133–150). Washington, DC: American Psychological Association. Retrieved October 31, 2008, from http://www.psyc.bbk.ac.uk/research/DNL/personalpages/Thomas_intell03.pdf

Towse, J. N., Redbond, J., Houston-Price, C. M. T., & Cook, S. (2000). Understanding the dimensional change card sort: Perspectives from task success and failure. *Cognitive Development, 15*(3), 347–365.

Usher, M., & McClelland, J. L. (2001). The time course of perceptual choice: The leaky, competing accumulator model. *Psychological Review, 108*(3), 550–592.

van der Maas, H. L. J., & Molenaar, P. C. M. (1992). Stagewise cognitive development: An application of catastrophe theory. *Psychological Review, 99*(3), 395–417.

Wallis, J. D., Anderson, K. C., & Miller, E. K. (2001). Single neurons in prefrontal cortex encode abstract rules. *Nature, 411*(6840), 953–956.

Wellman, H. M., Cross, D., & Bartsch, K. (1986). Infant search and object permanence: A meta-analysis of the A-not-B error. *Monograph of the Society for Research in Child Development*, No. 214, *51*(3), 1–67.

Yerys, B. E., & Munakata, Y. (2006). When labels hurt but novelty helps: Children's perseveration and flexibility in a card-sorting task. *Child Development, 77*(66), 1589–1607.

Zelazo, P. D., Muller, U., Frye, D., & Marcovitch, S. (2003). The development of executive function in early childhood. Monographs of the Society for Research in Child Development, 68(3), Serial No. 274.

Zelazo, P. D., Frye, D., & Rapus, T. (1996). An age-related dissociation between knowing rules and using them. *Cognitive Development, 11*(1), 37–63.

# CHAPTER 8

## Dynamical Insight into Structure in Connectionist Models

*Whitney Tabor*

This paper adds its voice to those that have pointed out that connectionist models are a subclass of dynamical systems. They are a particularly interesting subclass because of their psychological plausibility. Moreover, they have a valuable trait: one can make connectionist models that behave in psychologically plausible ways. However, connectionist models are, at least initially, difficult to understand. The difficulty arises in the fact that the plausible behaviors of connectionist models are achieved through a *learning algorithm*—a simple, low-level process that tunes the parameters (or *weights*) of the network on the basis of repeated interactions with a body of data. Although the principle of the learning algorithm is well understood—it says, in essence, at each step, make a small change that improves performance—the pattern of weights that results from training a network on an interesting problem is typically quite complex and difficult to interpret. Such inscrutability might seem like a negative point, but one can sometimes learn things by striving to see into what, at first, seems inscrutable. Dynamical systems theory is helpful in this regard; it provides conceptual constructs (e.g., trajectories, basins, attractors, bifurcations) and analytical methods (e.g., phase-space portraits, bifurcation analysis), which can help one understand relationships and causes in these psychologically relevant models. (For introductions to dynamics terms, see Abraham & Shaw, 1984, and Strogatz, 1994). Dynamical systems theory has been embraced by the field of ecological psychology as the best formal means of clarifying the notion that psychological behaviors are emergent phenomena in organism–environment systems (Kugler & Turvey, 1987).

In addition to the connectionist method of psychological modeling and the accompanying analytic framework of dynamical systems theory, it is useful, at this point, to mention a third element: *symbolic computation theory.* I use this term to denote theories which describe behavior using discrete symbols and what are called *well-formedness rules.* Observed behaviors are modeled as allowable symbol patterns under the well-formedness rules. The grammatical theories of natural language proposed by the field of generative grammar are examples of this kind of theory (e.g., Chomsky, 1957). Unfortunately, the symbolic approach has generally been treated as an enemy by both connectionists and ecological psychologists, including many ecological psychologists favorably oriented toward dynamical systems. While I agree with the rejection of symbolic models as a sufficient foundation for the science of organism–environment systems, I argue here that symbolic models are nevertheless particularly relevant to cognition and, in fact, to the relationship between the connectionist and dynamical approaches. In a nutshell, symbolic models correspond to certain idealized behaviors, which connectionist models can exhibit and which influence learning the way an attractor can be said to influence the trajectory of a dynamical system. This is so even though realistic connectionist systems rarely, if ever, spend time at these attractors. Symbolic models are thus inadequate as the sole basis for modeling in psychology because they are only a subpart of the broader dynamical framework which is needed to characterize cognition. On the other hand, they are an important proper subpart.

I'll make this discussion more concrete by focusing on one major domain: development. Several people have suggested that the dynamical notion of a phase transition is particularly relevant to development (Smith & Thelen, 1993; Thelen & Smith, 1994; van der Maas & Molenaar, 1992; van Geert, 1998). In the sense intended here, a phase transition is a qualitative change in the behavior of a system that is associated with the monotonic and continuous change of a parameter—a so-called *control parameter*. Phase transitions seem potentially relevant to modeling developmental stages because they correspond to situations in which behavior changes subtly and quantitatively for a long period of time and then suddenly changes dramatically and qualitatively. Indeed, the researchers mentioned above have pointed out plausible resemblances between developmental data and various mathematical models of phase transitions. Here, I go along this path and then make an additional suggestion: developmental phase transitions may involve a phase in which *well-formedness* rules fail to characterize the structure of behavior and shift to a phase in which these rules constitute a good approximation of the behavior. Thus, this account treats the symbolic models as relevant to characterizing stages but not sufficient for characterizing all states individuals go through in the process of growing up.

What is the evidence for these claims? It takes the form of a connectionist model called a fractal learning neural network (FLNN). I want to make clear at the outset that I offer no measured developmental data to motivate/promote this model. Instead, one might think of this work as an exploration of the abstract structure of information. The data that motivate the model are the linguistic observational data that reveal recursive patterning in every natural language. For two reasons, I believe that the results may be of interest to developmentalists. (1) The FLNN model is a connectionist learning model and thus has relatives which have made plausible empirical predictions about many domains, including development. (2) The FLNN model helps a person think clearly about appealing alternatives to some current theories of development (e.g., maturational explanations for stages, nativist explanations for grammatical complexity). The FLNN model's account is appealing because it (a) uses a simple learning mechanism and thus corresponds to a theory that links development and learning, (b) predicts stagewise behavior with progressive quantitative changes occurring within the stages, a phenomenon also observed in development, (c) derives phase transitions from the learning, thus predicting, rather than stipulating, the stages, (d) shows highly structured, domain-specific behavior arising from a mechanism which is capable of a wide variety of behaviors, not all of which are highly structured (thus, in some sense, deriving the structure where no structure was apparent), and (e) generalizes beyond its input in a way that is, at least partially, consistent with human generalization in the comprehension and production of recursive sentence structures.

I will describe the model and then return to these points.

## Description of the Model

### Background

Elman (1990, 1991) trained a network, which he dubbed the *simple recurrent network*, to process sequences of symbols, including sequences of words. Such a network (Fig. 8.1) has three layers of units (input, hidden, and output) plus a context layer, which stores a copy of the hidden unit activations from the previous time step. As depicted in Figure 8.1, connections feed forward from Input to Hidden to Output and from Context to Hidden. The copying operation associated with the context layer is a type of recurrence, and it permits the network to encode information about past inputs, allowing it to keep track of the temporal dependencies which are crucial to syntactic processing.

In one experiment, Elman (1991) trained his network on a corpus of words approximating the syntactic properties of English. The corpus included singular and plural nouns, as well as center-embedded relative clause structures [see

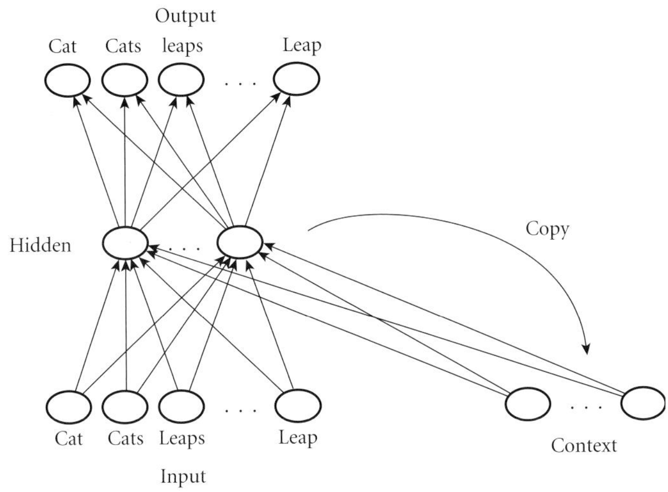


**Figure 8.1. Elman's simple recurrent network**
This figure presents a schematic of Elman's simple recurrent network (Elman, 1990, 1991). Connections feed forward from the Input layer to the Hidden layer to the Output layer and from the Context layer to the Hidden layer. The network uses the Context layer to store a copy of the activations in units of the Hidden layer; this copy dates from the previous time step.

example (1)].[1] (Articles such as *the* and *a* were left out for simplicity.)

(1) dogs/dog who girls chase run/runs.

When processing an example like (1), the network must store the number of the head noun of the subject of the sentence (*dog* or *dogs*) during the time that it processes the relative clause (*who girls chase*) in order to properly predict the number of the main verb (*run* vs. *runs*). Elman used a single-on-bit (typically referred to as *one-hot*) encoding for each word on the input or output layer. The codes for the words of the corpus were presented in order, one at a time, on the input layer of the network. The network was trained on the task of predicting the next word at each point.[2] Because the sequence of sentences in the corpus was approximately random, the network was not able to guess specifically what word was going to appear at any point. However, it could detect stable, syntactic properties of the language. In particular, once the network was fully trained, the activation of each output unit approximated the probability that the word corresponding to that output unit would occur next in the corpus. For example, after having encountered *dog* at the beginning of a sentence, output activation was concentrated mostly on singular verbs and the word *who*. While such local co-occurrence tendencies are fairly easy to model [e.g., using Markov (1960) processes], some kinds of longer distance dependencies, including those arising in center-embedded structures such as example (1), are more difficult. Thus, it is significant that the network was also successful at predicting the number on the verbs in relative clause structures like example (1).

Since Elman's initial experiments, several projects have successfully used the simple recurrent network and variants on it to learn complex grammatical dependencies in other human language-like corpora (e.g., Christiansen & Chater, 1999; Frank, Mathis, Gussine, Stowe, & Vindiola, 2005; Rohde, 2002; Rohde & Plaut,

[1] Example (1) is said to be a sentence with one level of center-embedding. In particular, it consists of a matrix sentence, *dogs run*, with a single subordinate sentence, essentially *girls chase dogs*, embedded in the middle of it.

[2] The network was trained using the sum-squared error cost function and back-propagation of the error signal through time (Pearlmutter, 1995; Rumelhart, Hinton, & Williams, 1986).

**Table 8.1 ABC grammar**

| 1.0 S → ABC 1.0A → a (S) 1.0 B → b (S) 1.0 C → c (S) |
|---|

*Note:* Parentheses denote optional constituents, which occur with a probability of .2 in every case. A probability, indicated by a decimal number to its left, is associated with each production. A sentence is generated by starting with the S node and iteratively replacing symbols according to the rules (and their probabilities) until all the symbols are lowercase. The resulting lowercase string is, by definition, a member of the ABC language. For example, abc, aabcbc, and abcabc are members of the ABC language, but bc, aabc, and acb are not. The least powerful type of symbolic device that is capable of stepping sequentially through any string of *a*'s, *b*'s, and *c*'s and determining whether it belongs to the ABC language is the pushdown automaton.

1999). Several projects have investigated the learning of abstract formal languages, somewhat related to human languages, in order to better understand the principles by which the network was succeeding at prediction (Boden & Wiles, 2002; Rodriguez, 2001; Rodriguez & Wiles, 1998; Rodriguez, Wiles, & Elman, 1999; Wiles & Elman, 1995). Schmidhuber and colleagues (Gers, Schmidhuber, & Cummins, 2000; Gers & Schmidhuber, 2001; Hochreiter & Schmidhuber, 1997) propose a different, but related, kind of recurrent network called *long short term memory* (LSTM), which could keep track of very long temporal dependencies (with up to thousands of intervening events).

I use the term *abstract formal language* to mean a set (finite or infinite) of finite-length strings of symbols drawn from a finite symbol alphabet. A foundational idea in the theory of computation is to ask, for a given formal language, what kind of computing device is needed to decide if a particular string is, or is not, a member of that language (Chomsky, 1956/1965; Hopcroft & Ullman, 1979; Kozen, 1997). Being able to do this for any finite string presented is called *recognizing the language*. The answer is uninteresting if the set of strings is a finite set because the device can simply be a big list of the strings of the language; when a string is presented for testing, the device simply runs through the list, comparing each list string to the test string, and accepts the test string if it finds a match. The situation becomes more interesting when infinite sets of strings are considered.[3] The list approach runs into problems with lengthy searches in this case, but for an important class of examples, one can build compact symbol-manipulating devices that take advantage of the structures of languages in order to recognize them. These come in several subclasses. The most basic subclass is the *finite-state machine*; it consists of a computer whose so-called brain can only be in a finite number of states. A somewhat more complex subtype is the *pushdown automaton*, which stores a sequence of symbols in memory and then recovers them in reverse order, retrieving each at just the moment when the environment indicates that it is time to recall it. The string of symbols is called a *stack*. The stack is a very useful device: it underlies *recursion*, the primary mechanism that gives modern computer programming languages their power; it has, arguably, formed the basis for cataloging the myriad complex patterns that occur in plant growth (Lindenmayer & Prusinkiewicz, 1989); it is related to fractal geometry, which is widely observed in nature (Mandelbrot & Ness, 1968); and, in one guise or another, it is a primary workhorse of all formalized linguistic theories (Gazdar, 1981). Another name for the class of languages recognized by pushdown automata is *context-free*. *Context-free languages* can be described either by pushdown automata or by finite lists of rules of the form X → β where β is a finite string of symbols. An example is shown in Table 8.1.

The projects mentioned above, on neural mechanisms that learn to recognize formal

[3] It may not seem plausible that real, physical creatures like human beings have infinite knowledge. However, they have something closely related: the ability to handle arbitrary new cases when they arise. One can understand the theory of infinite string languages as a hypothesis about the nature of this open-ended response capability.

[4] The language $a^nb^n$ consists of all strings of *a*'s and *b*'s in which some number of *a*'s is followed by the same number of *b*'s: ab, aabb, aaabbb, and so forth.

languages on the basis of experience with them, focus on the learning of context-free languages. Although the networks were successful on one simple case (the language $a^n b^n$)[4] and the analyses provided clear insight into the method by which the networks succeeded (see further discussion below), the networks were much less systematically successful with more complex cases (e.g., palindrome languages), which more closely resemble human languages. Meanwhile, work on the encoding of complex symbolic structures in connectionist units independent of the learning question (Moore, 1998; Tabor, 2000) has provided insight into how it is possible to encode complex syntax in connectionist architectures.

The construction of the FLNN (Tabor, 2003) was motivated by the combination of these research strands. The FLNN combines the representational results of Moore and Tabor with a hill-climbing algorithm that discovers weight settings embodying complex symbolic processes (Moore, 1998; Tabor, 2000, 2003). I'll describe the architecture of this device, along with some simulation results, and then make some comments about how this effort may shed light on development and on the relation between dynamical systems and connectionist approaches.

## Architecture and Mechanism

The world of stacks and grammars is a symbolic world where set membership is the natural currency, but quantitative measurement seems to have little meaning. On the other hand, canonical dynamical systems in general, and connectionist networks in particular, live in metric spaces where real-valued distances between states play a central role in the system dynamics. Where might these disparate worlds meet? A plausible answer is fractal sets, recursive structures that lie in metric spaces. A fractal set is a set that exhibits self-similarity over an infinite range of scales (i.e., scale invariance), including arbitrarily small scales. The best known example is the Cantor set, the fractal set that remains when one starts with a line segment, removes the middle third to produce two smaller line segments, removes the middle thirds of each of these, and so on ad infinitum. The sets of points in the left and right thirds of the original segment are called the left and right branches of the fractal, respectively. Numerous investigators (Levy, Melnik, & Pollack, 2000; Melnik, Levy, & Pollack, 2000; Moore, 1998; Tabor, 2000) have shown how to use generalizations of the Cantor set to make dynamical systems that recognize context-free languages. The system starts at a prespecified location on the fractal and jumps to another part of the fractal whenever a symbol of a string is processed; it uses the recursive spatial structure of the fractal to keep track of the recursive symbol embedding of the language.[5]

The FLNN is a learning connectionist network related to Elman's simple recurrent network, which is designed to interact with sentences from a context-free language and discover how to move around on an appropriate fractal and recognize strings from the language. The FLNN architecture has four layers (Fig. 8.2). The first layer is the input layer, where each unit stands for a word. The second layer has a linear activation function with linear recurrent connections that are gated by the input units. It is in this layer that the fractal arises. Cantor sets and some of their generalizations can be constructed with affine operators on vector spaces [see equation (1)] (Barnsley, 1988; Barnsley & Rising, 1993).

$$\mathbf{z} \leftarrow \mathbf{a}\mathbf{z} + \mathbf{b} \tag{1}$$

Here, $\mathbf{z}$ is the vector of activations on the second layer. The gating connections, by which the input units specify the weights of the recurrent self-connections in the hidden layer, allow the input pattern to specify parameter $\mathbf{a}$, which is set to a value smaller than one ($\mathbf{a} < 1$) for

[5] This work is closely related to the large body of work on variable binding with connectionist networks (e.g., Hinton, 1990; Jagota, Plate, Shastri, & Sun, 1999; Plate, 1995; Smolensky, 1990). Both problems are concerned with the challenge of productivity—how to allow a large number of combinations to be stored in a fixed-width vector; one may think of formal language modeling as a subcase of the variable binding problem. In fact, some variable binding proposals use fractal scaling to allow graceful packing of many patterns into one vector (e.g., Plate, 2003). The work of Moore (1998) and Tabor (2000) is distinguished by providing a general formalization of the fractal/grammar relationship, although it does not apply the method to variable binding in general.

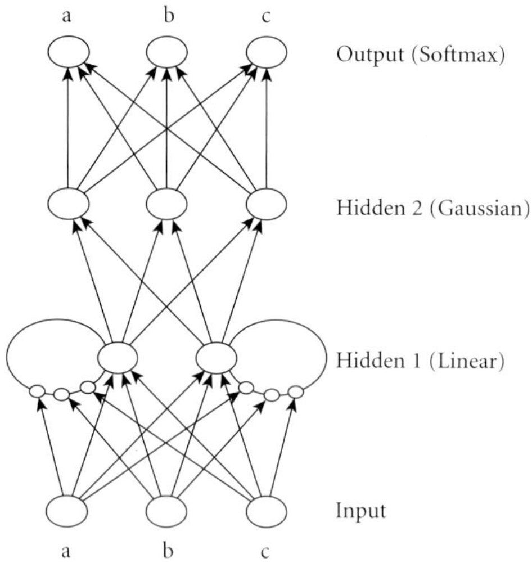


**Figure 8.2. A fractal learning neural network (FLNN)**

contraction (i.e., moving to a smaller scale on the fractal) and to a value greater than one (**a** > 1) for expansion (i.e., moving to a larger scale on the fractal). The linear connections from the input to the hidden layer allow the network to specify parameter **b**, which determines which branch of the fractal the system occupies at each point. The third layer of the FLNN has Gaussian units (i.e., radial basis functions) whose means are learned but whose variances are fixed (arbitrarily) at 0.25. Gaussian units are a kind of on-center, off-surround unit: they activate strongly when the vector of their inputs is in a particular spherical region centered at their mean and of radius specified by their variance. These third-layer units end up being used to detect which branch of the fractal the second layer is on at each point.[6] The fourth layer of the FLNN is the output layer. In the output layer, as in the input layer, each unit stands for a word. The output units, as a group, have the normalized exponential activation function (the so-called softmax activation function) because they specify probability distributions over words. In the fractal grammar recognizers of Moore (1998) and Tabor (2000), the *set of possible successor symbols* at any point is determined by the branch of the fractal that the system inhabits at that point. In the FLNN, as in Elman's simple recurrent network, the *set of possible successor symbols* is translated into a probability distribution over successor symbols with the probabilities approximating the actual occurrence rates.

There is an important question about the number of units in the hidden layers. I chose the number of units to be the minimal number that I knew were necessary to solve the task well (in this case two units in the recurrent hidden layer and three units in the Gaussian hidden layer). This constraint may play a role in the success of the generalization touted below. Further testing is needed to find out whether the network generalizes as well when it has an excess of hidden units.

### Training

I trained the network on formal languages that are simple but more complex than those studied in prior work that successfully interpreted the network symbol processors' hidden unit representations (Gers et al., 2000; Gers & Schmidhuber, 2001; Rodriguez, 2001; Rodriguez et al., 1999; Rodriguez & Wiles, 1998; Wiles & Elman, 1995). On the other hand, the training languages are not as complex as the corpora used by researchers focused on the details of the resemblance of the network's behavior to natural language (Elman, 1990, 1991; Christiansen & Chater, 1999; Rohde, 2002; Rohde & Plaut, 1999). See Tabor (2003) for a discussion of this complexity issue. The training data for the network were generated from the grammar shown in Table 8.1. In the simulation reported here, the training corpus consisted of one instance of each sentence from the grammar that has 9 or fewer words. Thus, the maximum number of center-embeddings was 2.

At the beginning of each sentence, the two recurrent hidden unit activations were set to 0.

[6] The analog of a sphere in two dimensions is a disk, and in one dimension, a line segment. Thus, in the case of a simple Cantor set, these Gaussian units would be activating in response to a particular segment of the real line. In the example network given here, there are two hidden units, so the Gaussian third-layer units are sensitive to disk-shaped regions in the space of the linear second-layer units (see Fig. 8.2).

**Table 8.2** Performance on training and test sets

| Corpus | RMSE | SE | % Cor. | SE | *n* | N points |
|---|---|---|---|---|---|---|
| Training | 0.013 | 0.001 | 100.000 | 0.000 | 9 | 129 |
| Test | 0.048 | 0.005 | 99.424 | 0.195 | 9 | 4755 |

*Note*: % Cor. = percent correct; *n* = number of networks that contributed to the computation of SE; N points = the number of words tested per network; RMSE = root mean square error; SE = standard error. All networks were trained using a hill-climbing algorithm.

The words were presented one at a time, in order, on the input layer of the network. The grammar was used to determine output probabilities following each word. In the simulation reported here, these probabilities were used as the targets of training.[7]

The means of the Gaussian units and the Gaussian→Output weights were set to random initial values (mean = 0, variance = 0.3). The gated self-connections in the hidden layer were set to initial value 1 (this is the unbiased choice; Tabor, 2000). The recurrent connections between different units (as opposed to the recurrent connections from each hidden unit to itself) were clamped to 0 throughout training and testing. This choice, also, was motivated by the knowledge that these connections would not be needed for the task at hand.

The network was trained by a hill-climbing algorithm with epoch-sized steps. The first of the hill-climbing steps was to present every sentence in the corpus to the network and compute the total error at the output layer. Kullback-Leibler (KL) divergence was used as the error measure.[8] Then, a variety of slight distortions of the current weight setting was examined (in particular, each weight was adjusted slightly, while keeping all the others fixed). For each of these distorted weight settings, the entire corpus was tested, and error was again accumulated. If one of these adjustments made a maximal improvement in the error, than it was adopted as the new current weight setting. (Ties for best were broken arbitrarily.) This process was repeated until total error dropped below a prespecified threshold (0.02).

### Test results

As noted, the network was trained on all sentences of length up to 9 words (with, at most, 2 levels of center embedding). Then, the network was tested on all sentences of length 10–15 words (with, at most, 4 levels of center embedding). The results (shown in Table 8.2) indicate that, while the network's performance on the test corpus was not perfect, it was nearly so.

### Analysis of Weights

In line with the scheme developed by Tabor (2000), the trained network has discovered how to move around on a fractal set to keep track of the dependencies in the ABC language. An illustration of the learned fractal is shown in Figure 8.3E. Tabor (2003; 2005) provides evidence that the learned weight set is organized on the same principles as the ideal fractal (Tabor, 2000).

### Code

MATLAB code for running these simulations is available at the following website: http://www.sp.uconn.edu/~ps300vc/Projects/FractalLearning/index.html

[7] The use of probability distributions rather than individual events as training data is not consistent with what people actually experience. However, this simplification is superseded by a second one—the use of batch error accumulation in the hill-climbing procedure described below; this second simplification was used because it is more efficient.

[8] The KL divergence of a probability distribution $p_2$ with respect to another probability distribution $p_1$ is as follows: $p_1 = p_2 \times \log(p_2/p_1)$ (Cover & Thomas, 1991). This is a natural way to compare probability distributions and has a motivated use in back-propagation learning when events correspond to single output units, as they do here (Rumelhart, Durbin, Golden, & Chauvin, 1995). I do not know, at present, how it compares with other error measures when used with the hill-climbing algorithm.

This code runs slowly but is fairly simple to interpret.

### Connectionist Networks and Dynamical Systems Theory

Rodriguez, Wiles, and Elman (Rodriguez, 2001; Rodriguez et al., 1999; Rodriguez & Wiles, 1998; Wiles & Elman, 1995) analyzed simple recurrent networks trained on simple, context-free languages related to the case discussed above. These investigators showed that standard dynamical systems constructs (e.g., trajectories, attractors, repellers) provided a foundation for understanding how the network successfully generalized from the finite sample from training sessions. For example, Rodriguez and colleagues (Rodriguez et al., 1999) showed that a network trained on the language $a^n b^n$ employed two quasi-linear dynamical objects: an attractor and a repeller. When the code for *a* was presented on the input layer, the network's hidden unit dynamics consisted of a single attractor basin with a stable, fixed point. At the beginning of processing, the hidden unit activation vector was relatively far from the attractor. Each successive presentation of an *a* caused the state to jump a step closer to the fixed point. When the input was switched to *b*, the result was a new dynamical regime, similar to the *a* regime, but with a repeller that was located on the other side of hidden unit space. The first presentation of *b* moved the state to a location close to the repeller. The distance between this point and the repeller was a decreasing function of the number of *a*'s that had been presented. Additional presentations of *b* caused the network to jump away from the repeller and head back to the location associated with the beginning of a sentence. The hidden state arrived back at this location when the number of *b*'s matched the number of *a*'s that had been presented. The Hidden→Output weights of the network implemented linear separations of the hidden unit space as follows: (1) when the network was near the periphery of the attractor's basin, it expected the start of a new sentence (i.e., only possible next word = *a*), (2) when the network was jumping toward the attractor, it predicted *a* and *b*, and (3) when the network was in the region of the repeller, it predicted only *b*'s until it arrived back at the start-of-sentence region. Thus, the balancing of the number of *a*'s with the number of *b*'s was accomplished by matching the contraction effect of the attractor with the expanding effect of the repeller.

Rodriguez's (Rodriguez et al., 1995) network moves on a simple fractal, a contractive geometric series, to keep track of levels of embeddings. A *contractive geometric series* is a series of numbers of the form $ar^n$, where $a$ is a nonzero constant, $r$ is nonzero between $-1$ and 1, and $n = 0,1,2\ldots$. At the end of training, the FLNN described above moves on a slightly more complex fractal, a two-dimensional Cantor Set, to simultaneously track levels and types of embedding. As in Rodriguez's case, the system trajectories are directed by a collection of attracting and repelling fixed points with carefully coordinated interactions.

If these cases are typical, they suggest the following relation between connectionist models, dynamical systems theory, and symbolic processes. Certain connectionist networks with certain weight settings mimic symbolic processes; moreover, certain dynamical objects are formally related, via the networks, to the symbolic entities (Crutchfield & Young, 1990; Moore, 1998). By way of example, when the attractor and repeller are in perfect coordination in the symbol prediction network (and thus travel around on a Cantor-set like fractal), the network generates the same formal language as a particular pushdown automaton; in fact, for every pushdown automaton language (i.e., context-free language), there exists a fractal traversing network that generates the language (Tabor, 2000). This suggests that dynamical systems theory is in a position to provide insight into the relation between symbolic and connectionist models. The next section explores the process by which the network approaches such a symbolically organized state and provides some evidence that the relation proposed here is not a fluke enforced by excessive training. The next section also suggests that not all states of the network's development

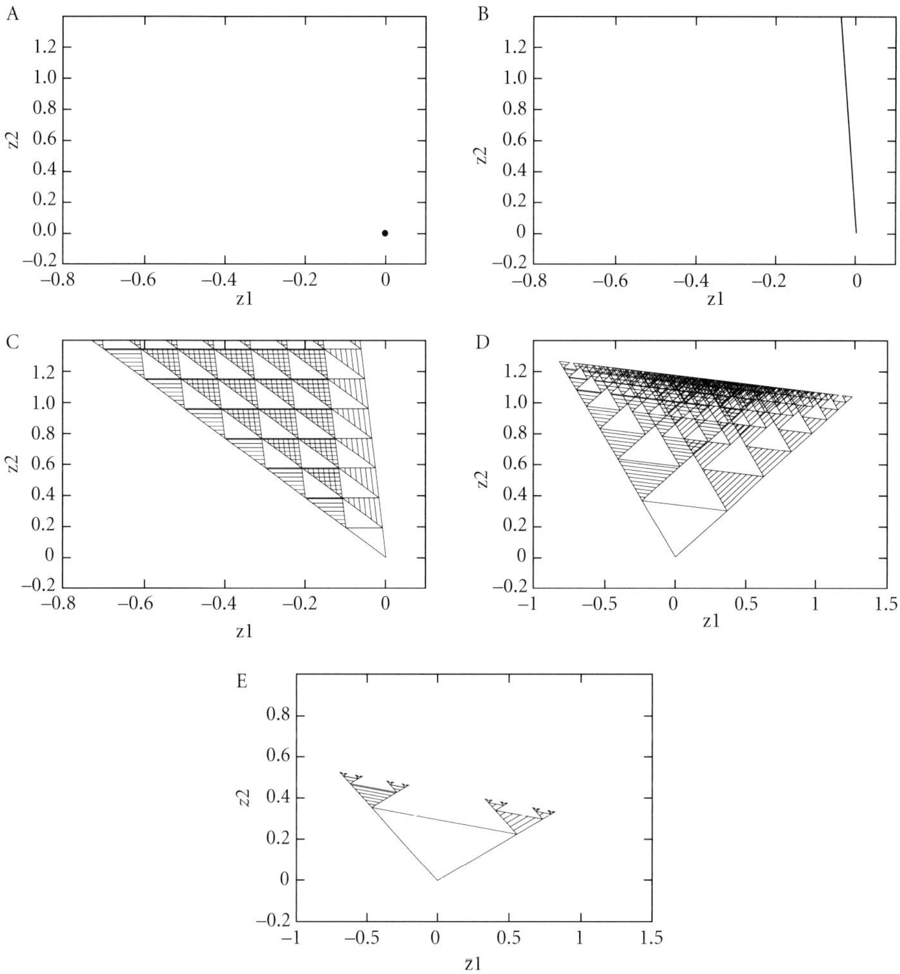


**Figure 8.3. Visited states in FLNN processing of grammatical strings**
This figure depicts the evolution of the set of visited states in the linear hidden unit layer during processing of grammatical strings over the course of FLNN training on the language of the grammar in Table 8.1. Here, z1 and z2 (the axes) are hidden unit activations. Only states generated by legal sequences of *a*'s and *b*'s are shown—that is, only states following sequences that increase/maintain the current level of embedding, not any that decrease it. **(Panel A)** At the beginning of training, the visited states consist of the single point [0, 0]. **(Panel B)** At 550 epochs, the visited states lie on an infinite ray with endpoint at [0, 0]. **(Panel C)** At 1000 epochs, the visited states lie on an infinite lattice (the visited states are at the vertices of triangles marked with horizontal and vertical hatching; the bottommost vertex of the whole lattice is the initial state). There is near-complete overlap between states in which *a b* (or *a*) is predicted next (horizontal hatching) and states in which *a c* (or *a*) is predicted next (vertical hatching). The network is jumping around on these triangle vertices, going further up in the plane of the figure when *a*'s are encountered, going right at *b*'s, and going down at *c*'s. **(Panel D)** At 10000 epochs, the lattice has become finite and the branches corresponding to *b*-prediction and *c*-prediction are spreading apart. **(Panel E)** At 51,174 epochs (the end of training), the branches have fully separated, and the set of visited states lies on a totally disconnected fractal in the sense of Barnsley (Barnsley, 1988; Barnsley & Rising, 1993). At this stage, the network's behavior is nearly identical to that of the corresponding pushdown automaton. It is not exactly identical because, as mentioned in the text, the expansion associated with *c* inputs only approximately inverts the contraction associated with *a* and *b* inputs.

correspond to symbolically describable behaviors.

### Phase Portraits and Review of Claims

Figure 8.3 is a succession of *phase portraits*[9] of the model over the course of training on the ABC language. The portraits show the points visited by the model when it processes sentences containing up to 7 levels of embedding. Only points visited during the processing of initial strings containing *a*'s and *b*'s are shown. If strings containing *c*'s were shown, the pictures would look similar but fuzzier because the model imperfectly approximates the symmetry between contraction and expansion.

Figure 8.3 illustrates several successive, qualitatively distinct stages of the model's behavior over the course of training. With this illustration in view, I review the claims about *appealing properties* made above and assess their validity:

(a) *The model learns.* The model clearly involves a tuning process, which brings it from domain-oblivious behavior to highly domain-sensitive behavior. The hill-climbing algorithm itself is not biologically plausible because it involves thorough exploration of the environment every time a weight is adjusted. Nevertheless, its success is a prerequisite for the success of currently known biologically plausible algorithms (e.g., Hebbian learning).

(b) *The model passes through qualitatively distinct stages.* Distinct stages can be observed in Figure 8.3. The dimensional expansions that have occurred between stage A (Fig. 8.3A) and stage B and between stage B and stage C, along with the *unfolding* that completes itself between stage D and stage E, appear to be associated with *bifurcations*, the mathematical analogues of phase transitions.[10] These bifurcations correspond to changes in the symbolic computational behavior of the system. For example, at stage A, the model generates the simplest finite state language on a 3-symbol alphabet: there is only 1 state, and all symbols are equally likely to occur at all points in time. Stages B and C approximate progressively more complex finite state devices in that the network becomes increasingly sensitive to differences between the symbols but exhibits no contraction or expansion; hence, the network cannot generate recursive structures. Stage D involves contraction and expansion, but the recursion is confused in a complex way which may correspond to *super-Turing computation* (Siegelmann, 1999). Stage E approximates a *pushdown automaton*. This apparent correspondence between bifurcations and grammar changes is one foundation of the claim that there is a link between the dynamical and symbolic computational perspectives (see Crutchfield & Young, 1990; Moore, 1998). The model exhibits *progressive quantitative change occurring during the stages*. For example, during the *unfolding* stage (illustrated in Fig. 8.3, panels C and D) the fractal scale at which the branches first overlap is continuously decreasing. Thus, the depth of embedding that the model can successfully parse is gradually increasing. This is a *progressive quantitative change* because it makes the model's behavior more like the fractal parser it eventually becomes, even though it has not yet crossed the threshold of becoming isomorphic to a pushdown automaton (see Fig. 8.3, crossing the threshold occurs between panels D and E).

What is meant by the claim that the model transits from a stage at which *well-formedness rules* are inadequate for describing its behavior

[9]Here, I use *phase portrait* to mean a diagram of the state space at one stage of training that gives enough information that, with knowledge of the system at hand, one can deduce the dynamics at that stage. Standard phase portraits of continuous systems display a sample of their continuous trajectories. Since FLNNs are both discrete (i.e., they jump from point to point) and stochastic (i.e., there are multiple futures of most points), plotting trajectories would make the figure very busy. See Tabor (2000) for more detailed trajectory portrayals.

[10]I have not yet formalized this claim, so it remains tentative. I am assuming a standard interpretation of bifurcation: a bifurcation has occurred if there does not exist a homeomorphism between the before and after systems (Guckenheimer & Holmes, 1983; Strogatz, 1994). Intuitively, two systems are homeomorphic if one can be put into one-to-one correspondence with the other by stretching and squeezing. For example, changes in dimensionality require bifurcations, so the transitions a→b and b→c appear to involve bifurcations.

to a stage at which the rules provide a good approximation? In Figure 8.3C, the region spanned by the model during the processing of all sentences of the ABC language has infinite area. This is because contraction has not developed yet and the inputs are causing incremental, equal-sized displacements. But the interpretation of the model as an approximation of a pushdown automaton depends on its use of the fixed-width Gaussian units in the second hidden layer to classify each grammatical condition it encounters. When the model does not reliably land within the *spotlights* of these Gaussian units, the basis of the isomorphism with a symbolic device is lost. Viewing the case from the perspective of self-organization, one might say that the model has not yet reached the stage where the constituents of the emergent structure are successfully coordinating.

(c) *The stages derive from the learning.* The qualitatively distinct stages come about through the iterative application of a single simple process: move to the best proximal weight setting. In this sense, the stages are derived from the learning. The derivation of the stages from the interaction of the simple, low-level learning process with data is interesting from a developmental standpoint. As noted under section (b), the different stages appear to correspond to different classes of computational devices.[11] Under classical maturational accounts, stages are generally stipulated; the causes of their distinct qualities are assumed to be buried in the history of evolution. The current model offers a view of what an explanatory developmental account might look like. We can study the model to find out why the stages occur, whether they will occur in the same order every time, and what types of deviations of the architecture and environmental conditions can stop them. More generally, the network may be a useful formal example to consider when thinking about the question of what it means for an organism to come up with something new.

In one regard, the current model offers an advance over catastrophe theory models (Thom, 1985), which identify traits that phase transitions can be expected to have but do not provide a great deal of insight into what causes the transitions. On the other hand, it is not clear that the bifurcations in the current model exhibit the full range of properties that catastrophe theory predicts (e.g., modality, sudden jumps, hysteresis, divergence, critical slowing down) and for which there is compelling evidence in the developmental literature (van der Maas & Molenaar, 1992; van Geert, 1998). If this discrepancy is present, it may stem, in part, from the fact that the FLNN does not model real-time processing; in this instance, the FLNN models what might be called *sentence-time processing* with its smallest grain being the word. In any case, further assessment of the relation between catastrophe theory and the current fractal learning approach may be fruitful.

(d) *Complex, domain-specific behavior from a general mechanism.* Expert evidence from currently practicing linguists on the types of grammatical patterns present in natural languages finds all of them best described with the symbolic paradigm. I'm skeptical that the sample reflects the true distribution. My skepticism stems from the fact that, until recently, people have not known how to look for anything else. Even within the symbolic paradigm, there is currently no satisfactory model of how a specific grammar is chosen from among the immense number of possibilities the paradigm affords. For these reasons, again recognizing that further empirical justification is needed, I treat it as desirable to adopt a model with very general computational capabilities and to let learning do significant work. This is in contrast to established linguistic treatments of native capacity (e.g., Gibson & Wexler, 1994), which assume that the complex machinery is prefabricated by the genes and only a finite number of

[11] One might be inclined to say that the model progresses monotonically from simplicity at birth (Fig. 8.3A) to complexity at adulthood (Fig. 8.3E). However, if my speculation above about the super-Turing character of the phase illustrated in Figure 8.3D is correct, then that phase, which presumably corresponds to some stage of childhood, is arguably more complex than the adult phase (Figure 8.3E).

discrete parameter settings must be selected in order to achieve adult behavior.

That connectionist learning is a very general induction system is well attested. Here, we appear to see it building a fairly complex kind of mechanism, a pushdown automaton, to suit the needs of a particular task. A question that needs to be asked is this: To what extent was this result contrived by the constraints put on the model? If the observed pushdown stack is the only complex behavior the model is capable of conducting (or if it is one of a small, finite number of possibilities), then it is not very revealing or remarkable that the learning algorithm discovered it. (Under these conditions, the treatment would be very similar to current nativist accounts.) On the other hand, if many behaviors are possible, then the model offers an appealing alternative to current versions of nativism in language learning theory. I know this much at present: close relatives of this model are capable of a great variety of behaviors, including all context-free languages that are free of all context (Tabor, 2000), a crossed serial dependency language (Tabor, 2001), several finite-state languages (Tabor & Terhesiu, 2004), and several chaotic processes (Tabor & Terhesiu, 2004). I do not yet know how many of these behaviors can be learned. Also, the question remains open as to whether these FLNN learning results will scale to more realistic natural language corpora.

(e) *Generalization.* When trained on shallow embeddings (e.g., down to 2 center embeddings) the model generalizes to greater depths (3+). In fact, its end-of-training behavior closely mirrors the behavior of the infinite-state pushdown automaton that inspired the training data.[12] The fact that the neural model generalizes to this well-known and very useful mechanism on the basis of a minute number of examples (i.e., 4 sentences in the depth 2 corpus for the ABC language) is surprising. Further tests show that when the model is trained on various other small subsets of the infinite state ABC language, it converges on the same infinite state behavior. This suggests that the infinite state behavior is a kind of attractor for the learning process. These observations constitute another piece of evidence in support of the view that there is a systematic relation between the symbolic and dynamical perspectives. They also imply a kind of causal role for symbolic structures in the developmental dynamics of connectionist models: the attractor steers the learning process.[13] This view is at odds with discussions that treat connectionism as a refutation of the relevance of symbolic models. On the other hand, it is also at odds with symbolic models that want to build in rules natively. On the present view, the recursive structure stems from a very general property of neurons (i.e., potential for fractal organization), which is closely related to the nature of physical space and thus is not an arbitrary property of particular organisms.

But how relevant are these infinite state languages to real human behavior? This is a point of long-standing controversy. Chomsky (1957) and many other people have argued that we ought to be studying ideal language ability—so-called competence—which is said to include the ability to process embeddings of arbitrary depth. That people cannot manage more than 3 levels of center-embedding easily (e.g., [The essay [that the guinea pig [someone [I knew 4] cared for 3] shredded 2] was about Duchamp 1].) is taken to be a performance limitation. It is supposed that a stack is limited, so we run out of space after only a few embeddings (Kimball, 1973). Others have argued that the distinction between competence and performance is dubious and that connectionist networks like Elman's simple recurrent network (SRN) predict the limited ability of people to process

[12] The network only approximates the symbolic device in two senses: (1) after finite training it has not completely converged and (2) the softmax units at the output layer take on activations in the open interval [0, 1], which does not include 0 and 1. Since these output activations are interpreted as probabilities, the network assigns a positive probability to every possible string. The network approximates symbolic behavior by assigning a relatively high probability to transitions that are deemed legal by the corresponding symbolic system.

[13] I am thankful to Denis Mareschal for suggesting this approach to thinking about the model.

center-embedded sentences without needing to adopt the full power of a pushdown automaton (Servan-Schreiber, Cleeremans, & McClelland, 1991) and without needing to make a seemingly untestable distinction between competence and performance (Christiansen & Chater, 1999). The proponents of the latter arguments offer connectionist models as examples of systems in which there is no distinction between competence and performance.

The present results, however, suggest a different interpretation of at least some connectionist models: the FLNN only exhibits perfect recursive computation if there is perfect balance between contraction and expansion and if weights and activation are computed with infinite precision. When any noise is added to the values, the system is computationally equivalent to a finite-state machine (Casey, 1996). This observation suggests a close parallel between Turing machines and FLNNs—the infinite precision is playing the role of the infinite tape (Pollack, 1987). Moreover, the possible attractor status of recursive weight settings suggests that even though real systems only approximate these settings, the idealized settings are relevant theoretical constructs which can help us understand the principles by which the device is succeeding at the task. Put this way, the fractal structure is very much like an element of *competence* in the sense of Chomsky (1957). This line of reasoning casts some doubt on the claim that connectionist models provide a basis for rejection of the competence–performance distinction (but see McClelland, 1993).

However, the argument does not stop here. It should be noted that there are important differences between the symbolic and FLNN realizations of the competence–performance issue. First, the current account is not a cavalier use of the distinction, in which data that do not match the theory are peremptorily labeled as *performance phenomena* and subsequently ignored. The model provides an explicit interpretation of both competence (i.e., a universe of topological structures) and performance (i.e., a particular location within that universe). Second, in order to predict that center-embeddings will be hard to process beyond 3–4 levels, the symbolic approach must bound the length of a stack to 3 or 4 elements, a stipulation which is not well motivated. By contrast, there is better motivation for claiming that noise is present in the neural implementation of a fractal computer; also, the combination of fixed-variance neural noise with fractal scaling correctly predicts that the growth of difficulty in center-embedding should have a roughly exponential structure (Miller & Isard, 1964). Third, the treatment of recursion as an emergent property rather than an architectural property gives the system a kind of flexibility that seems suited to generating the variety of recursive structures that natural languages exhibit (e.g., center-embedding vs. crossed-serial dependency; Savitch, 1987) and the mutability, over periods of decades and centuries, of grammatical structures (Hopper & Traugott, 1993). Fourth, the treatment of ideal devices as attractors rather than as mechanisms raises the possibility that multiple ideals could simultaneously shape behavior; this prediction was born out by psycholinguistic work (Tabor, Galantucci, & Richardson, 2004) and is possibly relevant to creole continua and other cases of language contact. Fifth, while the symbolic paradigm separates the ideal system from the perturbations of it caused by *performance factors* in a stark way—by defining the system as signal and the perturbations as noise—connectionist models claim that there is a continuum between the two. The latter position may be relevant for understanding how new structures arise in language change (Givón, 1971; Tabor, 1994, 1995) and for understanding why grammatical patterns in some languages parallel statistical patterns in others (Bresnan & Nikitina, 2003/2008-manuscript in preparation). (In such cases, noise may nourish metamorphosis and thus should not be segregated in the theory.) In these regards, the FLNN properties are consistent with the claim put forth by Servan-Schreiber and colleagues (Servan-Schreiber et al., 1991) that recurrent connectionist networks are a new kind of computational device that may naturally be called a *graded state machine*. Similarly, a fractal is aptly described as a *graded recursive structure*.

## Conclusion

The original motivation for developing FLNNs was to find a way for connectionist networks to effectively learn complex syntactic dependencies and to discern how connectionist treatments differ from symbolic treatments of these behaviors. It is important to clarify how much of this task has been accomplished and how much not, and to draw attention to other elements of language, not treated by this model, which may play an important role in the developmental picture.

First, it is known that connectionist models with these general characteristics (i.e., symbol predictors with recurrent hidden layers employing second order weights) are capable of correctly processing context-free languages (Moore, 1998; Tabor, 2000). I've only tested learning on two such languages, so it remains an open question how many of these parameterizations are reachable via learning.

Second, it is important to note that human language knowledge contains a good deal of systematic structure beyond that of phrase structure. One major element of this other structure is termed *pragmatic* or *semantic* knowledge. Some have argued that pragmatic knowledge plays a central role in bootstrapping syntactic understanding (e.g., Pinker, 1984). Others have argued for the reverse causal relationship (e.g., Fisher, Gleitman, & Gleitman, 1991). I am disinclined to believe that the causality is serial in either direction, but a mutual-causality view also attributes a significant role to pragmatic structure. Encouragingly, Elman (1990) found that pragmatic information reflected in the statistical properties of corpora was detected by a simple recurrent network and was encoded via refinement of the syntactic encoding. Servan-Schreiber and colleagues (Servan-Schreiber et al., 1991) found that subtle statistical biases of the type that plausibly stem from pragmatic knowledge can help an SRN learn syntactic structure. Weckerly (1995) provided sentence processing evidence that syntactic and pragmatic information are not independent but are intertwined in a way predicted by the SRN. I have not studied this kind of information structure in the FLNN yet, but it is clearly important, and a good next step will be to see if the fractal analysis can help elucidate the nature of the intertwining.

As a final remark, in some dynamical approaches to development as well as in some connectionist work, there is a tendency to view ideal concepts as misleading. For example, Thelen, Schöner, Scheier, and Smith (2001) remark:

> Does this [dynamical field theory] model have anything to say about Piaget's issue: when do infants acquire the object concept? We believe this question is ill-posed and cannot be answered because there is no such thing as an "object concept" in the sense of some causal structure that generates a thought or a behavior (Smith, Thelen, Titzer, & McLin, 1999). There is only "knowledge" of objects as embedded in the immediate circumstances and the history of perceiving and acting in similar circumstances. (p. 34)

In keeping with this view, the present work portrays a world in which knowledge is nothing more than experience-tuned action (Elman, 2004), there is considerable flexibility of behavior, and virtually any behavior can occur at any time in an organism's life if strong enough biases encourage it. And yet, according to the view put forth here, there is still value in referring to ideal, or absolute, forms; these are the limiting dynamical systems which exert a force-like influence on the learning process. An example in the present case is the infinite-state grammar which the network converges to in the limit of infinite training and the absence of noise. While it is important to recognize that a realistic network will never behave exactly like this infinite-state grammar, the grammar nevertheless provides insight into the nature of the system that the network uses to organize its knowledge of its training data. The current example thus encourages one to look for analogous attractors in other domains, in which development appears to cleave to, or consort with, coherent patterns of behavior.

## Acknowledgments

In addition to the very helpful interactions at the conference that prompted this book, I had enlightening discussions with James Dixon, James Magnuson, Jay Rueckl, Aaron Schultz, and Dalia Terhesiu.

## References

Abraham, R. H., & Shaw, C. D. (1984). *Dynamics, the geometry of behavior: Parts 1–4* [Series: Visual mathematics library: Vol. 1]. Santa Cruz, CA: Aerial Press.

Bharucha-Reid, A.T. *Elements of the Theory of Markov Processes and Their Applications*. NewYork: McGraw-Hill, 1960.

Barnsley, M. F. (1988). *Fractals everywhere*. Boston: Academic Press.

Barnsley, M. F., & Rising, H. (1993). *Fractals everywhere* (2nd ed.). Boston: Academic Press Professional.

Boden, M., & Wiles, J. (2002). On learning context-free and context-sensitive languages. *IEEE Transactions on Neural Networks, 13*(2), 491–493.

Bresnan, J., & Nikitina, T. (2008). The gradience of the dative alternation. In L. Uyechi & L.-H. Wee, *Reality exploration and discovery: Pattern interaction in language and life* [manuscript in preparation]. Stanford, CA: CSLI Publications. (Original work released 2003) Retrieved October 31, 2008, from Stanford University, Center for the Study of Language and Information website: http://www.stanford.edu/~bresnan/bresnan-nikitina.pdf

Casey, M. (1996). The dynamics of discrete-time computation, with application to recurrent neural networks and finite state machine extraction. *Neural Computation, 8*(6), 1135–1178.

Chomsky, N. (1956). Three models for the description of language. *IRE Transactions on Information Theory, 2*(3), 113–124. [Corrected version appears in R. D. Luce, R. R. Bush, & E. Galanter (Eds.). (1965). *Readings in Mathematical Psychology* (Vol. 2, pp. 105–124). New York: Wiley.]

Chomsky, N. (1957). *Syntactic structures* [Series: Janua linguarum (Gateway to languages), Vol. 4]. Hague, Netherlands: Mouton.

Christiansen, M. H., & Chater, N. (1999). Toward a connectionist model of recursion in human linguistic performance. *Cognitive Science, 23*(2), 157–205.

Cover, T. M., & Thomas, J. A. (1991). *Elements of information theory*. New York: Wiley.

Crutchfield, J. P., & Young, K. (1990). Computation at the onset of chaos. In W. H. Zurek (Ed.), *Complexity, entropy, and the physics of information* [Proceedings]. Redwood City, CA: Addison-Wesley.

Elman, J. L. (1990). Finding structure in time. *Cognitive Science, 14*(2), 179–211.

Elman, J. L. (1991). Distributed representations, simple recurrent networks, and grammatical structure. *Machine Learning, 7*(2–3), 195–225.

Elman, J. L. (2004). An alternative view of the mental lexicon. *Trends in Cognitive Science, 8*(7), 301–306.

Fisher, C., Gleitman, H., & Gleitman, L. R. (1991). On the semantic content of subcategorization frames. *Cognitive Psychology, 23*(3), 331–392.

Frank, R., Mathis, D., Gussine, E., Stowe, J., & Vindiola, M. (2005, September). *Question formation, neural networks, and the poverty of the stimulus*. Poster presentation at the annual Architectures and Mechanisms for Language Processing Conference, Ghent, Belgium. Retrieved October 31, 2008, from http://www.cog.jhu.edu/faculty/frank/

Gazdar, G. (1981). On syntactic categories [Special issue: The psychological mechanisms of language. A discussion organized jointly for the Royal Society and the British Academy]. *Philosophical Transactions of the Royal Society of London. Series B, Biological Sciences, 295*(1077), 267–283. Retrieve by subscription from http://journals.royalsociety.org/content/n57v56156693/

Gers, F. A., & Schmidhuber, J. (2001). LSTM recurrent networks learn simple context-free and context-sensitive languages. *IEEE Transactions on Neural Networks, 12*(6), 1333–1340.

Gers, F. A., Schmidhuber, J., & Cummins, F. (2000). Learning to forget: Continual prediction with LSTM. *Neural Computation, 12*(10), 2451–2471.

Gibson, E., & Wexler, K. (1994). Triggers. *Linguistic Inquiry, 25*(3), 407–454.

Givón, T. (1971). Historical syntax and synchronic morphology: an archaeologist's field trip. In *Proceedings of the Seventh Regional Meeting of the Chicago Linguistics Society* (pp. 394–415). Chicago: Chicago Linguistics Society.

Guckenheimer, J., & Holmes, P. (1983). *Nonlinear oscillations, dynamical systems, and bifurcations of vector fields* [Applied mathematical sciences]. New York: Springer.

Hinton, G. E. (Ed.). (1990). *Connectionist symbol processing*. Cambridge, MA: MIT Press–Bradford Books.

Hochreiter, S., & Schmidhuber, J. (1997). Long short-term memory. *Neural Computation, 9*(8), 1735–1780.

Hopcroft, J. E., & Ullman, J. D. (1979). *Introduction to automata theory, languages, and computation*. Menlo Park, CA: Addison-Wesley.

Hopper, P. J., & Traugott, E. C. (1993). *Grammaticalization*. New York: Cambridge University Press.

Jagota, A., Plate, T., Shastri, L., & Sun, R. (Eds.). (1999). Connectionist symbol processing: Dead or alive? *Neural Computing Surveys, 2*, 1–40.

Kimball, J. (1973). Seven principles of surface structure parsing in natural language. *Cognition 2*(1), 15–47.

Kozen, D. C. (1997). *Automata and computability* [Series: Undergraduate texts in computer science]. New York: Springer.

Kugler, P. N., & Turvey, M. T. (1987). *Information, natural law, and the self-assembly of rhythmic movement: Theoretical and experimental investigations*. Hillsdale, NJ: Erlbaum.

Levy, S., Melnik, O., & Pollack, J. B. (2000). Infinite RAAM: A principled connectionist basis for grammatical competence. In *Proceedings of the 22nd annual meeting of the Cognitive Science Society* (pp. 298–303). Mahwah, NJ: Erlbaum. Retrieved on October 31, 2008. from http://www.cs.wlu.edu/~levy/pubs/raam-cogsci00.pdf

Lindenmayer, A., & Prusinkiewicz, P. (1989). Developmental models of multicellular organisms: A computer graphics perspective. In C. G. Langton (Ed.), *Artificial Life* [Proceedings; Series: Santa Fe Institute studies in the sciences of complexity] (pp. 221–250). Redwood City, CA: Addison-Wesley.

Mandelbrot, B. B., & Ness, J. W. V. (1968). Fractional Brownian motions, fractional noises and applications. *SIAM review. Society for Industrial and Applied Mathematics, 10*(4), 422–437.

McClelland, J. L. (1993). The GRAIN model: A framework for modeling the dynamics of information processing. In D. E. Meyer & S. Kornblum (Eds.), *Attention and performance: Vol. XIV. Synergies in experimental psychology, artificial intelligence, and cognitive neuroscience* (pp. 655–688). Hillsdale, NJ: Erlbaum.

Melnik, O., Levy, S., & Pollack, J. B. (2000). RAAM for infinite context-free languages. *5Proceedings of the IEEE-INNS-ENNS International Joint Conference on Artificial Neural Networks*, (2000), 585–590. Silver Spring, MD: IEEE Press.

Miller, G. A., & Isard, S. (1964). Free recall of self-embedded English sentences. *Information and Control, 7*(3), 292–303.

Moore, C. (1998). Dynamical recognizers: Real-time language recognition by analog computers. *Theoretical Computer Science, 201*(1–2), 99–136.

Pearlmutter, B. A. (1995). Gradient calculations for dynamic recurrent neural networks: A survey. *IEEE Transactions on Neural Networks, 6*(5), 1212–1228.

Pinker, S. (1984). *Language learnability and language development* [Series: Cognitive science series]. Cambridge, MA: Harvard University Press.

Plate, T. A. (1995). Holographic reduced representations. *IEEE Transactions on Neural Networks, 6*(3), 623–641.

Plate, T. A. (2003). *Holographic reduced representation: Distributed representation for cognitive structures*. Stanford, CA: CSLI Publications.

Pollack, J. B. (1987). *On connectionist models of natural language processing*. Unpublished doctoral dissertation, University of Illinois.

Rodriguez, P. (2001). Simple recurrent networks learn context-free and context-sensitive languages by counting. *Neural Computation, 13*(9), 2093–2118.

Rodriguez, P., & Wiles, J. (1998). Recurrent neural networks can learn to implement symbol-sensitive counting. In M. I. Jordan, M. J. Kearns, & S. A. Solla (Eds.), *Advances in neural information processing systems, Vol. 10* (pp. 87–93). Cambridge, MA: MIT Press–Bradford Books.

Rodriguez, P., Wiles, J., & Elman, J. L. (1999). A recurrent neural network that learns to count. *Connection Science, 11*(1), 5–40.

Rohde, D. (2002). *A connectionist model of sentence comprehension and production*. Unpublished doctoral dissertation, Carnegie Mellon University.

Rohde, D. L. T., & Plaut, D. C. (1999). Language acquisition in the absence of explicit negative evidence: How important is starting small? *Cognition, 72*(1), 67–109.

Rumelhart, D. E., Durbin, R., Golden, R., & Chauvin, Y. (1995). Backpropagation: The basic theory. In Y. Chauvin & D. E. Rumelhart (Eds.), *Back propagation: Theory, architectures, and applications* [Series: Developments in connectionist theory] (pp. 1–34). Hillsdale, NJ: Erlbaum.

Rumelhart, D. E., Hinton, G. E., & Williams, R. J. (1986). Learning internal representations by error propagation. In D. E. Rumelhart, J. L. McClelland, & PDP Research Group (Eds.), *Parallel distributed processing: Explorations in the microstructure of cognition: Vol. 1. Foundations* (pp. 318–362). Cambridge, MA: MIT Press–Bradford Books.

Servan-Schreiber, D., Cleeremans, A., & McClelland, J. L. (1991). Graded state machines: The representation of temporal contingencies in simple recurrent networks. *Machine Learning, 7*(2–3), 161–193.

Siegelmann, H. T. (1999). *Neural networks and analog computation: Beyond the Turing limit.* Boston: Birkhäuser.

Smith, L. B., & Thelen, E. (Eds.). (1993). *A dynamic systems approach to development: Applications.* Cambridge, MA: MIT Press–Bradford Books.

Smith, L. B., Thelen, E., Titzer, R., & McLin, D. (1999). Knowing in the context of action: The task dynamics of the A-not-B error. *Psychological Review, 106*(2), 235–260.

Smolensky, P. (1990). Tensor product variable binding and the representation of symbolic structures in connectionist systems [Special issue]. *Artificial Intelligence, 46*(1–2), 159–216.

Strogatz, S. H. (1994). *Nonlinear dynamics and chaos: With applications to physics, biology, chemistry, and engineering* [Series: Studies in nonlinearity]. Reading, MA: Addison-Wesley.

Tabor, W. (1994). Syntactic innovation: A connectionist model [Doctoral dissertation, Stanford University]. *Dissertation Abstracts International, A: The Humanities and Social Sciences* 55(10), pp. 3178-A.

Tabor, W. (1995). Lexical change as nonlinear interpolation. In J. D. Moore & J. F. Lehman (Eds.), *Proceedings of the 17th Annual Cognitive Science Conference.* Mahwah, NJ: Erlbaum.

Tabor, W. (2000). Fractal encoding of context-free grammars in connectionist networks. *Expert Systems: The Journal of Knowledge Engineering, 17*(1), 41–56.

Tabor, W. (2001). Sentence processing and linguistic structure. In S. C. Kremer & J. F. Kolen (Eds.), *A field guide to dynamical recurrent networks.* Silver Spring, MD: IEEE Press.

Tabor, W. (2003). Learning exponential state-growth languages by hill climbing. *IEEE Transactions on Neural Networks, 14*(2), 444–446.

Tabor, W. (2005). *Fractal learning neural networks.* Manuscript submitted for publication. Retrieved October 27, 2008, from http://www.sp.uconn.edu/~ps300vc/Papers/tabor03flnn.pdf

Tabor, W., Galantucci, B., & Richardson, D. (2004). Effects of merely local syntactic coherence on sentence processing. *Journal of Memory and Language, 50*(4), 355–370.

Tabor, W., & Terhesiu, D. (2004). On the relationship between symbolic and neural computation. In AAAI Fall Symposium Technical Report (Report No. FS-04-03, pp. 63–68). Menlo Park, CA: AAAI Press. Retrieved by subscription from http://www.aaai.org/Press/Reports/Symposia/Fall/fs-04-03.php

Thelen, E., Schöner, G., Scheier, C., & Smith, L. B. (2001). The dynamics of embodiment: A field theory of infant perseverative reaching. *Behavioral and Brain Sciences, 24*(1), 1–34, commentary 34–70, authors' response 70–80.

Thelen, E., & Smith, L. B. (1994). *A dynamic systems approach to the development of cognition and action.* Cambridge, MA: MIT Press–Bradford Books.

Thom, R. (1985). *Structural stability and morphogenesis: An outline of a general theory of models* (2nd ed.). Reading, MA: Benjamin.

van der Maas, H. L. J., & Molenaar, P. C. M. (1992). Stagewise cognitive development: An application of catastrophe theory. *Psychological Review, 99*(3), 395–417.

van Geert, P. (1998). A dynamic systems model of basic developmental mechanisms: Piaget, Vygotsky, and beyond. *Psychological Review, 105*(4), 634–677.

Weckerly, J. (1995). *Object relatives viewed through behavioral, electrophysiological, and modeling techniques.* Unpublished doctoral dissertation, University of California, San Diego.

Wiles, J., & Elman, J. (1995). Learning to count without a counter: A case study of dynamics and activation landscapes in recurrent networks. In J. D. Moore & J. F. Lehman (Eds.), *Proceedings of the Seventeenth Annual Conference of the Cognitive Science Society* (pp. 482), Pittsburgh, PA. New York: Taylor & Francis–Psychology Press. Retrieved October 31, 2008, from http://crl.ucsd.edu/~elman/Papers/wiles_elman/counter.pdf

# CHAPTER 9

## The Robot as a New Frontier for Connectionism and Dynamic Systems Theory

*Matthew Schlesinger*

*By nature, men are nearly alike; by practice, they get to be wide apart.*
*Confucius (551–479 BCE)*

---

Are connectionism and dynamic systems theory (hereafter, DST) separate approaches to development or, instead, different manifestations of a unified theory of development? One way to answer this question is to place the two approaches side-by-side, and then to catalog all the similarities and differences, functionally, structurally, conceptually, and so on (e.g., Thelen & Bates, 2003). One difficulty with such an analysis, however, is that neither approach is a static entity. Both have an *evolutionary past* in cultural terms, that is, the intellectual history that has helped to shape each approach. In much the same way, both connectionism and DST also have their own unique developmental trajectories, influenced by the major figures who emerged and led their respective communities in a particular direction. More importantly, the simple act of dialog between the advocates (e.g., followers, practitioners) of connectionism and DST will inevitably influence the future of each approach, if only by the exchange of ideas and practices.

The compare-and-contrast strategy, it seems, offers limited prospects. As an alternative, I suggest a more holistic approach, in which connectionism and DST are viewed in organismic terms. What follows from this perspective? First, note that connectionism and DST share a similar historical heritage: both approaches became established in the latter quarter of the 20th century. In particular, proponents of both approaches were motivated, in part, by a reaction to the prevailing winds of *mentalism*. The spread of the cognitive revolution was also a factor. As a result, the two approaches share the view that knowledge is distributed and graded, rather than symbolic and centrally stored. Second, each has endeavored to formulate a broad, metatheoretical framework that identifies and investigates fundamental principles of development (e.g., continuity vs. discontinuity, endogenous vs. exogenous influences on development). Third, and perhaps most importantly, both connectionism and DST have constructed and utilized an impressive array of mathematical tools, methods, and models for describing, explaining, and predicting developmental processes.

To paraphrase Confucius, then: "By nature, connectionism and DST are nearly alike." In any case, their developmental trajectories over the past 20 years have often traveled in parallel directions (though not always at the same time), confronting many of the same questions and challenges. Over time, though, many salient differences have also taken root, almost certainly as a consequence of the specific mathematical techniques that each approach has exploited (e.g., feed-forward networks vs. differential equations, principal component analysis vs. catastrophe theory). So it is probably also true that "by practice, they've gotten to be wide apart."

My goal, then, is to provide an optimistic forecast for the future of connectionism and DST. In particular, I focus on the idea that—regardless of how similar or dissimilar connectionism and DST appear to be at this moment in their development—there are numerous signs that *hybridization* of the two approaches is not only possible, but also has begun to occur. In the next section, I first provide a context for my forecast by reviewing three major, crosscutting themes that are shared by connectionism and DST. Second, I highlight the evidence for an optimistic outlook by describing recent work in the field of adaptive behavior and robotics, which is illustrated by numerous examples of models that blend elements of connectionism and DST. Finally, in the conclusion, I return to the crosscutting themes and elaborate on each in light of the progress that robotics researchers have made toward a hybrid approach.

## Introducing Three Crosscutting Themes

To suggest that connectionism and DST are on converging paths is an admittedly hopeful prediction, especially in light of the fact that advocates for one approach hardly speak the same language as advocates for the other approach! For example, some of the jargon used by connectionists includes *feed-forward activation*, *supervised learning*, and *gradient descent*, while the DST community talks about *stability versus instability*, *attractors*, and *bifurcations*. Still, these differences really beg the question: are the differences only skin deep, and if so, how deep are the *threads* they share, which bind the two approaches together? In this section, I focus on three specific themes that cut across connectionism and DST, each helping to guide two separate developmental trajectories toward a common destination.

### Shared Concepts

There are a number of key ideas that are shared by both connectionism and DST. I highlight here two major concepts. First, a particularly strong link between the two approaches is through the concept of *embodiment*. While the term is broad enough to be used in many different ways, both approaches generally agree that knowledge should not—maybe even cannot—be divorced from the physical context in which it is acquired and manifested. DST has often pushed this notion to its logical limits, that is, *organism-environment coupling* in which the concept of embodied knowledge includes both the organism's physical body and the *outside* world in which that body acts (e.g., Smith & Thelen, 1993; Thelen, Schöner, Scheier, & Smith, 2001; Thelen & Smith, 1994).

While DST makes an explicit commitment to the concept of embodiment, connectionism appears at times to be less certain. In particular, throughout the 1990s a large number of neural network models were designed and studied that seemed to offer only a minimal, indirect role for the physical body and the physical environment during development (e.g., Munakata, McClelland, Johnson, & Siegler, 1997; Plunkett & Marchman, 1991; Quinn & Johnson, 1997; for a critical review, see Schlesinger & Parisi, 2001a). For example, consider the task of learning to convert present-tense verbs into their past-tense counterparts (i.e., verb inflection). By focusing on a small piece of the information-processing stream, a model of this task might assume that (1) sensible auditory signals comes from *somewhere out there*, (2) some disembodied perceptual system *somewhere else* in the nervous system parses the auditory stream into appropriate units (e.g., syllables, phonemes), and (3) after the morphology problem is solved, *some other part* of the nervous system will know what to do with that information (e.g., utter the correct form of the desired verb).

It would be incorrect, though, to suggest that connectionists are unaware or unconcerned with the concept of embodiment (e.g., Bechtel, 1997). Indeed, there are a number of recent models that explicitly incorporate both structural and functional elements of embodiment (e.g., Mareschal, Plunkett, & Harris, 1999; Mareschal & Shultz, 1993; Schlesinger & Parisi, 2001a; Schlesinger, Parisi, & Langer, 2000). For example, these models are more structurally complex than generic feed-forward models, reflecting the idea that neural embodiment requires a diverse range of neural structures and functions (e.g., *what* vs. *where* visual streams; see Mareschal,

Plunkett, & Harris, 1999; Schlesinger, 2006). In addition, these models pay careful attention to how sensory signals originate, sometimes simulating the transduction process itself (e.g., with a simulated retina; see Schlesinger, 2003). Finally, there are also a class of connectionist models—called *ecological neural networks*—that simulate *both* the organism and the environment, allowing the two to interact in exactly the tightly coupled way DST envisions (e.g., Parisi, Cecconi, & Nolfi, 1990).

As I emphasized earlier, a second, related concept shared by connectionism and DST is the idea of *distributed knowledge*. Again, while the language used by each approach is often quite different—for example, connectionism's term is *distributed representations* while DST's term is *soft assemblies*—the core concept is more or less the same. Each approach emphasizes the idea that *whatever knowledge is* (and surely, there are important differences between the epistemology of connectionism and DST),[1] it is not a passive store of information located in a central location waiting to be retrieved for a relevant situation. Insofar as connectionism and DST share a common view, it is that knowledge is fluid, dynamic, and somehow *spread out* over space and time.

Of course, advocates of DST may disagree with the idea that knowledge-as-distributed-representations are anything like the in-the-moment, ephemeral quality of a soft assembly (e.g., the convergence of motor neuron activity in a particular context, at a particular moment, that gives rise to a particular reaching movement). Once again, this skepticism is likely the product of an overabundance of standard feedforward network models in the literature (for a critical review of the so-called back-prop nets, including an overview of alternative modeling approaches, see Schlesinger & Parisi, 2004). To alleviate this unfortunate bias, consider the subset of connectionist models, which (for some unknown reason) are less-often used and written about; importantly, these models are intentionally designed so their internal states are always a function of (a) whatever sensory experience is occurring at that moment and (b) their internal state a moment before. For example, simple recurrent networks, Hebbian networks, and auto-associator networks, all exhibit this capacity for time- and state-dependent activity (e.g., Elman, 1990; Munakata & Pfaffly, 2004; Sirois, 2004). Watching one of these networks in action leaves the observer with the feeling that they are highly dynamic, *percolating*, *bubbling*, perhaps even *deliberating their options* in much the same way that a dynamic field model seems to do (e.g., Hock, Schöner, & Giese, 2003; Schutte, Spencer, & Schöner, 2003).

### Timescales

A second theme that links connectionism and DST is the dimension of time. In particular, both approaches agree with the idea that to understand a given behavior, it must be analyzed over multiple timescales. Even the somewhat mundane example of an infant's reaching movement illustrates a behavior that spans a surprisingly wide range of timescales. To begin, consider (a) microscopic timescales, including multi-millisecond events such as changes in neural and muscle activation patterns, versus (b) macroscopic, psychologically salient timescales during which overt behaviors occur (i.e., those visible to the naked eye, for instance, movement of the infant's hand toward a cup). Taken together, these first two timescales are conventionally referred to as *real time*. In addition, consider also (c) the timescales including and extending just beyond real time, the so-called *microgenetic* span of time (e.g., minutes,

[1] For example, many connectionists treat the mind as a *representational system* whose primary function is *information processing* (i.e., encoding, storing, and retrieving task-relevant knowledge). In contrast, DST often takes more of an antirepresentationalist stance, discarding the view of the mind (and the knowledge it contains) as a stable entity/structure, instead replacing it with a suite of ongoing perceptual, cognitive, and behavioral processes that exist only in the momentary realization of the soft assembly.

hours, days), (d) the *ontogenetic* history of the infant, spanning days, weeks, and months, which allows the cumulative outcomes of past reaches to influence the one that is occurring right now, and finally (e) the *phylogenetic* history of the species, which imposes many of the neural and biomechanical constraints that our young infant is learning to identify and exploit (e.g., the elbow and shoulder joints permit two-dimensional versus three-dimensional rotations, respectively).

A shared interest in multiple timescales of behavior is one thing. It is a different question, though, to ask how, *in practice*, connectionism and DST have addressed the issue. Indeed, DST makes a strong commitment to studying multiple timescales, while connectionism has focused less on this issue. Admittedly, it is probably a fair criticism to note that connectionist models of development have rarely investigated timescales at the *fast* end of real time (e.g., at the level of milliseconds), even though the connectionist *zeitgeist* itself is inspired by the concept of neural processing. In addition, there are also few connectionist models of behavior at the speed of naked-eye real time, though models of reaching are a notable exception (e.g., Kettner, Marcario, & Port, 1993; Schlesinger et al., 2000; Vos & Scheepstra, 1993).

Again, there may be an element of *déjà vu* at play here. The problem is not that connectionists are ambivalent about studying different timescales. Rather, it may follow from the pervasive practice of using generic feed-forward networks, which exist *out of time* insofar as they (more or less) instantaneously produce an output for a given input. In other words, their architecture lacks internal dynamics, and as a result, feed-forward networks are in a sense too blunt an instrument to study behavior as it unfolds (cf. cascade-activation models, e.g., Cohen, Dunbar, & McClelland, 1990).

To be fair however, there is a bright side to this story. While many connectionist models of development may treat the individual behavioral unit as a molar, indivisible whole, virtually all of these same models are successful at relating the timescale of that behavioral unit to longer timescales. To illustrate, let us return to the problem of learning to inflect English verbs. Plunkett and Marchman (1991) show how varying verb frequencies during learning can lead to the U-shaped acquisition pattern observed in English-speaking children (e.g., both the model and children produce *goed* after first learning to produce *went*). More generally, connectionist models of syntax acquisition, object permanence, and reasoning on the balance scale, just to name a few, have addressed a variety of long-term, developmental phenomena, including changes in knowledge representation, stages of development, and transition mechanisms (e.g., Elman, 1993; Munakata, 1998; Shultz, Mareschal, & Schmidt, 1994).

While in practice connectionist and DST models have tended to focus at different places on the behavioral timescale, there are, in principle, no fundamental obstacles to bridging these gaps. Indeed, it might be seen as a strength of the two respective approaches that when combined, a hybrid connectionist–DST model could conceivably span the entire timescale from milliseconds and minutes up to days, months, and years. However, as I highlight in the next section, parsimony is a key element of model building. In other words, just because an ambitious, across-all-timescales model *could* be built, does not necessarily mean it *should* be built.

## Constraints

It is impotant to note that all models, whether designed from the connectionist or DST perspective, are by necessity a simplified version of the behavioral phenomenon they are intended to represent. In other words, in order to capture a specific feature of development, a model must leave other things out. A model of perceptual categorization may ignore the fact that visual inputs include color and stereoscopic information. A model of semantics and word learning may ignore the fact that different cultures both speak and listen to their children in different ways.

While the choice of what to leave in is by no means an exact science, there are several meaningful metrics for deciding whether a model *has enough in it* to be a reasonable simulation of a real-world phenomenon. One, of course, is parsimony (e.g., Ockham's razor): if two models are equivalent in performance, the one with *less stuff inside* (i.e., fewer untested assumptions, lower number of free parameters) is to be preferred. A second metric concerns how closely the output of the model corresponds to the real-world behavior it is intended to represent. For example, a neural network model of reaching may produce as output a vector of joint angles, which are used to instantaneously move the simulated arm to a new position (e.g., Kettner et al., 1993). Alternatively, the output of the model might be interpreted as muscle activation values, which incrementally move a virtual arm in a particular direction (e.g., Schlesinger et al., 2000). Note that of the two approaches, it is the second, more embodied model that has the opportunity to exploit kinematic constraints that arise in real time as the arm is being moved (e.g., interactions between shoulder and elbow rotations). Finally, a third, related metric involves what happens on the input side. In particular, the more desirable model of the two is the one that requires as little preprocessing or *massaging* of the raw input as possible.

The common constraint that puts these metrics into a single, coherent package is *empirical data*. In other words, the story begins when a behavioral researcher gives the model-builder a behavioral dataset and asks, "Can you model this?" Three fundamental (but not exclusive) goals are to (a) capture performance differences in the data—with an emphasis on developmental patterns—while keeping the (b) input and (c) output of the model as similar (or analogous) as possible to the stimuli presented to, and the responses produced by, respectively, the participants in the behavioral study.

At first glance, DST models appear to have an advantage over connectionist models with regard to modeling specific behaviors. In particular, a model that does a better job of capturing behavior in real time is likely to have a close correspondence between its output and real-world behaviors (e.g., Schutte & Spencer, 2002; Thelen et al., 2001). Despite their strengths on the output side, however, these models have had less success in exploiting input representations that are naturalistic, or at least close analogs to the kinds of stimuli that are used in comparable behavior tasks (cf. Bicho, Mallet, & Schöner, 2000; Steinhage & Schöner, 1997).

In contrast, connectionist models have some ground to cover before they can adequately claim to capture empirical data patterns in more than a coarse, roughly analogous sense. Indeed, I have already criticized generic feed-forward models as (a) suffering from disembodiment while (b) lacking the capacity for internal dynamics. To the list we can now add a third deficit: many connectionist models of the last decade only weakly resembled, if at all, the behavioral phenomena they were designed to study. A good example of this tendency is offered by looking at models of the balance scale task, in which a child is shown a balance scale—with weights on each side of a fulcrum—and is asked to predict which side, if either, will tip. Connectionist models of this task *simplified* things by distilling the input into four numeric features (the two weights and their distances from the fulcrum, respectively); the output, meanwhile, is two dichotomous variables, scaled so that three outcomes can be *read off* the model (i.e., tip left, tip right, or balance; e.g., McClelland, 1989; Shultz et al., 1994).

Despite being somewhat impoverished analogs of real-world experience, these models remain a valuable research tool because they are able to account for a variety of developmental phenomena (e.g., four stages of reasoning on the balance scale task). In other words, they are often good models of the ontogenetic timescale. Nevertheless, there is an emerging consensus that if connectionist models of development are going to progress beyond their infancy—metaphorically speaking—they must undergo a number of qualitative changes, including (a) capturing behavioral patterns at the level of real time and (b) providing a closer correspondence (on both the input and output sides) to behavioral data (e.g., Christiansen & Chater, 2001; Schlesinger & Parisi, 2001a). In the next section, I survey a variety of research programs in the field of robotics and adaptive behavior, with the goal of demonstrating not only that the metamorphosis of connectionism is already well underway, but also that DST is coming along for the ride.

## LINKING CONNECTIONISM AND DST THROUGH ROBOTICS

Sheldon Kopp (1972), in a quirky blend of Buddhism and psychotherapy, writes in *If you meet the Buddha on the road, kill him!*: "The Zen way to see the truth is through your everyday eyes. It is only the heartless questioning of life-as-it-is that ties a man in knots" (p. 187). Perhaps he is right, or perhaps not. Nevertheless, the point here is to highlight a fundamental tension between a philosophical question—namely, are connectionism and DST one thing or two?—and a practical reality—that is, the scientific progress being made by those who are relatively unconcerned (or unfettered) by philosophical constraints. To be blunt, some of this progress has been achieved by researchers from other disciplines (e.g., engineers, computer scientists) who have been using the tools and ideas of connectionism and DST while our attention was focused on resolving philosophical matters.

To be fair, it is more than a little disingenuous to suggest that the underlying theoretical tenets or mathematical tools that play a central role in connectionism and DST *belong*, in any concrete sense, to either approach. Of course, they are borrowed or, more accurately, shared. And so it is somewhat inevitable that a brief survey outside psychology reveals a life for connectionism and DST that we might never have suspected or imagined. Indeed, engineers, computer scientists, roboticists, and others who study adaptive behavior have retooled/reshaped the two approaches, blending them into a variety of unique models, paradigms, and computational hybrids.

In this section, I sample from recent work in robotics and adaptive behavior, highlighting convergent and complimentary applications of connectionism and DST. This survey is organized to focus on three fundamental lessons for students of development. First, the robotic platform—whether real or simulated—is an ideal tool for studying both microgenetic and ontogenetic changes. Second, the field of *evolutionary robotics* supplements the developmental perspective with the means to study phylogenetic (i.e., inherited) changes. And finally, robotics provides a comprehensive platform for studying embodied knowledge.

### From Real time to Developmental Time

A convenient fact about most robots is that they are designed to "exist" in the here and now. In other words, their behaviors are often temporally scaled so that (a) they are observable to the naked eye and (b) they correspond to the behaviors of real organisms (e.g., a colony of ants, a mechanical dog, a humanoid robot). As a consequence, robots are interesting devices in their own right that elicit a wide range of social-interactive behaviors from curious humans (e.g., Adams, Breazeal, Brooks, & Scassellati, 2000). In more formal language, the *life* of a robot spans multiple timescales, ranging from real time up to ontogenetic or developmental time (i.e., from milliseconds and minutes to days and years).

#### Real Time

The task of designing a robot that can operate in real time is a fundamental challenge for a robot builder. At a minimum, a robot is a *sensorimotor system*: it has *sensors* that operate in one or more modalities (e.g., tactile, infrared, sonar, digital video), which are used to gather information about its current state, and it has *effectors* (e.g., motor systems), which enable it to move, orient its sensors in a new direction, and more generally, interact with its environment. In engineering terms, the physical device that is the robot is called the *plant* (e.g., akin to a power plant). Between the sensors and effectors, the robot builder designs and implements an input–output system called a *controller*. The controller converts sensory inputs into desired motor outputs through a mapping or function called a *policy*.

Researchers in the field of *machine learning* pursue methods for computing, discovering, or learning *optimal policies*; that is, policies that enable a robot (or other input–output device) to reach a desired goal as efficiently (or reliably, quickly, etc.) as possible. It is important to note that finding an optimal policy depends on whether or not the number of states and actions for a given task is within computationally tractable limits (e.g., tic-tac-toe vs. backgammon or chess). For simple tasks, an optimal policy can be computed in advance by mathematical analysis. For other tasks—particularly those with large, multidimensional state spaces (i.e., possible sensor values)—an approximation method is used to estimate the optimal policy. In particular, the latter case is often the starting point for researchers in the field of *adaptive behavior*; they use *incremental learning methods* to gradually improve the behavior of their robot (or other autonomous agent) as it interacts with a real or simulated environment. In other words, analytical methods correspond to *fixed policies*, while incremental methods correspond to *adaptive policies* that are modified through experience.

Adaptive policies often rely on sensory feedback as a behavior unfolds (e.g., visual and proprioceptive signals as a hand reaches toward a target). Thus, adaptive policies are also called *closed-loop* policies; a given behavior generates sensory feedback, which is then used to generate the next behavior, which produces new sensory feedback, and so on. In contrast, fixed policies are called *ballistic* or *open-loop* policies; once the behavior (or behavioral sequence) begins, ongoing sensory feedback is not used to modify or update the programmed behavior.

The use of adaptive or closed-loop policies to control the movement of a mobile robot raises two important issues. First, *kinematic control* involves moving a robot through two- or three-dimensional space (i.e., changing position over time). However, robots are composed of multiple, mechanical segments (e.g., wheels, arms, legs), which can move in a wide variety of ways. Because the segments of the robot are physically linked, the movement of one segment can lead to the movement of another (e.g., rotation of the shoulder joint moves not only the upper arm but also the forearm; e.g., Turvey, Fitch, & Tuller, 1982). Second, *dynamic control* is a higher level problem, as it involves the same challenges of kinematic control with the added task of moving the robot while also accounting for its mass and acceleration (i.e., changing velocity over time), while several forces act on the body (e.g., gravity, inertia). In other words, like the kinematic effects that moving one limb or mechanical segment has on another, there are also dynamic *reactive forces* that are generated when part of the robot's body is moved. For example, rotation of a humanoid robot's torso or body axis creates a centripetal force, which causes the arms to move outward, away from the body.

Thus, a fundamental challenge of employing adaptive control in real time requires dealing with unintended movements that are due to both the kinematic and dynamic interactions between body segments. A traditional strategy for dealing with this problem is to *linearize the plant*, that is, to remove or suppress the nonlinear, dynamic interactions by increasing the stiffness of the links or joints between each of the segments (this helps explain the slow, stiff movement of earlier humanoid robots).

In contrast to linear methods, there are a number of more recent approaches which seek to identify the nonlinear properties of robotic systems. For example, some researchers use artificial neural networks as a computational structure for both acquiring and storing the policy. In particular, an advantage of using multilayered neural networks is that they implement a form of *nonlinear function approximation*, in which complex or dynamic relationships between sensory input and desired motor output can be estimated with incremental methods. Comparable methods can also be used to design and train a *forward model*, which learns the mapping in the opposite direction, that is, the sensory input that is predicted to occur as the result of a planned action (see Jordan & Rumelhart, 1992).

A second, more general approach is the use of *dynamic control methods*, which not only identify but also exploit the intrinsic dynamics of the robot's body or mechanical system. For example, the trajectory of a humanoid robot's arm movement can be controlled by planning in high-dimensional coordinates that include (1) the current sensory input, (2) the optimal or desired trajectory, and (3) the posture at the end of the movement. Alternatively, dynamic control methods such as *equilibrium-point control* can be used to reduce the dimensionality of the planning problem (e.g., Mussa-Ivaldi, 1999). According to the *equilibrium point hypothesis*, movement of the arm can be generated by shifting the equilibrium point of the arm (i.e., the point at which opposing forces on each joint are balanced) from one stable configuration or posture to another. By this view, high-dimensional planning or control is both inefficient and unnecessary; an optimal reaching trajectory can emerge for free, in a sense, by allowing the nonlinear dynamic properties of the mechanical system to determine the real time characteristics of the movement.

Indeed, there is a growing community of researchers who advocate for exploiting, rather than suppressing, the intrinsic dynamics of robots. For example, they argue that dynamic control methods are more (a) cost efficient in energetic terms, (b) cost efficient in cognitive terms (e.g., planning at a low-dimensional, more abstract, level of behavior), (c) biologically plausible, (d) generalizable or transferable across sensorimotor systems, and (e) robust over noisy/uncertain/unpredictable environments (e.g., Bizzi, Mussa-Ivaldi, & Giszter, 1991; Flash, 1987; Grupen, 2003; Schaal, Kotosaka, & Sternad, 2000; Schweighofer, Arbib, & Kawato, 1998; Williamson, 1998).

### Developmental Time

Compared to the study of adaptive behavior in real time, robotics researchers have devoted less attention to studying behavioral changes that occur over days, weeks, months, and years. Nevertheless, the disciplines of robotics and developmental psychology have recently begun to interact and exchange ideas, giving rise to a hybrid field of study called *developmental robotics* (e.g., Lungarella, Metta, Pfeifer, & Sandini, 2003; Weng, et al., 2001). The two goals of developmental robotics are reciprocal: (a) to exploit basic principles of learning and development in order to build better robots and (b) to use robots and other artificial agents as a platform for studying developmental processes.

It should be stressed that while some developmental robotics research is, at best, loosely inspired by developmental psychology, there are numerous studies that take development seriously. For example, one active area of research focuses on the fundamental issue of development as a *sequence of qualitative changes* (i.e., stages). This issue has been investigated in a variety of ways, including robotic models and simulations of developmental phenomena such as self-recognition and hand–eye coordination (e.g., Michel, Gold, & Scassellati, 2004; Schlesinger et al., 2000). A related area of research focuses on *growth of the body* and corresponding developments in sensorimotor activity (e.g., Dominguez & Jacobs, 2003; Lungarella & Berthouze, 2003).

As a detailed example of this approach, consider the development of spatial orienting and navigation. Prior to the onset of crawling, young infants tend to employ an *egocentric orienting strategy*, which uses the infant's body as a reference point for localizing nearby objects (e.g., "on my left side"). As infants begin to crawl and walk, however, an *allocentric orienting strategy* emerges, which relies on stable, external landmarks for localizing objects (e.g., "next to the chair"; Acredolo, Adams, & Goodwyn, 1984). Hiraki, Sashima, and Phillips (1998) investigated the shift from egocentric to allocentric orienting in a mobile robot by presenting the robot with a three-cup shell game. Prior to testing, the robot experienced either passive locomotion (i.e., by being pushed) or self-produced locomotion. Mirroring the developmental pattern observed in human infants, experience in the passive locomotion condition resulted in egocentric responding on the shell-game search task, while experience in the self-produced locomotion condition resulted in allocentric responding.

As Lungarella and colleagues (Lungarella et al., 2003) note in their review, however, much of the robotics research that is inspired by developmental processes focuses on sensorimotor systems, that is, "competence in interacting with the local environment—in particular, basic visuo-motor competencies such as saccading, gaze fixation, joint attention, hand-eye co-ordination and visually-guided reaching" (pp. 175–176). Indeed, it might be argued that while developmental robotics is well suited to studying low-level problems such as motor control, there are numerous skills and abilities that might prove difficult, if not impossible, to develop in a machine at a level that rivals human performance (e.g., speech or face recognition). In other words, sometimes it might be better to use existing engineering knowledge to build in or hardwire a skill, rather than use adaptive processes to acquire it. How can we tell when it is best to use adaptive processes?

This question can be addressed in several ways. First, as I noted earlier, the use of analytical methods—in which a policy is engineered or computed by hand rather than learned—is limited to tasks or problems where the size of the state space is tractable. However, for problems with large state spaces, prior knowledge can be built into the search process, so that even when approximate or incremental methods are employed, less time is spent searching regions of the state space where suboptimal solutions reside. On a related note, hardwired systems tend to do better when the environment is *stationary* over time (i.e., once acquired, the optimal policy does not need to be adjusted with changes in the environment) and predictable or *deterministic* (i.e., each action always results in the same outcome). In contrast, systems that learn, adapt, or develop generally have an advantage over hardwired ones when the state space is large, the environment is nonstationary, and sensory data are imperfect, uncertain, noisy, or *stochastic* (i.e., each action has a range of probable outcomes).

Second, regardless of whether the performance of a developing robot is superior to that of a hardwired robot, many developmental researchers are more interested in the developmental process than in the outcome itself. For instance, there is growing interest in the notion of *developmental immaturity*, that is, the idea that limiting or constraining early developmental capacities can facilitate long-term development (e.g., Bjorklund & Green, 1992). This approach has been successfully applied to a variety of perceptual and motor skills, including learning to swing, to reach, and to exploit binocular disparity as a depth cue (e.g., Dominguez & Jacobs, 2003; Lungarella & Berthouze, 2002; Schlesinger et al., 2000).

### Evolutionary Time

Complementing the study of developmental robotics, there is a parallel community of researchers who investigate *evolutionary robotics* (e.g., Nolfi & Floreano, 2001). The field of evolutionary robotics includes two major areas of research (e.g., Schlesinger, 2004). First, some investigators focus on the use of *evolutionary computation* (i.e., neo-Darwinian principles of variation and selection) as a search strategy for exploring large state spaces. Research from this perspective tends to focus on issues central to engineering and mathematics, including efficient search for optimal policies, tuning of algorithms, and scaling of policies from simulation to real-world environments. Note that scaling—that is, converting a simulated or *toy-world* model so that it can operate in a more complex or real environment—is a fundamental issue. Some researchers argue that both training and testing should be conducted exclusively with real robots in a real environment (e.g., Brooks, 1991). Alternatively, others investigators suggest that, while a policy can be acquired in a simulation, the performance of the model should be confirmed by testing it in a real environment (e.g., Husbands, Harvey, Cliff, & Miller, 1997; for a detailed discussion, see Nolfi, Floreano, Miglino, & Mondada, 1994).

Evolutionary computation, which is primarily a set of tools for machine learning, is contrasted by a second area of evolutionary robotics research, in which the robot provides a platform for studying biological processes. In particular, many *artificial life* (Alife) researchers use the same neo-Darwinian algorithms, but with the goal of simulating all biological phenomena, both real and possible (Langton, 1995; Parisi, 1996).

The general research method of evolutionary robotics is to create a large population of agents, and then allow all of the agents to compete on a given task. Because the individuals who are most fit are chosen to reproduce (i.e., determine the members of the next generation of robots), the performance of the population as a whole tends to improve over time. This computational strategy illustrates the concept of *hill-climbing*, in which small changes to the model are made that seek to incrementally improve performance. In contrast, *gradient-descent* algorithms like back-propagation-of-error seek to minimize or reduce errors, which are represented as the difference between observed and ideal performance.

It should be stressed that by including the evolutionary timescale, evolutionary robotics expands the range of issues that developmental researchers can investigate. For example, how does the evolutionary history of a species influence the development of a particular ability (e.g., motor skill, foraging and navigation, language acquisition; see Hinton & Nolan, 1987)? What precursors or components of the ability tend to be encoded genetically and inherited versus learned anew by each generation? How are different levels of adaptation—genetic, neural, behaviour, and so forth—related during both evolution and development (see Cangelosi, Parisi, & Nolfi, 1994)? Consider the various levels of biological adaptation: genetic, neural, behavioral, and so forth (see Cangelosi, Parisi, & Nolfi, 1994). More generally, evolutionary robotics is an ideal platform for investigating *epigenetic processes*, that is, the role of genes and environment in the interaction between learning, maturation, development, and evolution.

An additional perspective offered by evolutionary robotics—one that is often overlooked by developmental researchers—is the population-based approach. In other words, evolutionary robotics also provides a comparative perspective, in which a variety of developmental patterns or trajectories can emerge within a population of artificial agents. As a concrete example, let us return to the development of *allocentric spatial orientation* (i.e., the use of external landmarks and spatial cues; e.g., Newcombe & Huttenlocher, 2000; Wang, Hermer, & Spelke, 1999). Using a mobile robot within the evolutionary framework, Miglino and Lund (2001) studied this capacity and found that a number of component behaviors, including wall-following and orienting toward landmarks, emerged independently in a population of simulated robots, and then became coordinated, giving rise to successful navigation skills.

Of particular relevance to connectionism and DST is work in evolutionary robotics that synthesizes elements of both approaches. For example, Beer and colleagues (e.g., Beer, 1996; Slocum, Downey, & Beer, 2000) have studied populations of simulated agents whose nervous systems are represented by continuous-time, recurrent neural networks (CTRNNs), which evolve over multiple generations. Like conventional recurrent networks, CTRNNs have a dynamic internal state that is time dependent. In other words, their response to the same stimulus at two different points in time will vary as a function of their internal state (i.e., recent experience). From a functional perspective, then, CTRNNs are able to maintain a short-term internal representation that can serve as a memory trace of recent experience. Beer exploits this feature to show how a population of evolving CTRNNs can solve a number of *minimally cognitive* problems, such as catching falling objects, self-object discrimination, and selective attention (e.g., Slocum, Downey, & Beer, 2000). As highlighted above, it should be noted that approaches like these—that investigate behavior at both population and individual levels—provide an integrated computational framework for relating changes within and between the evolutionary and developmental timescales.

## Embodied Knowledge

To conclude the survey of robotics and adaptive behavior, I return to the issue of *embodied knowledge* (i.e., the issue of robots as *embodied knowers*). In particular, it is interesting to note that while there is wide variation in the types of platforms that robotics researchers use, and even wider variation in the questions they investigate, there is a broad consensus that the physical structure of the robot and the environment that it inhabits collectively exert a fundamental influence over how the robot learns, develops, and evolves (e.g., Brooks, 1991; Grupen, 2003; Lungarella et al., 2003; Weng et al. 2001).

This issue highlights a critical difference between connectionist models that are disembodied (i.e., have only an implied connection to the environment) and those that are embodied and situated (i.e., occupy and interact with an explicitly defined environment). The strategy of *classic connectionism* is to present input (sensory) patterns in random sequences to minimize the influence of experiencing one input pattern before another. Alternatively, networks that have a body and occupy an environment (i.e., autonomous agents, ecological networks) *actively sample* or *self-select* their sensory inputs (e.g., Nolfi & Parisi, 1993; Parisi, 1997). In other words, by virtue of their own behavior, embodied and situated networks are able to (a) limit or constrain the kinds of sensory data that reach their nervous system at any given moment and, more importantly, (b) modulate the pattern of data flow over time (e.g., Schlesinger & Parisi, 2001a).

The study of embodied networks raises two important questions. First, does self-selection of sensory input facilitate or simplify any particular developmental tasks? Second, what similarities, if any, are there between the developmental trajectories of embodied networks and real organisms?

A recent model of hand–eye coordination in infants can be used to address both of these questions (Schlesinger & Parisi, 2001b; Schlesinger et al., 2000). The model employs a population-based approach, in which a collection of artificial neural networks are used to represent a population of motor units that control the movement of an eye and arm. The position of the arm endpoint is determined by three joint angles: (1) the orientation of the body axis, (2) the shoulder joint, and (3) the elbow joint. Each of the arm joints, in turn, is controlled by two muscles that pull in opposite directions (i.e., flexor–extensor pairs). Note that because the arm reaches along a two-dimensional surface, only two of the three joint angles (i.e., the body axis and the elbow joint) are needed to reach an arbitrary point. In effect, then, the model is placed in a two degrees-of-freedom (DOF) workspace, leaving it with an extra or redundant DOF.

The model rapidly acquires a set of control strategies that exploit its particular physical design (Schlesinger et al., 2000). For example, it quickly learns to lock out the shoulder joint by rotating the shoulder to one side of its body, while it keeps the elbow joint away from the fully flexed or extended position. In addition, the model also shifts from a pattern of coactivating the muscle pair that controls the shoulder (which results in little or no net movement of the upper arm) to a pattern of *reciprocal activation*, in which one muscle pulls while the other remains in a relaxed state. Interestingly, the shift from coactivation to reciprocal activation in opponent muscle pairs is not explicitly selected by the model, but instead appears to emerge as a byproduct of rewarding the model for reaching the target as quickly as possible (i.e., a minimum-time reaching constraint).

These findings bear directly on both the questions raised earlier. First, by freezing or locking out the movement of the shoulder joint, the model effectively eliminates one dimension of the three-dimensional joint search space. In other words, rather than attempting to exhaustively sweep through the space of all possible torso-shoulder-elbow combinations, the model simplifies the search problem by holding the position of the shoulder constant, while exploring various torso-elbow configurations. Second, and more importantly, the strategy of *reducing the search space* does not appear to be an isolated phenomenon, but instead may be ubiquitous both during the development of reaching in infants (i.e., freezing degrees of freedom, muscle coactivation; e.g., Berthier, Clifton, McCall, & Robin, 1999; Spencer & Thelen, 2000) and during motor learning throughout the life span (e.g., Vereijken, Van Emmerik, Whiting, & Newell, 1992).

To conclude the discussion on embodied knowledge in robots, it should be emphasized that research in this area is not limited to the development of motor control or visual-motor skill. Indeed, there is a broad community of researchers in adaptive behavior who agree that sensorimotor activity is a fundamental starting point for cognition, in general. Examples of this work include spatial navigation, intercepting or catching falling objects, and perception of time (e.g., Nolfi & Parisi, 1999; Slocum, Downey, & Beer, 2000; Tuci, Trianni, & Dorigo, 2004).

To provide a detailed illustration of how roboticists define and study *sensorimotor cognition*—the idea that all forms of knowing are grounded in actively sensing and exploring the world—consider the related problems of learning to perceptually discriminate a set of objects and forming a set of categories, within which (either functionally or perceptually) similar objects can be grouped. Scheier and Pfeifer (1995) addressed this question by studying a mobile robot that develops a functional classification of graspable and nongraspable objects. Interestingly, Scheier and Pfeifer designed the robot's environment so that passively viewing the target objects results in ambiguous perceptual information (i.e., *perceptual aliasing*). In particular, two different objects may appear the same when viewed from certain perspectives; however, by approaching these objects in systematic ways and observing how their appearances change over time, they can be discriminated. As a result, the robot learns to exploit regularities in its own sensorimotor activity—that is, contingencies between how it moves and how its infrared sensors are affected—as the basis for differentiating and grouping objects.

While the example of using sensorimotor patterns as a way of dividing the world into categories may not illustrate high-level cognition (e.g., inductive or deductive logic, metaphorical or analogical reasoning), it does provide an example of how a foundation for sensorimotor cognition can be established. More generally, this approach is not only historically rooted in Piaget's *genetic epistemology* (1952) but is also supported by work in linguistics, philosophy, and other areas of cognitive science (e.g., Johnson, 1987; Keijzer, 1998; O'Regan, 1992; Thelen, 2000; Varela, Thompson, & Rosch, 1992).

## Conclusion

This chapter began with an important question: Are connectionism and DST separate approaches to development, or instead, different manifestations of a unified theory of development? My strategy for addressing this question was to propose an analogy, in which the two approaches are interpreted loosely from an organismic perspective (i.e., as lines of thought that exist within and across evolutionary and developmental timescales). A number of interesting consequences follow from this perspective. First, the systems of interrelated concepts, scientific methods, and empirical data that comprise connectionism and DST are not viewed as static structures, but rather, as complex entities that change over time. A second consequence of the organismic analogy—and one that is clearly highlighted throughout this volume—is that there is no single definition or description that captures either approach. Surely there are prevailing *ways of thinking* for DST or connectionist researchers and common practices that complement these separate ways of thinking, but there are also a number of important and meaningful differences among the advocates that subscribe to each approach.

It might also prove helpful to think of connectionism and DST as part of a large family of distinct but related instances that have emerged from a common set of ideas. What might be gained by this view? First, the fact that there are multiple flavors/adaptations of connectionism and DST is not a weakness or deficit, but instead, a healthy indication that each approach has both grown and diversified as a result of adaptive pressures. Second, and perhaps more importantly, the idea that connectionism and DST belong to a multigenerational family argues against the likelihood of an ultimate, grand unification. Like a metaphorical Noah's ark, which pair of approaches should be chosen to define a unified theory? Which specific instances (or particular features) of DST and connectionism should be selected (or excluded) to unify the two?

My goal has been to argue *against* the strategy of distilling connectionism and DST to their fundamental essences, and then attempting to compare and contrast the two. Instead, consider the following thought experiment. Imagine for a moment that you were a farmer, looking for a way to cross two strains of tomatoes into a new, more vigorous breed. With literally hundreds of different breeds to choose from, you would focus on those with the most appropriate or desirable features, such as color, shape, resistance to pests, shelf life, and so on. I would argue, using the same logic, that rather than asking if connectionism and DST are one thing or two, we instead should focus on creative and novel ways of blending the two approaches into theoretical and methodological hybrids that bring together the best features of each approach, as appropriate for the particular empirical question, developmental task, or theoretical issue being studied. Indeed, an example of recent work that bridges connectionism and DST is work by Toussaint (2004), which combines elements of Hebbian learning and dynamic field theory.

It should be stressed, however, that hybridization as a method for conducting developmental science is not constrained solely by pragmatics (e.g., the likelihood that a particular blend of DST and connectionism will perform well). Two other issues are just as important. First, failures are instructive. An engineer may discard a model due to its inability to solve a particular task. A developmental researcher, however, may discover that the so-called unsuccessful behavior patterns generated by a model, in fact, correspond to what infants or children actually do at a particular stage of development (e.g., search for hidden objects at an incorrect location, generalize the word *mama* to all women). Second, hybrid models should not arbitrarily combine elements or features of different approaches, but should instead select from the available options with both common sense and theoretically motivated constraints as guides.

If indeed hybridization is a metaphor that will enable the developmental trajectories of connectionism and DST to align and perhaps even converge, then as I have already illustrated, there is ample evidence from the field of adaptive behavior that several existing connectionism-DST hybrids are both viable and self-sustaining. In the final section, I conclude by returning once more to the crosscutting themes I discussed earlier and pulling together several of the issues I highlighted in previous sections.

### Crosscutting Conceptual Theme: Shared Concepts

At the beginning of the chapter, I focused on two shared, core concepts that play a fundamental role in connectionism and DST. One of these is the concept of *embodiment*. While connectionism and DST may agree that the physical body and the physical environment constrain (e.g., influence, shape) developmental processes, the two approaches have also explored different ways for investigating constraints. For example, DST research has often focused on changes in behavior during real-time performance, while connectionist research has often focused on long-term changes on the developmental timescale. These differences are exacerbated by the fact that many connectionist models have provided only a vague, implicit role for the physical body and environment.

Fortunately, studying development in a robot (real or simulated) collapses all of these issues onto a single, coherent platform. First, a robot is, by definition, an embodied object that occupies a real or virtual environment. Second, it acts and exists through both real time and developmental time (and perhaps even evolutionary time). Third, at the robot builder's discretion, a number of physical properties can be incorporated into the robot's design that provide a basis for generating nonlinear forces, including elasticity, gravity, inertia, and friction (i.e., the robot *is* a dynamical system). Fourth, and also at the robot builder's discretion, an artificial neural network can be used as the control system for acquiring and storing the policy. Clearly, a robot provides connectionist and DST researchers with a unique opportunity to combine many of their ideas and techniques in an embodied, situated, and physically grounded system.

I also highlighted the concept of *distributed knowledge* as a theme or bridge connecting connectionism and DST. In particular, both approaches agree that knowledge is not something that exists apart from action and perception, but instead is tied to the here and now of experience. In terms of potential structures for storing knowledge, however, it should be stressed that a variety of techniques for nonlinear function approximation are widely used in the fields of robotics and adaptive behavior; such techniques include fuzzy logic, cellular automata, and machine-based classifier systems. These techniques, and others like them, are obviously an ideal platform for connectionist and DST researchers to facilitate a dialog that focuses on defining and investigating distributed knowledge.

### Crosscutting Conceptual Theme: Timescales

A second major theme that cuts across both connectionism and DST is the issue of learning and development over multiple timescales. As I have already suggested, because robots are physical entities that exist through time, they provide a natural framework for studying change both within and among the real time, microgenetic, ontogenetic, and phylogenetic timescales. Indeed, as robots acquire the capacity to design or repair themselves (i.e., robot self-replication), the distinction between simulated or modeled changes in artificial systems, and biological or psychological changes in real organisms, may begin to break down. If ongoing research with robots indicates a future trend, the robots of the future may have a life of their own!

### Crosscutting Conceptual Theme: Constraints

The final crosscutting theme that I described is the need for computational models of learning and development to map more directly onto situations and behaviors developmental psychologists actually study. For computational models—whether motivated by connectionism or DST—to become fully relevant and accessible to good old-fashioned developmental researchers, they must be able to generate datasets that are more or less comparable to those generated by human subjects.

Unfortunately, the practice of using a model to replicate the performance of an infant or child in a particular study may be the exception, rather than the norm (cf. Simon, 1998). Note that by *replicate* I don't mean to *roughly* or *approximately* simulate the situation and the behavior. Instead, I suggest that we raise the standard as high as possible with the aim of placing robots (or simulated agents) in situations that are (more or less) the same as those experienced by human subjects. The aim would be to enable the agents to respond to those situations in behaviorally realistic or meaningful ways (e.g., looking, grasping, vocalizing). Although this particular approach is comparatively new, there are a few promising examples. For instance, robots (both real and simulated) have recently been used to replicate/simulate infants' preferential looking or visual attention during violation-of-expectation studies (e.g., Lovett & Scassellati, 2004; Schlesinger, 2006; Weng, Zhang, & Chen, 2003). Hopefully, as this trend continues to grow, researchers using both the connectionist and dynamic systems approaches will not only find common ground on which to stand, but also a common language with which to communicate, share ideas, and collaborate.

## References

Acredolo, L. P., Adams, A., and Goodwyn, S. W. (1984). The role of self-produced movement and visual tracking in infant spatial orientation. *Journal of Experimental Child Psychology, 38*(2), 312–327.

Adams, B., Breazeal, C., Brooks, R., & Scassellati, B. (2000). Humanoid robots: A new kind of tool. *IEEE Intelligent Systems, 15*(4), 25–30.

Bechtel, W. (1997). Embodied connectionism. In D. M. Johnson & C. E. Erneling (Eds.), *The future of the cognitive revolution* (pp. 187–208). New York: Oxford University Press.

Beer, R. D. (1996). Toward the evolution of dynamical neural networks for minimally cognitive behavior. In P. Maes, M. J. Mataric, J.-A. Meyer, J. Pollack, & S. W. Wilson (Eds.), *From animals to animats: Vol. 4. Proceedings of the Fourth International Conference on Simulation of Adaptive Behavior* (pp. 421–429). Cambridge, MA: MIT Press–Bradford Books.

Berthier, N. E. Clifton, R. K., McCall, D. D., & Robin, D. J. (1999). Proximodistal structure of early reaching in human infants. *Experimental Brain Research, 127*(3), 259–269.

Bicho, E., Mallet, P., & Schöner, G. (2000). Target representation on an autonomous vehicle with low-level sensors. *International Journal of Robotics Research*, *19*(5), 424–447.

Bizzi, E, Mussa-Ivaldi, F. A., & Giszter, S. (1991). Computations underlying the execution of a movement: A biological perspective. *Science*, *253*(5017), 287–291.

Bjorklund, D. F., & Green, B. L. (1992). The adaptive nature of cognitive immaturity. *American Psychologist*, *47*(1), 46–54.

Brooks, R. A. (1991). Intelligence without representation. *Artificial Intelligence*, *47*(1–3), 139–159.

Cangelosi, A., Parisi, D., & Nolfi, S. (1994). Cell division and migration in a 'genotype' for neural networks. *Network*, *5*(4), 497–515.

Christiansen, M. H., & Chater, N. (2001). Connectionist psycholinguistics: Capturing the empirical data. *Trends in Cognitive Sciences*, *5*(2), 82–88.

Cohen, J. D., Dunbar, K., & McClelland, J. L. (1990). On the control of automatic processes: A parallel distributed processing account of the Stroop effect. *Psychological Review*, *97*(3), 332–361.

Dominguez, M., and Jacobs, R. A. (2003). Developmental constraints aid the acquisition of binocular disparity sensitivities. *Neural Computation*, *15*(1), 161–182.

Elman, J. L. (1990). Finding structure in time. *Cognitive Science*, *14*(2), 179–211.

Elman, J. L. (1993). Learning and development in neural networks: The importance of starting small. *Cognition*, *48*(1), 71–99.

Flash, T. (1987). The control of hand equilibrium trajectories in multi-joint arm movements. *Biological Cybernetics*, *57*(4–5), 257–274.

Grupen, R. A. (2003, August). A developmental organization for robot behavior. In C. G. Prince, L. Berthouze, H. Kozima, D. Bullock, G. Stojanov, & C. Balkenius (Eds.), *Proceedings of the Third International Workshop on Epigenetic Robotics: Modeling Cognitive Development in Robotic Systems* (pp. 25–36), Boston, MA. Lund, Sweden: Lund University Cognitive Studies. Retrieved October 31, 2008, from http://www-robotics.cs.umass.edu/Papers/epirob03.pdf

Hinton, G. E., & Nolan, S. J. (1987). How learning can guide evolution. *Complex Systems*, *1*, 495–502.

Hiraki, K., Sashima, A., & Phillips, S. (1998). From egocentric to allocentric spatial behavior: A computational model of spatial development. *Adaptive Behavior*, *6*(3–4), 371–391.

Hock, H. S., Schöner, G., & Giese, M. A. (2003). The dynamical foundations of motion pattern formation: Stability, selective adaptation, and perceptual continuity. *Perception and Psychophysics*, *65*(3), 429–457.

Husbands, P., Harvey, I., Cliff, D., & Miller, G. (1997). Artificial evolution: A new path for artificial intelligence? *Brain and Cognition*, *34*(1), 130–159.

Johnson, M. (1987). *The body in the mind: The bodily basis of meaning, imagination, and reason*. Chicago: University of Chicago Press.

Jordan, M. I., & Rumelhart, D. E. (1992). Forward models: Supervised learning with a distal teacher. *Cognitive Science*, *16*(3), 307–354.

Keijzer, F. A. (1998). Doing without representations which specify what to do. *Philosophical Psychology*, *11*(3), 269–302.

Kettner, R., Marcario, J., & Port, N. (1993). A neural network model of cortical activity during reaching. *Journal of Cognitive Neuroscience*, *5*(1), 14–33.

Kopp, S. B. (1972). *If you meet the Buddha on the road, kill him! The pilgrimage of psychotherapy patients* [Series: Bantam new age books]. New York: Bantam.

Langton, C. G. (Ed.). (1995). *Artificial life: An overview.* Cambridge, MA: MIT Press–Bradford Books.

Lovett, A., & Scassellati, B. (2004). Using a robot to reexamine looking time experiments. *Proceedings of the Third International Conference on Development and Learning* (pp. 284–291). Cambridge, MA: IEEE Press. Retrieved October 27, 2008, from http://www.cs.yale.edu/homes/scaz/papers/Lovett-ICDL-03.pdf

Lungarella, M., & Berthouze, L. (2003). On the interplay between morphological, neural, and environmental dynamics: A robotic case-study. *Adaptive Behavior*, *10*(3–4), 223–241.

Lungarella, M., Metta, G., Pfeifer, R., & Sandini, G. (2003). Developmental robotics: A survey. *Connection Science*, *15*(4), 151–190.

Mareschal, D., Plunkett, K., & Harris, P. (1999). A computational and neuropsychological account of object-oriented behaviours in infancy. *Developmental Science*, *2*(3), 306–317.

Mareschal, D., & Shultz, T. R. (1993). A connectionist model of the development of seriation. *Proceedings of the Fifteenth Annual Conference of the Cognitive Science Society* (pp. 170–175). Hillsdale, NJ: Erlbaum.

McClelland, J. L. (1989). Parallel distributed processing: Implications for cognition and development. In R. G. M. Morris (Ed.), *Parallel distributed processing: Implications for psychology and neurobiology* (pp. 8–45). New York: Oxford University Press.

Michel, P., Gold, K., & Scassellati, B. (2004, September). Motion-based robotic self-recognition. In *Proceedings of the IEEE/RSJ International Conference on Intelligent Robots and Systems* (pp. 2763–2768). Cambridge, MA: IEEE Press.

Miglino, O., & Lund, H. H. (2001). Do rats need euclidean cognitive maps of the environmental shape? *Cognitive Processing, 4*(2001), 1–9.

Munakata, Y. (1998). Infant perseveration and implications for object permanence theories: A PDP model of the AB task. *Developmental Science, 1*(2), 161–184, commentary 184–204, author's response 205–211.

Munakata, Y., McClelland, J. L., Johnson, M. H., & Siegler, R. S. (1997). Rethinking infant knowledge: Toward an adaptive process account of successes and failures in object permanence tasks. *Psychological Review, 104*(4), 686–713.

Munakata, Y., & Pfaffly, J. (2004). Hebbian learning and development [Special section/issue]. *Developmental Science, 7*(2), 141–148.

Mussa-Ivaldi, F. A. (1999). Modular features of motor control and learning. *Current Opinion in Neurobiology, 9*(6), 713–717.

Newcombe, N. S., & Huttenlocher, J. (2000). *Making space: The development of spatial representation and reasoning.* Cambridge, MA: MIT Press–Bradford Books.

Nolfi, S., & Floreano, D. (2001). *Evolutionary robotics. The biology, intelligence, and technology of self-organizing machines.* Cambridge, MA: MIT Press–Bradford Books.

Nolfi, S., Floreano, D., Miglino, O., & Mondada, F. (1994). How to evolve autonomous robots: Different approaches in evolutionary robotics. In R. Brooks & P. Maes (Eds.), *Artificial life: Vol. IV. Proceedings of the Fourth International Workshop on the Synthesis and Simulation of Living Systems* (pp. 190–197). Cambridge Mass: MIT Press–Bradford Books.

Nolfi, S., & Parisi, D. (1993). Self-selection of input stimuli for improving performance. In G. A. Bekey & K. Y. Goldberg (Eds.), *Neural networks in robotics* (pp. 403–418). Boston: Kluwer.

Nolfi, S., & Parisi, D. (1999). Exploiting the power of sensory-motor coordination. In D. Floreano, J.-D. Nicoud, & F. Mondada (Eds.), *Advances in artificial life: Proceedings of the Fifth European Conference on Artificial Life* (pp. 173–182). New York: Springer.

O'Regan, J. K. (1992). Solving the "real" mysteries of visual perception: The world as an outside memory. *Canadian Journal of Psychology, 46*(3), 461–488.

Parisi, D. (1996). Computational models of developmental mechanisms. In R. Gelman & T. K.-F. Au (Eds.), *Perceptual and cognitive development* (pp. 373–412). San Diego: Academic Press.

Parisi, D. (1997). Active sampling in evolving neural networks. *Human Development, 40*, 320–324.

Parisi, D., Cecconi, F., & Nolfi, S. (1990). Econets: Neural networks that learn in an environment. *Network: Computation in Neural Systems, 1*(2), 149–168.

Piaget, J. (1952). *The origins of intelligence in children* (M. Cook, Trans.). Madison, CT: International Universities Press.

Plunkett, K., & Marchman, V. (1991). U-shaped learning and frequency effects in a multi-layered perceptron: Implications for child language acquisition. *Cognition, 38*(1), 43–102.

Quinn, P. C., & Johnson, M. H. (1997). The emergence of perceptual category representations in young infants: A connectionist analysis. *Journal of Experimental Child Psychology, 66*(2), 236–263.

Schaal, S., Kotosaka, S., & Sternad, D. (2000). Nonlinear dynamical systems as movement primitives. In *Proceedings of the First IEEE-RAS International Conference on Humanoid Robotics.* Cambridge, MA: IEEE Press. Retrieved October 31, 2008, from http://www-slab.usc.edu/publications/S/schaal-ICHR2000.pdf

Scheier, C., & Pfeifer, R. (1995). Classification as sensorimotor coordination: A case study on autonomous agents. In F. Morán, A. Moreno, P. Chacón, & J. J. Merelo (Eds.), *Advances in artificial life: Proceedings of the Third European Conference on Artificial Life* (pp. 657–667). New York: Springer.

Schlesinger, M. (2003). A lesson from robotics: Modeling infants as autonomous agents. *Adaptive Behavior, 11*(1), 97–107.

Schlesinger, M. (2004). Evolving agents as a metaphor for the developing child [Special section/issue]. *Developmental Science, 7*(2), 158–164.

Schlesinger, M. (2006). Decomposing infants' object representations: A dual-route processing account. *Connection Science, 18*(2), 207–216.

Schlesinger, M, & Parisi, D. (2001a). The agent-based approach: A new direction for computational models of development. *Developmental Review, 21*, 121–146.

Schlesinger, M., & Parisi, D. (2001b). Multimodal control of reaching: Simulating the role of tactile feedback. *IEEE Transactions on Evolutionary Computation: Special Section on Evolutionary Computation and Cognitive Science*, *5*(2), 122–128.

Schlesinger, M., & Parisi, D. (Section Eds.). (2004). Beyond backprop: Emerging trends in connectionist models of development: An introduction [Special section/issue]. *Developmental Science*, *7*(2), 131–132.

Schlesinger, M., Parisi, D., & Langer, J. (2000). Learning to reach by constraining the movement search space. *Developmental Science*, *3*(1), 67–80.

Schutte, A. R., & Spencer, J. P. (2002). Generalizing the dynamic field theory of the A-not-B error beyond infancy: Three-year-olds' delay- and experience-dependent location memory biases. *Child Development*, *73*(2), 377–404.

Schutte, A. R., Spencer, J. P., & Schöner, G. (2003). Testing the dynamic field theory: Working memory for locations becomes more spatially precise over development. *Child Development*, *74*(5), 1393–1417.

Schweighofer, N., Arbib, M. A., & Kawato, M. (1998). Role of the cerebellum in reaching movements. I. Distributed inverse dynamics control. *European Journal of Neuroscience*, *10*(1), 86–94.

Shultz, T. R., Mareschal, D., & Schmidt, W. C. (1994). Modeling cognitive development on balance scale phenomena. *Machine Learning*, *16*(1–2), 57–86.

Slocum, A. C., Downey, D. C., & Beer, R. D. (2000). Further experiments in the evolution of minimally cognitive behavior: From perceiving affordances to selective attention. In J.-A. Meyer, A. Berthoz, D. Floreano, H. L. Roitblat, & S. W. Wilson (Eds.), *From animals to animats: Vol. 6. Proceedings of the Sixth International Conference on Simulation of Adaptive Behavior* (pp. 430–439). Cambridge, MA: MIT Press–Bradford Books.

Simon, T. J. (1998). Computational evidence for the foundations of numerical competence. *Developmental Science*, *1*(1), 71–78.

Smith, L. B., & Thelen, E. (Eds.). (1993). *A dynamic systems approach to development: Applications*. Cambridge, MA: MIT Press–Bradford Books.

Sirois, S. (2004). Autoassociator networks: Insights into infant cognition [Special section/issue]. *Developmental Science*, *7*(2), 133–140.

Spencer, J. P., & Thelen, E. (2000). Spatially specific changes in infants' muscle coactivity as they learn to reach. *Infancy*, *1*(3), 275–302.

Steinhage, A., & Schöner, G. (1997). Self-calibration based on invariant view recognition: Dynamic approach to navigation. *Robotics and Autonomous Systems*, *20*(2–4), 133–156.

Thelen, E. (2000). Grounded in the world: Developmental origins of the embodied mind. *Infancy*, *1*(1), 3–28.

Thelen, E., & Bates, E. (2003). Connectionism and dynamic systems: Are they really different? [Special section/issue] *Developmental Science*, *6*(4), 378–391.

Thelen, E., Schöner, G., Scheier, C., & Smith, L., B. (2001). The dynamics of embodiment: A field theory of infant perseverative reaching. *Behavioral and Brain Sciences*, *24*(1), 1–34, commentary 34–70, authors' response 70–80.

Thelen, E., & Smith, L. B. (1994). *A dynamic systems approach to the development of cognition and action*. Cambridge, MA: MIT Press–Bradford Books.

Toussaint, M. (2004). Learning a world model and planning with a self-organizing dynamic neural system. *Advances in Neural Information Processing Systems*, *16*, 929–936. Retrieved October 31, 2008, from http://books.nips.cc/papers/files/nips16/NIPS2003_CS04.pdf

Tuci, E., Trianni, V., & Dorigo, M. (2004). 'Feeling' the flow of time through sensorimotor co-ordination. *Connection Science*, *16*(4), 301–324.

Turvey, M. T., Fitch, H. L., & Tuller, B. (1982). The Bernstein perspective: Vol. I. The problem of degrees-of-freedom and context-conditioned variability. In J. A. S. Kelso (Ed.), *Human motor behavior: An introduction* (pp. 239–252). Hillsdale, NJ: Erlbaum.

Varela, F. J., Thompson, E. T., & Rosch, E. (1992). *The embodied mind: Cognitive science and human experience*. Cambridge, MA: MIT Press.

Vereijken, B., van Emmerik, R. E. A., Whiting, H. T. A., & Newell, K. M. (1992). Free(z)ing degrees of freedom in skill acquisition. *Journal of Motor Behavior*, *24*(1), 133–142.

Vos, J. E., & Scheepstra, K. A. (1993). Computer-simulated neural networks: An appropriate model for motor development? *Early Human Development*, *34*(1–2), 101–112.

Wang, R. F., Hermer, L., & Spelke, E. S. (1999). Mechanisms of reorientation and object localization by children: A comparison with rats. *Behavioral Neuroscience, 113*(3), 475–485.

Weng, J., McClelland, J., Pentland, A., Sporns, O., Stockman, I., Sur, M., et al. (2001). Artificial intelligence: Autonomous mental development by robots and animals. *Science, 291*(5504), 599–600.

Weng, J., Zhang, Y., and Chen, Y. (2003, July). Developing early senses about the world: "Object permanence" and visuoauditory real-time learning. *Proceedings of the International Joint Conference on Neural Networks* (pp. 2710–2715), Portland, OR. Cambridge, MA: IEEE Press. Retrieved October 32, 2008, from http://www.cse.msu.edu/~weng/research/IJCNN2003ss. df

Williamson, M. M. (1998). Exploiting natural dynamics in robot control. In R. Trappl (Ed.), *Proceedings of the Fourth European Meeting on Cybernetics and Systems Research*. Austria: Austrian Society for Cybernetic Studies.

# PART IV

# Hybrid Models: Case Studies

# CHAPTER 10

# Combining Connectionist and Dynamic Systems Principles in Models of Development: The Case of Analogical Completion

*Denis Mareschal, Robert Leech, and Richard P. Cooper*

## Introduction: There Are Two Sides to Any Coin

In 2002, the journal *Developmental Science* devoted an entire special issue to exploring two new theoretical approaches for understanding development: dynamic systems theory and connectionism [*Developmental Science,* 6(4), 2003; Spencer & Thelen, 2003]. What makes these two approaches stand out from other contemporary theories of development in the literature is the central role played in the approaches by mechanisms of change. In other words, both approaches are fundamentally interested in understanding *how* change occurs in developmentally relevant phenomena.

To some outsiders, both approaches appear to share a number of similarities. First, they both emphasize the use of computer modeling as a research tool and use mathematics to describe the internal processes of their models. Although such methods are far more common in psychology now than they were even 15 years ago (especially in the study of adult cognition and perception), they are still relatively rare in the study of developmental psychology. Second, both approaches eschew the sequential processing assumptions of cognition that underlie many of the early information processing theories. That is, instead of viewing cognition as the outcome of a sequential series of steps (e.g., Pylyshyn, 1984; Sternberg, 1969), the emphasis is on parallel processing of multiple sources of information (e.g., Rumelhart, McClelland, & PDP Research Group, 1986). Third, both approaches reject the fundamental notions of the *language of thought hypothesis* (Fodor, 1975). That is, they reject the idea that cognition consists in the syntactic manipulation of symbolic tokens.

However, a closer look reveals what appear to be fundamental differences between the two approaches. Connectionist systems are deeply committed to the notion of representations, and emphasize how *knowledge* that is represented in active or latent form (e.g., Munakata & McClelland, 2003) constrains emergent behaviors. Indeed, the history of connectionist and neural network modeling shows how much effort has gone into exploring how information about the world is internally represented in such systems (e.g., Feldman & Ballard, 1982). This may be because the first applications of neural networks—that would later become the mainstay of connectionist approaches—were intended to model memory systems (e.g., Hinton & Anderson 1981; Rosenblatt, 1958; for a detailed history of connectionist neural computation research, see also Anderson & Rosenfeld 1989). In fact, it is precisely this emphasis on the acquisition of task-appropriate representations that first drew developmental psychologists to connectionism (e.g., Plunkett & Sinha, 1991). Because of their history, connectionist systems have emphasized the importance of stable states (e.g., Rosenblatt, 1958). A memory was said to have been retrieved when the network had stabilized in a state that was consistent with the information contained in the cues. Dynamic processes were involved in the retrieval of the memory, but what was really important was the end state.

In contrast, dynamic systems approaches have their origin in the study of control, in particular how neural networks can be used as control systems for locomotion (e.g., Barto, Sutton, & Anderson, 1983). Through the 1970s and 1980s, the problem of motor control was approached in the same way that many other cognitive problems were approached; that is, through the use of symbolic serial processing (Boden, 1988). However, it soon became clear that, at least in the domain of motor control, serial processing was, in fact, an illusion; stable states were only fleetingly present if ever. In reality, it was argued, the sensorimotor system was a continuously adapting system, closely coupled to the environment (Ballard et al., 1997). This emphasis on a close coupling between control systems and environment was immensely successful in producing models of sensorimotor responses that worked both as engineering projects (e.g., they could control a robot in the real world) and as explanatory frameworks in accounting for human locomotion (e.g., Jordan, 1989). These successes led developmentalists in the dynamic systems camp to deny a role for representations and instead to focus on process as the key factor in explaining behavior (e.g., Smith & Samuelson, 2003; Smith & Thelen, 1993; Thelen & Smith, 1994).

So, while connectionism and dynamic systems share some fundamental similarities when viewed from a distance, there are at least two core elements of the approaches that seem to put them in opposition. First, connectionism emphasizes representations, whereas historically the dynamic systems approach has denied that representations play a role in the generation or control of behavior. Second, connectionism emphasizes the stable states of a system, whereas dynamic systems has often emphasized the transition states of a system.

In fact, we believe that while these are, theoretically, fundamental differences, a detailed examination of the formal descriptions and properties of each approach suggests that these differences are differences in emphasis rather than differences in kind, and that, at least in the context of cognitive development, connectionism and dynamic systems theory can profitably be viewed as two sides of the same coin. To illustrate this, consider the equations used to implement the common dynamical field theory version of dynamic systems theory (e.g., Thelen, Schöner, Scheier, & Smith, 2001). These are generalized continuous versions of the equations used to describe interactive activation networks (Amari, 1977, 1980; Grossberg, 1980; Rumelhart, McClelland, & PDP Research Group 1986). Thus, while the conceptual emphasis is placed differently, the two approaches are in fact speaking of the same thing.

In our view, the visions are complementary, and both can contribute to a better understanding of behavior as it unfolds through both moment-to-moment and developmental time. To illustrate, this chapter describes a recent connectionist model of the development of analogical completion in early childhood that draws on (a) the gradual learning and distributed representations of connectionist networks and (b) the notion of relatively rapid settling into changing attractor states of dynamic systems theory to capture the fine-grained performance of children at different ages.

## Development of Analogical Completion

Analogical reasoning has traditionally been the preserve of high-level models that use symbolic representations and explicit mapping of structured relations between an analogy's source domain and its target domain (French, 2002). Thus, in completing an analogy such as "*kitten* is to *cat* as *puppy* is to *what*?," traditional models have postulated a process of identifying a relation between the terms in the source domain (*kitten* and *cat*) and a process of mapping this relation to the target domain (*puppy* and *what*) to yield the answer, *dog*. Such models have tended to view development as something of an afterthought. The more developmentally inspired model presented here serves not only to emphasize the importance of development as a major constraint on cognitive theories, but also to illustrate the compatibility of connectionist and dynamic systems accounts of cognitive development. Within the model, analogy arises both from how connection weights change over long timescales following exposure to instances of relations and, importantly, how this interacts with the way in which the network settles over short timescales into stable attractor states. Equally importantly is how these attractor states subsequently change when perturbed by a different input.

Our model is designed to simulate young children's performance on "*a* is to *b* as *c* is to *d*" analogies involving simple causal domains, for example, *apple* is to *cut apple* as *bread* is to *what?* (*cut bread*). This type of analogy has been extensively studied in the developmental literature (e.g., Goswami & Brown, 1990; Rattermann & Gentner, 1998; see additional references below). In the model, analogy is viewed as a byproduct of the priming of relations within a simple memory system. Relations are represented implicitly as transformations between states (e.g., the relation *cutting* equates to the transformation from *apple* to *cut apple*) (Rogers & McClelland, 2004; Thomas & Mareschal, 1997). Under our account, analogical completion occurs though the *a* and *b* terms of the analogy priming a relation which then transforms the *c* term so that the pattern completion properties of the network produce the analogically appropriate *d* term. The model demonstrates the complementarity of connectionist and dynamic systems approaches by simultaneously employing both gradual, long-term, connection-weight-based acquisition of relational knowledge and more immediate, short-term settling into attractor states, which lead to analogical completion. The model stands up well to comparison with a set of key behavioral phenomena from the developmental analogical reasoning literature, as listed below. Several of these phenomena are currently beyond the scope of any nondevelopmental accounts. The key phenomena are:

1. One of the central findings in previous experimental research is the strong relationship between the development/accretion of relational knowledge and the acquisition of analogical reasoning (e.g., Goswami & Brown, 1990; Rattermann & Gentner, 1998). That is, children can only form and comprehend analogies involving relations with which they are, to some degree, familiar.
2. A corollary of the importance of relational knowledge accretion is that acquisition of analogical reasoning develops at different paces in different knowledge domains. Analogical reasoning ability does not arise from a domain-general change (Goswami & Brown, 1990).
3. The existing developmental literature suggests that the ability to reason by analogy develops spontaneously. Children regularly form analogies without the need for any formal tuition (Inagaki & Hatano, 1987; Pauen & Wilkening, 1997; Tunteler & Resing, 2002).
4. Gentner and colleagues (Gentner, 1988; Gentner & Toupin, 1986; Rattermann & Gentner, 1998) have repeatedly found evidence that a *relational shift* occurs over development. Children gradually change from initially making analogical judgments on the basis of object similarity (i.e., superficial or perceptual features shared by objects in the analogy) to increasingly relying on relational knowledge (i.e., knowledge of relations between objects).
5. Young children can solve so-called cross-mapping analogies, in which there is a direct conflict between relational and object-similarity responses (Gentner, Rattermann, Markman, & Kotovsky, 1995; Kotovsky & Gentner, 1996). In a cross-mapped analogy, the same object appears in both the base and the target but with a different role (e.g., *cat* in "*dog* is to *cat* as *cat* is to *mouse*"), so the correct response involves disregarding the identity of the object shared in the source and target domains of the analogy and focusing on the relation in which that object participates. Children find cross-mapped analogies harder and solve them later in development than other analogies.
6. The developmental literature (e.g., Kotovsky & Gentner, 1996; Loewenstein & Gentner,

2005; Rattermann, Gentner, & DeLoache, 1990) has repeatedly shown that the analogical abilities of children can be improved by prior training with or exposure to relational labels. Thus, Loewenstein and Gentner (2005) found that children performed better on a search task (in which a target was hidden in an analogically equivalent location to a source object) when the locations were described in relational terms (using prepositions such as *in*, *on*, and *under*) than when they were not.

7. Hosenfeld and colleagues (Hosenfeld, van der Maas, & Van den Boom, 1997) have directly compared the development of analogical reasoning with predictions derived from dynamic systems theory. They argue that dynamic systems theory supports the existence of a discontinuous change in performance at some point in development. These investigators found evidence consistent with three markers of such a discontinuous change: (1) a sudden jump in response accuracy; (2) a critical slowing in response time in the neighborhood of the sudden jump; and (3) a decrease in response consistency accompanying the sudden jump.

## A Developmental Model

Figure 10.1 depicts the general architecture of a connectionist network proposed by Leech, Mareschal, and Cooper (2001) as a model of both the acquisition of relational information and the development of analogical completion. The model implements the idea that analogical completion can arise as a byproduct of relational priming in a semantic network. Although the network is only trained on relational knowledge, it can, nevertheless, capture a range of empirical effects found in the developmental analogical completion literature.

The network contains bidirectional and symmetrical weights. This allows activation to flow in all directions within the network. The weights embody constraints on the coactivation states of the units within the network. As a result, such networks can carry out a form of pattern completion. When a subset of units is activated, the activation will circulate throughout the network via the weights; this reactivates an overall pattern that is consistent with the initial partial activation (for an excellent discussion of pattern completion, see Rumelhart, McClelland, & PDP Research Group, 1986). These stable states are sometimes referred to as the *stable attractors* of the network.

The current network can be conceptualized as consisting of three broad layers of units. The bottom (or input) layer is split into two banks of units representing the presentation of two objects in a *before*, or pretransformation, state (t1). Similarly, the upper (or output) layer is also split into two banks, representing the same two objects in an *after*, or posttransformation, state (t2). The *before* and *after* states can be thought of as two temporally contiguous, causally related states of the world. For example, one might see some bread and a knife, followed by some cut bread and a knife.

In the model, a causal relation is conceived of as the transformation from a *before state* to an *after state* within the world. Specifically, the Object (t1) layer corresponds to some object (e.g., *bread*) and the CA (t1) layer corresponds to a causal agent (e.g., *knife*), which is the object that causes the transformation (e.g., *cutting*) to occur.

During training, contrastive Hebbian learning is used to change the connection weights so that stable attractor states across the output units come to approximate the desired target output states of the network (O'Reilly, 1996; O'Reilly & Munakata, 2000). The network learns about causal relations in the world from a series of observations of pre- and posttransformation states of pairs of objects (e.g., bread and knife, stone and knife, bread and fire, paper and fire). Analogical completion is then assessed against this growing background knowledge.

The network is tested in the same way as children were in the original Goswami and Brown tasks (Goswami & Brown, 1990). The network is presented with the initial part of the analogy (e.g., *bread* is to *cut bread*), then with the first half of the final part (*apple* is to ?). If the appropriate knowledge is present, the network is able

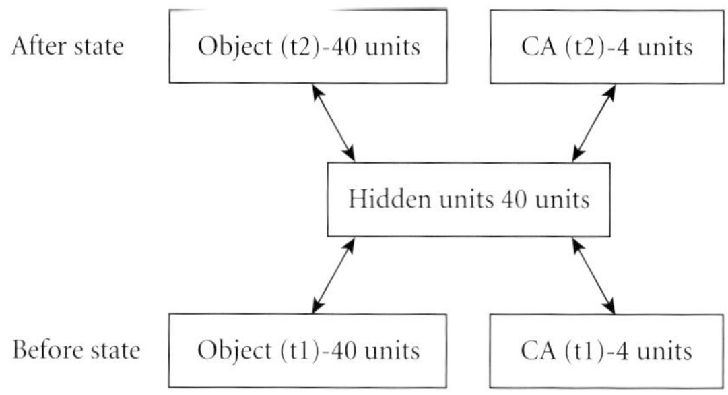


**Figure 10.1.** **Schema of the general architecture of a connectionist model (CA stands for causal agent)**

to solve this analogy because the presentation of "*bread* is to *cut bread*" primes the network with (i.e., puts it in an activation state consistent with) the relation *cut*. When subsequently presented with *apple*, the network completes this later item by applying the constraints from its current state (*cut*) with the new input (*apple*) to settle on *cut apple*. Both the training and testing of the network are described in more detail below.

### Training the Network

The network was trained with input patterns produced on the fly by adding Gaussian noise (mean = 0.0, variance = 0.1) to 20 object prototype patterns and 4 causal agent (CA) prototype patterns. This was intended to capture the fact that although two instances of cutting an apple with a knife are similar, they are never identical. Four transformation vectors were also randomly generated. The transformation vectors encode the relation between the pre- and post-transformation states of the object. In fact, the transformed state of the object, Object (t2), is obtained by adding a transformation vector to an object at Object (t1). For example, Object (t1) (*apple*): [.5 .2 .8 .2 .4], might be transformed by the vector (*cutting*): [−.4 .0 .0 .7 .0] resulting in Object (t2) (*cut apple*): [.1 .2 .8 .9 .4]. Note that the transformation vector remains implicit during training. Only Object (t1) and a transformed Object (t2) are ever presented to the network, not the actual transformation vector itself. The causal agent remained untransformed; that is, CA (t2) was always identical to CA (t1). Different objects (e.g. *bread* or *apple*) transformed by the same relation (e.g. *cutting*) were transformed by the same vector. Thus, the network generalizes across objects on the basis of shared transformations tied to particular casual agents.

Training proceeded according to the standard 2-step contrastive Hebbian learning procedure (O'Reilly & Munakata, 2000). This consisted in randomly selecting an object and a causal agent, clamping Object (t1) and CA (t1) inputs to these patterns respectively and then letting activation cycle round the network 5 times. The process was then repeated with the additional Object (t2) and CA (t2) target patterns clamped on. The weights were then updated according to the difference in Hebbian terms corresponding to the activations in these 2 phases. Because both the pre- and post-transformation states of the network could be obtained by direct observation of the world, learning of relational information constitutes a form of self-supervised learning (Japkowicz, 2001).

In this model, each of the 20 possible objects was affected by 2 of the 4 causal agents, and so 2 of the 4 transformations. Thus, there were 360 potential analogies for the network testing. However, when an object was presented in conjunction with 1 of the 2 causal agents that did not transform it, the target Object (t2) output was the same as the untransformed Object (t1) input (i.e., there was no transformation). Thus, the presence of the causal agent alone did not predict that a transformation will occur. Finally, the activation of any unit varied according to a sigmoidal activation function ranging from 0 to 1. The initial weights were uniformly randomized between ±0.5. The learning rate was uniformly set to 0.1, and activations were updated for 5 cycles between each weight update.

## Testing the Network

Analogical completion was tested separately from the learning of relation information. During analogical completion the activation pattern corresponding to the initial 2 elements of the analogy (e.g., *bread* and *cut bread*) were clamped on at Object (t1) and Object (t2), respectively, while CA (t1) and CA (t2) were initially set to 0.5, their resting values. This is analogous to presenting the network with *bread:cut bread* (i.e., the initial half of an *a:b::c:d* analogy). The network was then allowed to settle into a stable attractor state by filling in CA (t1) and CA (t2) and arriving at hidden unit activations that are consistent with the transformation *cutting*. Following this, the Object (t1) and Object (t2) units were unclamped, and a second pattern, corresponding to the first of the final 2 elements of an analogy (e.g., *apple*), was presented to Object (t1) with nothing presented to Object (t2). This is analogous to presenting the network with the activation pattern *apple:?* (i.e., the final probe-half of the *a:b::c:d* analogy). At this point, the network was no longer stable and settled into a new attractor state. However, the prior priming of the transformation *cutting* in the context of *bread* biased the network to settle into the attractor state corresponding to the transformation *cutting*, resulting in the *cut apple* pattern at Object (t2) (i.e., the analogically appropriate response).

## Evaluating the Model

How the network arrives at a response illustrates the important role that dynamic attractors play in analogical completion in this model. Consider the example of completing the following analogy: *bread:cut bread::apple:?*. Children are tested by being shown a series of pictures that define the first 3 terms and then given 4 further pictures to choose from in order to complete the analogy (e.g., Goswami & Brown, 1990). Figure 10.2 is an illustrative example of the network responding to one of these analogies at 3 points in training. Here, the sum-of-square distance (SSD) between the actual output of the network when tested on this analogy and 4 possible outputs corresponding to the 4 pictures that would be shown to the children are plotted. The lower the *y* axis value, the closer the actual activation is to that possible output pattern. Thus, Figure 10.2 shows which of the 4 objects is closest to the network's actual response after different amounts of training. The one with the lowest SSD can be taken as the network's response in this forced choice paradigm.

After 100 epochs of training (Fig. 10.2, panel A), the network fails to complete the analogy appropriately. Instead its output is closest to that of a perceptual similarity match (i.e., *apple*). After 1500 epochs of training (Fig. 10.2, panel B), the network's response is ambiguous. Finally, after 5000 epochs of training (Fig. 10.2, panel C), the network settles into the appropriate state that completes the analogy (i.e., *cut apple*). However, to infer correct analogical completion, it is also important that the network does not produce an analogical response when it is *not* appropriate. Indeed, it could be the case that the attractor basin corresponding to *cut apple* is so wide that *cut apple* is always the network's response. However, this is not the case as Figure 10.2, panel D, indicates. Here, the network is presented with a non-analogy (*bread:bread: apple:?*) and produces the appropriate nontransformed response (*apple*).

The model was also assessed against the seven key phenomena from the developmental literature listed earlier. These are considered in turn:

1. *Relational knowledge accretion.* We can assess how well the model has mastered the causal knowledge domain by examining its ability to produce the correct Object (t2) response when presented with Object (t1) and CA (t1). Figure 10.3 plots the mean sum-of-squares error at Object (t2) as well as the percentage of analogies completed appropriately by the network as a function of training. The proportion of analogies correct and the sum-of-squares error are strongly negatively correlated ($r = -0.97$, $p < .001$). This shows that that there is a strong association

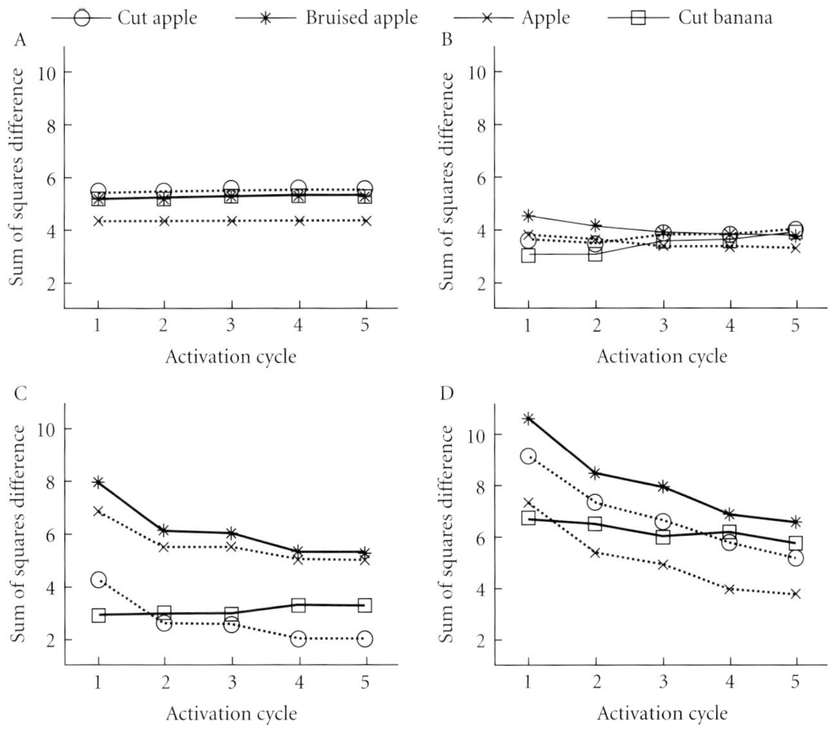


**Figure 10.2. Analogical completion over development**
The *y* axis shows the sum-of-squares distance calculated between the actual output of the network and 4 target patterns (note the legend for the lines at the top of the figure). The lower the *y* axis value, the closer the actual activation is to that target pattern. Panels A–C depict the network's response to an analogy: for example, *bread* is to *cut bread* as *apple* is to *what*. Panel D depicts an example of the network's response to a non-analogy: for example, *bread* is to *bread* as *apple* is to *what*.

between the network's *knowledge* of causal relations and its performance on analogies requiring this knowledge. The interrelatedness between analogical completion and relational knowledge can also be seen when considering how well the network recovers the appropriate causal agent when presented only with the untransformed (*apple*) and transformed (*cut apple*) objects. This ability is important because it demonstrates that the analogical completion observed in the network is not simply a matter of forming a straightforward stimulus–response input–output link. The successful production of the appropriate causal agent correlates strongly with analogical completion ($r = 0.93$, $p < .001$). This finding mirrors the results obtained from a similar experimental comparison by Goswami and Brown (1990) with young school-age children.

2. *Domain specificity.* In a related vein, there is no general or global shift that enables the network to solve all analogies. Rather, as with children, the network's analogy performance depends on underlying specific relational knowledge (Goswami & Brown, 1990).
3. *Spontaneity.* The priming effects described in the network would occur spontaneously in a broad range of task contexts and are not just limited to *a:b::c:d* formal contexts. This is consistent with the reported spontaneous use of analogy by young children in a broad range of informal contexts (Inagaki & Hatano, 1987; Pauen & Wilkening, 1997; Tunteler & Resing, 2002).

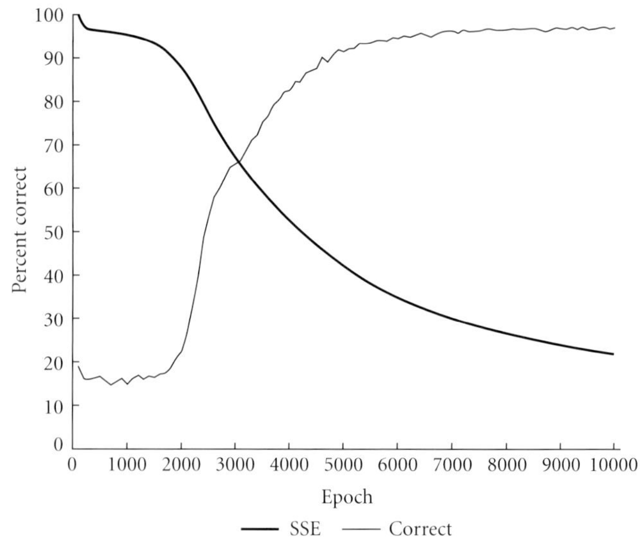


**Figure 10.3. Percentage of analogically appropriate responses**
This figure depicts the percentage of analogically appropriate responses at Object (t2) (thin line) and the normalized mean sum of squared error (SSE, thick line) over training. These results are the average of 50 replications.

4. *Relational shift.* The network, like children, shows behaviors consistent with a relational shift, in which children move from completing analogies on the basis of surface similarity to completing analogies on the basis of structural or functional similarity. Table 10.1 compares the response profile of children of different ages (Rattermann & Gentner, 1998) on the target *a:b::c:d* task with the response profile of the network at different points in training. Note the close parallel between network and child error rates, both early and later in training. This profile suggests that the network, like children, undergoes an apparent change from responding on the basis of object similarity to responding on the basis of relational similarity.
5. *Cross-mapped analogies.* Both children (Gentner et al., 1995) and the network model can solve cross-mapping analogies. These are analogies in which the same item plays different roles in the two parts of the analogies (e.g., *small:medium::medium:?*). In such analogies, there is a strong conflict between object and relational similarity. The network requires more training before solving these analogies, reflecting the finding that these analogies are harder for children than analogies where object similarity aids the relational analogy (Gentner et al., 1995).
6. *Effect of relational labels.* We contrasted analogical completion in our original model with a modified version that included an additional orthogonal label cue, presented to the network in addition to the causal agent. The analogical completion performance of the modified model follows a very similar developmental trajectory to the original model, but shifted earlier in training. Thus the model mirrors the observed augmenting effect of training with relational labels on children's analogical development (Kotovsky & Gentner, 1996; Loewenstein & Gentner, 2005
7. *Indicators of a discontinuous change.* Figures 10.4, panels A–C, indicate that our model also demonstrates developmental markers of discontinuous change (Hosenfeld et al., 1997) similar to those witnessed in children's development. Figure 10.4, panel A, shows that a *sudden jump* (i.e., a marked speeding up in the network's analogy completion across training epochs) occurs at around 2400 epochs of training. Figure 10.4, panel B, suggests that concurrent with the sudden jump, there is a peak in inconsistent responses (as measured by presenting the same network twice with different noisy versions of the same input). This spurt in inconsistency was observed in the neighborhood of the sudden jump, as is also found in the

**Table 10.1. The profile of responses made by children and the network**

| Response type | Children | | Network | |
|---|---|---|---|---|
| | 4-year-olds | 5-year-olds | 2300 epochs | 2800 epochs |
| Correct analogy | 35% | 67% | 39% | 63% |
| Wrong analogy | 35% | 28% | 16% | 9% |
| Object-similarity error | 22% | 3% | 22% | 3% |

*Note*: Network's responses are the average of 50 replications.
*Source*: Rattermann and Gentner (1998) for data on children's responses.

children's data. Figure 10.4, panel C, similarly suggests that there is an increase in the number of activation cycles for the network to produce its response. Again—and consistent with children's reaction times reported by Hosenfeld and colleagues (Hosenfeld et al., 1997)—this slowing down was strongly correlated with the sudden jump in successful analogy completion.

## DISCUSSION

We have seen that the development of simple analogical completion abilities, as demonstrated by children's changing performance on analogies of the *a:b::c:d* type, may be simulated by a model that employs relational priming to arrive at the *d* term of an analogy given the *a:b::c:_* parts of the analogy (Leech, Mareschal, & Cooper, 2003). These simulations capture seven key phenomena apparent in children's performance over time (see section Development of Analogical Completion). The model (and underlying theory) differs substantially from other models (and theories) of analogical reasoning. We therefore begin this discussion with a more detailed consideration of the relation of our theory and model to other theories and models of analogical reasoning. In the second part of the discussion, we turn our attention to five crosscutting themes raised by the comparison of connectionism and dynamic systems theory.

### Relational Priming and Other Approaches to Analogical Reasoning

Traditionally, accounts of analogical reasoning have been grounded in symbolic approaches to cognitive processing. Thus, according to structure mapping theory (SMT: Gentner, 1983, 1988), one of the dominant approaches, analogical reasoning involves first selecting base and target domains for an analogy and then mapping the object in the base domain to the target domain based on purely structural properties of the objects. For example, the representation *apple* in the predicate representation cut (*apple, cut apple*) would map onto *bread* in cut (*bread, cut bread*). Predicates or relations between objects are preserved across this structural mapping, so that if a predicate holds an object in the base domain, then by analogy, the predicate will hold the corresponding object in the target domain. Many other accounts of analogical reasoning adopt a similar basic assumption: that the core to analogical reasoning lies in mapping or aligning the structural relations between elements in the base domain with corresponding elements in the target domain. This is even the case in hybrid connectionist/symbolic accounts such as analogical mapping by constraint satisfaction (ACME; Holyoak & Thagard, 1989) and LISA (Hummel & Holyoak, 1997). The relational priming account of analogy differs substantially from all of these accounts in that the fundamental process is held not to be one of structural alignment, but one of priming. Relational priming therefore constitutes a genuine distinct alternative theory of analogical reasoning.

The difference in mechanism between relational priming and structural alignment accounts of analogical reasoning reflects differences in origins of the theories and differences in the types of analogical reasoning that the theories take as core data. While SMT has been applied to developmental data (e.g., Gentner, 1988), the theory was originally conceived in the context of adult analogical competence (Gentner, 1983). Other structural alignment

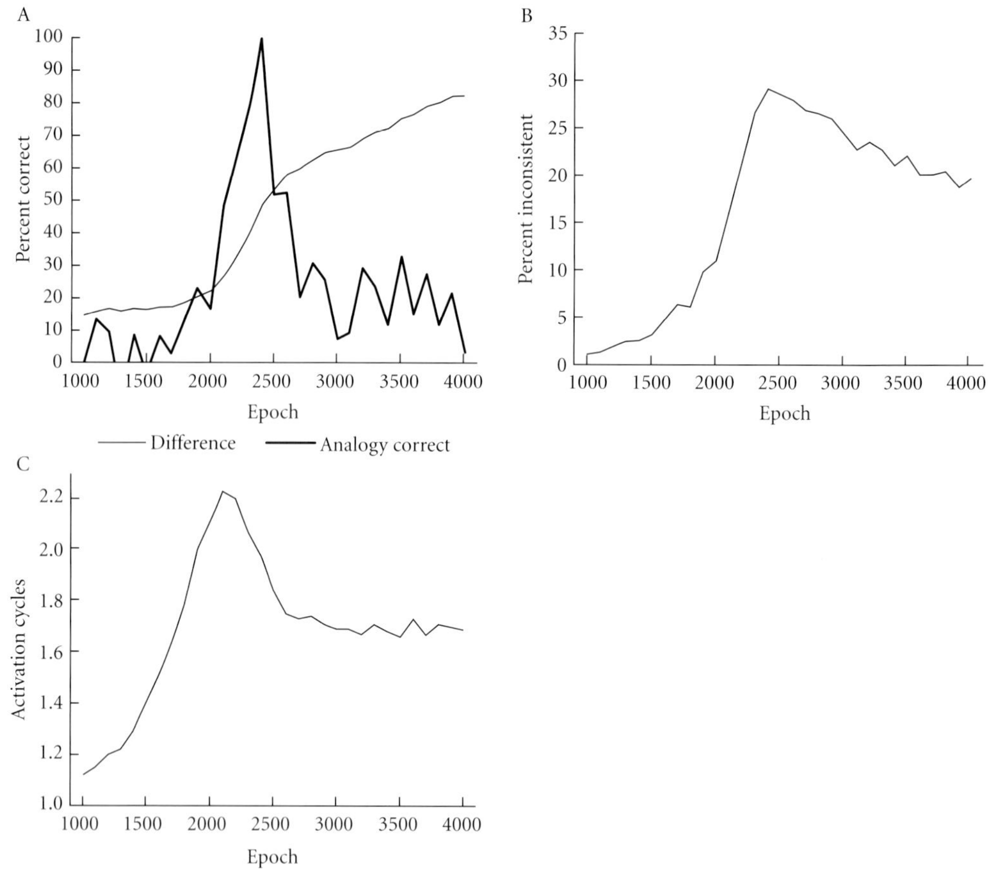


**Figure 10.4. Indicators of a discontinuous change**
(Panel A) This graph provides evidence for a *sudden jump* (i.e., a marked speeding up) in the network's analogy completion across training epochs. The thin line is the percentage of analogies completed correctly. The thicker line is the normalized difference in correct analogies across time, demonstrating a sudden jump centered around 2400 epochs of training. (These results are the average of 50 replications.) (Panel B) This graph provides evidence for inconsistency in the percentage of inconsistent responses across training. There is a sharp increase around the timing of the sudden jump in the network's analogy completion. (These results are the average of 50 replications.) (Panel C) This graphs provides evidence that the number of activation cycles required for the network to produce its response changes across training. There is a sharp increase around the timing of the sudden jump in the network's analogy completion. (These results are the average of 100 replications.)

accounts of analogical reasoning have largely tended to follow this lead. Correspondingly, the data that such theories have been primarily concerned with is that of the adult literature (e.g., Gick & Holyoak, 1980). Complex, multirelational analogies, such as that between the solar system and the Rutherford atom,[1] are central within the adult literature. In presenting the relational priming account of analogical reasoning, we take the primary research to be

[1] The Rutherford atom is a stylized representation of the generic atom fashioned by Ernst Rutherford (1871–1937). Rutherford's design is reminiscent of the solar system; that is, tiny electrons are depicted (with visible tracks/orbits) circling around a dense nucleus, somewhat like planets circling the sun. While misleading on several scientific points, the Rutherford atom attracted the attention of the lay pubic and is frequently used as a symbol of atomic energy (e.g., in the emblem of the U.S. Atomic Energy Commission).

that involving children's developing abilities at relatively simple analogies (i.e., involving a single relation). Such research is represented by the seven key phenomena described in our review of the behavioral data. Relational priming (and its instantiation in the model presented above) is successful to the extent that it can account for each of these phenomena.

It remains to be demonstrated whether relational priming can account for the more complex analogical abilities of adults (e.g., multirelational analogies). In our view, complex analogies are not beyond the scope of relational priming (for further discussions of this point, see Leech, Mareschal, & Cooper, 2008). Rather, such analogies require relational priming applied either in an iterative fashion or with more complex relations (e.g., ternary relations, relations between relations).

### Crosscutting Themes

The behavioral data highlighted in our review of the literature include some that might be regarded as characteristic of connectionism (e.g., relational knowledge accretion) and others that might be regarded as characteristic of dynamic systems theory (e.g., indicators of discontinuous change). Crucially, our model exhibits key features of both connectionism approaches and dynamic systems approaches. We now consider the implications of this.

### Development

How does the model view development? At the cognitive level, development is driven by the gradual acquisition of relational knowledge through continuous exposure to exemplar transformations; development is not driven by biophysiological maturation. At the computational level, what changes with development is the attractor structure of the network. Initially, this attractor structure, if present at all, is entirely random. Early in training, the network learns that features from the before state are likely to be preserved in the transformation to the after state (to a first approximation across the whole training set). Knowledge is encoded in the initial, weak attractor structure; at this point, transformations of the objects are dominated by perceptual similarity. Only through additional learning does the attractor structure change to reflect the relational nature of causal agents (so that *apple*, *knife*, and *cut apple* correspond to a stable state of the network).

The model also demonstrates that there is no inconsistency between knowledge accretion (even in an account in which the only change over time results from knowledge accretion) and indicators of discontinuous change. Knowledge accretion sits comfortably within the dynamic systems approach; in concert, dynamic systems account for apparent discontinuities in behavior. Moreover, the discontinuities are emergent features of our model. There is no sense in which discontinuities are programmed in the model. In our opinion, the dynamic systems approach is essential in accounting for Hosenfeld's data (Hosenfeld et al., 1997). We believe that the data would not easily be captured by a model based on purely feed-forward connectionist techniques, or by one based on serial symbolic processing; this is, in part, because some of Hosenfeld's dependent variables are dynamical in nature (e.g., response reaction time).

### Core Concepts

As should be clear from the above, core concepts of both connectionism and dynamic systems theory play central roles in our account of the development of analogical reasoning. From a representational perspective, the account (and subsequent model) is committed to *featural representations* of objects and learned, nonsemantically transparent *hidden layer representations*. These representations are the product of constraint satisfaction implemented in the dynamics of the settling network. At this level, the model is committed to the core concept of *settling* to an *attractor state* as exhibited by dynamic systems theory.

The activation pattern of the hidden units effectively represent a relation as a qualified transformation of the before state. The transformation is qualified because *knife* does not transform all objects (e.g., *stone*). (In this example, *knife* and *stone* correspond to a stable state of the fully trained network.) At the same time, hidden unit representations are the product of a nonsymbolic learning algorithm and are not semantically transparent. The model thus shares with many connectionist models the characteristic that there is no way in which individual hidden units may be mapped to macro-level features.

Solving analogies of the *a:b::c:d* type within the model actually involves two attractor states: (1) one state relating *a*, *b*, and an unspecified causal agent and (2) another state involving *c*, the unspecified causal agent, and *d*. The transition from one attractor to another is typical of dynamic systems theory.

### Timescales of Change

Although connectionist and dynamic systems principles work together in our account, they work on different timescales of change. Connectionist principles operate on developmental timescales, with learning and the accretion of knowledge shaping the structure of attractors over timescales corresponding to months and years. Principles of dynamic systems theory on the other hand, function over a much shorter time window (e.g., solving individual analogies in terms of settling to attractor states and priming transitions between attractor states).

### Levels of Explanation

We have argued that our relational priming model may be viewed as both a dynamic systems model and a connectionist model. Given this, it is tempting to think that the difference in views reflects a simple difference in levels of explanation, with, for example, the dynamic systems view being at Marr's *implementational level* (Marr, 1982) and the connectionism view being at a somehow higher level. This is not an accurate depiction of the relation between the two views.

The fundamental philosophical claim behind levels of explanation is that while a system's behavior may be explained at multiple levels, explanations within a level are, in some way, complete. Thus an algorithm can be described independent of its implementation, and a computational function can be specified independent of its method of evaluation (i.e., independent of an algorithm for the function). This kind of separation cannot be given for the dynamic systems and connectionist elements of our model. They are highly interrelated, and—regardless of whether one adopts Marr's (1982) traditional tripartite approach or more fine-grained approaches such as Newell's (1990) approach—higher level explanations require principles from neither connectionism nor dynamic systems theory, and lower level explanations make reference to principles from both. Given this, the relation between dynamical systems principles and connectionist principles within our model is better characterized in terms of differing timescales (as described above) rather than in terms of differing levels of explanation.

### Constraints

The theoretical assumption of relational priming as the core process underlying simple analogical reasoning requires some computational mechanism to establish and hold a representation during the several seconds taken to solve an analogy. Either dynamic systems theory or connectionism might provide such a mechanism via recurrent connections within a network of nodes. At the same time, empirical constraints (particularly those suggesting discontinuous change) point to the involvement of other principles grounded in dynamic systems theory (e.g., descriptions in terms of phase changes). Yet models founded purely on dynamic systems principles cannot easily account for the empirical facts of change over developmental timescales (i.e., learning). Learning, particularly as it relates to relational knowledge accretion and domain specificity, argues against a purely dynamic systems (i.e., antirepresentationalist) approach. The theoretical and empirical constraints, therefore, point to a system that combines principles of both dynamic systems theory and connectionism.

With regard to the neural implementation of the model, there are no strong constraints from physiology. At best, one might argue that the neural evidence points to the distributed representation of semantic knowledge in the temporal cortices (for example, see Shelton & Caramazza, 2001). The model is consistent with this in that it involves distributed representations of semantic knowledge. Beyond this, the model makes no neural claims.

## Conclusion

We have presented a model of the development of simple analogical reasoning and have shown how the model accounts for seven characteristics of children's developing abilities to reason analogically. We have further argued that the model's success is critically dependent upon principles of both connectionism and dynamic systems theory. Thus, our model demonstrates complementarity between the approaches. Connectionism and dynamic systems theory need not be viewed as mutually exclusive, alternative conceptions of cognitive development. Rather, they may by viewed as alternative, but mutually supportive, perspectives on a single system. While this argument has been made with a specific model, there is nothing unique about the model's relation to connectionism and dynamic systems theory. The connectionist approach to representation and the dynamic systems approach to settling and constraint satisfaction can be (and have been) combined in a range of domains. It is our opinion that this combination of approaches is likely to continue to yield advances in our understanding of the computational foundations of cognitive development.

## Acknowledgments

This work was supported by an ESRC studentship awarded to Robert Leech, and by European Commission NEST-Pathfinder grant 029088 (ANALOGY).

## References

Amari, S.-I. (1977). Neural theory of association and concept formation. *Biological Cybernetics, 26*(3), 175–185.

Amari, S.-I. (1980). Topographic organization of nerve fields. *Bulletin of Mathematical Biology, 42*(3), 339–364.

Anderson, J. A., & Rosenfeld, E. (Eds.) (1989) *Neurocomputing: Foundations of research.* Cambridge, MA: MIT Press–Bradford Books.

Ballard, D. H., Hayhoe, M. M., Pook, P. K. & Rao, R. P. N. (1997). Deictic codes for the embodiment of cognition. *Behavioral and Brain Sciences, 20*(4), 723–742, commentary 743–758, authors' response 758–767.

Barto, A. G., Sutton, R. S., & Anderson, C. W. (1983). Neuron-like adaptive elements that can solve difficult learning control problems. *IEEE Transactions on Systems, Man and Cybernetics, 13*(5), 834–846.

Boden, M. A. (1988). *Computer models of mind: Computational approaches in theoretical psychology* [Series: Problems in the behavioural sciences]. New York; Cambridge University Press.

Feldman, J. A., & Ballard, D. H. (1982). Connectionist models and their properties. *Cognitive Science, 6*(3), 205–254.

Fodor, J. A. (1975) *The language of thought.* Cambridge, MA: MIT Press.

French, R. M. (2002). The computational modeling of analogy-making. *Trends in Cognitive Sciences, 6*(5), 200–205.

Gentner, D. (1983). Structure-mapping: A theoretical framework for analogy. *Cognitive Science, 7*(2), 155–170.

Gentner, D. (1988). Metaphor as structure mapping: The relational shift. *Child Development, 59*(1), 47–59.

Gentner, D., Rattermann, M. J., Markman, A. B., & Kotovsky, L. (1995). Two forces in the development of relational similarity. In T. J. Simon & G. S. Halford (Eds.), *Developing cognitive competence: New approaches to process modeling* (pp. 263–313). Hillsdale, NJ: Erlbaum.

Gentner, D., & Toupin, C. (1986). Systematicity and surface similarity in the development of analogy. *Cognitive Science, 10*(3), 277–300.

Gick, M. L., & Holyo ak, K. J. (1980). Analogical problem solving. *Cognitive Psychology, 12*, 306–355.

Goswami, U., & Brown, A. L. (1990). Melting chocolate and melting snowmen: Analogical reasoning and causal relations. *Cognition, 35*(1), 69–95.

Grossberg, S. (1980). How does the brain build a cognitive code? *Psychological Review, 87*(1), 1–51.

Hinton, G. E., & Anderson, J. A. (Eds.). (1981). *Parallel models of associative memory.* Hillsdale, NJ: Erlbaum.

Holyoak, K. J., & Thagard, P. (1989). Analogical mapping by constraint satisfaction. *Cognitive Science, 13*(3), 295–355.

Hosenfeld, B., van der Maas, H. L. J., & van den Boom, D. C. (1997). Indicators of discontinuous change in the development of analogical reasoning, *Journal of Experimental Child Psychology, 64*(3), 367–395.

Hummel, J. E., & Holyoak, K. J. (1997). Distributed representations of structure: A theory of analogical access and mapping. *Psychological Review, 104*(3), 427–466.

Inagaki, K., & Hatano, G. (1987). Young children's spontaneous personifications as analogy. *Child Development, 58*, 1013–1020.

Japkowicz, N. (2001). Supervised and unsupervised binary learning by feedforward neural networks. *Machine Learning, 42*(1–2), 97–122.

Kotovsky, L., & Gentner, D. (1996). Comparison and categorization in the development of relational similarity. *Child Development, 67*(6), 2797–2822.

Leech, R., Mareschal, D., & Cooper, R. P. (2003, August). A connectionist account of analogical development. In R. Alterman & D. Kirsh (Eds.). *Proceedings of the Twenty-Fifth Annual Conference of the Cognitive Science Society* (pp. 710–715), Boston, MA.

Leech, R., Mareschal, D., & Cooper, R. P. (2008). Analogy as relational priming: A developmental and computational perspective on the origins of a complex cognitive skill. *Behavioral and Brain Sciences, 31*(4), 357–378, 409–414.

Loewenstein, J., & Gentner, D. (2005). Relational language and the development of relational mapping. *Cognitive Psychology, 50*(4), 315–353.

Marr, D. (1982). *Vision: A computational investigation into the human representation and processing of visual information*. San Francisco, CA: Freeman.

Munakata, Y., & McClelland, J. L. (2003). Connectionist models of development. *Developmental Science, 6*(4), 413–429.

Newell, A. (1990). *Unified theories of cognition*. Cambridge, MA: Harvard University Press.

O'Reilly, R. C. (1996). Biologically plausible error-driven learning using local activation differences: The generalized recirculation algorithm. *Neural Computation, 8*(5), 895–938.

O'Reilly, R. C., & Munakata, Y. (2000). *Computational explorations in cognitive neuroscience: Understanding the mind by simulating the brain*. Cambridge, MA: MIT Press–Bradford Books.

Pauen, S., & Wilkening, F. (1997). Children's analogical reasoning about natural phenomena. *Journal of Experimental Child Psychology, 67*(1), 90–113.

Plunkett, K., & Sinha, C. (1992). Connectionism and developmental theory. *British Journal of Developmental Psychology, 10*, 209–254.

Pylyshyn, Z. W. (1984) *Computation and cognition: Toward a foundation for cognitive science*. Cambridge, MA: MIT Press–Bradford Books.

Rattermann, M. J., & Gentner, D. (1998). More evidence for a relational shift in the development of analogy: Children's performance on a causal-mapping task. *Cognitive Development, 13*(4), 453–478.

Rattermann, M. J., Gentner, D., & DeLoache, J. (1990). The effects of familiar labels on young children's performance in an analogical mapping task. *Proceedings of the Twelfth Annual Conference of the Cognitive Science Society*, (pp. 22–29). Hillsdale, NJ: LEA. Retrieved by subscription from http://www.questia.com/search

Rogers, T. T., & McClelland, J. L. (2004). *Semantic cognition: A parallel distributed processing approach*. Cambridge, MA: MIT Books.

Rosenblatt, F. (1958). The perceptron: A probabilistic model for information storage and organization in the brain. *Psychological Review, 65*(6), 386–408.

Rumelhart, D. E., McClelland, J. L., & PDP Research Group. (1986). *Parallel distributed processing: Explorations in the microstructure of cognition. Vol. 1: Foundations*. Cambridge, MA: MIT Press–Bradford Books.

Shelton, J. R., & Caramazza, A. (2001). The organization of semantic memory. In B. Rapp (Ed.) *The handbook of cognitive neuropsychology: What deficits reveal about the human mind* (pp. 423–443). Philadelphia, PA: Psychology Press.

Smith, L. B., & Samuelson, L. K. (2003). Different is good: Connectionism and dynamic systems theory are complementary emergentist approaches to development. *Developmental Science, 6*(4), 434–439.

Smith, L. B., & Thelen, E. (Eds.). (1993). *A dynamic systems approach to development: Applications*. Cambridge, MA: MIT Press–Bradford Books.

Spencer, J. P., & Thelen, E. (Section Eds.). (2003). Connectionist and dynamic systems approaches to development [Special section/issue]. *Developmental Science, 6*(4).

Sternberg, S. (1969). Memory-scanning: Mental processes revealed by reaction-time experiments. *American Scientist, 57*(4), 421–457.

Thelen, E., Schöner, G., Scheier, C., & Smith, L. B. (2001). The dynamics of embodiment: A field theory of infant perseverative reaching. *Behavioral and Brain Sciences, 24*(1), 1–34, commentary 34–70, authors' response 70–80.

Thelen, E., & Smith, L. B. (1994). *A dynamic systems approach to the development of cognition and action*. Cambridge, MA: MIT Press–Bradford Books.

Thomas, M. S. C., & Mareschal, D. (1997). Connectionism and psychological notions of similarity. In *Proceedings of the Nineteenth Annual Conference of the Cognitive Science Society*, (pp. 757–762). Mahwah, NJ : Erlbaum.

Tunteler, E., & Resing, W. C. M. (2002). Spontaneous analogical transfer in 4-year-olds: A microgenetic study. *Journal of Experimental Child Psychology, 83*(3), 149–166.

# CHAPTER 11

## Integrating Connectionist Learning and Dynamical Systems Processing: Case Studies in Speech and Lexical Development

*Bob McMurray, Jessica S. Horst, Joseph C. Toscano, and Larissa K. Samuelson*

The goal of experimental psychology is to use observable behavior to understand mental and biological processes responsible for cognition, perception, and language. This task is doubly hard for the developmental scientist. Developmental science must strive not only to understand these unobservable processes, but it must also make sense of the (typically unobservable) processes that cause them to emerge and change over the life span. This study of *developmental mechanism* often appears intractable—there are too many degrees of freedom, too many points of inference, and too few data.

Computational theory, however, can make the search for mechanism more tractable. Computational approaches provide a check on the logical consistency of theories, a method for examining the prediction space of a theory, a way to determine what factors of an adult processing model are susceptible to developmental change. Importantly, they also offer a set of biological, mathematical, and physical constraints on theories. In particular, connectionism (Elman et al., 1996) and dynamical systems theory (Smith & Thelen, 1993; Thelen & Smith, 1994) both offer compelling paradigms for modeling development. However, they differ on the mechanisms they highlight and the timescales over which they operate.

This chapter examines the interplay of dynamical systems and connectionism at the level of both theory and computation. We start with a discussion of developmental mechanism focusing on two particular processes—one typically associated with connectionist approaches and one typically associated with dynamical systems. We then illustrate potential hybrid approaches with case studies from the field of language acquisition. The first—a model of speech category learning—illustrates how a dynamical systems perspective may inform a classic connectionist mechanism (i.e., statistical learning). The second—a model of early word learning—combines connectionist and dynamical systems principles. Both suggest that we can no longer treat these two paradigms as independent.

### Developmental Mechanism, Connectionism, and Dynamical Systems Theory

As a rough working definition, we define *developmental mechanism* as a fundamental process or capability within the developing system that accounts for long-term developmental change (but may also account for short-term processes) in a particular domain or developmental problem. While developmental mechanism often refers to specific singular mechanisms (e.g., the Piagetian notion of accommodation), development itself is clearly multiply determined. Development represents the totality of these many causes of change, but for ease of analysis, developmental science has historically peeled off components of this larger process for study. Ultimately, however, it is probably not appropriate to assume that these unary mechanisms are independent—interactivity is likely the norm rather than the exception.

With this idea in mind, the purpose of this chapter is explicitly to examine the interaction of two such mechanisms offered by connectionism and dynamical systems theory: statistical learning proposed by connectionism, and online in-the-moment processing (or constraint satisfaction) proposed by dynamical systems. We suggest that these mechanisms may be profoundly interrelated and, thus, that both approaches may provide useful views on the same problems. The purpose of this chapter, then, is to move toward such interaction. In the special issue of *Developmental Science* (Spencer & Thelen, 2003) that preceded this volume and the conference that inspired it, Smith and Samuelson (2003) asked:

> *Are these in-principle differences between connectionist and dynamic systems theories? That is, could you track the trajectory of change in the weights of a connectionist model and gain insight into the process of learning or development in the modeled system? Or, could a dynamical system learn the statistical regularities inherent in the world and use them to direct and change its behavior? Surely both are possible.*

This chapter will (modestly) do both by examining the interrelationship of developmental mechanisms proposed by each paradigm.

## Connectionism and Statistical Learning

Connectionism rests, in large part, on the demonstration that complex behavior can emerge out of simple local interactions in networks of neuron-like units. The emergence of this complexity is the cornerstone of *developmental connectionism*, which suggests that an examination of the *ongoing* emergence of structure and the ways in which the environment influences it provides an explanatory framework for development. These local interactions and their emergent properties arise out of environmentally tuned learning, allowing connectionist models to provide a platform for testing this theoretical mechanism of development.

Connectionist networks are models of cognitive and perceptual processes that transduce some input into some output, usually in a goal-specific manner. Typically, connectionist models simplify this process in two ways. First, the models transform input to output in a small number of discrete steps (although there are many examples that do not so; for instance, Bharucha, 1987; Dell, 1986; McClelland & Elman, 1986) leaving out the real-time, moment-by-moment dynamics. Second, connectionist networks rarely directly model the raw sensory inputs or motoric outputs (although, see Nakisa & Plunkett, 1998; Schlesinger, Parisi, & Langer, 2000). These simplifications create a resemblance to information processing models (e.g., a focus on functional input–output mappings), but this is only superficial. Connectionism's commitment to task-specific processing means that intermediate processing states (hidden units) are simply that; they are not the abstract, domain-independent, symbolic representations posited by information-processing approaches.

These two simplifications are made largely because of an underlying commitment to a single developmental mechanism that may apply across domains: task-dependent statistical learning. Processing in a network is fundamentally a matter of connection weights. Activation is passed between neuron-like units along variable-strength weights, and the specificity of processing in a network is largely due to its particular weights. Most learning rules alter these weights, encoding statistical relationships between features of the input and between those same features and the output.

Under this view, the environment (or at least the inputs to a given system) contains rich statistical structure that, once encoded in the matrix of weights, enables the system to perform tasks such as word segmentation, face identification, or edge detection with minimal domain-specific architecture. That is, the same simple processing units and learning rules can handle many different types of inputs. Thus, differences between domains arise because of the statistical structure of the task, not the underlying neural hardware. This simple approach has been surprisingly successful. Networks can learn to use transitional probabilities between phonemes (Elman, 1990), eigenvectors of face images (Turk & Pentland, 1991), and regularities in edges and their orientation in visual scenes (Olshausen & Field, 1996) to perform common cognitive and perceptual tasks. These instances differ in terms of learning principles and architecture, but they make the common points (1) that the information needed to solve a particular problem is available in the world and (2) that development can be accounted for by the accumulation of this information as biases on processing (i.e., connection weights).

While connectionist networks are *capable* of solving many sorts of problems, an individual network only learns the specific statistical structure relevant for the particulars of the task it is performing. Given the same input and a different goal, a different structure may be extracted. In fact, the actual statistics that are directly encoded in a network's weights reflect only those necessary to relate inputs to outputs. This structure is not encoded as a separate knowledge store, but rather, it is embedded directly within the processing framework. As such, the connection weights themselves contain the information necessary to accomplish what the network has been trained on for the current task. In this sense, the information represented in a connectionist network is more like procedural knowledge (i.e., how to accomplish a particular goal) than any explicit representation of the structure of the input.

A substantial body of work on goal-directed statistical learning has examined the sorts of statistics generated from and the properties of various learning schemes. However, a small number of studies have taken an implicitly dynamical systems approach. Many of the learning rules posited by connectionism work by gradient descent, that is, the model learns by gradually adjusting its weights to minimize an error signal or maximize a likelihood function (Barto, 1995). Importantly, the magnitude of change (i.e., learning) is proportional to the distance from the goal—once the system arrives at its goal-state, this change (quite logically) becomes zero. If we assume that the statistical structure in the environment is then stable, this gradual change in weights can be seen from a dynamical systems perspective as movement toward an attractor-state in weight-space along the dynamics specified by the learning rule. In this case, the attractor state is the set of weights that allows the network to achieve its goal (although some networks never do) since no further change is possible once the system has arrived at the goal—it has settled.

By framing learning in terms of dynamical systems concepts, we can ask about the stability and dynamics of learning itself. For instance, learning via back-propagation is susceptible to catastrophic interference (McCloskey & Cohen, 1989). Training a network on one batch of data followed by switching to a different training set can result in a reduction in the ability to process the first training set. Put another way, manipulating the stability of the input causes the model to gravitate toward a new attractor state, causing it to partially *forget* the prior input set. Similarly, Hebbian learning models that have adopted an attractor state (in weight-space) for a given input are often unable to adapt to new statistical structure—forcing new data into the old hole. Thus, Hebbian learning, in many cases, does not have the ability to leave an attractor state, once established.

This approach to learning dynamics, focusing on whether the model learned or not, is cursory. Clearly, a closer analysis of learning in a particular network would reveal more subtleties about the process. Nonetheless, discussion at this level does suggest a means for solving this particular problem: slow learning (for an excellent discussion, see McClelland, McNaughton, & O'Reilly, 1995). This prevents the network from settling into an incorrect attractor state and maximizes the quantity of data received by the model prior to settling. This then makes it more likely that a model would be exposed to multiple data sets prior to stabilizing. While beneficial, the use of slow learning creates two additional concerns. First, the necessity of slow learning implicitly suggests that we'll need additional mechanisms to account for phenomena like short-term perceptual learning (e.g., Kraljic & Samuel, 2005; Norris, McQueen, & Cutler, 2003). Second, slow learning also does not fit with the fact that, occasionally, the ability to acquire knowledge appears rapid and complete with only a few exposures. These concerns are well illustrated in the case studies that follow.

## Dynamical Systems Theory and Online Processing

As a theory of development, dynamical systems theory offers two important additions to our cache of mechanisms: embodiment and online processing. Under the embodiment view (Samuelson, 2002; Thelen, 2000), the mechanics of the body, sensory and perceptual processes, and the role of the child as an active learner and creator of his or her environment, all shape motor behavior and underpin cognitive and linguistic behavior.

We agree on the importance of embodiment. It is important to note, however, that embodiment is distinct from the fundamental *computational* issues that divide connectionism and dynamical systems (which are largely framed around slow and fast timescales). Thus, there are dynamical systems models that make no reference to embodiment (e.g., Spivey, 2007; van Geert, 1998) and connectionist approaches that embrace it (e.g., Bechtel, 1997). Nonetheless, for many accounts, this factor is indeed a key difference: connectionists tend to lean toward explanations based on information (broadly construed), and dynamical systems theorists lean toward those based on embodiment.

The second major contribution is that the dynamical systems paradigm views time as continuous and fundamental, using the simple, but powerful formalism expressed in equation (1).

$$S_t = F(S_{t-1}) \tag{1}$$

The state of the system at the current time ($S_t$) is some function (F) of the state at the previous moment in time ($S_{t-1}$). F(S) then represents development itself—how one describes or explains the change that is unfolding over time. Of course, a more radical (and realistic) instantiation of this formalism is expressed in equation (2).

$$S_t = F(S_{t-\Delta t}) \tag{2}$$

Again, the state of the system depends on its state at the previous moment in time; however, equation (2) treats time as continuous (since $\Delta t$ is infinitely small). The implication is that there is no principled place to divide time for measurement. It makes no sense to assume the momentary state of a 6-month-old infant is a function of the momentary state of a 3-month-old infant—that would require causality to leap into the future, ignoring the many state changes that have occurred in the intervening months. Instead, to understand development, we must look at adjacent times and examine moment-by-moment processing. That is, the state of the 6-month-old infant is the product of many moment-by-moment processes leading from birth (and before) to later states. This does not preclude the idea of multiple mechanisms operating at different timescales (e.g., memory vs. learning, adaptation vs. development). Rather, it entails that within such mechanisms, development and processing is continuous in time.

This formalism admits no clear division between the long spans of developmental time and the short spans of in-the-moment behavior and action, although there may be differences in the processes at each timescale (as we will demonstrate). Dynamical systems theory argues that the proper study of development must be concerned with in-the-moment behavior. Crucially, by examining time at much shorter scales, dynamical systems theory has also revealed a host of mechanisms operating in-the-moment. Simmering, Spencer, and Schöner (2006) used a dynamic field model to capture drift and precision in short-term memory representations over the course of several seconds. In this model, memory drift is the emergent product of the online interaction of perceptual input (i.e., reference frames) and long-term spatial memory. Spivey and Dale (2004) modeled taxonomic categories emerging from the dynamic interaction of multiple object features. Both models make use of attractor dynamics by which the model evolves toward a stable representation that satisfies the constraints from multiple domains. In the model of Simmering and colleagues (Simmering et al., 2006), these constraints are provided by the current perceptual input and the long-term memory representations; in the model of Spivey and Dale (2004), these constraints are provided by multiple banks of features. In both cases, the ultimate representation is one that is partially consistent with all of the input. We loosely term this notion of assembling an attractor on the basis of multiple sources of input *online constraint satisfaction*.

The dynamical systems' focus on the smaller (and more continuous) timescale motivates a view of development in which in-the-moment behavior is a crucial determinant of longer-term change. Thus, behavior is not learned or acquired in any kind of static sense, but rather, it is assembled online on the basis of constraints provided by the context, the developing systems' abilities, and the developmental history.

This, in a sense, takes the connectionist emphasis on goal-directed processing to task by pointing out that any behavioral task is the product of multiple constraints and influences, over and above what may be learned by the system. The connectionist concept of process, under this view, is too simple. It ignores the behavioral context and active nature of the child in the environment, while it focuses on the underlying information processing. This may be true, but it does not negate the power of the connectionist argument; for many perceptual, cognitive, and linguistic problems, the statistical structure in the input and in the tasks that a child encounters can be a powerful source of information for generating behavior.

## Toward an Integration

Given the description above, connectionism and dynamical systems provide complementary views of development that often appear to describe different timescales (see also Smith & Samuelson, 2003). While connectionism focuses on slower learning processes, dynamical systems argues that moment-by-moment behavior cannot be ignored. Moreover, while connectionist models use learning to argue that there is sufficient information in the input to accomplish some task, dynamical systems focus more on the temporal dynamics of learning rather than the information and mechanisms behind it.

Both paradigms are dominated by mathematical formalism as a key component of understanding developmental mechanism. Moreover, connectionist systems are a special class of dynamical systems (see Smolensky, Mozer, & Rumelhart, 1996), and many dynamical systems adopt connectionist-like architectures (e.g., dynamic field theory: Thelen, Schöner, Scheier, & Smith, 2001). Despite this, there have been few attempts to integrate both developmental mechanisms into a single computational framework. Connectionist networks that embrace moment-by-moment processing (e.g., McClelland & Elman, 1986) abandon a continuous notion of learning, assuming that processing occurs over a settled (learned) system. Other models (e.g., Tabor & Tanenhaus, 1999) implement a dynamical system on top of a connectionist network—in a sense, treating the moment-by-moment processing as a separate module from the connectionist learning. In contrast, dynamical systems that embrace learning (e.g., van Geert, 1991, 1998) do so by focusing only on the slow developmental time course and by simplifying the nature of learning and relevant information.

Integrating these approaches has proven to be practically difficult because learning and online processing each represent dynamical systems on their own; thus, each contributes its own sources of instability. Including both approaches in a single model, then, multiplies the sources of instability. This should ultimately be a central question of concern to developmental scientists: how are the brain and the body constrained to avoid these multiple sources of instability (see Schöner & Dineva, 2007)? However, in the short term, this difficulty has slowed the integration of the two developmental mechanisms into a single model.

For instance, consider the classic dynamical system described by the logistic function expressed in equation (3) (Perko, 1991; Spivey, 2007; Ward, 2001).

$$y_{t+1} = \eta \cdot y_t \cdot (1 - y_t) \quad (3)$$

This simple dynamical system is stable, with a tendency to trend (over some number of iterations) toward a value between 0 and 1 (its attractor states) regardless of the initial values of *y*. However, this stable pattern of behavior only exists when $0 < \eta < 3$; it oscillates when $3 < \eta < 4$; and it behaves chaotically when $\eta > 4$.

If one were to add learning to this simple system, the obvious place would be to use a learning rule to optimize $\eta$. However, the behavior of the system is dependent on $\eta$. Thus, the learning rule itself must have an attractor between 0 and 4. This may appear simple in this instance; however, the massively parallel and interactive dynamical systems and learning rules used to model actual development will contain many more parameters—thus, ensuring stable optimization may be quite difficult.

These problems have led to an assumption of *additivity*. That is, while many connectionists acknowledge the importance of online dynamics, they seem to assume (1) that they live on top of an underlying connectionist network (e.g., Tabor & Tanenhaus, 1999), (2) that learning proceeds from the results of an already settled process, or (3) that the connectionist network itself is a distillation of a much more complex, underlying dynamical system and the simplification doesn't matter.

This further motivates an assumption of *independent timescales*. Connectionist models often simplify time to a series of discrete input/output events (much like a series of trials in a standard experiment). However, given the traditionally slow pace of connectionist learning, trial-by-trial analyses are not common (although, see Mareschal & French, 2000). Rather, connectionism often focuses on long-term *developmental time*. Dynamical systems, in contrast, focus on time as continuous, often placing more emphasis on timescales that occur *within* a given input/output event—even at the scale of milliseconds, the short-term *processing time*. These complementary timescales have furthered the assumptions that the mechanisms that operate on each timescale are independent and additive.

However, it may not be safe to assume that these developmental mechanisms are additive—their interactions may have crucial implications for development. Thus, we turn to our investigation to these interactions. The two case studies described below examine the interaction of developmental change at short and long timescales, and they demonstrate that understanding processing at both timescales is crucial to understanding development/learning. The first case study looks at this interaction in the area of speech perception, and the second looks at it in word learning.

## Case Study 1: Statistical Learning and Speech Category Acquisition

The acquisition of speech categories presents an ideal domain in which to study the dynamic properties of statistical learning processes because the input is well understood, the developmental time course has been clearly mapped out over many studies, and we have empirical evidence for statistical learning. As a result, we have now modeled this process in a number of ways that illustrate how dynamic systems–style processing/analysis may be critically important to understanding what appears to be a classic statistical learning problem. In particular, these models demonstrate that (1) statistical learning alone is *insufficient* to solve the problem—some further processing such as competition is needed; (2) the particular statistics being learned can force development to follow very specific dynamics; (3) the same learning that accounts for developmental phenomenon can stably play a role at the scale of online processing; and (4) when the required competition is implemented as a simple dynamic system, it can illustrate the linking between largely online behavior and developmental history.

### Empirical Results

Speech input can be characterized by a variety of graded acoustic cues, such as formant frequencies, amplitude of components (e.g., vocal chord vibration or aspiration), and relative timing of such components. This continuous acoustic/phonetic cue-space is carved into many categories by a given language (e.g., phonemes), and the number of categories and their locations (in cue-space) differ cross-linguistically. One of the first tasks that infants are faced with, then, is determining the relevant categories of speech sounds for their languages.

Empirical work demonstrates that infants are initially capable of discriminating a range of phonetic contrasts, many of which will not be used in their native languages. Werker and Tees (1984), for instance, found that 6-month-old infants raised in an English-speaking environment (as well as those raised in a Hindi-speaking environment) could discriminate alveolar /d/ and dental /d/ (dental stops are used in Hindi but not in English), as well as /b/ and /d/, and /k'/ and /q'/ (ejective phonemes used in Thompson, a Native American language). By 12 months of age, however, infants were only able to discriminate sounds from their own languages. These findings have been extended to a range of phonetic contrasts and the pattern that has emerged is clear: infants initially are able to discriminate more sounds than they will need, and this ability is lost over the first year of development (for a review, see Werker & Curtin, 2005).

Statistical learning has been proposed as a mechanism for this change (Maye, Werker, & Gerken, 2002). Under this view, infants track the frequency of specific cue-values (picture the generation of something like a histogram) and identify the number, location, and size of clusters in acoustic-phonetic space in order to learn categories. This is supported by three major arguments. First, a growing body of evidence in phonetics has shown that speech cues are not distributed uniformly, but rather, that they statistically cluster in a way that reflects underlying categories. Lisker and Abramson (1964) measured voice onset times (VOTs) in a large number of languages and found that VOTs were distributed in a multimodal distribution with categories appearing as separate Gaussian clusters. This clustering has also been found for vowels and approximants (Espy-Wilson, 1992; Peterson & Barney, 1951). Thus, many continuous speech cues form Gaussian statistical clusters—perfect fodder for statistical learning.

Second, speech categories reflect a corresponding Gaussian structure. A number of studies by Joanne Miller and her colleagues (reviewed in Miller, 1997) have examined adults' goodness ratings for wide ranges of VOTs and found that these ratings form a roughly Gaussian distribution with the best exemplars located at the most statistically frequent exemplars. These graded categories have also been seen in infants. Miller and Eimas (1996) showed that infants who were habituated to a non-prototypical /k/ (with an extreme VOT of 200 ms) dishabituated when exposed to a more prototypical one (VOT of 110 ms). Finally, McMurray and Aslin (2005) demonstrated that the voicing categories of 8-month-old infants exhibit a gradient pattern which was apparent over small differences (i.e., as small as 10 ms VOT).

The third argument, and the linchpin in the case, comes from a study by Maye and colleagues (Maye et al., 2002). Infants were exposed to a series of tokens from a /d/-/t/ VOT continuum. One group of infants heard tokens whose frequencies formed a bimodal distribution (i.e., two categories) and a second group heard tokens from a unimodal distribution (i.e., one category). After a brief (2.5 minutes) exposure to these statistical distributions, both groups were tested for discrimination. Infants in the bimodal group discriminated pairs of sounds consistent with their having learned two categories. However, infants in the unimodal group did not discriminate these sounds (see also, Maye, Weiss, & Aslin, 2008).

A number of connectionist models have been proposed to account for this process. However, each model embodies a number of additional components that may affect the developmental trajectory of the model while not necessarily being related to statistical learning. For instance, Guenther and Gjaja (1996) include topographic architectures and competition, Elman and Zipser (1988) include logistic activation functions and supervised learning, Nakisa and Plunkett (1998) use unique architectures created by a genetic algorithm, and McClelland (see Chapter 1) uses competition. Since the goal of the present chapter is to understand the dynamics of learning itself, it is necessary to distill this learning process to the simplest possible learning device using the fewest number of architectural assumptions. Thus, we developed a model of statistical speech category learning based on the mixture of Gaussians (MOG) model architecture (McMurray, Aslin,

& Toscano, in press). This model uses a classic statistical technique that is well understood and models learning with simple gradient descent techniques. This simplicity, combined with the lack of strong paradigmatic commitments, enables a deeper understanding of the underlying dynamics of statistical learning.

### Architecture of the Model

Each speech category (e.g., voiced or voiceless) is modeled by Gaussian distributions [Box 11.1, equation (4)] along a continuous acoustic dimension (e.g., VOT). A phonetic dimension (e.g., voicing), then, is modeled as the sum of the

**Box 11.1 Formulas for the mixture of Gaussians model of speech categorization**

Note that while VOT is used in place of X, it is only used for the sake of concreteness of description—this model is intended as a general model for distributions along any single speech dimension.

$$G_i(\text{vot}) = \frac{\phi_i^2}{\sum_{j=1}^{k}\phi_j^2}\frac{1}{\sqrt{2\pi\sigma_i^2}}\exp\left(\frac{(\text{vot}-u_i)^2}{2\sigma_i^2}\right) \tag{4}$$

Equation (4): Probability of a given VOT for model $i$. K represents the number of total models in the mixture, $\varphi_i^2/\Sigma\varphi^2$ represents the likelihood of model $i$, normal against the likelihood of the other models. Here, $\mu_i$ is the mean VOT (category center) of model $i$; $\sigma_i$ is the width.

$$M(\text{vot}) = \sum_{i=1}^{k} G_i(\text{vot}) \tag{5}$$

Equation (5): Probability of a given VOT for all models ($i$). The likelihood of any given VOT is the sum of the likelihoods of each model.

$$\Delta\phi_i = \eta\cdot\left(\frac{G_i(\text{vot})}{M(\text{vot})}\cdot\frac{2}{\phi_i} - \frac{2\phi_i}{\sum_{j=1}^{k}\phi_j^2}\right) \tag{6}$$

Equation (6): Learning rule for adjusting $\varphi_i$ (from which the prior likelihood of model $i$ is computed) for a given VOT. Note that this rule only operates on a single model (the one with the highest likelihood) to create a winner-take-all form of competition. Here, $\eta$ represents a learning rate parameter.

$$\Delta\mu_i = \eta\cdot\left(\frac{G_i(\text{vot})}{M(\text{vot})}\right)\frac{(\text{vot}-u_i)}{\sigma_i^2} \tag{7}$$

Equation (7): Learning rule for adjusting $\mu_i$ (the center of the category) given a single VOT. Here, $\eta$ represents a learning rate parameter.

$$\Delta\sigma_i = \eta\cdot\left(\frac{G_i(\text{vot})}{M(\text{vot})}\right)(\sigma_i^{-3}(\text{vot}-u_i)^2 - \sigma_i^{-1}) \tag{8}$$

Equation (8): Learning rule for adjusting $\sigma_i$ (the width of the category) given a single VOT. Here, $\eta$ represents a learning rate parameter.

probabilities that a given VOT arose from some number (K) of Gaussians [equation (5)]. For instance, a translation of Lisker and Abramson's (1964) analysis into the MOG framework would suggest that voicing in English is the sum of two (K = 2) Gaussians with equal prior likelihoods (φ), means (μ) at 0 ms (voiced) and 50 ms (voiceless), and standard deviations (σ) of around 10 ms (voiced) and 20 ms (voiceless). Given a particular parameter set, this model categorizes a given VOT by computing the likelihood of that VOT for each of the component Gaussians, with the most likely Gaussian representing its voicing category.

To model the learning process, the model is exposed to a series of individual VOTs. After each one, three sets of parameters are adjusted via the learning rules given in Box 11.1 [equations (6)–(8)]. These learning rules are derived from standard gradient descent procedures to maximize the likelihood of the entire model (the sum of all the individual Gaussians) given a particular data point. The learning rule for each parameter is the derivative of the log-likelihood (i.e., how probable the current model is) given the current input (VOT), with respect to the parameter being updated (e.g., μ). These learning rules are unsupervised—there is no error signal; the rules simply push the model to find the most likely parameter set.

Most mixture models are trained on batches of data (not on a single data point at a time), using more powerful methods such as expectation maximization (EM) or supervised techniques. However, supervised methods lack the psychological plausibility to study development, and batch-mode methods (like EM) do not permit a moment-by-moment analysis of learning. Gradient descent allows the learning rules to be applied iteratively (i.e., one input at a time) and without a teacher.

Because children do not know in advance how many categories may exist on a given dimension (languages like English have two voicing categories, while Thai has three), the model should not have that knowledge either. Instead, the model, like children, should be capable of learning any distribution. Thus, prior to training, the model is initialized with a large number of Gaussians (more than will be necessary). Each of these Gaussians starts with equal likelihoods ($\varphi = 1/K$) and standard deviations ($\sigma \approx 2$). The locations (μ's) of each category in VOT-space are randomized to provide some variation for the learning rules to act on. The model is then exposed to a single VOT, all of the parameters are updated, and the next VOT is presented. These inputs are sampled from a data set that is generated from the actual distribution of VOTs in a given language.

## Testing the Model

### Competition

Initial tests were conducted to verify that the model could successfully learn the distribution of speech sounds. One hundred models were initialized with K = 20, random μ's, and σ = 5. These models were exposed to a random distribution of VOTs (taken from data on English reported by Lisker & Abramson, 1964). These models were initially tested to see if they learned the correct number of categories (the number of Gaussians for which $\varphi_i > 0.01$). Surprisingly, none of the models learned two categories—most had substantially more ($M = 10.25$). In fact, each category was typically represented by multiple overlapping Gaussians (see Fig. 11.1). While one could argue for a distributed representation in which a given speech category is represented by a number of overlapping Gaussians, this model did not even approximate that well. On average, 67% of the Gaussians were 5 ms of VOT or more away from a correct μ for the data set, and the number of Gaussians that was closest to a given category did not reflect the true base rate (φ): the voiced category was heavily overrepresented ($M = 52\%$ of the φ-space) compared to the voiceless ($M = 17\%$ of the φ-space).

This lack of success was perhaps not surprising—fitting a mixture of Gaussians with an unknown K and priors on each Gaussian (φ) has been a historically difficult problem (for a discussion, see De Boer & Kuhl, 2003). To solve it, we adopted a competition scheme loosely based on the competitive learning network of

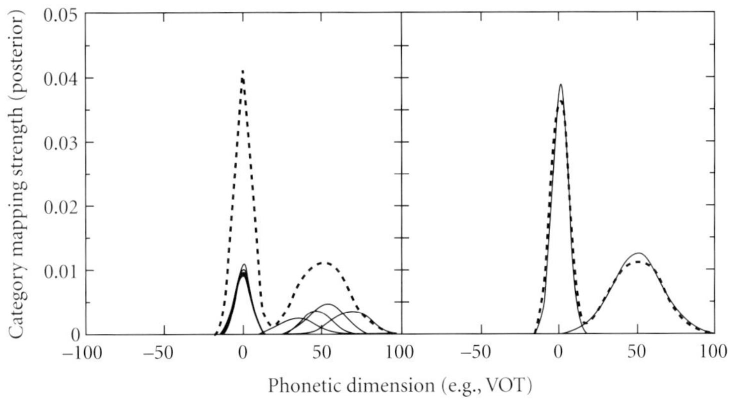


**Figure 11.1. Testing the MOG model: Comparison of models after 10,000 learning trials**
This figure depicts comparisons of MOG models after 10,000 learning trials with and without winner-take-all learning competition. Shown is the posterior probability of each VOT for each Gaussian (an estimate of the strength of the category mapping). (Left panel) Model without winner-take-all learning. Multiple overlapping Gaussian curves are present for each category of the input (dashed line). (Right panel) Model with winner-take-all learning. Only one Gaussian curve remains for each category.

Rumelhart and Zipser (Rumelhart & Zipser, 1986; see McMurray, Aslin, & Toscano, in press). Winner-take-all learning was adopted as a component of the learning rule for φ; only the most likely Gaussian in the mixture would alter its φ. Since the effective likelihood of a Gaussian is normalized ($\varphi^2/\Sigma\varphi^2$), an increase in the winning φ resulted in an effective decrease of the losing φ's. The winner-take-all metric is psychologically plausible; φ is analogous to frequency, so updating the frequency of only one category after a single exposure to a speech sound makes sense.

This simple change to the learning procedure resulted in dramatically improved performance. On a new batch of 100 models, 95% learned the correct (two category) solution, and 100% of these successful models learned the correct μ's and σ's. While this may seem like a minor detail, the simplicity of this model highlights its importance. Statistical learning mechanisms alone lack the power to learn multimodal distributions of speech cues; an additional competition mechanism is needed. While this has long been implicit in the connectionist literature (e.g., Rumelhart & Zipser, 1986), this approach makes it explicit. Thus, even the simplest statistical learning problems require a minimal amount of processing to be successful. While our winner-take-all metric is simple and relatively non-dynamic, as we shall demonstrate shortly, a dynamical systems approach may offer a method of implementing this more realistically.

### Dynamics of Learning

The more successful version of the model described above was next examined developmentally. The first goal was to determine if it displayed the same developmental time course as has been demonstrated empirically. The second, more important goal, was to examine learning from a dynamical systems perspective, asking what happens moment-by-moment.

In terms of the developmental time course, we first point out that the initial random distribution of μ's (Fig. 11.2) ensures that newborn models will be able to discriminate a large number of phonetic contrasts—very few pairs of VOTs are assigned to the same category. Of course, as this model learns the training distribution, it eliminates unneeded Gaussians and tunes the remaining ones to the parameters found in its language, whether it is a two-category language such as English, or a three-category language such as Thai (Toscano & McMurray, 2005). This

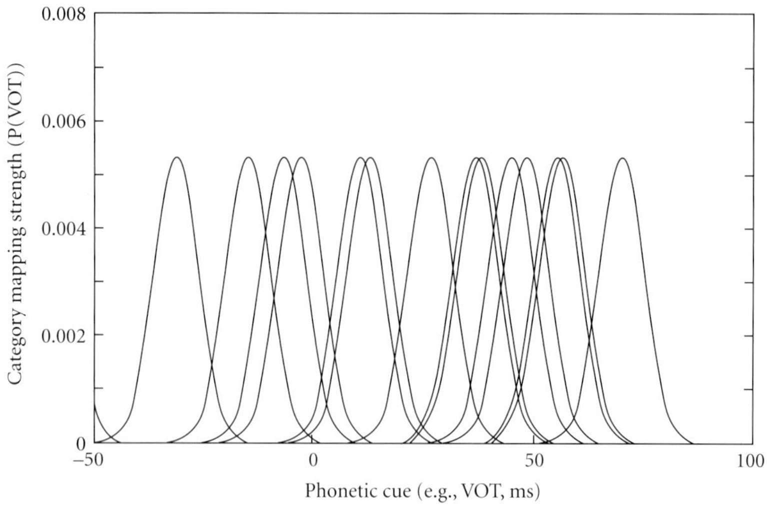


**Figure 11.2. One initialization for a K = 15 model**
Here, φ and σ are constant across Gaussians (.067 and 5, respectively) while μ's are randomly distributed.

is exactly what infants do between 6 and 12 months of age (see Werker & Curtin, 2005, for a review).

To test this, we evaluated 100 models on discrimination of pairs of VOTs over development (see McMurray, Aslin, & Toscano, in press). Discrimination was determined by a simple metric. For each VOT, we computed which Gaussian had the highest posterior probability using equation (4) (Box 11.1). If the same Gaussian was most likely for both VOTs, the model responded *same*, otherwise it responded *different*. During exposure, within-category VOT pairs (−40 vs. 0) were initially categorized as belonging to different categories and later as a single category (see Table 11.1). The between-category VOT pair (0 vs. 40) was classified as different throughout development. A last comparison between a prototypical voiceless sound (40 ms) and a non-prototypical one (80 ms) yielded moderate discrimination throughout training (analogous to Miller & Eimas, 1996).

This attunement should be part of any model that learns the statistics of the input. However, a richer empirical test will require a subtler analysis of the developmental time course. In particular, we wanted to determine whether development in the model is marked by simple gradual improvement, or if the computational problem posed by statistical learning is marked by more dramatic changes in development. To probe this statistical learning mechanism more deeply, we conducted a trial-by-trial analysis of the trajectory of learning for a new set of 150 models. These models were initialized with K = 5, random μ's, and starting σ's of 5, 10 or 15 (50 models for each starting σ). These models were exposed to a simple distribution of VOTs (μ's at [0, 50], σ's at [10, 10] and φ's at [.5, .5]) and their parameters extracted every 50 training cycles. Models were run until they had stabilized (either successfully or not), which usually took around 6000 training cycles. Overall, only 79% of these models succeeded in learning the distribution, although this was due to small K's and

[1]This success rate was also related to the starting σ. While models starting with σ's of 5 or 15 did fairly well (σ = 5, 42/50 succeeded; σ = 15, 44/50 succeeded), models starting with σ = 10 (the correct σ) were largely responsible for the low success rate (33/50 succeeded). This was reliable [$\chi^2(2) = 8.37$, p = .015]. It is unclear why starting with the correct σ resulted in worse performance than staring with incorrect ones. However, it is important to note that the learning rule for σ is only optimal if μ has been learned correctly. Thus, early in development, an accurate estimate of σ may not be optimal—something that should be examined in future simulations.

**Table 11.1. Discrimination as a function of training cycle**

| Comparison (ms VOT) | Proportional discrimination after X training cycles | | | | | Description |
|---|---|---|---|---|---|---|
| | 1000 | 2000 | 3000 | 4000 | 5000 | |
| −40 vs. 0 | 0.88 | 0.81 | 0.60 | 0.52 | 0.37 | Spanish prevoiced vs. voiced |
| 0 vs. 40 | 0.99 | 1.00 | 0.99 | 1.00 | 1.00 | English voiced vs. voiceless |
| 40 vs. 80 | 0.76 | 0.71 | 0.65 | 0.61 | 0.70 | English voiceless vs. extreme |

*Note:* Discrimination as a function of training cycles is shown for within-category (−40 vs. 0), between-category (0 vs. 40), and extreme (40 vs. 80) comparisons; units = ms VOT (millisecond voice onset time).

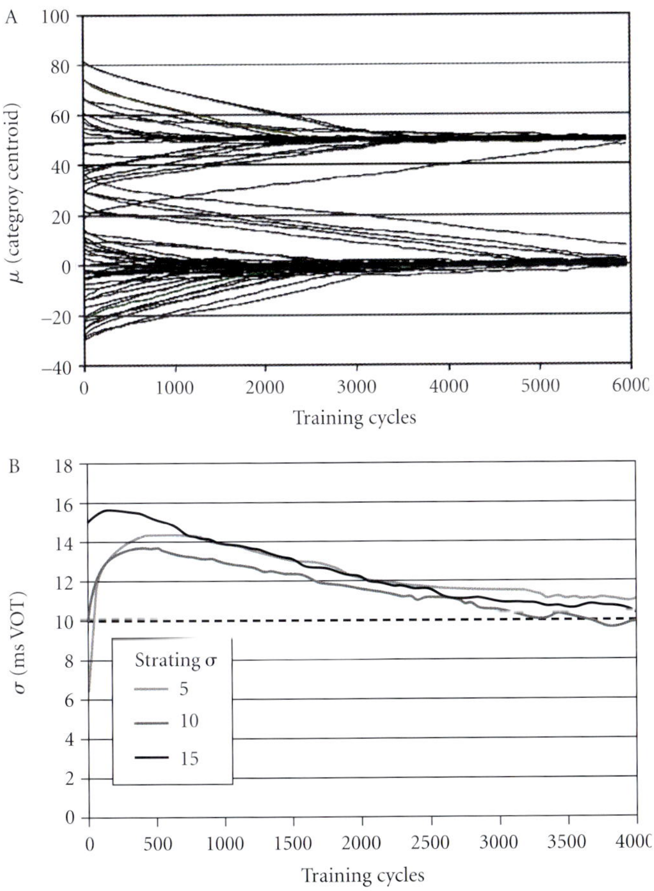


**Figure 11.3. Results of an analysis of learning in the mixture of Gaussians framework** (A) Trajectories of μ's for individual models. (B) Mean σ as a function of training cycles and starting σ. The dashed line represents the σ of the data set used in training.

larger starting σ's.[1] A trial-by-trial analysis of changes in μ across individual models is shown in Figure 11.3A. Not unexpectedly, there was a large amount of variance in μ initially (since they were set to random values), and rather quickly they approached the values in the training distribution (0 and 50) following a characteristic hyperbolic curve (also not surprising given the formula for Δμ). Thus, learning the locations of the categories may simply reflect a gradual improvement in the estimate.

The analysis of σ was more intriguing. Figure 11.3B, displays σ as a function of training cycles for models with a starting σ of 5 (smaller than that of the training set), 10 (the same size), or 15 (larger). In all three cases, σ very quickly grew to be much larger than would eventually be required (even if it was already larger) and then gradually settled toward the correct solution. This was a consequence of the learning rules. A large mismatch between the current data point and μ resulted in both a shift in μ and an increase in σ (at least until μ settles, at which point σ starts to decrease). Thus, even if the model started with the correct σ, it momentarily increased σ until μ was learned.

This predicts a significant increase in the size (in phonetic space) of speech categories followed by a slower decrease, particularly during the period of 6 to 12 months of age when speech categories are changing the most. This cannot be tested with head-turn and habituation methods because they do not allow enough repeated measurements; however, the two-alternative, anticipatory eye-movement paradigm (McMurray & Aslin, 2004) may offer a solution. Importantly, this analysis of development in the width of categories suggests that there may be regularities in the dynamics of learning based on the statistics to be learned. In this case, an accurate estimate of σ requires an estimate of μ, but the converse is not true. Thus, while μ shows no systematic effects (other than convergence on the true values), σ shows early growth. An analysis of both the dependencies among statistics and their relationship to learning processes, thus, has the potential to reveal systematic dynamics over developmental time.

### Perceptual Learning

An oft neglected facet to connectionist learning is that learning is always happening—every input forces a model to evaluate the relationship between its current parameters and the input, and adjust the parameters if necessary. While this is obvious, it has important implications for current work on perceptual learning. A number of recent studies demonstrate that adult listeners are capable of shifting category boundaries of speech sounds with very few exposures (Kraljic & Samuel, 2005; Norris et al., 2003). These studies hypothesize a secondary mechanism to account for short-term plasticity. For instance, Mirman, McClelland, and Holt (2006) add Hebbian learning to TRACE (McClelland & Elman, 1986) to model short-term plasticity, but do not use it to model the acquisition of phoneme categories or words. This is largely due to the assumption that (developmental) learning must be slow. However, an alternative is that the same developmental mechanism (statistical learning) that accounts for initial speech category learning can also yield short-term perceptual plasticity. If true, this would suggest that connectionist-style learning may play an important role over the shorter timescales addressed in dynamical systems theory.

To test this possibility, 25 models were exposed to a bimodal distribution ($\mu$ = [0, 50]; $\sigma$ = [10, 10]; $\varphi$ = [.5, .5]). After 10,000 training cycles (sufficient for the model to settle), a new data set was introduced with rightward-shifted categories

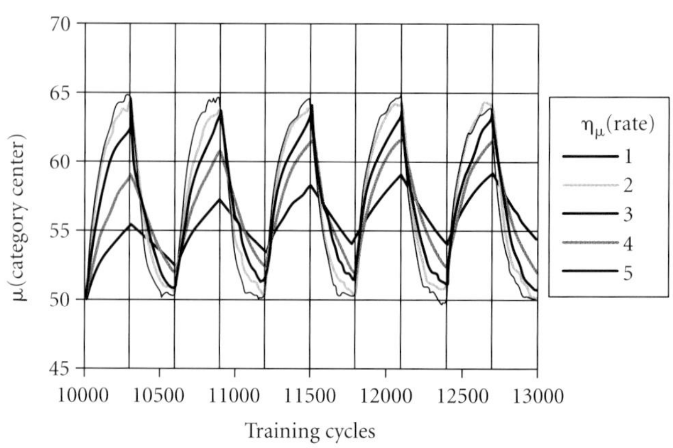


**Figure 11.4. Plasticity simulations** This figure depicts plasticity simulations with $\mu$ plotted as a function of time and $\eta_\mu$ representing the voiceless category. After 10,000 cycles of training on categories centered at [0 and 50], the training set alternated every 300 epochs (vertical lines) between [0 and 50] and [10 and 65]. Note that only the voiceless category is shown for clarity of presentation.

($\mu$ = [15, 65]). Training alternated between these two data sets every 300 cycles, while $\mu$ and $\sigma$ were recorded, for 6 alternations. This was repeated with 5 different values for $\eta_\mu$ (the rate of change in $\mu$). Figure 11.4A, shows $\mu$ (for the rightmost category) as a function of time and $\eta_\mu$. It is clear that the model can quickly adapt to the new means. By the third alternation, some models can adapt within 30–40 training cycles (which is faster than what was required by human listeners in Clark & Luce, 2005). This was particularly true for the models with a high $\eta_\mu$. Models with a low $\eta_\mu$ were not able to shift their means all the way to either end and tended to drift toward the average of the two values.

To assess this short-term plasticity, we computed the time it took for the model to switch from within 5 ms of the old mean to within 5 ms of the new one. A hierarchical regression analysis examined switch time as a function of $\eta_\mu$, target (original $\mu$'s or shifted), and alternation number. It confirmed that this switch time was highly correlated with $\eta_\mu$ ($R^2_{partial}$ = 51, $\beta$ = −.696, t(756) = 28.1, p < .0001)—faster learning rates led to faster switch times. The alternation number (1st, 2nd, 3rd) was not correlated with switch time ($R^2_{partial}$ = .001)—the model's plasticity did not change over the alternations. However, the model was significantly faster at shifting toward the original $\mu$'s ([0, 50]) than the novel ones ($R^2_{partial}$ = .08, $\beta$ = −.21, t(756) = 8.5, p < .0001), suggesting that it may have not shifted entirely to the new values. This was particularly the case for low $\eta_\mu$'s, shown by an interaction of $\eta_\mu$ and target ($R^2_{partial}$ = .012, $\beta$ = −.27, t(753) = 3.011, p = .003).

At least in principle, then, the same learning mechanism that accounts for initial category learning could operate quickly enough (upon established categories) to account for short-term plasticity. However, it is possible that we have accomplished this by artificially increasing $\eta_\mu$. As mentioned above, slower learning is generally thought to yield a more stable model and a greater likelihood of success; increasing the learning rate to aid in short-term processing may hinder basic developmental processes (if the increased $\eta_\mu$ exceeds that needed for stable development). An analysis of these simulations revealed this was not the case at these learning rates. Overall, 88% of the simulations correctly learned the two-category solutions and a logistic regression confirmed that success was not correlated with $\eta_\mu$ (Wald(1) = .096, p > .2). While increasing $\eta_\mu$ beyond about 20 led to instability and failure at learning, switch time appears to flatten at $\eta_\mu$'s above 16. Thus, there is no disadvantage to high $\eta_\mu$'s, and no additional plasticity advantage for very high $\eta_\mu$'s. The slower learning that is important for development need not be so slow that it prevents perceptual learning on short timescales. Therefore, the same statistical learning mechanism that accounts for developmental change can also account for short-term perceptual learning, without any loss of efficacy in the basic developmental process.

This seems to predict quite a bit of plasticity in even the adult perceptual system. If this is true, why is it that adults cannot easily acquire non-native speech categories? In answering this question, it is important to bear in mind that most work done on non-native adult acquisition has examined difficult cases (e.g., involving Japanese language) where listeners are not simply shifting existing categories, but must abandon a single-category solution in favor of a two-category solution (e.g., McCandliss, Fiez, Protopapas, Conway & McClelland, 2002). Even with the higher learning rate, the MOG model struggles here. This was seen in a series of 10 models that were trained on a unimodal distribution ($\mu$ = 10, $\sigma$ = 20) that shifted (after 15,000 epochs) to a bimodal distribution ($\mu$ = [−90, 90], $\sigma$ = [20, 20]). All 10 models failed to learn the two-category solution.

This occurred because prior history of the model caused it to settle into a one-Gaussian solution. Further learning simply could not act upon this new state to achieve a two-category solution. For $\varphi$, this remaining single category was generally the winner on any given trial (so only it could adjust $\varphi$). For $\mu$ and $\sigma$, both are adjusted as a function of $\varphi$ (which was small for all Gaussians but one). Thus, the theoretical notion of plasticity is complex even in a simple model like this—it can be a function of both the history of the system, and the statistical parameter being learned.

### A Network Approach

The MOG approach provides a distillation of statistical learning processes. In doing so, it reveals the dynamics of learning and also links short-term plasticity and developmental mechanism. Crucially, it points out that statistical learning alone is not sufficient for successful learning; competition among categories is required. Our one-step, discrete winner-take-all competition, however, represents a simplified, implausible form of competition. A more realistic framework—in which competition is spread out over time—would allow us to implement statistical learning in a more dynamical framework.

Thus, we developed a Hebbian normalized recurrence (HNR) network (McMurray & Spivey, 1999) for speech categorization (see Fig. 11.5). The 130 input nodes formed a topographic map of the input dimension (e.g., VOT). Each node had a Gaussian tuning curve whose mean (i.e., characteristic VOT) was organized from low (–40 ms of VOT) to high (80 ms). Input nodes were initially randomly connected to output nodes. After receiving an input, activation was weighted and added to the current output [Box 11.2, equation (9)]. Output nodes compete with each other with a squared-normalization rule [equation (10)]. This formalism approximates winner-take-all competition in the limit (i.e., if run repeatedly). However, at any given time step competition is partial and reflects the intermediate states leading up to the winner-take-all climax. Output nodes pass activation back to the input nodes via a set of feedback connections [equation (11), Fig. 11.5]. Importantly, feedback only affects *active* input units, which prevents the model from activating perceptual input on the basis of top-down evidence alone (i.e., a *hallucination*). Since the feedback is multiplicative, it also introduces some nonlinearity into the system. After the activation provides feedback to the input layers, the input activations are normalized [equation (12)], and then activation cycles in this way until the output nodes settle (i.e., the summed derivative gets close to zero). Importantly, at every cycle a small amount of Hebbian learning takes place [equation (13)] strengthening the weights connecting active inputs with outputs and decaying weights for which either is inactive.

Preliminary tests of this architecture showed that 48 of 50 models trained on the distribution of VOTs from Lisker and Abramson (1964) learned two categories. Moreover the model showed a similar developmental pattern to the empirical data and the MOG model; displaying multiple stable categories early in development ($M$ = 4.2 stable categories halfway to the end of learning). Finally, the quadratic normalized competition in the output layer was required for learning. Without it, none of the 50 models learned the distributions.

Thus, this model has largely the same features as the MOG model in terms of learning and competition. However, the recurrent feedback/competition loop also enables us to examine the time course of processing. Figure 11.6A, displays a trained model's activation of voiced and voiceless output nodes as a

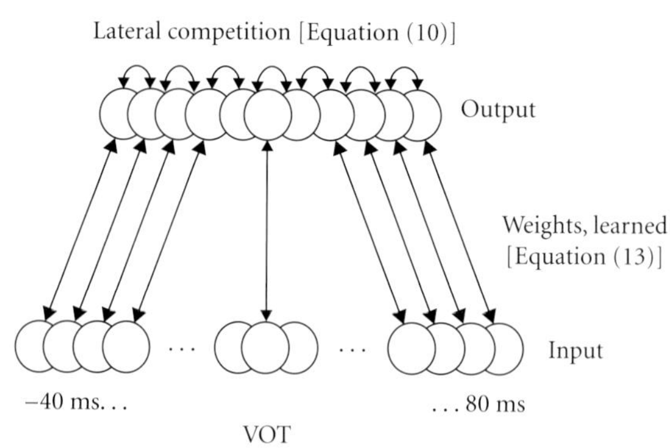


**Figure 11.5. Hebbian normalized recurrence network used for speech categorization**

**Box 11.2 Formulas for Hebbian normalized recurrence network for speech categorization**

$$o_x = o_x + \sum_{z \in i} w_{xz} \cdot i_z \tag{9}$$

Equation (9): Activation in an output node ($o_x$) is the sum of activation from the input ($i_y$) nodes multiplied by the strength of the connection between them ($w_{xy}$).

$$o_x = o_x^2 \Big/ \sum_{z \in o} o_z^2 \tag{10}$$

Equation (10): Output activation is squared and normalized to sum to 1 as a form of non-linear competition. If this step is repeated many times (for any pattern of activation), this formula always yields a single output node active at 1, and all of the others at 0 (unless two nodes are perfectly tied). Thus, its limit behavior approximates Rumelhart and Zipser's (1986) competitive learning, while its moment-by-moment behavior is continuous. Also note that this is provably equivalent to lateral inhibition in which a given node inhibits all other nodes as a function of the proportion of total activation it holds (see McMurray, Horst, & Samuelson, 2008, in preparation) and is similar to Heeger's (1993) findings of normalization in population codes in the visual system.

$$i_y = i_y + i_y \cdot \sum_{z \in o} o_z w_{zy} \tag{11}$$

Equation (11): Output activation spreads back to input layer (along the same weights). Multiplying feedback by activation in the input layer prevents hallucination—feedback cannot affect an input node unless it is already active.

$$i_y = i_y \Big/ \sum_{z \in i} i_z \tag{12}$$

Equation (12): Input layer is normalized to sum to 1.

$$\Delta w_{xy} = \eta \cdot \left( i_y o_x - w_{yx} \right) \tag{13}$$

Equation (13): Weights are adjusted at each cycle through the model by a modified Hebbian rule. Here, $\eta$ represents the learning rate and is very low (~1e-6).

function of time after receiving a voiced input. The model temporarily considers both alternatives, but eventually settles on the correct target. Figure 11.6B, illustrates the graded nature of this model's speech categories. Here, if the network received a voiced VOT (e.g., *bear*), the activation for the voiceless competitor (e.g., *pear*) systematically increases as the VOT approaches the category boundary—even for VOTs that are ultimately categorized by the network as voiced. Thus, as in similar attractor networks (e.g., Spivey, 2007; McClelland & Elman, 1986), early on activation for competitors reflects gradiency in the input, while later it settles into a more discrete representation.

This provides a good qualitative fit to eye-movement data from McMurray, Tanenhaus, and Aslin (2002). This study used the visual world paradigm (Allopenna, Magnuson, & Tanenhaus, 1998) to examine subjects' moment-by-moment word recognition as a function of within-category VOT variation. Here, subjects heard an auditory target such as *bear*, whose VOT was manipulated in small steps (as in the model). A subject would click on the correct target from a screen containing the *bear*, a voicing

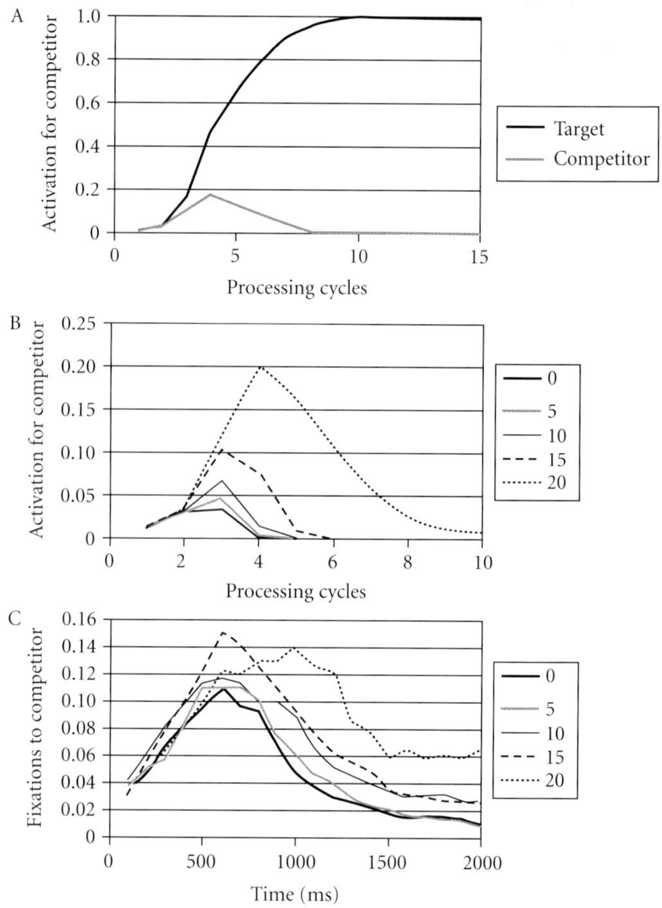


**Figure 11.6. Temporal dynamics in the Hebbian normalized recurrence network for speech categorization**
(A) Activation for targets and competitors for a single VOT is depicted as a function of time (i.e., processing cycles). (B) Activation of voiceless competitor is depicted as a function of time (i.e., processing cycles) and VOT (0–20 ms). (C) Probability of fixating the voiceless competitor is depicted as a function of time (as per McMurray, Tanenhaus, & Aslin, 2002). Each curve indicated looks to the voiceless competitor after hearing a voiced target with a VOT of 0–20 ms for trials in which the subject eventually selected the voiced target. Fixations here are analogous to activation (e.g., Allopenna, Magnuson, & Tanenhaus, 1998).

competitor (*pear*), and two unrelated objects. We measured activation for the competitor word by examining looks to the pear, and we found that the pattern of fixations over time revealed systematic gradiency (see Fig. 11.6C). As VOT approached the category boundary, subjects were more likely to fixate a competitor (e.g., *pear* when the target was *bear*). An analysis of the model revealed that the gradient pattern of responding derives from the graded statistical structure of the category. Thus, the gradiency in lexical activation may be the result of statistical learning of speech cues. Given emerging results that this gradiency may be crucial to normal adult online comprehension mechanisms (McMurray, Munson, & Gow, 2008, submitted for publication; McMurray, Tanenhaus, & Aslin, 2008, submitted for publication), this developmental mechanism may prepare the system for efficient online processing.

### Case Study 1: Discussion

The first case study examined what appears to be a relatively straightforward case of statistical learning, the acquisition of speech categories. Preliminary simulations with the mixture of Gaussians model, however, suggested that statistical learning alone is insufficient—an additional competition mechanism is needed. In a typical connectionist model, it is easy to lose sight of the theoretical import of such mechanisms as they are often the default assumption of the architecture (e.g., logistic activation functions in an auto-associator, or winner-take-all competition in a competitive learning network). In contrast, we built a learner up from minimal paradigmatic commitments, in order to highlight the role of mechanisms like competition. Further, when this was expanded to a connectionist learning paradigm with dynamical processes, it demonstrated how developmental history (the statistical distribution) can explain a nondevelopmental online-processing phenomenon (gradiency). Further, a consideration of the dynamics of learning itself led to new insights into this mechanism. In particular, it suggested that the width of phonetic categories may increase early in development—even if the categories are already larger than those of the input. This suggests that the statistical structure to be acquired can play an important role in constraining the dynamics of development. That is, statistical learning may not always yield incremental gradual improvement; it can also yield some interesting systematic dynamics.

The mixture of Gaussians model was able to simulate short-term plasticity without a secondary adaptation or priming mechanism. This suggests that learning itself is best thought of as an in-the-moment process, something that must also be considered by accounts focusing on processing time. This is not a unique point. Mareschal and French (2000) applied this idea to infant habituation; Mirman and colleagues (Mirman et al., 2006) demonstrate how TRACE (McClelland & Elman, 1986) can be augmented for short-term learning. However, the mixture of Gaussians approach does provide an important counterpoint to arguments that connectionism focuses only on long timescales of learning.

## Case Study 2: Word Learning

The second case study offers a different view of the relationship between learning and in-the-moment processes. Here we start with what appears to be entirely an online, in-the-moment process, and offer a simulation that demonstrates how connectionist learning principles can benefit from and influence this process. This case study focuses on early word learning, an ideal domain for this purpose because there are clear components of long-term learning (vocabulary growth) and in-the-moment behavior (identifying referents) that may interact. Our simulations of this suggest processes that operate in both online and developmental time, and suggest that they may be critically interrelated.

### Empirical Results

Between the ages of 1 and 5 years, children acquire new words at an extraordinary rate, learning as many as 50,000 words by adulthood (Carey, 1978). This immense rate of learning has led many to posit a need for specialized (perhaps innate) mechanisms that are directed toward forming sound/meaning mappings. One such proposed mechanism has become known as fast-mapping. While the term *fast-mapping* has been used as a catchphrase to describe any word learning that occurs with minimal exposure, we use it here to describe a specific referent selection task based on mutual exclusivity. In such tasks, children quickly link a novel name and referent based on the assumption that a novel word will refer to something for which they do not already know a name. In the seminal demonstration of fast-mapping, Carey and Bartlett (1978) asked children to retrieve the *chromium* tray from a set containing a chromium (olive-green) tray and a blue one. Children retrieved the correct tray, suggesting that they had inferred that *chromium* referred to the color for which they did not have a name. This form of fast-mapping has been shown to be operative in children as young as 17 months of age (Halberda, 2003).

It is not clear, however, that this sort of fast-mapping represents learning. While children can identify the referent of novel words in this sort of task, this may actually represent in-the-moment problem solving, not learning. Assessing this requires the researcher to determine if the mappings formed in a fast-mapping task are retained. Existing experimental results are not clear on this fact. Some studies have tested retention by presenting children with a subsequent trial of the same sort (the novel words paired with a known word), raising the possibility that children are simply solving the problem again (e.g., Mervis & Bertrand, 1994). Other studies review the name–object linkages before testing (e.g., Goodman, McDonough, & Brown, 1998). None of these studies rule out the possibility that fast-mapping represents a form of in-the-moment problem solving, and they do not provide evidence that children are truly learning the words (for a review, see Horst & Samuelson, 2008).

A recent study by Horst and Samuelson (2008) provides convincing evidence that children do not retain word–referent mappings formed in fast-mapping contexts. In this study, children were initially given a series of typical fast-mapping trials in which they were presented with two familiar objects and one novel object. Children were asked to retrieve the *fode* and correctly selected the novel object 68% of the time. Following a 5-minute break, the children were then presented with the *fode*, another recently fast-mapped object, and a novel object that had previously been seen but not named. When asked to retrieve the *fode* in this context, children performed at chance (see Fig. 11.7A). This suggests that children's behavior on the fast-mapping trial did not result in much (if any) learning. Subsequent experiments showed that this was the case even when the number of novel names presented in the fast-mapping task was reduced to one, when the number of total trials was reduced to three, and when the experimenter provided labeling after the child made a selection. This evidence suggests that fast-mapping may represent an in-the-moment problem solving behavior, but not a specialized learning mechanism (or a learning mechanism at all). This also suggests that fast-mapping could be modeled by online constraint-satisfaction and competition alone. Further, it could be that while fast-mapping does not offer the *one-shot* learning that has been trumpeted in the literature, it is still related to word learning. Here we explore a subtle version of this hypothesis. Our simulations demonstrate that in-the-moment problem solving may (1) provide a training signal to permit slow connectionist learning and (2) allow the system to behave as if it knows a word, despite relatively little knowledge. By so doing, these simulations illustrate the deep relationship between online processing and learning.

## Model Rationale

The HNR architecture has a number of properties that make it well suited to model this process. As we have shown, HNR models are capable of learning subtle regularities in the environment using simple unsupervised mechanisms. Thus, they may be able to capitalize on statistical associations (between sounds and visual stimuli) to learn words. We propose an associationist account: consistent co-occurrence between auditory and visual forms permits word learning. For instance, every time *dog* is heard, the only visual stimulus that is *consistently* present is also a *dog*. Although other visual stimuli may be available, *dog* will be more consistent over time. This invokes behaviorist explanations of word learning which were largely ruled out by arguments that unambiguous reference is not possible in most visual scenes without innate constraints (although, see Smith, 2000). While this may be true in a single presentation, word learning does not have to happen in a single presentation if co-occurrence between words and referents is consistent and reliable over the long term (Horst, McMurray, & Samuelson, 2006; see Yu & Smith, 2007, for evidence that adults can learn words in such circumstances).

However, learning is not the whole story. Children are engaging in a complex constraint-satisfaction process when determining the referent of a novel word in the context of the contents of their lexicons and the visual context. This is where the domain of dynamical systems and the HNR architecture (as representative of a class of interactive settling systems) offers a solution. In the previous case study, the competition/settling algorithm of this model only captured competitive dynamics. However, nonlearning versions of this model (e.g., Spivey, 2007) have been shown to solve a host of problems in which multiple, graded constraints must be probabilistically integrated in real time. These include complex multidimensional categorization and visual search. Thus, this implementation of online constraint-satisfaction should be able to determine the correct word/object mappings in the moment. That is, the HRN will find the optimal solution (the referent) given partial input

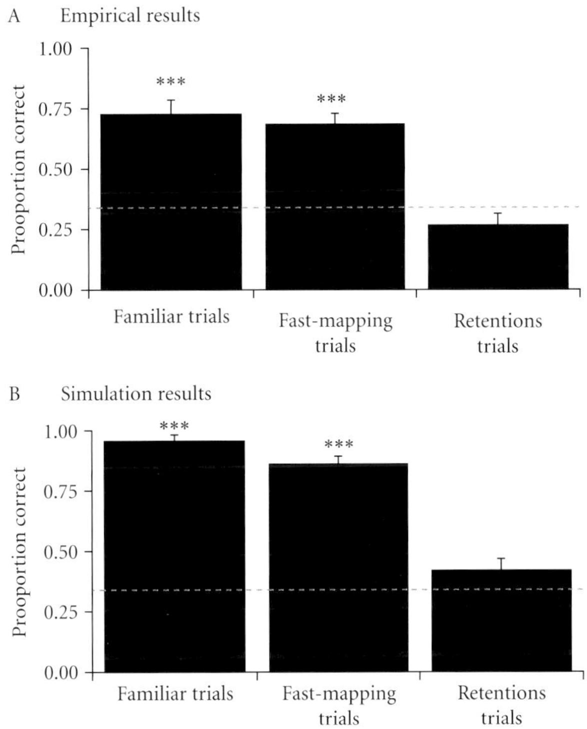


**Figure 11.7. Fast-mapping versus retention**

(the word and the set of visual competitors) and partial knowledge (prior lexical mappings)—much as infants appear to do.

Thus, the current model is built around an associationist statistical learning core which implements referent selection using a simple, dynamic probabilistic constraint satisfaction device. The present simulations will show that these two components interact at multiple timescales to generate learning and development that exceed what either could do alone.

## Architecture

The HNR network used for word learning is shown in Figure 11.8 (for more details, see McMurray, Horst, & Samuelson, 2008, in preparation). Two banks of localist input units (auditory and visual) were used. In each array, a single node was designated for each possible spoken word and for each visual referent. Auditory and visual arrays are independent; they cannot directly pass activation to each other. On a given trial, nodes corresponding to the auditory and visual inputs are activated (including nodes for visual competitors). Input arrays then pass activation to the decision layer [equivalent to the output layer in the prior simulation; see Box 11.3, equation (14)]. As before, this decision layer is squared and normalized [equation (15)], and activation is fed back to the two input layers [equation (16)], normalized [equation (17)], and the cycle continues.

As in the prior network, all weights were initially random. On each cycle of processing, a small amount of Hebbian Learning is applied to the weights [equation (18)]. Note that this learning rule differs slightly from that of the prior network, specifically in the way weights decay. While the previous rule (based on Rumelhart & Zipser, 1986) decayed all of the weights equally, the current rule (adapted from ART, Grossberg, 1976) only decays weights that are connected to either an active input unit or an active decision unit. This restricted weight-decay function maintains plasticity in a large number of weights (those connected to both inactive input and output nodes) allowing the system to learn new words.

**Box 11.3 Formulas for Hebbian normalized recurrence word learning simulation**

$$d_x = d_x + \sum_{z\in a} w_{xz}\cdot a_z + \sum_{z\in v} u_{xz}\cdot v_z \tag{14}$$

Equation (14): Decision node activation ($d_x$) (similar to output units) is the sum of the weighted ($w_{xz}$) activation of the auditory input ($a_z$) nodes plus the weighted ($u_{xz}$) activation of the visual input nodes ($v_z$).

$$d_x = d_x^2 \Big/ \sum_{z\in d} d_z^2 \tag{15}$$

Equation (15): Decision activation is squared and normalized to sum to 1 as a form of nonlinear competition.

$$a_y = a_y + a_y \cdot \sum_{z\in d} d_z w_{zy}$$
$$v_y = v_y + v_y \cdot \sum_{z\in d} d_z u_{zy} \tag{16}$$

Equation (16): Output activation spreads back to the two input layers (along same weights). Note that the only way that auditory and visual input arrays can affect each other is via this feedback mechanism (through the decision layer)—there are no direct connections.

$$a_y = a_y \Big/ \sum_{z\in a} a_z \quad v_y = v_y \Big/ \sum_{z\in a} v_z \tag{17}$$

Equation (17): Both input layers are independently normalized to sum to 1.

$$\Delta w_{xy} = \eta\cdot\left(a_y d_x - a_y w_{yx} - d_x w_{yx}\right)$$
$$\Delta u_{xy} = \eta\cdot\left(v_y d_x - v_y u_{yx} - d_x u_{yx}\right) \tag{18}$$

Equation (18): Weights are adjusted at each cycle through the model by a modified Hebbian rule. Here, $\eta$ represents the learning rate. Note that the weight decay function (adapted from adaptive resonance theory, Grossberg, 1976) insures that weights for which either the decision or input (*a* or *v*) units are inactive (0) do not decay. This preserves plasticity so that these weights may be involved in learning other words.

Crucially, learning occurs continuously throughout processing (as opposed to occurring only at the end of the settling process). This sets the HNR architecture apart from other dynamical connectionist models. It may represent a slight improvement in plausibility as there is no executive (homuncular) control over when learning can occur—the dynamics of learning simply occur during processing.

## Testing the Model

### Strength of Associative Learning

The first set of simulations with this architecture revealed that it could learn a lexicon even

[2]This was more decision units than the model would eventually use. However, by using a large number, we ensure that initially (when the weights are random) the model is more likely to activate different decision units for different auditory and visual inputs. This helps minimize the chance of overgeneralization (in which a single decision unit maximally responds to two words). Moreover, this large number of units (and the resulting sparse representation) allows the network to be built with no a priori notion of how many output units (categories) will be needed.

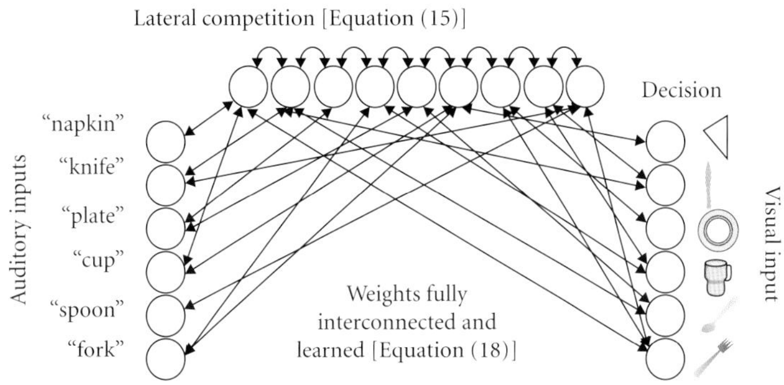


**Figure 11.8. The Hebbian normalized recurrence network used for word learning**
This figure presents a comparison of empirical results from Horst and Samuelson (2008) (panel A) and model performance (panel B). Error bars represent SEM and *'s indicate significant deviation from chance (33%, the dashed line).

if there were a large number of visual competitors present for any given auditory input. To test this model, 240 models were trained on a small 20-word lexicon for 30,000 epochs. We manipulated the number of competitors by randomly activating some of the visual competitors with a competitor-likelihood ranging from 0 to 0.95 (thus anywhere from 0 to 95% of the lexicon could be co-present on any given trial). Table 11.2 shows the model's performance at each competitor-likelihood. The model performed above 90% until a competitor-likelihood of 0.4, and was close to 70% with competitor-likelihoods of 0.75 (in which three-quarters of the lexicon was visually present on any trial during learning). Surprisingly, even at 0.95 competitor-likelihood, the model's performance was better than chance, $M = .405$, $t(19) = 3.01$, $p = .006$, and a single model trained in this condition for additional 90,000 epochs performed at 88.5% correct. Thus, simple statistical learning mechanisms can extract word/object pairings from extremely noisy input, particularly when combined with online constraint-satisfaction principles.

### Fast Mapping and Word Learning

The next simulations tested the model on an explicit fast-mapping task. Twenty models were initialized with 15 auditory and visual nodes and 90 decision units.[2] Input units were divided into three classes of five words each: familiar words, novel words (for novel fast-mapping trials), and held-out words (which were never activated and would be used as foils to examine weight-changes during retention). The models were initially trained on only the familiar words, to simulate an infant who has learned a few words prior to the fast-mapping experiment. Next, the model was tested in 15 testing trials. These trials consisted of three types. First, the five familiar words were tested to establish learning. Next, the five novel words were tested twice, once in a novel fast-mapping trial and once in a retention trial, to replicate Horst and Samuelson (2008).

Familiar words were tested by exposing the model to a familiar word (auditorily), along with the visual target, plus one familiar and one novel visual competitor. After presentation of the input, the model was cycled until decision units stopped changing (the squared derivative summed across output units was close to zero). Because the feedback/competition cycle suppressed activation for visual competitors, this allotted more attention to the target. Thus, the visual unit that had the most activation was the object that the model selected. On each novel fast-mapping trial (randomly interspersed among the familiar trials), the model was tested on one of the five novel words paired with the (experimenter-defined) correct (novel) referent, and two familiar visual competitors. Again, selections were based on the visual array. Retention trials were conducted

**Table 11.2. Average performance of models as a function of competitor-likelihood**

| Competitor-likelihood | % Correct | Competitor-likelihood | % Correct |
|---|---|---|---|
| 0.00 | 100.0 | 0.50 | 79.8 |
| 0.05 | 99.8 | 0.55 | 79.3 |
| 0.10 | 99.3 | 0.60 | 75.8 |
| 0.15 | 99.5 | 0.65 | 75.0 |
| 0.20 | 100.0 | 0.70 | 70.3 |
| 0.25 | 98.8 | 0.75 | 69.8 |
| 0.30 | 96.3 | 0.80 | 58.8 |
| 0.35 | 91.8 | 0.85 | 52.3 |
| 0.40 | 90.0 | 0.90 | 49.5 |
| 0.45 | 86.5 | 0.95 | 40.5 |

*Note:* Competitor-likelihood refers to the likelihood that each visual competitor is visually present with the target during training. For example, a competitor-likelihood of 0.5 means that, on average, half of the words in the lexicon were co-present with the target on any given training trial.

*after* the fast-mapping trials and were designed to determine if the model retained anything about the novel words during the fast-mapping trials. The model was tested on each of the five novel items paired with a second novel item and one of the held-out objects. Here, if the model had learned during the fast-mapping trials, it should be able to select the correct referent. Importantly (and unlike many connectionist models), learning was not turned off during testing, allowing the model to potentially learn during the test period (as infants might do in such an experiment).

As can be seen in Figure 11.7, the model's performance was close to behavioral results from Horst & Samuelson (2008, A). While the model's overall performance was better than that of the infants ($M_{model} = 96\%$, $M_{infants} = 72\%$ for familiar words), the pattern of results was identical. On the familiar and fast-mapping trials, the model performed quite well (familiar: $M = .96$, $SD = 0.10$; fast-mapping: $M = .86$, $SD = 0.16$) and both were better than the chance score of 1/3. However, like the infants, the model performed very close to chance on the retention trials ($M = .42$, $SD = 0.24$).

It would be tempting to conclude from the performance of the model in the fast-mapping trials (or prior literature) that the model (or the infant) had complete linkages between auditory and visual input (i.e., had learned the words). Conversely, it would be tempting to conclude from the model's (or infant's) failure to complete the retention trials that no true learning had occurred at all—only in-the-moment constraint satisfaction. However, a subtler possibility is that some learning occurred, but it was not enough to show retention of the fast-mapped words. To evaluate this hypothesis, we conducted an analysis of the weight-matrix to determine how much learning actually occurred.

The complete weight-matrices had been stored at three points during training for the above simulations: (1) prior to any exposure, (2) after the initial exposure but before test, and (3) after testing on the fast-mapping trials. We wanted to examine how much learning had occurred for a given class of items (familiar, novel, and held-out) at each point in time. Recall, that because the weight matrix describes connections between *localist* units, it is quite easy to carve it up and examine the weights for known words separately than those for fast-mapped words or for held-out words that received no input. The weights for familiar words reveal how much change we should expect for a word to go from unknown to known. Weights for the held-out words serve as a control, and reveal the amount of change when there is no direct input but only weight-decay. Thus, a comparison

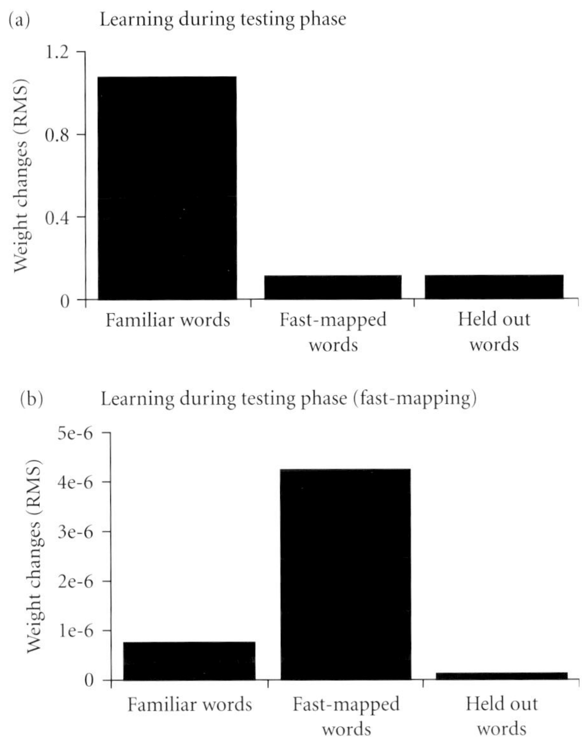


**Figure 11.9. Learning during training phase versus testing phase**
This figure presents the RMS (root-mean-square) weight change between the initialization of the model and the end of familiarization (panel A) and between the end of familiarization and the end of the testing phase (panel B). Note that the *y* axis scale in panel B is 1,000,000 times smaller.

of the weights for fast-mapped words to these groups will reveal how much learning has occurred over the fast-mapping trials relative to behaviorally understood benchmarks. Thus, we computed the RMS difference (the mean of the squared change in each weight) within each portion of the weight matrix at a window in time (e.g., the difference in the known portion between initialization and test).

Figure 11.9A, compares the amount of learning that occurred during the initial 3000 training epochs for the familiar, fast-mapped, and held-out words. Not surprisingly, very little learning occurred for the fast-mapped and held-out words ($RMS_{novel} = .08$; $RMS_{heldout} = .08$). At this point the model had not been exposed to any of these words, and the learning rule only decays weights connected to active nodes. Figure 11.9B, shows the same amount of change over the testing trials (both familiar and fast-mapping). First notice the *y* axis scale—there was notably less learning overall here than during training. Nonetheless, a comparison of the familiar and the fast-mapped words to the RMS of the held-out words (the control condition, since RMS [a squared value] cannot be directly compared to 0) revealed that both classes of words did receive statistically significant learning (familiar: $t(19) = 13.1$, $p < .001$; fast-mapped: $t(19) = 39.0$, $p < .001$).

Thus, the model learned something about the fast-mapped words during fast-mapping trials. However, this overall small amount of learning ($M = 1.9e\text{-}6$), did not even approach the amount of change that constituted learning for the familiar words ($M = .76$). Given this, we would be hesitant to call the weight changes for the fast-mapped words actually word learning. Rather, this analysis, suggests three things. First, online integration mechanisms permit the model to fast-map novel words in some situations, but this represents largely in-the-moment processing, not learning. Second, the fact that the model was able to capture the pattern of performance seen in children in this task suggests that it provides a good account of the processing underlying children's behavior in this task. Third, the model suggests that while a *single* fast-mapping trial creates insufficient weight change to constitute learning, it may drive learning over *many* trials. This is based on the fact that the same training regime was identical to the fast-mapping trials (presentation of a single auditory token coupled with a number of possible visual referents), and we found evidence of some learning on fast-mapping trial.

So how does the model accomplish this referent selection behavior, if not via learning mechanisms? Our analysis of the interaction dynamics revealed that it was the mismatch of the visual and auditory units on the known words that was important. On a fast-mapping trial, the decision units for the two known competitors received only partial bottom-up support. They received activation from the visual units, but not from the auditory units (and feedback could not activate the auditory units since there was no bottom-up support). For the novel word, however, there was typically at least one decision unit that happened to have connections to both the visual and auditory units (which one, of course, was random). As a result, bottom-up support from both input arrays was able to out-compete and suppress the other decision units. Feedback then caused these units to gradually shut off their corresponding visual units. Thus, both interactive feedback and lateral inhibition are crucial for this purely in-the-moment fast-mapping behavior.

### Interaction of Learning and Online Processing

The analyses above suggest that both inhibition and feedback are essential to fast-mapping. However, it is unclear if they are required for learning. Thus, the final simulations sought to understand the relationship between online processing (fast-mapping) and learning. Eighty networks were trained on a small (20-word) lexicon[3] and tested every 25 training epochs for 8000 cycles. In 20 of these networks both the recurrent feedback and competition (in the decision layer) were turned off to see if Hebbian learning alone was sufficient to learn a lexicon (the NULL model). An additional 20 were run with the competition alone (NORECURR), and 20 were run with the feedback alone (NOCOMP). The remaining 20 were run with both processes, a fully functional model (FULL).

In the previous two simulations, testing was conducted by analysis of which visual units were most active at the end of cycling (simulating the focusing of attention to the correct object). However, without recurrence the visual units did not change in response to the auditory input—selection based on them would be meaningless. Therefore, we adopted a different measure of success that was based on the decision units. A single auditory input was activated and the decision unit with maximal activation was computed. Next, the corresponding visual unit was activated and the decision unit with maximal activation was determined. If these two were the same decision unit, then the model had mapped the visual and auditory units onto the same decision unit—it knew the correct association, so to speak.

In Figure 11.10A, it is clear that, early in training, the fully functional model (FULL) underwent more learning than the other models. However, during this initial phase the model achieved only about 60% correct. While it eventually got quite close to 100%, the later phase of learning was markedly slower—in essence it learned the first 10–11 words quite quickly and then took more time to attain the remaining words. The model with recurrence but without competition (NOCOMP) was slightly delayed and did little learning until around 1000 epochs; however at this point it learned the complete lexicon quite fast. The model without recurrence but with competition (NORECURR) was seriously hindered and took much longer than the

[3]Larger lexica of 50 and 100 words have also been examined and are learnable (although with many more epochs). This simplification was adopted for ease of execution and analysis.

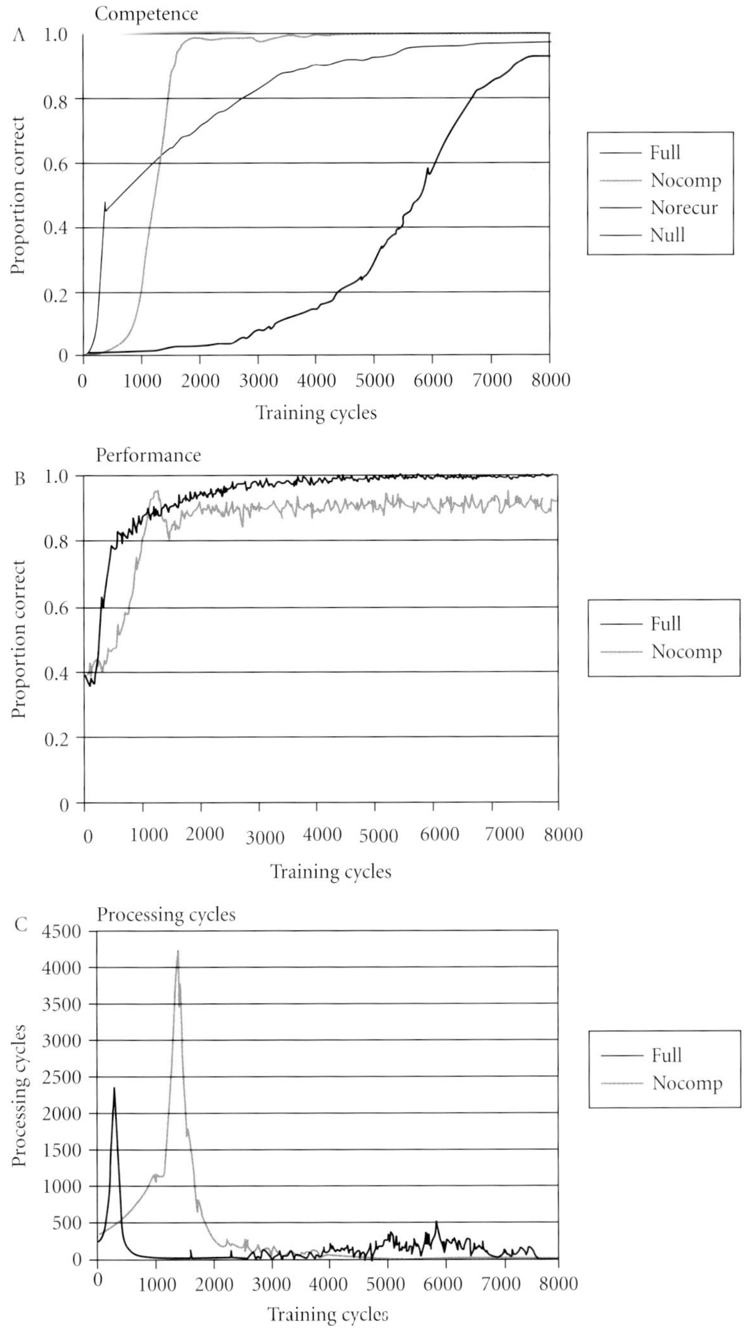


**Figure 11.10. Performance of the word learning model as a function of training and model components**

(A) Performance measured by connections to the decision units. Note, the NULL line stays constant at 0. (B) Performance measured in the visual array. (C) The average number of processing cycles required for each model to settle during the training sequence is depicted. Definitions: FULL = fully functional model, NOCOMP = model with recurrence but without competition, NORECUR = model without recurrence but with competition, NULL = model with both recurrent feedback and competition turn off. Refer to text for additional details.

other models. The NULL model was unable to learn at all.

It was a bit surprising that, of the two models with recurrence, NOCOMP ultimately outperformed FULL, increasing asymptotically to near 100% compared to 96% for FULL. Figure 11.10B, however, sheds a new light on this performance. This figure displays the percent correct using the prior method (based on activation in the visual array) for the two classes of models with recurrence. As before, FULL learned faster than NOCOMP. However, this time, it is FULL's performance that increases asymptotically to near 100%, while NOCOMP never reaches perfect performance.

This suggests a number of things. First, it is interesting that the NOCOMP model does not reach perfect performance despite apparently better weights. However, our measure of competence is based solely on the connection strength between the target and the decision units (i.e., it ignores the connections of the competitors). Thus, while it may have all of its targets associated properly, weights to the competitors may be quite strong as well, and without competition these cannot be suppressed. Competition in the FULL model, on the other hand, forces the decision units to reach a single active unit. This in turn forces competitor weights to decay, and allows the model to ultimately learn a better solution.

Second, the fact that the FULL model's performance seems to exceed the NOCOMP model's performance despite having less-than-perfect weights is probably the result of competition that allows the network to figure out the correct solution in-the-moment. Imperfect weights are sufficient to recognize a word, given the powerful constraint-satisfaction mechanisms.

Third, and most importantly, the lack of learning by the NULL model and very slow learning in the NORECURR models suggests that the online processing (fast-mapping) is essential to the learning process. Competition is clearly an asset (and without recurrence, required), but recurrent feedback loops among audition, decision, and visual units play a more critical role in aligning the auditory-decision mappings with the visual-decision mappings. This active comparison of decision-unit representations with available auditory and visual inputs is the fast-mapping process: in-the-moment integration of multiple constraints. This processing is, in a sense, serving as a teaching signal for the Hebbian learning. By forcing the model to arrive at the correct solution given the constraints, the model now has a more accurate representation upon which to learn—in-the-moment processing is generating a solution, and this is entrained in the weights.

Having shown that online processing is critical for learning, our final analyses looked at the converse: how learning affects online processing. Specifically, we examined the amount of time the model spent processing a single word. Figure 11.11C shows the average number of processing cycles it took each model to settle during the training sequence. For both the FULL and NOCOMP models (the two largely successful models), processing time had peaks at around the time the greatest change in performance was occurring. This would suggest that intermediate developmental states can destabilize the online settling process. Thus, over developmental time, the stability of the online process can be a function of learning. At both early and late points in the learning trajectory, the settling process is quite stable, while at intermediate points this is not so (paralleling our discussion of stability in section Toward an Integration). While the strong stability of the lateral inhibition can reduce the instability (from reducing the cycles from a peak of 4221 to 2348 cycles) it does not eliminate it.

### Case Study 2: Discussion

This model contributes an important theoretical explanation to the results of Horst and Samuelson (2008). It suggests that while learning occurs during fast-mapping, this learning could be relatively small—the seemingly good performance is largely due to online processing dynamics. However, on the flip side, it also demonstrates the possibility that online processing dynamics can significantly augment slow connectionist-style learning—it is possible to have slow learning and good performance, too. Thus, in explaining word learning, the timescales and processes of connectionism and dynamical systems may be intertwined. That is, learning necessarily operates over a slow timescale, but can be buttressed by fast, in-the-moment processes. Conversely, these momentary processes yield large observable behavior (referent selection), but their ability to be transformed into long-term change (via learning) is small due to the relatively slower learning dynamics. Thus, while both processes occur at the same time, the relative magnitude of the change they cause (and the persistence of them) make them look somewhat independent.

At a computational level, it also corroborates a finding from the first case study—the fact that in unsupervised learning schemes, competition (and in this case competition plus recurrent feedback) is essential for learning of this sort. Together these mechanisms are powerful enough that basic associative mechanisms were able to extract word–object mappings even with half of the lexicon visually co-present. In a sense, then, this online, dynamical constraint satisfaction provides a teaching signal for associative Hebbian learning. This parallels the role of the hippocampal system in the model of McClelland and colleagues (McClelland et al., 1995)—it speedily sets up weights that can then be learned using slower mechanisms. However, unlike McClelland's model, this competition/recurrence contributes to more than just learning. It also provides useful in-the-moment behavior (e.g., selecting a referent). This reinforces a view from dynamic systems approaches to development: that the task of the child is not simply to learn, but also to communicate, behave, and act in the moment.

At a broader level, this model suggests some powerful ways in which online processing and learning interact to drive development. Here, online processing forces the model to arrive at a correct referent, a solution that can subsequently be entrained in the weights. Importantly, statistical learning alone (as instantiated in this network) may not learn words as effectively as a hybrid approach combining online constraint satisfaction with learning. Complementarily, theoretical approaches based solely on online processing fail to relate fast-mapping behavior to learning. However, the combination also raises challenges—our analysis of the processing time of the model suggests that, particularly at intermediate states of learning, online dynamic processes may in fact be quite unstable. Using processes like lateral inhibition to create strong (almost obligatory) attractors can help, but do not eliminate this problem. This question, however, would not even be raised by a unitary connectionist or dynamical systems account. An interactive approach that considers both mechanisms thus offers a compelling alternative.

## Conclusion

Our case studies focused on two mechanisms offered by connectionism and dynamical systems theory: statistical learning and moment-by-moment processing. We have shown that not only are these mechanisms complementary, but that a tighter integration has the potential to significantly extend our understanding of development.

Our discussion of speech categorization revealed first that statistical learning alone could not account for development. Some form of competition is also needed. This was extended with a more dynamic form of competition in the Hebbian normalized recurrence network enabling it to fit the temporal dynamics of online processing, and to explain the underlying gradiency seen there as the product of a developmental process, the graded statistical distribution of the input. We also demonstrated that a dynamic approach to learning can yield novel insights, particularly with respect to the mixture of Gaussians model. The parameterization of learning (in terms of parameters of a Gaussian distribution) enabled a clearer understanding of different aspects of categorization change over time. Finally, and most importantly, our analyses of perceptual learning phenomena revealed that in some cases statistical learning *can be* online processing. The same statistical learning mechanism that accounts for developmental change may also be an integral part of short-term change. Thus, an understanding of in-the-moment behavior requires an understanding of developmental mechanism.

Our discussion of word learning extended the Hebbian normalized recurrence architecture in a way that clearly illuminated the complex relationship of in-the-moment behavior and long-term learning. Here we modeled the ability of children to quickly evaluate multiple constraints and select the correct referent, as well as the seemingly contradictory failure in their ability to retain such knowledge (Horst & Samuelson, 2008). Our simulations captured both findings but provided an explanation that is relevant to understanding the relationship between dynamical and connectionist mechanisms. We demonstrated that even if learning is slow (in a connectionist sense), in-the-moment behaviors can buttress partially formed representations, allowing models (and children) to behave in a way that exceeds the knowledge stored in their weights.

More importantly, we also demonstrated the converse: online processing is crucial for learning. Learning was substantially improved when competition and recurrent feedback were included as part of the model. In effect, these online processes (the probabilistic integration of auditory and visual input with knowledge stored in the weights) pushed the model to the correct answer and served as a teaching signal for learning—self-supervised learning. It was this powerful learning that accounted for the model's ability to learn word–object mappings even when 90% of the lexicon was visually co-present with the target during training. This argues for basic associative learning mechanisms, when supported by simple constraint satisfaction. More importantly, it shows that these processes are not additive, but interactive. While both types of processes are continually engaged during learning and development, their relative speed creates a tension in which in-the-moment processes continually prod learning, and as learning accumulates, it in turn can bias these processes.

As a package, these simulations provide only the beginnings of a framework for combining dynamical systems and connectionist paradigms. They illustrate that a combination of such mechanisms promises to yield a richer view of development than either alone. In particular, while individually, statistical learning and online processing provide only partial solutions to developmental problems, when combined, they introduce a new way to understand behavior, both moment-by-moment and the ways in which it changes over development. Our models divide development into mechanisms such as statistical learning and online processing for ease of study, and to some extent they can be factored out computationally. However, we must acknowledge a singular developmental process characterized by the moment-by-moment interaction of such mechanisms and the interdependence of change at multiple timescales. Our examination of statistical learning and online processing suggests they are two highly interactive sides of a multifaceted coin. Or, perhaps, they are two regions of a smooth sphere.

## Acknowledgments

The authors would like to thank Michael Spivey, Robbie Jacobs and Richard Aslin for discussions leading to the computational architectures presented here. Portions of this work were funded by NIH F31DC006537-01 to BM and NIH 5R01HD045713 to LKS.

## References

Allopenna, P. D., Magnuson, J. S., & Tanenhaus, M. K. (1998). Tracking the time course of spoken word recognition using eye movements: Evidence for continuous mapping models. *Journal of Memory and Language*, 38(4), 419–439.

Barto, A. G. (1995). Learning as hill-climbing in weight space. In M. Arbib (Ed.). *The handbook of brain theory and neural networks* (2nd ed., pp. 531–533). Cambridge, MA: MIT Press–Bradford Books.

Bharucha, J. J. (1987). Music cognition and perceptual facilitation: A connectionist framework. *Music Perception, 5*(1), 1–30.

Bechtel, W. (1997). Embodied connectionism. In D. M. Johnson & C. E. Erneling (Eds.). *The future of the cognitive revolution* (pp. 187–208). New York: Oxford University Press.

Carey, S. (1978). The child as word learner. In M. Halle, J. Bresnan, & G. A. Miller (Eds.), *Linguistic theory and psychological reality* (pp. 264–293). Cambridge, MA: MIT Press.

Carey, S., & Bartlett, E. (1978). Acquiring a single new word. *Proceedings of the Stanford Child Language Conference, 15*, 17–29.

Clark, C., & Luce, P. (2005, June). Perceptual adaptation to speaker characteristics: VOT boundaries in stop voicing categorization. *Proceedings of the ISCA Workshop on Plasticity in Speech Perception* (pp. 23–26), London. Bonn, Germany: International Speech Communication Association.

Dell, G. (1986). A spreading activation theory of retrieval in sentence production. *Psychological Review*, 93(3), 283–321.

De Boer, B., & Kuhl, P. K. (2003). Investigating the role of infant-directed speech with a computer model, *Auditory Research Letters On-Line (ARLO)*, 4, 129–134.

Elman, J. L. (1990). Finding structure in time. *Cognitive Science, 14*(2), 179–211.

Elman, J. L., Bates, E. A., Johnson, M. H., Karmiloff-Smith, A., Parisi, D., & Plunkett, K. (1996). *Rethinking innateness: A connectionist perspective on development.* Cambridge, MA: MIT Press–Bradford Books.

Elman, J. L., & Zipser, D. (1988). Learning the hidden structure of speech. *Journal of the Acoustical Society of America, 83*(4), 1615–1626.

Espy-Wilson, C. (1992). Acoustic measures for linguistic features distinguishing the semivowels /w j r l/ in American English. *Journal of the Acoustical Society of America, 92*(2 I), 738–757.

Goodman, J. C., McDonough, L., & Brown, N. B. (1998). The role of semantic context and memory in the acquisition of novel nouns. *Child Development, 69*(5), 1330–1344.

Grossberg, S. (1976). Adaptive pattern classification and universal recoding. I. Parallel development and coding of neural feature detectors. *Biological Cybernetics, 23*(3), 121–134.

Guenther, F. H., & Gjaja, M. N. (1996). The perceptual magnet effect as an emergent property of neural map formation. *Journal of the Acoustical Society of America, 100*(2 I), 1111–1112.

Halberda, J. (2003). The development of a word-learning strategy. *Cognition, 87*(1), B23–B34.

Heeger, D. J. (1993). Modeling simple-cell direction selectivity with normalized, half-squared, linear operators. *Journal of Neurophysiology, 70*(5), 1885–1898.

Horst, J. S., McMurray, B., & Samuelson, L. K. (2006, July). Online processing is essential for learning: Understanding fast mapping and word learning in a dynamic connectionist architecture. In R. Sun, and N. Miyake (Eds.) *Proceedings of the 28th Annual Conference of the Cognitive Science Society* (pp.339–344) Austin, TX: Cognitive Science Society..

Horst, J. S., & Samuelson, L. K. (2008). Fast mapping but poor retention in 24-month-old infants. *Infancy, 13*(2), 128–157.

Kraljic, T., & Samuel, A. G. (2005). Perceptual learning for speech: Is there a return to normal? *Cognitive Psychology, 51*(2), 141–178.

Lisker, L., & Abramson, A. S. (1964). A cross-language study of voicing in initial stops: Acoustical measurements. *Word, 20*, 384–422.

Mareschal, D., & French, R. M. (2000). Mechanisms of categorization in infancy. *Infancy, 1*(1), 59–76.

Maye, J., Weiss, D. J., & Aslin, R. A. (2008) Statistical phonetic learning in infants: Facilitation and feature generalization. *Developmental Science, 11*(1), 122–134.

Maye, J., Werker, J. F., & Gerken, L. (2002). Infant sensitivity to distributional information can affect phonetic discrimination. *Cognition, 82*(3), B101–B111.

McCandliss, B. D., Fiez, J. A., Protopapas, A., Conway, M., & McClelland, J. L. (2002). Success and failure in teaching the [r]-[l] contrast to Japanese adults: Predictions of a Hebbian model of plasticity and stabilization in spoken language perception. *Cognitive, Affective and Behavioral Neuroscience, 2(2)*, 89–108.

McClelland, J. L., & Elman, J. L. (1986). The TRACE model of speech perception. *Cognitive Psychology, 18*(1), 1–86.

McClelland, J. L., McNaughton, B. L., & O'Reilly, R. C. (1995). Why there are complementary learning systems in the hippocampus and neocortex: Insights from the successes and failures of connectionist models of learning and memory. *Psychological Review, 102*(3), 419–457.

McCloskey, M., & Cohen, N. (1989). Catastrophic interference in connectionist networks: The sequential learning problem. In G. H. Bower (Ed.), *The psychology of learning and motivation: Advances in research and theory Vol. 24* (pp. 109–165). San Diego, CA: Academic Press.

McMurray, B., & Aslin, R. N. (2004). Anticipatory eye movements reveal infants' auditory and visual categories. *Infancy, 6*(2), 203–229.

McMurray, B., & Aslin, R. N. (2005). Infants are sensitive to within-category variation in speech perception. *Cognition, 95*(2), B15–B26.

McMurray, B., Aslin, R. N., & Toscano, J. (in press). Statistical learning of phonetic categories: Insights from a computational approach. *Developmental Science.*

McMurray, B., Horst, J., & Samuelson, L. (2008). The interaction of learning and online constraint-satisfaction: a computational approach to early word-learning. Manuscript in preparation.

McMurray, B., Munson, C., & Gow, D. (2008). Lexical competition can delay progressive facilitation during the perception of assimilated speech. Manuscript submitted for publication.

McMurray, B., & Spivey, M. (1999). The categorical perception of consonants: The interaction of learning and processing. *Proceedings of the Chicago Linguistics Society, 35*, 205–219.

McMurray, B., Tanenhaus, M. K., & Aslin, R. N. (2002). Gradient effects of within-category phonetic variation on lexical access. *Cognition, 86*(2), B33–B42.

McMurray, B., Tanenhaus, M. K., & Aslin, R. N. (2008). Gradient sensitivity to sub-phonemic detail facilitates lexical ambiguity resolution. Manuscript submitted for publication.

Mervis, C. B., & Bertrand, J. (1994). Acquisition of the novel name–nameless category (N3C) principle. *Child Development, 65*(6), 1646–1662.

Miller, J. L. (1997). Internal structure of phonetic categories. *Language and Cognitive Processes, 12*(5–6), 865–869.

Miller, J. L., & Eimas, P. D. (1996). Internal structure of voicing categories in early infancy. *Perception and Psychophysics, 58*(8), 1157–1167.

Mirman, D., McClelland, J. L., & Holt, L. L. (2006). An interactive Hebbian account of lexically guided tuning of speech perception. *Psychonomic Bulletin and Review, 13*(6), 958–965.

Nakisa, R. C., & Plunkett, K. (1998). Evolution of a rapidly learned representation for speech. *Language and Cognitive Processes, 13*(2–3), 105–127.

Norris, D., McQueen, J. M., & Cutler, A. (2003). Perceptual learning in speech. *Cognitive Psychology 47*(2), 204–238.

Olshausen, B. A., & Field, D. J. (1996). Emergence of simple-cell receptive field properties by learning a sparse code for natural images. *Nature, 381*(6583), 607–609.

Perko, L. (1991). *Differential equations and dynamical systems.* New York: Springer.

Peterson, G. E., & Barney, H. L. (1951). Control methods used in the study of the vowels. *Journal of the Acoustical Society of America, 24*(2), 175–184.

Rumelhart, D. E., & Zipser, D. (1986). Feature discovery by competitive learning. In D. E. Rumelhart, J. L. McClelland, & PDP Research Group (Eds.), *Parallel distributed processing: Explorations in the microstructure of cognition. Vol. 1: Foundations* (pp. 151–193). Cambridge, MA: MIT Press–Bradford Books.

Samuelson, L. K. (2002). Statistical regularities in vocabulary guide language acquisition in connectionist models and 15-20-month-olds. *Developmental Psychology, 38*(6), 1016–1037.

Schlesinger, M., Parisi, D., & Langer, J. (2000). Learning to reach by constraining the movement search space. *Developmental Science, 3*(1), 67–80.

Schöner, G., & Dineva, E. (2007). Dynamic instabilities as mechanisms for emergence. *Developmental Science, 10*(1), 69–74.

Simmering, V. R., Spencer, J. P., & Schöner, G. (2006). Reference-related inhibition produces enhanced position discrimination and fast repulsion near axes of symmetry. *Perception and Psychophysics, 68*(6), 1027–1046.

Smith, L. B. (2000). Avoiding associations when it's behaviorism you really hate. In R. Golinkoff, K. Hirsh-Pasek., L. Bloom, G. Hollich., L. Smith, A. L. Woodward (Eds.). *Becoming a word learner: A debate on lexical acquisition* (pp. 169–174). New York: Oxford University Press.

Smith, L. B., & Thelen, E. (Eds.). (1993). *A dynamic systems approach to development: Applications.* Cambridge, MA: MIT Press–Bradford Books.

Smith, L. B., & Samuelson, L. K. (2003). Different is good: Connectionism and dynamic systems theory are complementary emergentist approaches to development. *Developmental Science, 6*(4), 434–439.

Smolensky, P., Mozer, M. C., & Rumelhart, D. E. (Eds.). (1996). *Mathematical perspectives on neural networks* [Series: Developments in connectionist theory]. Mahwah, NJ: Erlbaum.

Spencer, J. P., & Thelen, E. (Section Eds.). (2003). Connectionist and dynamic systems approaches to development [Special section/issue]. *Developmental Science, 6*(4).

Spivey, M. J. (2007). *The continuity of mind.* New York: Oxford University Press.

Spivey, M. J., & Dale, R. (2004). On the continuity of mind: Toward a dynamical account of cognition. In B. H. Ross (Ed.), *The psychology of learning and motivation: Advances in research and theory* (Vol. 45, pp. 87–142). San Diego, CA: Elsevier Academic Press.

Tabor, W., & Tanenhaus, M. K. (1999). Dynamical models of sentence processing. *Cognitive Science, 23*(4), 491–515.

Thelen, E. (2000). Grounded in the world: Developmental origins of the embodied mind. *Infancy, 1*(1), 3–28.

Thelen, E., & Smith, L. B. (1994). *A dynamic systems approach to the development of cognition and action*. Cambridge, MA: MIT Press–Bradford Books.

Thelen, E., Schöner, G., Scheier, C., & Smith, L. B. (2001). The dynamics of embodiment: A field theory of infant perseverative reaching. *Behavioral and Brain Sciences, 24*(1), 1–34, commentary 34–70, authors' response 70–80.

Toscano, J., & McMurray, B. (2005, November). Statistical learning, cross-linguistic constraints, and the acquisition of speech categories: A computational approach. Talk presented at the 11th Midcontinental Workshop in Phonology, University of Michigan, Ann Arbor, MI.

Turk, M., & Pentland, A. (1991). Eigenfaces for recognition. *Journal of Cognitive Neuroscience. 3*(1), 71–86.

van Geert, P. (1991). A dynamic systems model of cognitive and language growth. *Psychological Review, 95*(1), 3–53.

van Geert, P. (1998). A dynamic systems model of basic developmental mechanisms: Piaget, Vygotsky, and beyond. *Psychological Review, 105*(4), 634–677.

Ward, L. M. (2001). *Dynamical cognitive science*. Cambridge, MA: MIT Press–Bradford Books.

Werker, J. F., & Curtin, S. (2005). PRIMIR: A developmental framework of infant speech processing. *Language Learning and Development, 1*(2), 197–234.

Werker, J. F., & Tees, R. C. (1984). Cross-language speech perception: Evidence for perceptual reorganization during the first year of life. *Infant Behavior and Development, 7*(1), 49–63.

Yu, C., & Smith, L. B. (2007). Rapid word learning under uncertainty via cross-situational statistics: Research article. *Psychological Science, 18*(5), 414–420.

# PART V

# Reactions from the Outside

# CHAPTER 12

## Soft-Assembled Mechanisms for the Unified Theory

*Heidi Kloos and Guy C. Van Orden*

Do connectionist and dynamic-systems models converge on a unified theory? The answer depends on the kind of mechanism that the two models attempt to unravel. Two views of mechanism avail themselves to contemporary scientists. One view of mechanism represents cognitive activity as reducible to a cognitive architecture of separate cognitive components. Another view sees cognitive activity as emergent and highly dependent on fine details of the contexts in which behavior emerges. We will argue in this chapter that connectionist and dynamic systems models complement each other and collectively move toward a unified theory of development if they subscribe to the second view of mechanism, one that treats behavior as soft assembled in the immediate context.

The chapter organization is as follows. In section Where Models Converge in Stalemate, we address why models aimed at reducing behavior to cognitive components cannot make clear headway. The argument is that a reduction of behavior requires human performance to be relatively context free. Yet, as we show with the example of balance task performance, human performance is highly context dependent, even in the sterile laboratory context of balance experiments. In section Taking Context Effects Seriously, we elaborate on what such context dependence could mean. While not conclusive on its own, strong context dependence is consistent with the idea that cognitive activity is softly assembled to suit the immediate task environment. Soft assembly offers a plausible alternative to hard-assembled cognitive functions—functions that exist prior to and independently of the task context. We review more pointed evidence for soft assembly and discuss why models that take soft assembly seriously—connectionist or dynamic systems—anticipate the unified theory.

### Where Models Converge in Stalemate

At the center of the argument is strong context dependency in human performance. We develop this argument around well-studied examples from developmental psychology, in particular children's performance on balance scale tasks. However, our points pertain to cognitive modeling more generally.

#### Balance Scale Performance and Associated Models

Picture a child in a balance scale experiment. The balance scale straddles a fulcrum and has pegs along its surface on which to set weights (see Fig. 12.1). The experimenter sets some weights and asks the child to predict the behavior of the scale. Will it stay balanced, will the right side tip, or will the left side tip? To perform correctly in every case, the child must take into account the number of weights on each side of the bar, the distance of each weight from the fulcrum, and their product.

Different children give different kinds of answers to balance scale problems of this sort, but four kinds of answers seem reliable and systematic (Siegler, 1981). (I) One class of answer appears to reflect an exclusive focus on

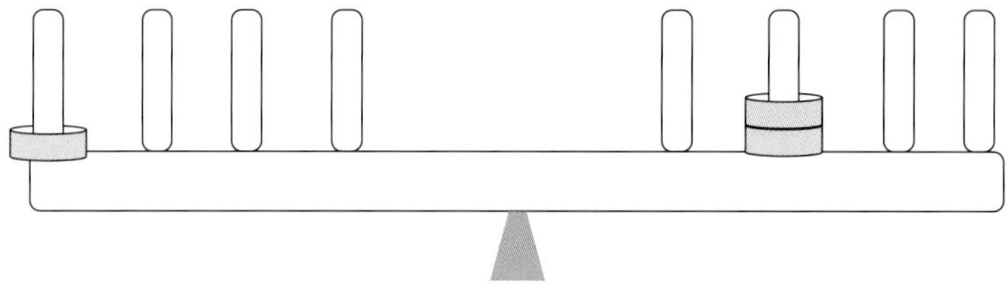

**Figure 12.1. Schematic illustration of a balance scale apparatus**
The balance scale apparatus depicted has four pegs on each side; the number of weights on each side varies during the balance scale task.

the amount of weight on each side of the fulcrum. Children who answer this way consistently expect the side with the most weights to tip. When there are equal numbers of weights on each side, children in this group expect the balance scale to stay level, ignoring distance from the fulcrum. (II) A second class of answer appears to take into account distance from the fulcrum, but only when weights are equally distributed on both sides. Children who produce this kind of answer predict that the heavier side will tip. But when both sides have the same numbers of weights, they predict that the side with weights furthest from the fulcrum will tip. (III) A third class of answer takes into account both weight and distance from fulcrum, but does not suggest a systematic integration of the two dimensions. Children who produce this class of answers correctly predict the effect of weight when the distances are the same, and they correctly predict the effect of distance from fulcrum when the number of weights is equal. However, when both weight and distance from fulcrum differ, then these children simply guess. (IV) A final, fourth class of answers appears to take into account both relevant dimensions (number of weights and distance from the fulcrum) and integrate dimensions appropriately when the two dimensions are pitted against each other.

A variety of models have been proposed to explain the four kinds of answers and the transition from one type of answer to the next more sophisticated. Production-rule models suggest that children's performance derives from rule-like algorithms or strategies that can change rather suddenly as a result of experience (e.g., Klahr & Siegler, 1978; Langley, 1987; Sage & Langley, 1983; Schmidt & Ling, 1996). Connectionist models, on the other hand, suggest that children's performance is not a function of explicit rules but rather a gradual adaptation to statistical relations between balance scale appearance and response (e.g., McClelland, 1989, 1995; Shultz, Mareschal, & Schmidt, 1994; Shultz, Schmidt, Buckingham, & Mareschal, 1995). Finally, dynamical systems models capture developmental changes as sudden jumps in the cusp catastrophe—a rule-like response looses stability and changes suddenly in a phase transition or bifurcation to a new rule (e.g., Van Rijn, Van Someren, & van der Maas, 2003).

The condensed overview of these models indicates some disagreement among them about the causal mechanism that could underlie balance performance. One solution to the disagreement has been to evaluate the degree to which the output of a model can capture human performance overall (Van Rijn et al., 2003). For instance, one could find a way to calculate overall fit across models in a single score and conclude that the model with the best score must be using the correct mechanism—the mechanism that best mimics the underlying cognitive architecture. On the grounds of this logic, production-rule models might receive a low score because they have difficulty capturing the developmental transition to a more sophisticated rule. And connectionist models might receive a low score because they do not capture rule-like human performance unless the rules are part of the statistical relations in the training set (Raijmakers, van Koten, & Molenaar, 1996).

We suggest, however, that such a solution is premature because it does not adequately consider how context figures into the disagreement. The models that presently compete emphasize the axiom that behavior can be the same across different laboratory contexts. They emphasize that psychological constructs are motivated by "behavioral consistency over varying contexts" (Embretson, 2006, p. 50; Cronbach & Meehl, 1955). Unquestioning adherence to this axiom rules out context-sensitive behavior as a topic of study. Or, at least, it ascribes context effects to secondary performance limitations that come and go with task, rather than to a participant's primary competence. Thus, the models that would compete for best overall fit tend to assume that scientifically informative behavior stands outside of context. This stance would beg the question of context effects, accepting before the fact that only context-free behavioral effects adequately pick out underlying mechanisms. But as we argue next, context effects actually distinguish between the models.

### Balance Scale Performance Is Context Sensitive

Adults, children, and even toddlers move through uncountable varieties of physical and social contexts with fluid ease—from outdoor play on the slippery slide to indoor play on the paintball court, from being a peer to being a child or parent, and from quietly drawing to arguments about bedtime. Even in the artificial and somewhat sterile environments of cognitive laboratories, context effects are paramount.

Take again the example of balance scale studies. The four types of balance scale behaviors described above—rules I, II, III, and IV—are well known and often cited. But it would be grossly misleading to suggest that this typology captures children's full range of balance behaviors. In fact, it captures only a limited set of options allowed within the particular task context (Hardiman, Pollatsek, & Well, 1986). Limits come from a variety of methodological choices that are rationalized in terms of good experimental design. In this section, we discuss some of these choices in more detail.

One methodological choice is to give or hold back feedback during children's test performance. In the standard balance scale task, a child is presented with a balance scale scenario in each measurement trial, and after predicting the behavior of the balance scale, the child progress immediately to the next trial. But feedback during testing changes the likelihood that a child's performance will adhere to one of the four rules, especially when feedback is tailored to a child's beliefs (Hardiman et al., 1986; Kliman, 1986). Siegler and Chen (1998), for example, carefully constrained the feedback provided to preschoolers to stop them from forming the rule *greater distance goes down* instead of rule I (*ignore distance*) or rule II (*pay attention to distance only if weight is equal*).

Another choice is the kinds of response options that children are allowed to give. In the standard task, children are allowed only three answer options: *left side down*, *right side down*, or *balance*. If children are provided with more flexible options, as when they are allowed to adjust the number of weights and the position of weights on the scale, children's performance can follow an entirely different rule called the *addition rule*. Under these circumstances, children integrate both the number of weights and their distance from the fulcrum according to an additive algorithm (Normandeau, Larivée, Roulin, & Longeot, 1989; Wilkening & Anderson, 1982). That is to say, they do not focus on one dimension only (rule I–III), nor do they integrate dimensions according to the correct multiplication algorithm (rule IV).

Yet another choice of method is the rather limited range of possible balance scenarios presented to participants. Only four pegs are fitted on each side of the fulcrum and only four weights can be placed on the pegs. This yields rather small discrepancies between the left and right arms of the balance scale. As it turns out, the magnitude of discrepancy between left and right side matters—an effect known as the *torque effect*. Children perform better with larger differences in torque when predicting the behavior of the balance scale (Ferretti & Butterfield, 1986; Ferretti, Butterfield, Cahn, & Kerkman, 1985; Jansen & Van der Maas, 1997).

The list goes on. Another design choice balances out the number of trials of each type. In the standard task, the dimension number-of-weights is pitted against the dimension distance-from-fulcrum in half of the trials, while either one or both dimensions are held constant in the other half. Tasks that allow a greater and unbalanced variety of configurations of weights do not elicit performance of the four rules (Kliman, 1986). Children no longer form rules on the basis of differences in weights or distance from the fulcrum; they form rules about mathematical relations, specific to their narrow experience. For example, a child will abstract from examples like "*one* weight on the *ninth* peg balances *nine* weights on the *first* peg" to become "*one* weight on the $n^{th}$ peg balances *n* weights on the *first* peg." Yet the child will generalize this belief no further, not even to accommodate two or more weights on the *n*th peg.

Finally, a fifth methodological choice puts limits on proprioceptive information available to the child. Children in the standard task are provided with proprioceptive information about weight (children are encouraged to hold the weights), but not about how distance of weights affects balancing. The lack of proprioceptive information about how distance affects balance selectively increases the difficulty for children to correctly take into account distance (Hardiman et al., 1986). In particular, exclusive proprioception of weight could bias children's performance toward rule I (*greater weight goes down*) and rule II (*if weight is the same, greater distance goes down*), but against a rule based on distance alone. In fact proprioceptive information about distance leads children to perform as though the rule is (*greater distance goes down*) but completely ignore differences in weight (Karmiloff-Smith & Inhelder, 1974). And children will sustain this rule for some time despite contradictory evidence, suggesting that this *distance rule* is at least as stable as rules I–IV.

Strong context dependence is of course not limited to balance scale tasks. Quite the opposite is the case. Context-dependent performance is the rule in developmental psychology, not the exception (e.g., Gigerenzer & Richter, 1990; Lawton, 1993). One could even argue that the most salient outcome of American-style post-Piagetian research has been to demonstrate extreme context sensitivity. Indeed, changes in task design can yield more and more sophisticated competence in younger children, making contextual support an important causal variable (c.f. Keen, 2003). Nonetheless, despite omnipresent context effects, the effect of context is rarely studied in its own right.

### Explaining Away Context Effects

One common argument is that differences in performance due to contexts can be ignored because they do not reflect children's true competence. Instead they are mere performance limitations due to underdeveloped language skills or memory limitations, for example. Yet where does performance end and competence begin? To illustrate this dilemma, take children's performance in another integration task, one that tests children's integrative understanding of distance, velocity, and travel time (Wilkening, 1981). In one task context, children estimated the distance an object traveled, after observing its velocity and travel time. In a second task context, children estimated the time an object traveled, after observing its velocity and travel distance. And in the third task context, children estimated an object's velocity, after observing the time and distances traveled. Each of these three contexts is a permutation of the relations among distance, velocity and travel time. And each permutation asks the child for a prediction of one factor based on the other two factors as antecedents. The important finding for our argument was that each permutation yielded a qualitatively different outcome. For example, 5-year-old children integrate speed and time multiplicatively to estimate the distance traveled, but they integrate speed and distance additively to estimate the time traveled, and they used distance exclusively to estimate speed—three qualitatively different outcomes.

Another way to treat task effects is to assume that they are a simple add-on to other factors, notably to cognitive functions. For instance, if children simply performed better in one balance task than another, then, all other things being equal, task effects are superposed on other effects. In this case, one could legitimately partition out task context as an independent source of variation in data. In a model, task context would then become a fixed parameter. But task and context effects rarely combine so straightforwardly, in balance performance or any other developmental phenomenon. As our review illustrates, children tested in a balance task do not simply perform better or worse. They exhibit different qualities of performance in different task contexts.

Still another way to treat contexts effects is to equate different effects with separate components of the mental architecture. Successful performance in one context might reflect implicit knowledge, for example, while performance in a different context reflects explicit knowledge. This solution may appear clear-cut when tasks differ conceptually. For instance, tasks that require a physical action such as balancing a rod can appear conceptually different from tasks that require a verbal judgment of balance (c.f. Kirst, Fieberg, & Wilkening, 1993; Levin, Siegler, & Druyan, 1990). A physical action might shed light on implicit knowledge, while verbal action might shed light on explicit knowledge—two distinct forms of representation. However, even if we ignore stalemates that have emerged over supposedly clear-cut distinctions (Cleeremans, 1997; Farah, 1994), this approach runs aground when tasks are conceptually alike. For example, it is not clear how a balance scale task with a small number of weights and pegs differs conceptually from balance scale tasks with a large number of weights and pegs. Both should entail the same knowledge; so finding a difference in performance leaves one guessing intuitively about what the separate components might be. But to merely equate effects with components, what has been called the *effects = structure* fallacy (Gibbs, 2006; Lakoff, 1987), yields a theoretical enterprise that is unpredictable, circular, and likely to end in stalemates among competing intuitions.

Finally, the most widely practiced response to task effects is to argue about which context is more transparent to mental functions, or equivalently which task context produces more *pure* data than another. In the example of the balance scale performance, Siegler (1981) defends the highly structured and methodologically precise task context of his assessment procedure, and Wilkening and Anderson (1982) justify the legitimate expansion of task contexts to explore the task space. But in truth, there is simply no empirical basis on which to decide which task context is best. All task effects refer equally to changes in outcome measures of performance. Distinctions among task effects are supported only by intuitions about competence based on convention or esthetics, but not evidence. Consider, however, that task effects might not be superficial aspects of performance, but rather that context is always and fundamentally constitutive of children's performance.

Arguments about the purity of data have not fared well in other domains. Conventional studies of adult cognition have virtually run to stalemate on the question of which task's data best reveal the architecture of cognition. Details of apparent stalemates have been described for perception (Uttal, 1990, 1997), language and reading (Goldinger & Azuma, 2003; Van Orden & Kloos, 2005; Van Orden, Pennington, & Stone, 2001), and memory processes (Watkins, 1990; Weldon, 1999). Similarly, functional neuroimaging of adult cognition shows signs of running to stalemate because subtle changes in task context cause cognitive functions to be in different parts of the brain (c.f. Cabeza & Nyberg, 2000).

In sum, we have reviewed how task context effects determine our laboratory pictures of children's knowledge. Such context effects demonstrate that a child's performance is an interaction of the child's knowledge and the specific task constraints within which they act. The solutions discussed so far hold onto the idea that context-free performance exists and should be given priority for a reduction to specific cognitive components. But these solutions have also, so far, led to stalemates about which tasks can separate context from components most successfully. In the next section, we discuss a more conservative solution to context sensitivity—one that takes context sensitivity at face value and accepts that context is constitutive of human behavior.

## Taking Context Effects Seriously

In the remainder of this chapter, we describe a path to circumvent stalemates about cognitive components and task contexts and thereby situate connectionist and dynamical systems models within a unified theory. The argument rests on the distinction between soft-assembled and hard-assembled cognition introduced by Turvey and Carello (1981). In what follows we describe this distinction in more detail and show how soft assembly provides a new way to think about context and behavior. We then return to the issue of modeling and discuss how connectionist and dynamical systems models can complement each other and point toward a new unified theory.

### Soft- Versus Hard-Assembled Mechanisms

Most of the research discussed so far is grounded in the assumption that cognitive activity is based on hard-assembled mechanisms. These mechanisms exist off-line in some form of inactive or dormant state and are activated in a particular task. Going back to the balance scale research, a hard-assembled mechanism could be a child's rule about what makes a balance beam tip, a recurring strategy that a child pursues in order to figure out the answer, or simply a child's knowledge about the effects of relevant dimensions. Hard-assembled mechanisms are independent of the immediate task context, they reveal themselves across multiple contexts, and are therefore discovered in context-independent performance.

By contrast, soft-assembled mechanisms emerge in contextually constrained, collective action of the brain and body. They come into existence with enaction, and they are only realized within the immediate context of enaction. An example of a soft-assembled system is the kinematics of a limb in a particular action. The mind and body in context will together create unique kinematics, and if the movement is repeated, each repetition will reveal unique kinematics (Berkinblit, Feldman, & Fukson, 1986; Bernstein, 1967). Looking across repeated movements, one sees a family resemblance, but no context-free mechanism exists to tie these movements together. Of course muscles and tendons and neuropil continue to exist throughout, but the instantaneous play of emergent control is realized in the movement itself, an enacted limb movement that is unique in each instantaneous context (Turvey, 1990).

*Self-organized criticality* is proposed as the mechanism that underlies soft-assembled cognitive activity (Juarrero, 1999; Turvey & Moreno, 2006; Van Orden, Holden, & Turvey, 2003), a concept borrowed from physics (Bak, 1996; Jensen, 1998). Criticality refers to a preparatory state of a system, also referred to as critical state, that emerges immediately before a response occurs. This preparatory state consists of several potential responses, all of which are contextually appropriate, although maybe not accurate. In the balance scale example, the potential responses will include the kinematics of indicating whether the scale will stay balanced, tip to the right, or tip to the left.

Self-organized criticality is brought about by local interactions among processes of the system. Those interactions that satisfy contextual constraints are strengthened, and thereby recruit other processes to their configuration. As a result, these local interactions extend to the periphery of the body and create interdependence among all component processes. It is this interdependence that allows components to act together, to express one of a potential set of contextually appropriate outcomes. Thus soft assembly creates poised, situated, state dynamics across the brain and body. Immediately prior to action, the final contextual contingencies of the trial, including the stimulus, will collapse the critical state to one response option, the response that the child will enact (c.f. Järvilehto, 1998).

Two features make self-organized criticality ideal to explain context effects. First, as noted, a critical state is a situated state, meaning that it is directly linked to the immediate constraints of the task context. Preparative cognition stays in the loop, so to speak, to create continually updating, contextually appropriate, critical states ready for action. This situated cognitive activity ensures that the child will respond appropriately (though not always accurately) with the response options that the experimenter allows. A cognitive act within an arbitrary context requires a situated preparatory state to anticipate the situated future—for instance, a child poised to stay on task in novel or familiar stimulus conditions and to make a balance scale response. Perpetually changing relations among context, brain, and body situate cognitive activity within an oncoming task.

Second, self-organized criticality ensures coordination across multiple scales of time and space. Even aspects of the artifactual balance scale environment change at different rates. Trial-by-trial changes in the distribution of weights occur on a relatively fast timescale, whereas the static laboratory backdrop changes on a much slower timescale. In another example, millisecond changes in the acoustics of a conversation co-occur with second-by-second, minute-by-minute, and more drawn out scales of change in structure, content, contextual backdrop, shared knowledge, turn-taking, and other facets of the conversation. Likewise there are many scales of change entailed in the optical flow of a walker, from nearby bumps in the road to landmarks or scenery, flowing by at different rates depending on relative size and proximity, to meandering detours and even changes of destination, to name just a few. Perpetual coordination of cognition is necessary because the environment changes perpetually on multiple timescales.

The brain, the body, and other biological systems are likewise organized on hierarchies of timescales (Soodak & Iberall, 1987). Each hierarchy spans faster and slower changes in neural activity and other physiological processes, timescales of change in limb and torso movements, and in fascia and skeletal movements. Self-organized criticality organizes the processes of the brain and body to act together, simultaneously, across their various scales of space and time. This organization across the brain and body allows context to work simultaneously at all scales, to fully and subtly situate the brain and body within the changing environment (Kello, Beltz, Holden, & Van Orden, 2007).

Local interactions among embodied processes on different timescales weave the intrinsic fluctuations of the component processes into a coherent fabric of flux, despite inherent tendencies of the different processes to vary at their own different rates (on their own timescales). Competitions among local rates of change strike a precise balance with globally emerging cooperative activity. In the precise balance of (or near) the critical state, they produce a long-range correlated, aperiodic pattern of change or flux in behavior—a complex fractal pattern of long-range correlations. The aperiodic flux is called *1/f noise*, *pink noise*, *1/f scaling*, *fractal time*, and other names, and is a generic prediction of systems in critical states.

### Evidence for Soft Assembly

What is the evidence that human cognition is soft assembled? As discussed above, soft-assembled processes have two characteristics that come from self-organized criticality: (1) pre-preparedness of critical states, rather than a dormant system that is merely reactive to a stimulus, and (2) long-term coordination and correlation of processes, rather than independent and static components. We present evidence next for these two features.

*Pre-prepared critical states.* Findings of ultrafast cognition provide evidence that cognition is pre-prepared. Take for example the results of Grill-Spector and Kanwisher (2005): The time it takes a participant to know that a picture was flashed on the screen is sufficient to know whether the picture showed a bird or a car. Perceivers apparently required no more time "for object categorization than for object detection" (Grill-Spector & Kanwisher, 2005, p. 157). Such ultrafast cognition has been found even when participants are asked to categorize novel pictures on the basis of animacy (i.e., *animate* versus *inanimate*), an archetype of high-level cognitive activity (Fabre-Thorpe, Delorme, Marlot, & Thorpe, 2001).

Ultrafast cognition is surprising and unexpected from the perspective of hard-assembled cognition. If cognition begins only after a stimulus onset, then substantial information processing remains to be completed prior to a categorization response. If cognition were one part of a hard-assembled chain of events that unfolds across a single timescale, then some measurable time should pass between successful object detection and successful object categorization. We are led to expect a longer time delay between stimulus and response than is observed in ultrafast cognition (Fabre-Thorpe et al., 2001; Kirchner & Thorpe, 2006). Nevertheless, ultrafast cognition is possible when the mind and body are pre-prepared to act immediately in one cascading lunge. Given pre-prepared cognition, an impulsion favoring one particular action trajectory can sometimes realize its action in an ultrafast cascade.

*Long-term Coordination.* Evidence for long-term coordination of embodied processes comes from ubiquitous 1/*f* scaling in human performance. Recall that long-term coordination among processes that change at different timescales predicts a pattern of long-range correlation or 1/*f* scaling in repeated measurements of an organism's behavior. Figure 12.2 illustrates the 1/*f* pattern in simple reaction time data. The time series is about 8000 trials worth, collected in one sitting, for about 3 hours. The upper right of the figure shows the wavy, aperiodic, fractal pattern of variation in reaction time from trial to trial. Below this trial data graph is a spectral portrait of 1/*f*. It is derived by artificially parsing the aperiodic pattern into multiple periodic component frequencies, usually sine waves. Examples of such sine waves are shown on the left hand side of the figure, the top left graph showing slow, lower-frequency oscillations, the left bottom graph showing rapid higher-frequency oscillations, and the middle two graphs showing intermediate frequencies (y axes are adjusted to make the higher-frequency sine waves visible). One can artificially segregate component waves of variation to yield higher frequencies plus intermediate frequencies plus lower frequency oscillations.

Emergent properties of the fractal pattern dictate the relation between amplitudes and frequencies. The remarkable finding is the lawful scaling relation between amplitudes of variation and frequencies of variation (on log scales). In this particular example, the relation between

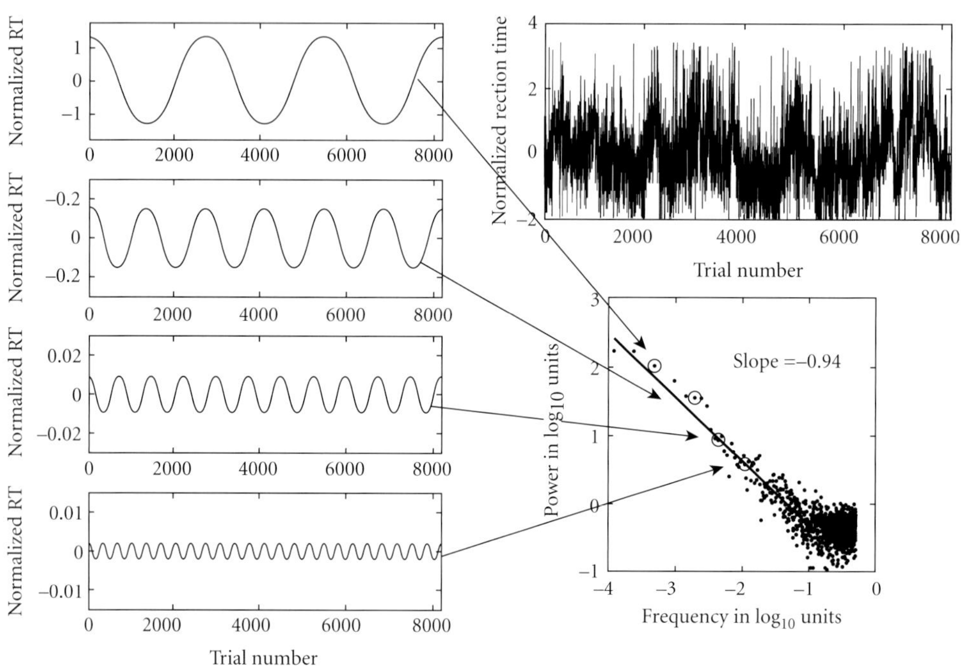


**Figure 12.2. Spectral portrait of 1/*f* scaling**

amplitudes and frequencies has a linear slope of −0.94. Amplitudes of oscillations with periods of tens of trials, for example, find their values on the same line that captures amplitudes for oscillations across periods of hundreds or thousands of trials.

Similar patterns of fractal long-term correlations are ubiquitous in physiology and human behavior (Bassingthwaighte, Liebovitch, & West, 1994; Gilden, 2001; Kello et al., 2007; Raichle & Gusnard, 2005; Riley & Turvey, 2002; van Orden et al., 2003; West, 2006). In fact, it begins to appear, as Machlup (1977, p. 157) suggested decades ago, if you have not yet found the fractal pattern, you have not taken enough data, "you have not waited long enough. You have not looked at low enough frequencies."

The initial response to $1/f$ scaling patterns has been a cautious reluctance to interpret these findings as evidence of soft assembly and self-organization (e.g., Wagenmakers, Farrell, & Ratcliff, 2005). Any complex pattern in data can be modeled by a linear decomposition into smaller patterns—that is how the spectral analysis works. This means that it will always be possible to model a particular $1/f$ pattern of a particular data set (Beran, 1994). The crippling paradox for hard assembly, however, is that a model's success becomes tied too closely to arbitrary details of how data are collected (Mandelbrot & Wallis, 1968/2002). Success depends too literally on the number of data points collected. One need only collect more data to, in effect, falsify the model.

As more data are collected, variation will grow in amplitude because the $1/f$ pattern will extend outside of the limits of sampled data points, and a larger sample will pick up more of the scaling relation. A longer data set reveals more of the lower frequencies, frequencies that are associated with much larger amplitudes of variation. So variation will grow in the larger sample. Consequently, the hard-assembled model must invent additional new components every time a longer data set is collected, out to natural limits of data collection (van Orden, Holden, & Turvey, 2005). As a consequence, no successful hard-assembled alternative has yet been put forward (Thornton & Gilden, 2005).

In addition, success of mimicking a $1/f$ process depends too literally on the particular kind of measurement taken for the observed behavior. One need only collect different kinds of measured values in the same behavior to, in effect, falsify the model. For example, a spoken word can be measured in parsing its acoustic spectrum into frequency bins and measuring the spectral intensity of the component frequency bins (300 Hz wide, evenly spaced up to 13.5 kHz, in Kello, Anderson, Holden, & Van Orden, 2008). If this same word is repeatedly spoken and an acoustic spectrum is computed for each repetition, then the repeatedly measured intensity values of the component frequency bins will all fluctuate in the pattern of $1/f$ scaling. Thus each arbitrary bin, 90 total, can be equated with a separate stream of variation, and each varies in the pattern of $1/f$ scaling. Thus a hard-assembled model must invent 90 distinct sources of $1/f$ scaling, one for every repeatedly measured value that is examined (see also Kello et al., 2007; Wijnants, Bosman, Hasselman, & Cox, 2007).

A hard-assembled model must invent a source of $1/f$ scaling for each measured value that fluctuates in the pattern of $1/f$ scaling. Yet $1/f$ scaling may extend outside the sampled values under consideration and appear in an indefinite set of different measured values that have not yet been considered. The absurdity descends from the idea that variation in measured values comes from independent hard-assembled sources, that measured values are transparent to causal properties of the component sources. The consequent paradoxical, endless additions to ad hoc models also stem from thinking that variation in measured values can be equated with components of cognition—that one's measured values are context-free and thereby transparent to state variables of cognitive components.

The paradoxes disappear, however, once we allow that cognitive activities are soft assembled. In other words, one moves past the paradox in recognizing that measured values are emergent products of dynamic linkages among component systems, including the system that comprises the context of measurement (Flach, Dekker, & Stappers, 2007; Turvey & Moreno, 2006). In soft assembly, context is causally entwined with the measurement of behavior (Van Orden, Kello, & Holden, in press). Consequently, measured outcomes differ in quality from the embodied processes from which they emerge. On-coming contingent details of task demands and immediate task contexts are constitutive of behavior. Contingent details, as fluid changes in constraints, change the interaction among the component processes from which behavior emerges.

### Context Constrains the Body

We have argued in previous sections that context effects are not simple add-ons to so-called real effects. Instead, pervasive context effects indicate different design principles altogether. This argument expands upon a famous description of development by connectionists—namely, that development entails "interactions all the way down" (Elman et al., 1996). Elman and colleagues use a newly hatched duck to illustrate how preparative constraints set up the potential for imprinting that is subsequently determined by the interaction with the environment. Beyond nature and nurture, a duck, or a child, is pre-prepared for developmental milestones, and the milestones are realized in interactions with the environment. Development reflects local details of the environment as a consequence. Likewise cognition itself is preparative and realized in local interactions with the environment (Turvey & Fitzpatrick, 1993).

Currently, dynamical systems and connectionist models address context effects by accounting for children's performance in more than one context, usually by additions to their architectures. For example, the dynamical systems model of Van Rijn and colleagues (Van Rijn et al., 2003) simulates rules I–IV, as well as the torque effect described by Ferretti and colleagues (e.g., Ferretti et al., 1985). McClelland's (1989) connectionist model can account for a difference in salience between weight and distance that could explain the discrepancy in performance when children are given proprioceptive information about weight versus distance. These attempts are insufficient in the present light, however. And this not only because they fail to capture the addition rules proposed by Wilkening and Anderson (1982) or the idiosyncratic rules discussed by Kliman (1986). Even if a model could be rigged to account for children's behavior in all the discussed contexts, it will not anticipate the inevitable next round of context effects.

Context effects are not exhausted at the level of the larger task contexts illustrated in the first part of this chapter. Context effects permeate the brain and body well below the level of trial judgments in particular tasks. That is to say, the context effects discussed in previous sections are just tips of icebergs, so to speak, and soft-assembled icebergs are context sensitive all the way down. Below the iceberg tips, each instantaneous muscle flex and each pattern of rhythmic cortical firing creates a context for every other muscle flex and every other pattern of neural firing (c.f. Belen'kii, Gurfinkel, & Pal'tsev, 1967; Freeman, Holmes, Burke, & Vanhatalo, 2003; Marsden, Merton, & Morton, 1983; Raichle & Gusnard, 2005).

Take, for example, the coordination of speech after an unexpected pull to a person's jaw (Kelso, Tuller, Vatikiotis-Bateson, & Fowler, 1984; Shaiman & Gracco, 2002). Articulation compensates with movements in the upper and lower lips to preserve the flow of speech such that a listener cannot distinguish between perturbed and unperturbed speech. The compensation entails cortical interactions, the fluid matrix of neuromuscular interactions in the lips, modulation of the force of breath and the pace of respiration, and all else that makes up speech. Most important, the fluid compensation stays within the limits of contextual constraints. Context in this case equals the unfolding of a spoken word as co-articulated speech. This context of constraints, specific to the particular co-articulation, exists in a particular configuration at the point at which the experimenter perturbs the jaw. This configuration constrains the fluid compensation, limiting potential compensations to those that insure intelligibility, all the way down.

The example illustrates how global and local context are embodied in local interactions as limits (or constraints) on cascading interactions among excitable neuromuscular media. These limits are demonstrated widely in motor coordination and also in the neuroscience of perception and action. In laboratory experiments, brain and body demonstrably reconfigure, in an instant, to accommodate local changes in the task environment. For example, a slight change in a motor coordination task—increasing the stimulus pace of the coordinated movements—leads to a new pattern in behavior in a virtually instantaneous phase transition across brain and body (Kelso, 1995). To accomplish this, active constraints must anticipate the phase transition, and must define a potential set of changes in concurrent interactions among excitable media. Active constraints maintain perpetually updated, context-specific potential sets of actions, which anticipate which actions are appropriate, necessary, and possible. Human and nonhuman animals prefigure how to act to satisfy their contexts of action. They must do so to keep apace of perpetually changing relations between actors and environments. In effect, brain, body, and context combine constraints to poise the actor perpetually ready for action.

## Implications for Modeling

Contemporary models do not usually represent context as a source of constraints. Context is usually implemented as activation or some other causal force. The pace at which activation is updated defines the primary timescale of a model's dynamics, and activations from context and other sources are usually integrated on this single timescale. A typical connectionist model will also include a second timescale of change in connection weights, but the model is always limited to a definite (usually small) number of timescales. This is true of dynamical systems models as well if they equate parameters and variables with hard-assembled mechanisms. Parameters and variables amount to a few explicit timescales of hard-assembled dynamics. However, for all their strengths, contemporary models grossly underestimate the number of temporal scales on which cognitive activity is actually assembled.

Actual cognitive activity unfolds across an indefinite number of timescales in a coordinated fractal pattern that hard assembly does not anticipate. We have explained this fact keeping in mind a picture of cognition that has a primary preparative function (c.f. Raichle & Gusnard, 2005). In the concrete terms of a laboratory experiment, cognition situates a person to participate. Contemporary models do not capture the situated behavior of participants and thereby fail to reveal the situated mechanisms of behavior. Instead, the most widely practiced modeling strategies have emphasized regularities across participants, central tendencies in data, or gross features of developmental change. These phenomena, though legitimate, do not reliably define a mechanistic level of explanation.

A different target for theory, modeling, and explanation is a level of emergent control, above the component details of enacted mechanisms. This *strategic reduction* captures causal properties of systems that are not transparent in component causes. A theory of emergent control makes progress so long as there actually are general principles of control to be discovered. This point about general principles takes a lesson from dynamical systems models. The cusp catastrophe, for example, is a very general account of control and qualitative change (Gilmore, 1993). And the search for empirical flags of the cusp catastrophe illustrates how one goes about establishing that cusp principles of control actually apply (van der Maas & Molenaar, 1992). However, if one grounds the ensuing model in hard-assembled components, then the paradoxes we have described come along for the ride.

At the level of emergent control, task context effects are equated with task constraints in a model's control parameters (Van Orden, Holden, Podgornik, & Aitchison, 1999). A control parameter is a ratio among constraints. Values of the ratio will favor one or another of the probabilistic outcomes. Control parameters are most often associated with dynamical systems models, but they are also discussed in the context of connectionist models (Kello, 2003; Kello, Sibley, & Plaut, 2005; Rueckl, 2002). For example, the ratio of weight to distance trials in a balance scale task could be conceived as a control parameter. The values of the ratio that favor distance trials yield different rules than the values of the ratio that favor weight trials. In a connectionist model, the ratio of weight versus distance trials is made explicit in the training regime and implicit in the weight matrix. In a dynamical systems model, this ratio could appear explicitly as a parameter in a system of equations. In both cases, the ratio controls competing outcomes that live on opposite sides of a critical value. The critical value defines the point of equally distributed constraints, an imaginary point of no decision, a state of criticality, and a precisely balanced tug of war between equally compelling rules.

This is a different view of modeling and control compared to previous schools of psychology. Models do not stand outside of history, in the sense of a cognitive architecture, except in the principles of their design. Most important, they do not capture phenomena outside of history. Models capture and make explicit the control of behavior emerging in time. Previous schools of psychology relegated control of behavior to relatively static loci in the environment (*behaviorism*) or the organism (*cognitivism*). The new unified theory will locate control in the perpetually changing interaction of child and environment.

## Conclusion

Hard-assembled cognition must inevitably treat context effects after the fact or as methodological problems of experimental control. Context effects are either something to be explained later, once the basic architecture is in place, or something that undermines data that could otherwise be equated with cognitive components. Nonetheless, cognitive performance is unduly dependent on the particulars of context. A unified theory of cognition and cognitive development must find its beginnings in the particulars of context sensitive phenomena. As for modeling, mimicking the control structure of human behavior captures the available causal basis of behavior. Control of behavior, even qualitative changes in control structure, can be simulated in both connectionist and dynamical system models. Therefore, a unified theory of soft-assembled cognition and development may embrace both dynamic systems and connectionist models. These attractive possibilities, perhaps inevitabilities, can be achieved by discarding hard assembly, by accepting that performance is not transparent to hard-assembled competence. Fluid soft assembly of cognition is the essential human competence and performance is transparent to this competence. Competence as context sensitivity and performance as sensitivity to context are two sides of the same coin.

## Acknowledgments

We thank Michael Riley, Kevin Shockley, and the reviewers for comments and questions that improved this chapter. We also acknowledge funding from the National Science Foundation to Heidi Kloos (DRL # 723638) and to Guy Van Orden (HSD #0728743; BCS # 0642716).

## References

Bak, P. (1996). *How nature works: The science of self-organized criticality.* New York: Springer.

Bassingthwaighte, J. B., Liebovitch, L. S., & West, B. J. (1994). *Fractal physiology.* New York: Oxford University Press.

Belen'kii, V. Y., Gurfinkel, V. S., & Pal'tsev, Y. I. (1967). Elements of control of voluntary movements. *Biophysics, 12*(1), 154–161.

Beran, J. (1994). *Statistics for long-memory processes.* New York: Chapman & Hall.

Berkinblit, M. B., Feldman, A. G., & Fukson, O. I. (1986). Adaptability of innate motor patterns and motor control mechanisms. *Behavioral and Brain Sciences, 9*(4), 585–599, commentary 599–626, authors' response 626–638.

Bernstein, N. A. (Ed.). (1967). *The co-ordination and regulation of movements.* London: Pergamon Press.

Cabeza, R., & Nyberg, L. (2000). Imaging cognition II: An empirical review of 275 PET and fMRI studies. *Journal of Cognitive Neuroscience, 12*(1), 1–47.

Cleeremans, A. (1997). Principles for implicit learning. In D. C. Berry (Ed.), *How implicit is implicit learning?* (pp. 196–234), New York: Oxford University Press.

Cronbach, L. J., & Meehl, P. E. (1955). Construct validity in psychological tests. *Psychological Bulletin, 52*(4), 281–302.

Elman, J. L., Bates, E. A., Johnson, M. H., Karmiloff-Smith, A., Parisi, D., & Plunkett, K. (1999). *Rethinking innateness: A connectionist perspective on development.* Cambridge, MA: MIT Press–Bradford Books.

Embretson, S. E. (2006). The continued search for nonarbitrary metrics in psychology. *American Psychologist, 61*(1), 50–55.

Fabre-Thorpe, M., Delorme, A., Marlot, C., & Thorpe, S. (2001). A limit to the speed of processing in ultra-rapid visual categorization of novel natural scenes. *Journal of Cognitive Neuroscience, 13*(2), 171–180.

Farah, M. J. (1994). Neuropsychological inference with an interactive brain: A critique of the "locality" assumption. *Behavioral and Brain Sciences 17*(1), 43–61, commentary 61—90, author's response 90—100.

Ferretti, R. P., & Butterfield, E. C. (1986). Are children's rule-assessment classifications invariant across instances of problem types? *Child Development, 57*(6), 1419–1428.

Ferretti, R. P., Butterfield, E. C., Cahn, A., & Kerkman, D. (1985). The classification of children's knowledge: Development on the balance-scale and inclined-plane tasks. *Journal of Experimental Child Psychology, 39*(1), 131–160.

Flach, J. M., Dekker, S., & Stappers, P. J. (2007). Playing twenty questions with nature (the surprise version): Reflections on the dynamics of experience. *Theoretical Issues in Ergonomic Science, 9*(2), 125–154.

Freeman, W. J., Holmes, M. D., Burke, B. C., & Vanhatalo, S. (2003). Spatial spectra of scalp EEG and EMG from awake humans. *Clinical Neurophysiology, 114*(6), 1053–1068.

Gibbs, R. W. (2006). *Embodiment and cognitive science.* New York: Cambridge University Press.

Gigerenzer, G., & Richter, H. R. (1990). Context effects and their interaction with development: Area judgments. *Cognitive Development, 5*(3), 235–264.

Gilden, D. L. (2001). Cognitive emissions of 1/*f* noise. *Psychological Review, 108*(1), 33–56.

Gilmore, R. (1993). *Catastrophe theory for scientists and engineers.* New York: Dover.

Goldinger, S. D., & Azuma, T. (2003). Puzzle-solving science: The quixotic quest for units in speech perception. *Journal of Phonetics, 31*(3–4), 305–320.

Grill-Spector, K., & Kanwisher, N. (2005). Visual recognition: As soon as you know it is there, you know what it is. *Psychological Science, 16*(2), 152–160.

Hardiman, P. T., Pollatsek, A., & Well, A. D. (1986). Learning to understand the balance beam. *Cognition and Instruction, 3*(1), 63–86.

Jansen, B. R. J., & van der Maas, H. L. J. (1997). Statistical test of the rule assessment methodology by latent class analysis. *Developmental Review, 17*(3), 321–357.

Järvilehto, T. (1998). The theory of the organism-environment system: I. Description of the theory. *Integrative Physiological and Behavioral Science, 33*, 321-334.

Jensen, H. J. (1998). *Self-organized criticality: Emergent complex behavior in physical and biological systems* [Series: Cambridge lecture notes in physics: Vol. 10]. Cambridge, England: Cambridge University Press.

Juarrero, A. (1999). *Dynamics in action: Intentional behavior as a complex system.* Cambridge, MA: MIT Press–Bradford Books.

Karmiloff-Smith, A., & Inhelder, B. (1974). If you want to get ahead, get a theory. *Cognition, 3*(3), 195–212.

Keen, R. (2003). Representation of objects and events: Why do infants look so smart and toddlers look so dumb? *Current Directions in Psychological Science, 12*(3), 79–83.

Kello, C. T. (2003). The emergence of a double dissociation in the modulation of a single control parameter in a nonlinear dynamic system. *Cortex, 39*(1), 132–134.

Kello, C. T., Anderson, G. G., & Van Orden, G. C. (2008). *Intrinsic fluctuations of speech reveal the coordinative basis of cognition.* Manuscript submitted for publication.

Kello, C. T., Beltz, B. C., Holden, J. G., & van Orden, G. C. (2007). The emergent coordination of cognitive function. *Journal of Experimental Psychology: General, 136*(4), 551–568.

Kello, C. T., Sibley, D. E., & Plaut, D. C. (2005). Dissociations in performance on novel versus irregular items: Single-route demonstrations with input gain in localist and distributed models. *Cognitive Science, 29*(4), 627–654.

Kelso, J. A. S. (1995). *Dynamic patterns: The self-organization of brain and behavior.* Cambridge, MA: MIT Press–Bradford Books.

Kelso, J. A. S., Tuller, B., Vatikiotis-Bateson, E., & Fowler, C. A. (1984). Functionally specific articulatory cooperation following jaw perturbations during speech: Evidence for coordinative structures. *Journal of Experimental Psychology: Human Perception and Performance, 10*(6), 812–832.

Kirchner, H., & Thorpe, S. J. (2006). Ultra-rapid object detection with saccadic eye movements: Visual processing speed revisited. *Vision Research, 46*(11), 1762–1776.

Kirst, H., Fieberg, E. L., & Wilkening, F. (1993). Intuitive physics in action and judgment: The development of knowledge about projectile motion. *Journal of Experimental Psychology: Learning, Memory, and Cognition, 19*(4), 952–966.

Klahr, D., & Siegler, R. S. (1978). The representation of children's knowledge. In H. W. Reese & L. P. Lipsitt (Eds.), *Advances in child development and behavior* (Vol. 12, pp. 61–116). New York: Academic Press.

Kliman, M. (1986). Children's learning about the balance scale. *Instructional Science, 15*(1), 307–340.

Lakoff, G. (1987). *Women, fire and dangerous things: What categories reveal about the mind.* Chicago: University of Chicago Press.

Langley, P. (1987). A general theory of discrimination learning. In D. Klahr, P. Langley, & R. T. Neches (Eds.), *Production system models of learning and development.* Cambridge, MA: MIT Press–Bradford Books.

Lawton, C. A. (1993). Contextual factors affecting errors in proportional reasoning. *Journal of Research in Mathematics Education, 24*(5), 460–466.

Levin, I., Siegler, R. S., & Druyan, S. (1990). Misconceptions about motion: Development and training effects. *Child Development, 61*(5), 1544–1557.

Machlup, S. (1977). Earthquakes, thunderstorms, and other 1/*f* noises. *Proceedings of the Sixth International Symposium on Noise in Physical Systems* [Special publication No. 614] (pp. 157–160). Washington, DC: National Bureau of Standards.

Mandelbrot, B. B., & Wallis, J. R. (2002). Noah, Joseph and operational hydrology. In B. B. Mandelbrot, *Gaussian self-affinity and fractals: Globality, the earth, 1/f noise and R/S* (pp. 236–251). New York: Springer. (Original work published 1968)

Marsden, C. D., Merton, P. A., & Morton, H. B. (1983). Rapid postural reactions to mechanical displacement of the hand in man. *Advances in Neurology, 39*, 645–659.

McClelland, J. L. (1989). Parallel distributed processing: Implications for cognition and development. In R. G. M. Morris (Ed.), *Parallel distributed processing: Implications for psychology and neurobiology* (pp. 8–45). New York: Oxford University Press.

McClelland, J. L. (1995). A connectionist perspective on knowledge and development. In T. J. Simon & G. S. Halford (Eds.), *Developing cognitive competence: New approaches to process modeling* (pp. 157–204). Hillsdale, NJ: Erlbaum.

Normandeau, S., Larivée, S., Roulin, J. L., & Longeot, F. (1989). The balance-scale dilemma: Either the subject or the experimenter muddles through. *Journal of Genetic Psychology: Child Behavior, Animal Behavior, and Comparative Psychology, 150*(3), 237–250.

Raichle, M. E., & Gusnard, D. A. (2005). Intrinsic brain activity sets the stage for expression of motivated behavior. *Journal of Comparative Neurology, 493*(1), 167–176.

Raijmakers, M. E. J., van Koten, S., & Molenaar, P. C. M. (1996). On the validity of simulating stagewise development by means of PDP networks: Application of catastrophe analysis and an experimental test of rule-like network performance. *Cognitive Science, 20*(1), 101–136.

Riley, M. A., & Turvey, M. T. (2002). Variability and determinism in motor behavior. *Journal of Motor Behavior, 34*(2), 99–125.

Rueckl, J. G. (2002). The dynamics of visual word recognition. *Ecological Psychology, 14*(1–2), 5–19.

Sage, S., & Langley, P. (1983). Modeling cognitive development on the balance scale task. *Proceedings of the Eight International Joint Conference on Artificial Intelligence, 1*, 94–96.

Schmidt, W. C., & Ling, C. X. (1996). A decision-tree model of balance scale development. *Machine Learning, 24*(3), 203–230.

Shaiman, S., & Gracco, V. L. (2002). Task-specific sensorimotor interactions in speech production. *Experimental Brain Research, 146*(4), 411–418.

Shultz, T. R., Mareschal, D., & Schmidt, W. C. (1994). Modeling cognitive development on balance scale phenomena. *Machine Learning, 16*(1–2), 57–86.

Shultz, T. R., Schmidt, W. C., Buckingham, D., & Mareschal, D. (1995). Modeling cognitive development with a generative connectionist algorithm. In T. J. Simon & G. S. Halford (Eds.), *Developing cognitive competence: New approaches to process modeling* (pp. 205–261). Hillsdale, NJ: Erlbaum.

Siegler, R. S. (1981). Developmental sequences between and within concepts. *Monographs of the Society for Research in Child Development, 46*(2), 1–74.

Siegler, R. S., & Chen, Z. (1998). Developmental differences in rule learning: A microgenetic analysis. *Cognitive Psychology, 33*(3), 273–310.

Soodak, H., & Iberall, A. S. (1987). Thermodynamics and complex systems. In F. E. Yates (Ed.), *Self-organizing systems: The emergence of order* [Series: Life science monographs] (pp. 459–469). New York: Plenum Press.

Thornton, T. L., & Gilden, D. L. (2005). Provenance of correlations in psychological data. *Psychonomic Bulletin and Review, 12*(3), 409–441.

Turvey, M. T. (1990). Coordination. *American Psychologist, 45*(8), 938–953.

Turvey, M. T., & Carello, C. (1981). Cognition: The view from ecological realism. *Cognition, 10*(1–3), 313–321.

Turvey, M. T., & Fitzpatrick, P. (1993). Commentary: Development of perception-action systems and general principles of pattern formation. *Child Development, 64*(4), 1175–1190.

Turvey, M. T., & Moreno, M. (2006). Physical metaphors for the mental lexicon. *Mental Lexicon, 1*(1), 7–33.

Uttal, W. R. (1990). On some two-way barriers between models and mechanisms. *Perception and Psychophysics, 48*(2), 188–203.

Uttal, W. R. (1997). Do rhetorical bridges exist between perceptual experience and neurophysiology? *Perspectives in Medicine and Biology, 40*(2), 280–302.

Van der Maas, H. L. J., & Molenaar, P. C. M. (1992). Stagewise cognitive development: An application of catastrophe theory. *Psychological Review, 99*(3), 395–417.

Van Orden, G. C., Holden, J. G., Podgornik, M. N., & Aitchison, C. S. (1999). What swimming says about reading: Coordination, context, and homophone errors. *Ecological Psychology, 11*(1), 45–79.

Van Orden, G. C., Holden, J. G., & Turvey, M. T. (2003). Self-organization of cognitive performance. *Journal of Experimental Psychology: General, 132*(3), 331–350.

Van Orden, G. C., Holden, J. G., & Turvey, M. T. (2005). Human cognition and 1/*f* scaling. *Journal of Experimental Psychology: General, 134*(1), 117–123.

Van Orden, G. C., Kello, C. T., & Holden, J. G. (in press). Situated behavior and the place of measurement in psychological theory. *Ecological Psychology.*

Van Orden, G. C., & Kloos, H. (2005). The question of phonology and reading. In M. J. Snowling, & C. Hulme (Eds.), *The science of reading: A handbook* (pp. 61–78). Oxford, United Kingdom: Blackwell.

Van Orden, G. C., Pennington, B. F., & Stone, G. O. (2001). What do double dissociations prove? *Cognitive Science, 25*(1), 111–172.

Van Rijn, H., van Someren, M., & van der Maas, H. L. J. (2003). Modeling developmental transitions on the balance scale task. *Cognitive Science, 27*(2), 227–257.

Wagenmakers, E.-J., Farrell, S., & Ratcliff, R. (2005). Human cognition and a pile of sand: A discussion on serial correlations and self-organized criticality. *Journal of Experimental Psychology: General, 135*(4), 108–116.

Watkins, M. J. (1990). Mediationism and the obfuscation of memory. *American Psychologist, 45*(3), 328–335.

Weldon, M. S. (1999). The memory chop shop: Issues in the search for memory systems. In J. K. Foster & M. Jelicic (Eds.), *Memory: Systems, process, or function?* [Series: Debates in psychology]. New York: Oxford University Press.

West, B. J. (2006). *Where medicine went wrong: Rediscovering the path to complexity* [Series: Studies of nonlinear phenomena in life science: Vol. 11]. Hackensack, NJ: World Scientific.

Wijnants, M. L., Bosman, A. M. T., Hasselman, F., & Cox, R. F. A. (2007). *1/f noise in precision aiming: Issues of practice and task difficulty.* Interdisciplinary College 2007, March 9–16, Günne at Lake Möhne, Germany.

Wilkening, F. (1981). Integrating velocity, time, and distance information: A developmental study. *Cognitive Psychology, 13*(2), 231–247.

Wilkening, F., & Anderson, N. H. (1982). Comparison of two rule-assessment methodologies for studying cognitive development and knowledge structure. *Psychological Bulletin, 92*(1), 215–237.

# CHAPTER 13

## Are Dynamic Systems and Connectionist Approaches an Alternative to Good Old-Fashioned Cognitive Development?

*Lisa M. Oakes, Nora S. Newcombe, and Jodie M. Plumert*

The goal of this book, and the conference that motivated it, is to determine whether connectionism and dynamic systems are two distinctly different theories of cognitive development or whether together they represent a paradigm shift in the field of cognitive development toward a single new unified theory. The three authors of this chapter represent relative outsiders to this discussion. We each have studied cognitive development from a blend of traditional theoretical approaches, influenced by the theories of Piaget, Gibson, and Vygotsky, as well as by the broad range of theories that fall under the umbrella of information-processing approaches to cognitive development. We each represent a different blend of these theoretical frameworks, and we are all sympathetic to the general aims of and ideas behind connectionism and dynamic systems. We therefore have taken on the task of critically evaluating these new approaches from the perspective of good old fashioned cognitive development (GOFCD), with an eye toward understanding what connectionism and dynamic systems bring to the field as well as understanding the extent to which they differ from more traditional approaches to cognitive development.

We have organized this chapter into four sections. In the first section, we discuss what connectionism and dynamic systems bring to the study of cognitive development. Because many of the chapters in the book deal directly with this issue, this section is relatively brief. In the second section, we examine how connectionist and dynamic systems theories relate to other GOFCD theories of developmental change. In some sense, this section traces the historical roots of key ideas in connectionist and dynamic systems theories. In addition, we show through examples of systematic programs of work examining developmental change in cognitive processes how the ideas inherent in connectionism and dynamic systems are not unique, although, importantly, connectionist and dynamic systems approaches may make some of these ideas more explicit and central, in part because developing formal analytical methods for modeling change provides a new set of sharper tools to drive progress. Next, we evaluate the contribution of connectionism and dynamic systems in more depth by examining explanations of two historically important issues in cognitive development: infants' behavior in the A-not-B task and children's solutions to the balance scale problem. Here we evaluate how these new theories compare to more traditional explanations of children's developing behavior in these tasks. Finally, we consider how well connectionist and dynamic systems approaches address criticisms often leveled at other theories of cognitive development.

### Dynamic Systems and Connectionist Approaches to the Study of Change

The field of cognitive development is broadly concerned with how children's thinking evolves from the preverbal representations of an infant to the high-level conceptual abilities of a 16-year-old. The challenge, of course, is how to characterize and study such change and the causes of change. Dynamic systems and connectionist approaches to cognitive development explicitly focus on understanding change over time. Each is concerned with demonstrating through mathematical models and careful empirical studies *how* change occurs, not simply documenting *that* change occurs. As a result, they bring a focus on mechanism to the forefront. Proponents of the two viewpoints are concerned with how systems self-organize, with organization arising from a less organized (or sometimes unorganized) state through real-time processes and the dynamic activity of the system. The two approaches conceive of this self-organization differently. For dynamic systems theories, developmental change is an emergent product of interactions among multiple components, occurring on many different timescales (Smith & Thelen, 2003). Theories adopting this framework emphasize multicausality and self-organization emerging out of the real-time dynamics of the child's own activity in a structured environment (Smith & Thelen, 2003). For connectionist theories of development, reorganization emerges out of nonlinearities in learning (Marchman, 1997; Thelen & Bates, 2003), and new structures only emerge from the interaction of the existing structure and environmental input (Bates & Elman, 2002; Elman, 2005). From an outsider's perspective, it is very difficult to distinguish between these two ways of thinking about change; for both, self-organization and emergent structure are a key feature of change. Structural change emerges from activity that occurs in real time, and developing systems exhibit high levels of variability during the process of change. Central to both connectionist and dynamic systems theories of development, therefore, is the explicit idea that new structures and behaviors are emergent products of multiple, interacting components. Moreover, in both styles of theorizing, change at longer timescales necessarily emerges from change at shorter timescales because all behavior is linked together through time. Perhaps most important, both approaches involve developing formal mathematical models of developmental change that provide a detailed level of specificity about those interactions (although it is important to point out that the two most influential volumes on connectionist and dynamic systems approaches to development did not include formal models; Elman et al., 1996; Thelen & Smith, 1994).

One consequence of these core ideas is that studies conducted within connectionist and dynamic systems frameworks often involve repeated observations of behavior over time, although the timescale is often relatively short (e.g., within a single session or across sessions separated by a few days or weeks). Because real-time change (i.e., the changes that happen from moment-to-moment) is intimately tied to change at longer timescales, development can be understood by observing change over many trials or epochs within a single experimental session or over several sessions across several days or weeks (for examples, see Samuelson, 2002; Spencer, Vereijken, Diedrich, & Thelen, 2000; Thelen, Corbetta, & Spencer, 1996). Because the work from a GOFCD perspective often does not have an explicit goal of uncovering the mechanisms of change, studies involving repeated observations over time are relatively rare (for a notable exception, see Siegler, 1996). Instead, GOFCD theorists generally use cross-sectional studies to document cognitive changes that occur over relatively long timescales (e.g., months, years). As others have noted, it is very difficult to examine mechanisms of change with cross-sectional studies. Thus, one contribution of the connectionist and dynamic system movements in cognitive development is to put the focus back on repeated observations over time. It must be pointed out that this approach is not new to cognitive development—Heinz Werner (Werner & Garside, 1957) called it microgenesis and Bob Siegler (1996) has strongly advocated and practiced this style of research over the past decade or more (for empirical examples of this approach, see also Oakes & Plumert, 2002; Plumert & Nichols-Whitehead, 1996). It should also be noted that despite their interest in the connections between changes on different timescales, most studies from a dynamic systems perspective examine change over relatively short timescales (such as trials or minutes). Thus, one weakness of many studies adopting a connectionist or dynamic systems framework is that they do not often examine changes over long timescales (Thelen's work on reaching would be an exception, e.g., Spencer et al., 2000; Thelen et al., 1996).

Another key contribution of dynamic systems and connectionist approaches to cognitive development are the tools they provide for studying the emergence of new structures or behaviors (Bates et al., 1998; Thelen, Schöner, Scheier, & Smith, 2001). As is evident from several of the chapters in this book (e.g., Chapters 4, 6, 7, 8, 10, and 11), not only do connectionist and dynamic systems theorists value repeated observations over time, but they also call for studies that seek to understand the processes that give rise to emergent behaviors. Such studies typically include perturbations or supports that change the organism–environment interaction. Usually, this entails manipulations of environmental structure (e.g., changing the salience of location B and noting the effect on infants' reaching for the location A in the A-not-B task; Diedrich, Highlands, Spahr, Thelen, & Smith, 2001), but sometimes it entails manipulation of organism characteristics (e.g., teaching children shape-based categories in the laboratory, presumably changing how they approach the word-learning context in general, and noting the influence of this change on their rate of vocabulary acquisition outside the laboratory; Samuelson, 2000). Observing how manipulations of either the task or the organism changes the resulting behavior leads to a better understanding of the processes that give rise to behavior. Again, this type of approach is not unique to dynamic systems and connectionism. There are many examples from GOFCD and other approaches (for example, see Johnston & Lickliter, Chapter 14) that explicitly engineer changes to either the organism or the environment to gain insight into developmental process (e.g., Oakes & Plumert, 2002; Plumert & Hund, 2001; Robinson, 2005).

As the chapters in this volume also make clear, these two theoretical paradigms have provided new mathematical and computational tools that make it easier to examine organism–environment interactions (see Chapters 1 and 2; Bates et al., 1998). For example, the dynamic field theory allows researchers to directly test how hypothesized processes within the organism (e.g., memory, attention) and inputs from the environment (e.g., salience of perceptual information) interact to produce predictable patterns of behavior (e.g., Schutte, Spencer, & Schöner, 2003). Likewise, connectionist models of learning allow researchers to directly test how patterns of behavior emerge out of the interaction of simple processing units (e.g., Mareschal, Quinn, & French, 2002). Together, the conceptual and computational focus on how behavior emerges from interacting components offers a significant step forward in our understanding of developmental process.

In summary, a major contribution of these two approaches to the field of cognitive development is a focus on the mechanisms of change that lead to the emergence of new behaviors. Hence, any new theory from either a dynamic systems or a connectionist perspective would seek to outline general principles that govern how new ways of thinking or behaving arise from multiple, interacting components. Importantly, this new theory would make explicit this focus on the emergence of new behaviors through such interactions and have as a central goal understanding those interactions rather than describing age changes in cognitive skills. In the next section we evaluate whether this would indeed be a new theory and show how many GOFCD theories have also had this as a goal.

## Is This a New Way of Thinking about Developmental Change?

GOFCD has long been interested in change, and much of the research and theorizing in the field of cognitive development is ultimately motivated by understanding developmental change. For example, information-processing theories, which have been criticized for focusing on *what* develops rather than on mechanisms of developmental change (Thelen & Smith, 1994), have described change in terms of an increase in processing speed (Kail, 1986), the number of relations a child can keep in mind (Halford, Wilson, & Phillips, 1998), the capacity and duration of memory stores (Case, 1985), and the availability of strategies for solving problems (Siegler, 1996). Importantly, in each of these examples, mechanisms of changes are provided through increasingly thorough and detailed descriptions of what is developing.

Why then does the emphasis on change in dynamic systems and connectionist approaches seem to be so unique and novel? In the 1980s and 1990s, nativist theories dominated the study of cognitive development. Such theories focused on identifying early emerging capabilities and not on how change occurs. This focus reflects, in part, the influence of Chomsky's (1968) notions that environmental events simply trigger preexisting behaviors (i.e., those specified in biology). The highly publicized and influential work of investigators such as Elizabeth Spelke (Spelke, Breinlinger, Macomber, & Jacobson, 1992; Spelke & Newport, 1998), Karen Wynn (1998), and Rochel Gelman (Gelman & Gallistel, 1978) was aimed at showing high-level cognitive competence at an early age rather than documenting the mechanisms that produce changes in cognitive abilities. This led to the impression that the field of cognitive development had collectively lost an interest in understanding how change occurs.[1]

However, understanding the mechanisms that lead to the emergence of new behaviors has a long history in the study of cognitive development and was central to the theories of Piaget, Gibson, and Vygotsky (a fact that is acknowledged by theorists from both connectionist and dynamic systems perspectives [see Bates & Elman, 2002; Thelen & Bates, 2003]). Piaget, for example, proposed that new mental structures emerge through the dynamic interplay between the child's developing cognitive structures and input from the environment. Similarly, for Gibson, changes in the organism lead to increased sensitivity to environmental structure, which in turn leads to changes in the organism at both neural and behavioral levels. Hence, change emerges out of cyclical organism–environment interactions over both shorter and longer timescales (Gibson, 1988; Gibson & Pick, 2000). For Vygotsky, new skills emerge at times when children are sensitive to social experiences that allow them to try out new ways of thinking and acting, sometimes referred to as the *zone of proximal development* (Wertsch, 1985). The notion that adult guidance must be developmentally appropriate necessarily implies that cognitive change emerges out of the interaction of the child and the social environment. Thus, the idea that behaviors emerge through interactions between the organism and the environment has been central to our understanding of development for quite some time. These historical views on emergence have played an important role in the application of dynamic systems and connectionist frameworks to understanding development.

Importantly, however, ideas about emergence in the theories of Piaget, Gibson, and Vygotsky have largely been lost or ignored over time, even by theorists who came from these traditions. Modern theorists whose work originated in a Piagetian tradition shifted focus from understanding how cognitive structures emerge out of the interaction of the child and the world to an almost exclusive focus on the cognitive structures or concepts themselves (Flavell, 1970). Gibsonian theorists have shifted away from viewing affordances as an emergent property of the interaction between the organism and environment to viewing affordances as objective properties of the environment—"information available about surfaces, places, obstacles, and

[1] It should be pointed out that although Rochel Gelman's views have historically been nativist in flavor, her recent chapter with Lucariello on learning in the third edition of *Stevens' Handbook of Experimental Psychology* (Gelman & Lucariello, 2002) is remarkably non-nativist. Thus, the reference here is to the work in the 1980s and 1990s that contributed to the nativist movement in the field of cognitive development, rather than the most recent writings by Gelman or any of the cited authors.

things as well as about oneself" (Gibson, 2003, p. 293). Likewise, sociocultural approaches to cognitive development have almost exclusively focused on the social environment and have had very little to say about how organism characteristics interact with social structure to produce changes in thinking. We believe these shifts in theoretical perspectives over time have occurred because the notions of *interaction* and *emergence* are very difficult concepts. It is difficult to think about *behavior* or *thinking* as being simultaneously determined by organism characteristics and environmental structure. It is much easier to assign causal priority to one or the other, rather than to both at the same time.

An additional issue is that arguments about the emergence of new ways of thinking and behaving from a GOFCD perspective have been made primarily at a conceptual level. Although this theorizing has led to advances in our understanding of the mechanisms of development, the hypothesized mechanisms have often been difficult to test. One significant contribution of connectionist and dynamic systems theories of cognitive development is to illustrate how interaction and emergence can be simulated with sophisticated formal models. Indeed, one of the most exciting products from this movement is the collection of mathematical models that simulate the emergence of qualitatively different stages of behavior from multiple, interacting components. Although computational models can oversimplify the complexity of cognitive processes, they offer important steps forward in formalizing and testing ideas about cognitive change.

Because interaction and emergence are difficult to conceptualize, many studies of cognitive development from a GOFCD perspective may appear to be merely descriptive at first blush, rather than revealing processes of cognitive change. However, the rich descriptions of cognitive development amassed over the last several decades have actually provided considerable understanding of the mechanisms of change, although the explicit goals of these investigations may not have always been to uncover such mechanisms. Thus, we have made more progress in our understanding of mechanisms of change than we often credit ourselves for, and we have even advanced our understanding of how behavior and structure emerge from organism–environment interactions. This progress has not occurred *despite* a focus on description, but rather is intimately tied to the descriptive success. In part, the two enterprises are linked because the rich description of change strongly constrains the search for explanation. But even more than that, rich description focuses attention on plausible mechanisms and explanatory principles. These mechanisms can then be tested in studies that may themselves seem (on the surface) descriptive. That is, investigators often test explanatory ideas by evaluating predictions regarding descriptive questions, such as how manipulations of task characteristics or environmental input will affect children's behavior or cognitive abilities.

Consider an example of a cognitive change: the emergence of configural, as opposed to featural, processing of visual stimuli. We may simply ask: At what age do people process upright and inverted faces differently (an effect that has been interpreted as a hallmark of configural rather than merely featural processing)? This may seem a supremely descriptive question (although an improvement over a global question such as when can children process faces). Yet, the answer revealed by research asking this descriptive question turned out to be quite complex. On the basis of classic work by Carey and Diamond (1977), we would conclude that a change in processing faces occurs around 10 years of age, and what propels this change to configural processing are prolonged developmental changes—maturational and experiential factors that occur over many years. Evidence that configural processing is in place by 4 to 5 months of age (Bhatt, Bertin, Hayden, & Reed, 2005; Turati, Sangrigoli, Ruel, & De Schonen, 2004), in contrast, led to a very different account of what factors influence a change from featural to configural processing. If configural face processing emerges early, then infants' experience with faces during the period in which the visual system is organized becomes a potentially important factor contributing to that change. Indeed, early restriction of visual experience, occurring when infants are born with congenital cataracts, has an enduring influence on face processing (Le Grand, Mondloch, Maurer, & Brent, 2004)—an observation that might appear to be merely a descriptive fact but that is also a test of a mechanism.

Even further specificity turns out to be possible. It turns out that there are three types of configural processing, which mature at different rates (Maurer, Le Grand, & Mondloch, 2002). Although some aspects of configural face processing may be present in infancy, others may not reach adult levels until 14 years of age. In addition, effects of early visual deprivation can co-occur with surprising levels of adult plasticity (Maurer, Lewis, & Mondloch, 2005), as may arguably be seen in the advent of configural sensitivity to non-face stimuli (Gauthier & Tarr, 2002). Taken individually, each of these investigations could be characterized as descriptive—they illustrate how the behavior unfolds over time. Taken together, however, the data (at a minimum) strongly constrain the family of possible accounts of developmental mechanisms. This body of work captures an important feature of rich description: Determining the timing of environmental effects and residual plasticity allows us to move beyond minimal explanations of developmental change (i.e., when and how adult competence is acquired) to formulating sophisticated (and explanatory) accounts of how environmental inputs and neural substrates interact to induce developmental change.

In our view, the (more or less explicit) goal of work conducted from a variety of GOFCD perspectives has been to understand how such interactions contribute to the emergence of new behavior. Examples from our own work will serve as illustrations of this approach. Ours are not the only programs of research that exemplify this trend, but they illustrate how studies that do not have the same explicit goals of dynamic systems and connectionism can reveal a great deal about emergence and interactions. Importantly, the focus of many of these studies was revealing the mechanisms of developmental change—or testing hypotheses about how such mechanisms constrain children's emerging behaviors. Moreover, these programs of research, although not explicitly adopting a dynamic systems framework or a connectionist framework, are aimed at uncovering how child–environment interactions operate to allow new behaviors to emerge.

For example, Oakes and her colleagues have studied infants' categorization from an information-processing perspective. This approach focuses on how basic cognitive processes such as attention and memory operate to produce categorization behavior. Their work, which was also influenced by Vygotskian notions of the context dependency of cognitive abilities, has shown that infants' categories are highly flexible and dependent on interactions between the developmental state of the infant (i.e., age) and the task (see Oakes, Horst, Kovack-Lesh, & Perone, 2008). At 10 months of age, infants respond to a category of *people* that excludes other land mammals (such as horses) in an object-examining task but not in a less structured sequential-touching task (Oakes, Plumert, Lansink, & Merryman, 1996). By 13 months of age, infants attend to this distinction even in the less structured task (Oakes et al., 1996). Thus, whether infants recognize the significance of the distinction between people and animals is determined both by infant age and how the items within and across categories are encountered. Oakes and Ribar (2005) observed a similar effect for 4- to 6-month-old infants' attention to the category of dogs versus cats in visual familiarization tasks. At 4 months of age, infants responded to this distinction only when the task minimized memory and other cognitive demands, whereas at 6 months of age, infants responded to the distinction even in a more cognitively demanding familiarization task. The point is that, in each of these investigations, infants' ability to respond to a particular category is a function of the interaction between their cognitive abilities (such as their ability to encode and recall individual items) and the structure of the task. In demanding tasks, infants have difficulty recognizing subtle distinctions (such as that between dogs and cats). In tasks that place fewer demands on infants' cognitive abilities, or as those abilities develop, infants can more easily recognize subtle distinctions between categories. Infants' recognition of the category, therefore, is emergent: it resides in the interaction between their abilities and the task context. They may have knowledge and past experience relevant to recognizing the category in the moment, but their behavior in the experimental task reflects more than this existing knowledge and past experience. Similarly, infants clearly have the ability to form narrowly defined categories even at 4 months of age, but whether they actually form such categories at any given moment is complexly determined by the interaction of their existing knowledge, the cognitive abilities they use to form such categories, and the context in which they are forming those categories. Although not motivated by connectionist or dynamic systems theories of categorization, these findings are consistent with these approaches, and future work may represent a blending of GOFCD (e.g., information-processing approaches) and these new approaches to development.

In a very different domain, Plumert and her colleagues have studied categorical bias in memory for location by combining ecological and information-processing perspectives. The aim of this work was to discover how categorical bias emerges out of the interaction of the cognitive system and the available environmental structure (for a review, see Plumert, Hund, & Recker, 2007). Categorical bias refers to the tendency to remember locations in the same spatial group (e.g., same quadrant of a room) as closer together than they really are. The work by Plumert and her colleagues reveals the same general pattern as observed by Oakes and her colleagues: categorical bias is a function of developmental level (i.e., age) and environmental structure. When asked to learn the locations of 20 miniature objects in a large open box, memory for those locations is determined not only by age (e.g., 7-year-olds have less accurate memories than do adults) but also by task structure, such as the cues available for organizing the locations into groups during learning (Hund & Plumert, 2003; Hund, Plumert, & Benney, 2002; Plumert & Hund, 2001), by how long participants are required to maintain memory of locations over time (see Hund & Plumert, 2002, 2005), and by the cues available at test (Plumert & Hund, 2001). Importantly, this work demonstrates that we cannot explain patterns of categorical bias by referring only to task structure (e.g., presence or absence of boundaries) or by referring only to developmental differences in the cognitive system (e.g., strategic encoding of spatial groups). These studies have consistently shown that all age groups exhibit categorical bias under some task conditions but not under others. For example, adults always show significant categorical bias when at least one cue is available during learning, but they do not show bias when no cues are available during learning. Thus, it is impossible to predict categorical bias by referring to age alone. Likewise, these studies have consistently shown that children and adults differ in how they respond to cues for organizing the locations into groups, such as visible boundaries, spatiotemporal experience, or object relations. Clearly, children and adults extract different things from their experience with these tasks even though the task structure is identical for all participants. These variations in how the same age group responds to different task structure and how different age groups respond to the same task structure support the idea that categorical bias emerges out of the interaction of the cognitive system and the task structure.

A final example from work by Newcombe and her colleagues also illustrates how spatial thinking emerges out of the interaction of the child and the task. In a task in which children are required to find a hidden object after being disoriented (i.e., they are turned around with closed eyes), Newcombe and her colleagues have found that children's use of geometric properties (e.g., relative length of walls defining enclosures) and featural information (e.g., colors or markings on surfaces) to reorient is not solely a function of the child's underlying competence or developmental level. Rather, the exact mix of features children use to reorient depends on a confluence of factors such as the reliability, variability, and usefulness of the sources of information and the certainty with which the sources of information have been encoded, which contrasts with the nativist modular view of the origins of spatial knowledge. According to this view, various sources of spatial information are processed independently in separable cognitive processing units (e.g., Wang & Spelke, 2002). On the basis of the facts that (1) a wide variety of animal species can use geometric properties of enclosing spaces (e.g., relative length of walls defining enclosures) to reestablish spatial orientation after being disoriented (Cheng, 1986; Hermer & Spelke, 1996; for a review, see Cheng & Newcombe, 2005) and (2) both rats and human children fail to use disambiguating, nongeometric (i.e., featural) information (e.g., colors and markings on surfaces), some investigators have suggested that geometric processing constitutes a specialized cognitive module that is impenetrable to nongeometric information, even when that information has been processed (Gallistel, 1990). Moreover, some researchers have argued that this module is innately available (Hermer & Spelke, 1996).

Newcombe and Ratliff (2007) proposed, in contrast, that the existing data on integration of featural and geometric information can be best explained by an adaptive combination approach in which the likelihood of using the two kinds of information varies depending on factors such as uncertainty, cue validity, and cue salience (e.g., Ernst & Banks, 2002; Huttenlocher, Hedges, & Duncan, 1991). Newcombe and her colleagues have documented the conditions under which children use the colors of walls to find hidden objects, thereby showing that use of geometric versus featural information is not an all-or-none phenomenon at a given age. For example, younger children are more likely to use featural information in large rooms and under conditions in which they can move about the space (Newcombe & Ratliff, 2007). These differences in performance of children of the same age under different task conditions are difficult to explain from a modularity perspective. Hence, this program of research has provided the kind of data needed to differentiate between opposing theoretical views of the origins of fundamental spatial skills.

In summary, these three programs of research—by Oakes and her colleagues, Plumert and her colleagues, and Newcombe and her colleagues—were not motivated by a connectionist or dynamic systems theory of development and yet, nevertheless, examined the emergence of behavior as a function of multiple, interacting components. The research described here is compatible with connectionism and dynamic systems views, but has its roots in information-processing, Piagetian, Gibsonian, and Vygotskian perspectives. Some of this work may provide evidence for the kinds of mechanisms of change proposed by connectionist and dynamic systems theories of development—and could indeed be the foundation for future modeling and theorizing from these perspectives. The important point here is that ideas central to connectionist and dynamic systems approaches to development are compatible with (at least some) GOFCD approaches Although connectionism and dynamic systems approaches may provide new tools for examining these types of interactions in development, as may other modeling approaches, it is important to remember that these tools would all be used to support and further specify conclusions drawn from studies conducted from within GOFCD approaches.

## A Closer Look at Explanations of Change: How Do Dynamic Systems and Connectionist Theorists Explain Change on the A-not-B and the Balance Scale Tasks?

In this section we will further discuss the general approaches of dynamic systems theory and connectionism by examining specific examples of how these approaches have been used to explain developmental change in two domains: the A-not-B error and the balance scale (or balance beam) task. These two domains provide particularly good contexts for understanding the contributions of these approaches because they are classic problems that have been studied from the perspectives of GOFCD, connectionism, and dynamic systems theory. (Note that a formal dynamic systems model of the balance scale problem has not yet been proposed; van der Maas and colleagues have developed a cusp model of the balance beam task [see Chapter 15] that derives from dynamical principles but is not a pure dynamic systems model.) Thus, these domains allow us to consider dynamic systems and connectionist explanations of well-understood cognitive problems, as well as to consider how these approaches are (or are not) an advance over previous, more traditional cognitive development approaches.

### The Case of the A-not-B Error

In his classic work on the development of the understanding of object permanence during the sensorimotor period, Piaget noted that toward the end of the first year of life infants exhibit somewhat odd behavior toward hidden objects. Although infants can retrieve a hidden object by 8 or 9 months of age (when they enter stage IV of Piaget's sensorimotor period), they make the A-not-B error. In the prototypical error, infants observe an attractive object hidden underneath a cloth, and they successfully retrieve the object. After several such hidings (repeatedly in the same location, location A), the experimenter hides the object (while the infant watches) under a different cloth (or location B). Infants in Piaget's stage IV of sensorimotor development (from approximately 8 to 12 months of age) search not where they have just observed the object hidden, but rather where they have successfully uncovered it on previous trials. In other words, after seeing the object hidden at location B, infants search for the object in location A.

This puzzling behavior has long been the focus of study in infant cognitive development. A large number of studies have been conducted to replicate the conditions under which infants make this error and to understand why infants fail to search in the location they have just seen the object hidden. Piaget's classic interpretation is that this behavior indicates an incomplete understanding of object permanence. According to Piaget, the infant "seems to reason as if the place where the object was found the first time remains where he will find it when he wants to do so" (Piaget, 1954, pp. 46–47) even though the infant directly observes the object hidden in the new location! The infant has developed awareness that the object can be retrieved even when it is out of sight, but the infant does not fully understand the relation between the hiding place and where the object can be retrieved. Rather, according to Piaget, the infant conceives of the object *only* in the first place that he or she saw the object hidden and successfully retrieved it.

There have been many explanations for the A-not-B error, ranging from an appeal to innate or early emerging conceptual abilities (Baillargeon, 2004) to a discussion of the development of neuroanatomical structures responsible for inhibiting prepotent responses (Diamond, 1991). The dynamic systems and connectionist explanations of this behavior grew out of these previous theories (and indeed rely on the large corpus of data that has been collected examining the conditions under which the A-not-B error is obtained). The dynamic systems and connectionist explanations differ from those that came before them in that they focus directly on the error as an emergent product of the development of infants' underlying cognitive structures and the task demands. Previous explanations focused on what the error reveals about the development of the type of underlying representations. For example, context effects in infants' performance were viewed as masking infants' underlying competence, and as a result, it was proposed that the A-not-B error could be overcome by using tasks that more effectively tapped this competence (Baillargeon & Graber, 1988).

In contrast, dynamic systems and connectionist theories of the A-not-B error provide explanations of how the error emerges and changes in real time (i.e., in the context of a particular trial or experimental session) and how that emergence is a product of the task factors (e.g., salience of a particular hiding event) interacting with developmental changes in memory abilities, motor abilities, and other factors. For example, the dynamic systems account is that the error is actually an error in reaching, determined by the competition between the transient memory of the hiding event and the longer-term motor memory for past reaches (Smith, Chapter 4; Smith & Thelen, 2003; Thelen et al.,

2001). In clever and systematic studies, Smith, Thelen, and their colleagues have shown that the incidence of the error can be increased by increasing the strength of the motor memory for reaches at location A (e.g., increasing the number of reaches at location A) and decreasing the salience of the transient memory for the hiding event at location B (e.g., making location B more similar to location A). Furthermore, the incidence of the error can be decreased by reducing the strength of the motor memory (e.g., changing the infant's posture or arm weights before reaching on B-trials) or increasing the salience of the hiding event at location B (e.g., making location B very different from location A; see Smith, Chapter 4; Smith & Thelen, 2003). These types of manipulations are explicitly designed to perturb the components hypothesized to play a role in infants' behavior in the A-not-B task. In so doing, they reveal how organism–environment interactions give rise to the A-not-B error.

For Munakata and Morton's connectionist model of the A-not-B error, the strength of the underlying representation for the hiding event at location A competes with the underlying representation for the hiding event at location B. The system creates an *active representation* for the more recent hiding event (the working memory–like component) and a *latent representation* for previous hiding events (the long-term memory–like component; see Morton & Munakata, Chapter 7; Munakata, 1998; Munakata, Morton, & Stedron, 2003). Infants' representation of the location of the object is graded due to competition between latent and active memory for the hiding event. As was true for Smith and Thelen's dynamic systems model of the A-not-B error, early in development the transient memory for the hiding event at location B cannot compete with the stronger memory for the hiding event at location A that was built up over time. With development, the active memory for the hiding event at location B becomes more effective at competing, and as a result, the error is reduced. Moreover, Munakata and Morton can evaluate how this process works in real time by examining how connection weights change as the simulated child reaches several times to location A and then sees the object hidden at location B.

In both cases, online dynamics in context are the cause of the behavior. Both theories have received significant support, with empirical tests confirming predictions made by each model. Clearly, therefore, the two approaches are similar in many ways. They both describe the behavior in terms of the competition between longer term, more established memories and shorter term, more transient memories. This raises the central question of this volume: Are these actually separate theories, or are they two variations of the same unified theory? There are important differences in the details of the theories: a key component of Munakata and Morton's model (Chapter 7; Munakata, 1998; Munakata et al., 2003) is the internal representation (in terms of connection weights and recurrence) of the hiding events, whereas the hiding event and object are not represented in any obvious way in the dynamic systems theory of Smith and Thelen (Smith, Chapter 4; Smith & Thelen, 2003; Thelen et al., 2001). Similarly, the dynamic systems account is embodied, with the infant's memory for his or her own action playing an important role in behavior. The connectionist model does not have an obvious component for memories for actions carried out by a real body in actual space.

Are these approaches simply relabeling the same components? Perhaps the accounts of the A-not-B error are more similar than they are different. Moreover, these accounts have commonalities with more traditional ways of discussing the issues, such as in the discussion of errors relative to competing spatial frameworks that may each be used to code the locations of objects (Newcombe, 2001; Newcombe & Huttenlocher, 2000). Thus, this comparison may not be the best one for understanding the differences between the two approaches. But, considering these two approaches to the A-not-B task may be a good means of understanding what connectionism and dynamic systems have to offer beyond GOFCD. Although these two approaches to this task have similarities to some GOFCD theories of the error (Newcombe, 2001), clearly none of the approaches appeal to the qualitative changes in mental representation characteristic of a Piagetian approach or to innate competencies advocated by nativists. Thus, although the two explanations clearly derive from different theoretical stances—and as a result they have incorporated different components into their models and have argued for roles of different aspects of development in this error—they may have more in common than not. It may be that the truth is a blending of these two approaches, involving embodied cognition in which objects and locations are represented in long-term memory and actions are carried out by physical bodies in real space. Indeed, as is clear from several chapters in this book, in several domains, theorists are blending connectionism and dynamic systems to create new models of development (see Chapters 10 and 11). What is important for the present discussion is that both of these models of the A-not-B error have attempted to understand the error in terms of the real-time dynamics of the task combined with changes over a longer timescale.

### The Case of the Balance Scale Task

In a very different series of observations, Inhelder and Piaget (1958) documented children's responses to the *balance scale task*. In this task, children are presented with a balance scale and are asked to predict whether the scale will balance or tip when different configurations of weights are placed on each side of the fulcrum. To correctly solve this problem, children must consider not only the amount of the weight placed on each side, but the distance of the weights from the fulcrum as well. Based on the children's performance on this task, Inhelder and Piaget argued for qualitative changes in children's thought. Children with concrete, operational approaches were unsuccessful at solving this problem; they typically believe that weight is the only relevant factor. Children who had attained formal operations (and therefore were capable of propositional thought and hypothetical-deductive reasoning), in contrast, could successfully combine the two relevant variables in an accurate way, implicitly or explicitly computing torque.

As with the A-not-B task, there have been subsequent explorations of children's performance on the balance scale task from perspectives other than Piaget's. A well-known series of studies by Siegler (1976, 1981), adopting an information-processing approach to cognitive development, used rule-assessment methodology to delineate a succession of four modes of reasoning on balance scale problems, beginning with children 5–6 years of age. Importantly, Siegler's work clearly addressed questions about mechanisms of development; for instance, he showed that it was possible to diagnose which children would respond to environmental input regarding how the balance scale worked.

Subsequent work on the balance scale problem has, for the most part, fit within the GOFCD tradition. There has been discussion of issues such as whether rule-assessment methodology should be supplanted or supplemented by other techniques, such as latent class analysis (Jansen & van der Maas, 2002), and whether children younger than 5 years can also exhibit systematic (albeit one-dimensional) reasoning about these problems (Halford, Andrews, Dalton, Boag, & Zielinski, 2002). From these studies, a rich portrait of developmental change on this task has emerged, including an interesting suggestion that some aspects of change are abrupt and stage-like, whereas, at other periods, various rules or strategies coexist as children experiment with solutions (Jansen & van der Maas, 2002).

There have also been efforts to model children's behavior in this task. Some researchers have used connectionist approaches (e.g., McClelland, 1989, 1995; Shultz, Mareschal, & Schmidt, 1994), while others have not. Dynamic systems theorists have not dealt with the balance scale problem, arguably because thinking about the problem is a more symbolic and logical task than the A-not-B task and is also less intimately involved with motor experience. However, van der Maas and his colleagues have developed a cusp model of children's performance on the balance scale task that is derived from catastrophe theory and has some overarching points of contact with dynamic systems thinking (Jansen & van der Maas, 2001). These models have been less successful in making novel predictions and stimulating new research than connectionist or dynamic systems models of A-not-B, and, as argued at length by van der Maas and Raijmakers (Chapter 15), these models do not even do a complete job of accounting for known empirical phenomena. However, as in the case of models of the A-not-B error, such models represent a hybrid or mix of concepts from different traditions. Interestingly, van der Maas recently offered an ACT–R model of children's performance on the balance scale task (see van der Maas & Raijmakers, Chapter 15; van Rijn, van Someren, & van der Maas, 2003), which involves the postulation of explicit rules. This model seems to do the best job so far of capturing how children really behave. Although van der Maas and Raijmakers remain optimistic regarding the potential of neural networks with nonlinear properties, they suggest that more symbolic approaches seem to work better in the short run.

Overall, the history of dynamic systems and connectionist approaches to the balance scale problem suggests caution about how easy it will be to apply the techniques used so far in these traditions to higher cognitive functioning (van Geert [1998] provided a conceptual dynamic systems model of higher level cognitive processing, but there are no pure formal dynamic systems models of higher level cognitive processes). GOFCD has made good progress in delineating the progression of children's behavior in this realm and in suggesting reasons for developmental change that may differ at with age, including change in the encoding of relevant information, working memory, experimentation with strategies, and responses to direct instruction. It remains to be seen how successful nonlinear, dynamic models will be in explaining development in such domains and how these models will combine symbolic models with ideas from connectionism and dynamic systems approaches.

## Can Dynamic Systems and Connectionism Evade Classic Issues Facing the Study of Cognitive Development?

In this chapter, we have examined the contributions of dynamic systems and connectionist viewpoints and have argued that many of these contributions are not completely new ideas in the study of cognitive development. Rather, dynamic systems theory and connectionism bring ideas such as interaction and emergence to the forefront in our explanations of developmental change and provide formal models of these conceptions of development, therefore playing an important role in shaping how we think and talk about development. An important issue is whether these new frameworks avoid the problems of many previous theories of cognitive development; that is, how do these frameworks address the criticisms that have been leveled against many different theories of cognitive development? If dynamic systems and connectionism frameworks are in a position to replace traditional GOFCD frameworks, they must overcome those criticisms. We have identified at least four problems characteristic of many theories of cognitive development—some of these problems are solved by dynamic systems and connectionist approaches to cognitive development, while others are problematic for these approaches just as they are for GOFCD approaches.

First, all models of development must grapple with the difficult problem of characterizing the environment and the actual nature of the input and feedback that children receive. This is a central problem that was most clearly articulated and studied from an ecological perspective but was later minimized by the information-processing approach. The computational models developed from connectionist and dynamic systems approaches help bring the environment back into clear focus because they must describe the input and feedback the models receive. However, as was discussed in the section on the A-not-B error, it is not always transparent why a particular input to the model represents real, physical input in the environment. For instance, in Munakata's (1998) connectionist model of the A-not-B error, there are three input nodes, A, B, and C, corresponding to three potential hiding locations. However, it is not clear whether those nodes must refer to location (e.g., is the spatial relation between locations represented, and does it matter?) or whether they could equally well represent other features that differentiate the locations (e.g., colors of lids). When comparing models of the A-not-B error, we made the argument that these two models may be simply using different labels for essentially the same input. This is not characteristic of all connectionist and dynamic systems models, however. For instance, Mareschal and his colleagues (Mareschal, French, & Quinn, 2000) found that variability in the head and face regions of dog and cat stimuli predicted both how connectionist models and 3- to 4-month-old infants learned the categorical distinction between dog and cat. French et al. (2002) found that by blurring the visual input in a way that mimicked the visual acuity of young infants, a connectionist model would respond to the same kinds of categorical distinctions as infants do. Similarly, Smith and her colleagues (Yoshida & Smith, 2001) examined the structure of nouns to which children from different language environments are exposed and how those different kinds of inputs help create biases in children's assumptions about the referents of new words. Clearly, characterizing the nature of the input is extremely difficult. However, as recently pointed out by Saffran (2008; Saffran, Reek, Niebuhr, & Wilson, 2005), we cannot really understand the learning mechanism without understanding *what* is being learned.

A second problem that faces models of development is how to characterize the start and end states (and indeed, determining whether or not there is an end state) of the cognitive system with respect to a given ability. At minimum, a model must accurately characterize the cognitive system at the entry point of the period of development under study. Formal models from dynamic systems and connectionist perspectives accomplish this on one level because they must be very explicit about the starting state of the system (e.g., determining the weight or value of some parameter that refers to strength of memory trace). But, such determinations often seem arbitrary (e.g., how does a parameter value relate to a child's actual working memory capacity), and it is not often clear how they map onto actual developmental starting states in the child. In particular, often it is not clear where those start states came from (i.e., what developmental events occurred beforehand to create the starting state of the system), or whether and when development of a system ends. Some might argue that this is not the goal of these models, but the initial state of the system (e.g., the limited motoric and perceptual abilities of the newborn infant) places important constraints on how the system operates and, thus, what kinds of organism–environment interactions are possible.

Third, theories that focus on emergence must attempt to explain the nature of organism–environment interactions. Clearly, one advantage to connectionist and dynamic systems approaches is that they have brought the discussion of change to the forefront—in some sense, connectionist and dynamic systems theories are defined by their approach to emerging structure and their focus on understanding that emergence. In this way, these frameworks appear to address the criticism of GOFCD theories of development that such theories are merely descriptive. Although we believe that many cognitive developmentalists have indeed been interested in mechanisms of change (and have collected data relevant to understanding those mechanisms), we applaud the dynamic systems and connectionist frameworks for making this discussion explicit and for adding tools to our battery of methods for understanding change.

Finally, development is characterized by variability, and theories of cognitive development must come to terms with the high levels of variability that emerge with developmental change and task variation. For many, such variability is a nuisance; the goal is to develop sensitive tasks that eliminate sources of extraneous variability and reveal children's true competence. Dynamic systems and connectionist models embrace variability in behavior and use variability over time or with task variations as ways of explaining developmental changes in organism–environment interactions (see also Siegler, 1996). In fact, for the dynamic systems perspective, variability is a necessary precursor of change; systems must go through some sort of instability in order to change. Importantly, dynamic systems theory and connectionism provide an overall framework for thinking about the role of the environment or task, but understanding how specific tasks operate at specific points in development is a problem that can only be addressed by continued empirical and theoretical work in the field.

## Conclusion

Is this a new unified theory of development that represents a radical departure from good old-fashioned cognitive development? We don't think so. We think that, simply, but not at all trivially, this unified theory of development has helped to refine the explanations of development that have emerged in the field and has helped to enhance our search for the principles of emergence. Perhaps most importantly, these new approaches have provided us with new tools that will allow us to explore more deeply interactions and emergence in development. These new tools, in turn, have shifted the focus in our thinking about development to those processes. Thus, these new approaches have been and will continue to be important in shaping how we collectively think about developmental change. But, it is important to remember that the core notions underlying hypothesized mechanisms of developmental change have played important roles in our conceptualization and study of cognitive development for a long time and that dynamic systems theory and connectionism are not immune to the criticisms that have been raised about GOFCD theories.

## Acknowledgments

Preparation of this chapter was made possible by support from NIH grants HD49840 and HD49143 awarded to LMO, NSF grant BCS-0414302 awarded to NSN, and NSF grant BCS-0343034 awarded to JMP. We thank Bob Siegler for discussions about the state of work on the balance scale. Correspondences should be addressed to Lisa M. Oakes, Center for Mind and Brain, 267 Cousteau Pl, University of California, Davis, CA 95618; e-mail: lmoakes@ucdavis.edu.

## References

Baillargeon, R. (2004). Infants' physical world. *Current Directions in Psychological Science, 13*(3), 89–94.

Baillargeon, R., & Graber, M. (1988). Evidence of location memory in 8-month-old infants in a nonsearch AB task. *Developmental Psychology, 24*(4), 502–511.

Bates, E. A., & Elman, J. L. (2002). Connectionism and the study of change. In M. H. Johnson (Ed.), *Brain development and cognition: A reader* (2nd ed., pp. 623–642). Oxford: Blackwell.

Bates, E. A., Elman, J. L., Johnson, M. H., Karmiloff-Smith, A., Parisi, D., & Plunkett, K. (1998). Innateness and emergentism. In W. Bechtel & G. Graham (Eds.), *A companion to cognitive science* (pp. 590–601). Oxford: Basil Blackwood.

Bhatt, R. S., Bertin, E., Hayden, A., & Reed, A. (2005). Face processing in infancy: Developmental changes in the use of different kinds of relational information. *Child Development, 76*(1), 169–181.

Carey, S., & Diamond, R. (1977). From piecemeal to configurational representation of faces. *Science, 195*(4275), 312–314.

Case, R. (1985). *Intellectual development: Birth to adulthood.* Orlando, FL: Academic Press.

Cheng, K. (1986). A purely geometric module in the rat's spatial representation. *Cognition, 23*(2), 149–178.

Cheng, K., & Newcombe, N. S. (2005). Is there a geometric module for spatial orientation? Squaring theory and evidence. *Psychonomic Bulletin and Review, 12*(1), 1–23.

Chomsky, N. (1968). *Language and mind.* New York: Harcourt, Brace & World.

Diamond, A. (1991). Neuropsychological insights into the meaning of object concept development. In S. Carey & R. Gelman (Eds.), *The epigenesis of mind: Essays on biology and cognition.* Hillsdale, NJ: Erlbaum.

Diedrich, F. J., Highlands, T. M., Spahr, K. A., Thelen, E., & Smith, L. B. (2001). The role of target distinctiveness in infant perseverative reaching. *Journal of Experimental Child Psychology, 78*(3), 263–290.

Elman, J. L. (2005). Connectionist models of cognitive development: Where next? [Special issue on developmental cognitive neuroscience] *Trends in Cognitive Sciences, 9*(3), 111–117.

Elman, J. L., Bates, E. A., Johnson, M. H., Karmiloff-Smith, A., Parisi, D., & Plunkett, K. (1996). *Rethinking innateness: A connectionist perspective on development.* Cambridge, MA: MIT Press.

Ernst, M., O. & Banks, M. S, (2002). Humans integrate visual and haptic information in a statistically optimal fashion. *Nature, 415*(6870), 429–433.

Flavell, J. H. (1970). Concept development. In P. H. Mussen (Ed.), *Carmichael's manual of child psychology: Vol. 1.* (3rd ed.). New York: Wiley.

French, R. M., Mermillod, M., Quinn, P. C., Chauvin, A., & Mareschal, D. (2002). The importance of starting blurry: Simulating improved basic-level category learning in infants due to weak visual acuity. In W. Gray & C. Schunn (Eds.), *Proceedings of the 24th annual conference of the Cognitive Science Society.* (pp. 322-328). Mahwah, NJ: Erlbaum.

Gallistel, C. R. (1990). Representations in animal cognition: An introduction. *Cognition, 37*(1–2), 1–22.

Gauthier, I., & Tarr, M. J. (2002). Unraveling mechanisms for expert object recognition: Bridging brain activity and behavior. *Journal of Experimental Psychology: Human Perception and Performance, 28*(2), 431–446.

Gelman, R., & Gallistel, C. R. (1978). *The child's understanding of number.* Cambridge, MA: Harvard University Press.

Gelman, R., & Lucariello, J. (2002). Role of learning in cognitive development. In C. R. Gallistel (Ed.), *Stevens' handbook of experimental psychology: Vol. 3. Learning, Motivation, and Emotion* (3rd ed., pp. 395–443). New York: Wiley.

Gibson, E. J. (1988). Exploratory behavior in the development of perceiving, acting, and the acquiring of knowledge. *Annual Review of Psychology, 39,* 1–42.

Gibson, E. J. (2003). What psychology is about: Ruminations of an opinionated aged psychologists. *Ecological Psychology, 15*(4), 289–295.

Gibson, E. J., & Pick, A. D. (2000). *An ecological approach to perceptual learning and development.* New York: Oxford University Press.

Halford, G. S., Andrews, G., Dalton, C., Boag, C., Zielinski, T. (2002). Young children's performance on the balance scale: The influence of relational complexity. *Journal of Experimental Child Psychology, 81*(4), 417–445.

Halford, G. S., Wilson, W. H., & Phillips, S. (1998). Processing capacity defined by relational complexity: Implications for comparative, developmental, and cognitive psychology. *Behavioral and Brain Sciences, 21*(6), 803–831.

Hermer, L., & Spelke, E. (1996). Modularity and development: The case of spatial reorientation. *Cognition, 61*(3), 195–232.

Hund, A. M., & Plumert, J. M. (2002). Delay-induced bias in children's memory for location. *Child Development, 73*(3), 829–840.

Hund, A. M., & Plumert, J. M. (2003). Does information about what things are influence children's memory for where things are? *Developmental Psychology, 39*(6), 939–948.

Hund, A. M., & Plumert, J. M. (2005). The stability and flexibility of spatial categories. *Cognitive Psychology, 50*(1), 1–44.

Hund, A. M., Plumert, J. M., & Benney, C. J. (2002). Experiencing nearby locations together in time: The role of spatiotemporal contiguity in children's memory for location. *Journal of Experimental Child Psychology, 82*(3), 200–225.

Huttenlocher, J., Hedges, L. V., & Duncan, S. (1991). Categories and particulars: Prototype effects in estimating spatial location. *Psychological Review, 98*(3), 352–376.

Inhelder, B., & Piaget, J. (1958). *The growth of logical thinking from childhood to adolescence.* New York: Basic Books.

Jansen, B. R. J., & van der Maas, H. L. J. (2001). Evidence for the phase transition from rule I to rule II on the balance scale task. *Developmental Review, 21*(4), 450–494.

Jansen, B. R. J., & van der Maas, H. L. J. (2002). The development of children's rule use on the balance scale task. *Journal of Experimental Child Psychology, 81*(4), 383–416.

Kail, R. (1986). Sources of age differences in speed of processing. *Child Development, 57*(4), 969–987.

Le Grand, R., Mondloch, C. J., Maurer, D., & Brent, H. P. (2004). Impairment in holistic face processing following early visual deprivation. *Psychological Science, 15*(11), 762–768.

Marchman, V. A. (1997). Models of language development: An "emergentist" perspective. *Mental Retardation and Developmental Disabilities Research Reviews, 3*(4), 293–299.

Mareschal, D., French, R. M., & Quinn, P. C. (2000). A connectionist account of asymmetric category learning in early infancy. *Developmental Psychology, 36*(5), 635–645.

Mareschal, D., Quinn, P. C., & French, R. M. (2002). Asymmetric interference in 3- to 4-month olds' sequential category learning. *Cognitive Science, 26*(3), 377–389.

Maurer, D., Le Grand, R., & Mondloch, C. J. (2002). The many faces of configural processing. *Trends in Cognitive Sciences, 6*(6), 255–260.

Maurer, D., Lewis, T. L., & Mondloch, C. J. (2005). Missing sights: Consequences for visual cognitive development. *Trends in Cognitive Sciences, 9*(3), 144–151.

McClelland, J. L. (1989). Parallel distributed processing: Implications for cognition and development. In R. G. M. Morris (Ed.), *Parallel distributed processing: Implications for psychology and neurobiology* (pp. 8–45). New York: Oxford University Press.

McClelland, J. L. (1995). A connectionist perspective on knowledge and development. In T. J. Simon & G. S. Halford (Eds.), *Developing cognitive competence: New approaches to process modeling* (pp. 157–204). Hillsdale, NJ: Erlbaum.

Munakata, Y. (1998). Infant perseveration and implications for object permanence theories: A PDP model of the AB task. *Developmental Science, 1*(2), 161–184.

Munakata, Y., Morton, J. B., & Stedron, J. M. (2003). The role of prefrontal cortex in perseveration: Developmental and computational explorations. In P. T. Quinlan (Ed.) *Connectionist models of development: Developmental processes in real and artificial neural networks* (pp. 83–114). London: Taylor & Francis–Psychology Press.

Newcombe, N. S. (2001). A spatial coding analysis of the A-not-B error: What IS "location at A"?. *Behavioral and Brain Sciences, 24*(1), 57–58.

Newcombe, N. S., & Huttenlocher, J. (2000). *Making space: The development of spatial representation and reasoning.* Cambridge, MA: MIT Press.

Newcombe, N. S., & Ratliff, K. R. (2007). Explaining the development of spatial reorientation: Modularity-plus-language versus the emergence of adaptive combination. In J. M. Plumert & J. P. Spencer (Eds.), *The emerging spatial mind.* New York: Oxford University Press.

Oakes, L. M., Horst, J. S., Kovack-Lesh, K. A., & Perone, S. (2008). How infants categorize. In A. Needham & A. Woodward (Eds.), *Learning and the infant mind* (pp. 144–171). New York: Oxford University Press.

Oakes, L. M., & Plumert, J. M. (2002). Variability in thirteen-month-old infants' touching patterns in the sequential-touching task. *Infant Behavior and Development, 25*(4), 529–549.

Oakes, L. M., Plumert, J. M., Lansink, J. M., & Merryman, J. D. (1996). Evidence for task-dependent categorization in infancy. *Infant Behavior and Development, 19*(4), 425–440.

Oakes, L. M., & Ribar, R. J. (2005). A comparison of infants' categorization in paired and successive presentation familiarization tasks. *Infancy, 7*(1), 85–98.

Piaget, J. (1954). *The construction of reality in the child* (M. Cook, Trans.). New York: Basic Books.

Plumert, J. M., & Hund, A. M. (2001). The development of memory for location: What role do spatial prototypes play? *Child Development, 72*(2), 370–384.

Plumert, J. M., Hund, A. M., & Recker, K. M. (2007). Organism–environment interaction in spatial development: Explaining categorical bias in memory for location. In J. M. Plumert & J. P. Spencer (Eds.), *The emerging spatial mind.* New York: Oxford University Press.

Plumert, J. M., & Nichols-Whitehead, P. (1996). Parental scaffolding of young children's spatial communication. *Developmental Psychology, 32*(3), 523–532.

Robinson, S. R. (2005). Conjugate limb coordination after experience with an interlimb yoke: Evidence for motor learning in the rat fetus. *Developmental Psychobiology, 47*(4), 328–344.

Saffran, J. R. (2008). What can statistical learning tell us about infant learning? In A. Woodward & A. Needham (Eds.), *Learning and the infant mind.* New York: Oxford University Press.

Saffran, J. R., Reek, K., Niebuhr, A., & Wilson, D. (2005). Changing the tune: The structure of the input affects infants' use of absolute and relative pitch. *Developmental Science, 8*(1), 1–7.

Samuelson, L. K. (2000). *Statistical regularities in vocabulary guide language acquisition in 15- 20-month-olds and connectionist models.* Unpublished doctoral dissertation, Indiana University, Bloomington.

Samuelson, L. K. (2002). Statistical regularities in vocabulary guide language acquisition in connectionist models and 15-20-month-olds. *Developmental Psychology, 38*(6), 1016–1037.

Schutte, A. R., Spencer, J. P., & Schöner, G. (2003). Testing the dynamic field theory: Working memory for locations becomes more spatially precise over development. *Child Development, 74*(5), 1393–1417.

Shultz, T. R., Mareschal, D., & Schmidt, W. C. (1994). Modeling cognitive development on balance scale phenomena. *Machine Learning, 16*(1–2), 57–86.

Siegler, R. S. (1976). Three aspects of cognitive development. *Cognitive Psychology, 8*(4), 481–520.

Siegler, R. S. (1981). Developmental sequences within and between concepts. *Monographs of the Society for Research in Child Development, 46*(2), 1–74, 84.

Siegler, R. S. (1996). *Emerging minds: The process of change in children's thinking.* New York: Oxford University Press.

Smith, L. B., & Thelen, E. (2003). Development as a dynamic system. *Trends in Cognitive Sciences, 7*(8), 343–348.

Spelke, E. S., Breinlinger, K., Macomber, J., & Jacobson, K. (1992). Origins of knowledge. *Psychological Review, 99*(4), 605–632.

Spelke, E. S., & Newport, E. L. (1998). Nativism, empiricism, and the development of knowledge. In W. Damon & R. M. Lerner (Eds.), *Handbook of child psychology: Vol. 1. Theoretical models of human development* (5th ed., pp. 275–340). New York: Wiley.

Spencer, J. P., Vereijken, B., Diedrich, F. J., & Thelen, E. (2000). Posture and the emergence of manual skills. *Developmental Science, 3*(2), 216–233.

Thelen, E., & Bates, E. (2003). Connectionism and dynamic systems: Are they really different? *Developmental Science, 6*(4), 378–391.

Thelen, E., Corbetta, D., & Spencer, J. P. (1996). Development of reaching during the first year: Role of movement speed. *Journal of Experimental Psychology: Human Perception and Performance, 22*(5), 1059–1076.

Thelen, E., Schöner, G., Scheier, C., & Smith, L. B. (2001). The dynamics of embodiment: A field theory of infant perseverative reaching. *Behavioral and Brain Sciences, 24*(1), 1–34.

Thelen, E., & Smith, L. B. (1994). *A dynamic systems approach to the development of cognition and action.* Cambridge, MA: MIT Press.

Turati, C., Sangrigoli, S., Ruel, J., & de Schonen, S. (2004). Evidence of the face inversion effect in 4-month-old infants. *Infancy, 6*(2), 275–297.

van Geert, P. (1998). A dynamic systems model of basic developmental mechanisms: Piaget, Vygotsky, and beyond. *Psychological Review, 105*(4), 634–677.

van Rijn, H., van Someren, M., & van der Maas, H. L. J. (2003). Modeling developmental transitions on the balance scale task. *Cognitive Science, 27*(2), 227–257.

Wang, R. F., & Spelke, E. S. (2002). Human spatial representation: Insights from animals. *Trends in Cognitive Sciences, 6*(9), 376–382.

Werner, H., & Garside, E. B. (1957). *Comparative psychology of mental development* (rev. ed.). New York: International Universities Press.

Wertsch, J. V. (1985). *Vygotsky and the social formation of mind.* Cambridge, MA: Harvard University Press.

Wynn, K. (1998). Numerical competence in infants. In C. Donlan (Ed.), *The development of mathematical skills* (pp. 3–25). London: Taylor & Francis–Psychology Press.

Yoshida, H., & Smith, L. B. (2001). Early noun lexicons in English and Japanese. *Cognition, 82*(2), B63–B74.

# CHAPTER 14

# A Developmental Systems Theory Perspective on Psychological Change

*Timothy D. Johnston and Robert Lickliter*

The two theories of psychological development that are the focus of this volume demonstrate how quantitative precision can be brought to bear on the problem of explaining psychological change. In any theoretical endeavor, precision often comes at the expense of generality and simplification—some features of the phenomena being explained must be ignored so as to create a more limited domain within which to achieve precision. Our aim in this chapter is to draw attention to some aspects of development that we believe are underrepresented in both dynamic systems theory (DST) and connectionism. We intend this not as a criticism of either theoretical perspective but as a reminder that these aspects of development must be addressed in any eventual synthesis.

The approach we take is that of developmental systems theory, which shares with DST a focus on the whole organism in its environment and with connectionism a focus on the biological support for psychological change. It adds to those perspectives, a concern with more levels of biological analysis (including genetics), the incorporation of comparative, animal-based research, and attention to evolutionary considerations in the analysis of psychological change. Although both DST and connectionism have gone further than developmental systems theory in generating formal (mathematical) models of development and learning, developmental systems theory provides a broader context within which to pose questions about development and extends the analytic framework of the other two theories.

Nonetheless, there are broad compatibilities between developmental systems theory and the other theoretical positions under examination in this volume. In particular, all three perspectives reject the dichotomy between learned and innate features of the organism that has been such a persistent obstacle in developmental theory over the years and resist the tendency to ascribe features of behavior directly to genetic influences. We think the developmental systems perspective can enrich the accounts provided by DST and by connectionism in the following ways: by explicitly including gene expression in the explanation of development, by broadening the domain of environmental influences on development to include social factors, and by connecting development with evolutionary change. The expanded and enhanced perspective on development provided by this account helps to situate the issues specifically addressed by DST and connectionism in a broader theoretical context, and also illuminates a persistent problem for developmental theory, namely the relationships among the various timescales over which change occurs.

## The Developmental Systems Perspective

Developmental systems theory originated in opposition to accounts of development that assumed a dichotomous relationship between genetic and environmental influences on development, such as the ethological theories of Konrad Lorenz (1937, 1965). Early architects of developmental systems theory include Lehrman (1953, 1970), Schneirla (1956), Bateson (1976), and Gottlieb (1976), and their writings have continued to exert an important influence on the field. Susan Oyama's *Ontogeny of Information* (1985) was a landmark synthesis of the motivating concerns of the theory, which has since been developed by a number of other authors (Oyama, Griffiths, & Gray, 2001; Robert, 2004; West & King, 1987). Developmental systems theorists point out that all features of the organism—anatomical, physiological, behavioral, and psychological—emerge out of a set of interactions among a large number of factors. These include gene activity, hormones, social stimulation, food intake, and environmental temperature, among many others. Whereas some accounts of development focus on partitioning the organism's phenotypic characters among those that are genetically determined and those that are produced by the environment, developmental systems theory asserts that no such partitioning is possible, even in principle. Instead, the goal of developmental analysis should be to identify the ways in which various factors (including genetic and environmental ones) interact to produce all of the organism's phenotypic characters (e.g., Johnston & Edwards, 2002).

The specific factors that influence the development of any particular character of an organism (such as the mature song of a swamp sparrow or the perceptual abilities of a human infant) depend upon the ecological setting in which the organism normally develops. From a developmental systems perspective, understanding *what* influences are available in the ecological setting is critical to explaining *how* development occurs (Johnston, 1985; Lickliter, 2005). This ecological emphasis underlies another feature of developmental systems theory, namely its assertion of the continuity between developmental and evolutionary change. All phenotypic traits arise during ontogeny as products of individual development, and evolutionary change occurs as a result of variation in the patterns of development that give rise to individual development. Because the course of development is determined in part by the ecological factors to which the organism is exposed, changes in the phenotype of the mature organism may result from changes in those factors over long periods of time, as well as from alterations in the genetic makeup of the population resulting from natural selection, genetic drift, and other factors more usually considered in evolutionary analyses (Johnston & Gottlieb, 1990).

## Resources for Development

Griffiths and Gray (1994) refer to the various factors that influence development as *developmental resources.* These resources include all of the environmental factors shown to influence development, as well as factors that exert their influence within the organism, such as hormones and the products of gene expression. This terminology was proposed by Griffiths and Gray to emphasize that none of the factors that affect development are privileged and to eliminate the implication that the genes, in particular, are uniquely capable of specifying or encoding aspects of the phenotype directly. As discussed in detail by many authors (Gottlieb, 1998; Johnston, 1987; Lickliter & Honeycutt, 2003; Oyama, 1985), the idea that the genes contain instructions specifying the construction of the phenotype (in the form of a plan or blueprint) has, in some form or another, been widespread in the literature on development for many years. Neither DST nor connectionism espouses that nativist position and we will not reiterate the arguments against it here. Developmental systems theory also rejects the nativist position and is instead concerned with the ways and means by which the assembly of developmental resources is repeated within and across generations to produce behavioral phenotypes. Gene expression is, of course, a critical resource for behavioral development and is treated explicitly and in detail by developmental systems theory (Gottlieb, 1995; Johnston & Edwards, 2002; see further below).

The experience of the developing individual is another important resource for development. For example, children's skills develop in large part on the basis of their experiences with the world, and both DST and connectionism offer detailed accounts of the ways in which such experiences contribute to development. In many species, including humans in particular, the social group (e.g., parents, siblings, and nonfamily members) is a very important source of experience for the developing individual. Parents provide experience for the young directly, by supplying food, warmth, and tactile stimulation, and also indirectly by modifying the young animal's interactions with other features of its environment through mechanisms such as social facilitation, imitation, and scaffolding. The stabilizing and modifying roles of such factors both within and across generations are discussed further below.

Experience that constitutes a developmental resource for some behavioral or cognitive skill need not bear an obvious relationship to that skill. Much of developmental theory focuses on cases in which skill development is based on direct experience with the skill (i.e., practice) or with the situation in which the skill is used (i.e., observation or imitation). For example, the entire field of learning theory assumes that a skill is learned as a function of experience with the task to which the skill is applied. Similarly, connectionism is explicitly based on matching functions between a skill such as word recognition and experience with speech sounds that include the words being learned. Developmental systems theorists have pointed out that not all relevant experience bears such a direct and obvious relationship to the skill or capacity under consideration. For example, Gottlieb (1997) has shown that the following response of a young duckling to the call emitted by its mother depends not on prior experience with the call, but on exposure to the vocalizations produced by the embryonic duckling itself in the days immediately preceding hatching. These embryonic calls differ acoustically in terms of frequency and note composition from the maternal call, and so this is not an example of learning or familiarization. Exposure to the embryonic call is an essential, though nonobvious, resource for the normal development of the duckling's perceptual and behavioral specificity.

Another example of a nonobvious developmental resource is the importance of maternal licking and grooming of pre-weanling rat pups to their subsequent hormonal, physiological, and behavioral responsiveness to stress (Francis, Diorio, Liu, & Meaney, 1999). Pups receiving increased levels of maternal licking and grooming show enhanced synaptogenesis in several brain regions (Liu, Diorio, Day, Francis, & Meaney, 2000) and a decrease in emotional reactivity to novelty in adulthood. These types of findings suggest that the developmental consequences of an organism's varied encounters with its environment are not always as obvious or straightforward as we might expect.

In the remainder of this chapter, we will discuss some aspects of development that have not been emphasized by either DST or connectionism, but that are the focus of considerable attention within developmental systems theory. These aspects include the role of gene expression, the contributions of social experience, and the relationship between developmental and evolutionary change. In a final section, we also address the challenge of how to deal with developmental change as it occurs on multiple timescales simultaneously.

## Genes in Developmental Context

The idea that some behavior (*instinctive* or *innate*) can be directly attributed to instructions provided by the genes is a long-standing feature of much developmental theory (Johnston, 1987, 2001). With the discovery of the genetic code by Watson and Crick in the middle of the last century, this idea seemed to gather greater weight and authority, and the metaphor of the genetic blueprint, a set of precise and unequivocal instructions for the construction of the organism, gained wide currency. Numerous writers have criticized the blueprint metaphor and related ideas (e.g., Griffiths & Gray, 1994; Lewontin, 2000; Lickliter & Honeycutt, 2003; Moore, 2001; Morange, 2001; Oyama, 2000). One important aim of developmental systems theory has been to replace this misleading but seductive metaphor with a more accurate account of genetic influences on development that incorporates our growing understanding of how genes make their essential contributions to the development of behavior.

Genes are segments of DNA that are transcribed in a complex process to produce RNA molecules, which move out of the cell nucleus to serve as templates for the construction (translation) of protein molecules. This process, from DNA to protein, is generally referred to as *gene expression*. It used to be thought that proteins are coded by contiguous stretches of DNA arranged on the chromosome like beads on a string, but in fact, the picture has turned out to be far more complex. DNA consists of *exons* and *introns* that are initially transcribed sequentially into a primary RNA transcript. The intron sequences are then removed and discarded and the exons are spliced together into messenger RNA (mRNA) from which the protein molecule is translated. However, various kinds of alternative splicing can occur to produce different mRNA molecules from the same primary RNA transcript (Ast, 2005). Thus, the same stretch of DNA (the same gene, in conventional terms) can produce many different mRNA molecules, and hence many different proteins. The alternative splicing is carried out by a structure called the spliceosome and is regulated by specialized proteins produced within the cell in response to both intracellular and extracellular signals as development proceeds. This newly discovered one-to-many relationship between genes and proteins contradicts a long-standing dictum of molecular genetics (*one gene, one protein*) and explains how the relatively small number of genes in the human genome (fewer than 25,000) can give rise to more than 90,000 different proteins.

This complex relation between genes and protein synthesis illustrates and underscores the complexity of the various roles that genes play in the development of behavior—and the complexity of the various relationships between the genes and other developmentally important factors. Figure 14.1 depicts the place of the genes in the context of a developing organism, illustrating the

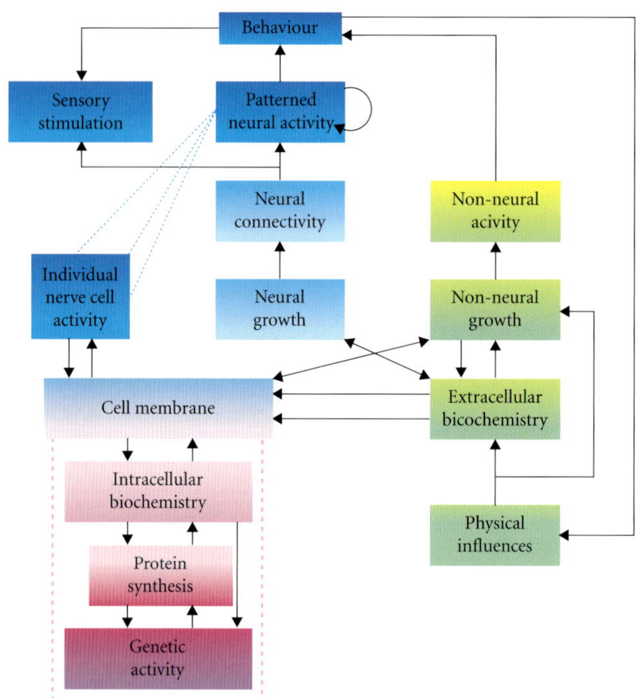


**Figure 14.1. Model of the relationship among the various factors that interact to produce developmental change in behavior (developmental systems theory)**
*Source:* Johnston & Edwards, 2002. Copyright 2002 by the American Psychological Association. Reprinted with permission of the authors.

influence of other factors, including experience (represented as sensory stimulation), the physical effects of the outer environment, and various internal elements such as individual cells and neural networks (Johnston & Edwards, 2002). As this depiction shows, genes do not affect behavior directly—genes are located within the nuclei of cells and affect the development of behavior only indirectly, through the regulated production of proteins and other gene products. Unlike many accounts of genetic contributions to development, this model shows that genes do not stand outside the developing system, acting on it as independent agents. Rather, they are embedded within the system, influencing its development through their effects on a complex cascade of interactions.

In the development of behavior, gene activity within the nervous system plays an especially important role. Genes, of course, exist within the nuclei of individual nerve cells, but nerve cells are organized into more or less complex neural circuits that control behavior. In explaining the neural control of behavior, it is the patterns of neural activity in those circuits that is most relevant. Figure 14.1 represents the nervous system in two different ways—as a collection of neural circuits that exhibit the patterns of activity controlling behavior and as a set of individual cells containing genes that help to regulate development. The dotted lines in the figure show that the activity of individual cells is contained within the activity of the neural circuits that they make up. This allows us to depict two important features of behavioral development that are sometimes hard to keep in focus: although behavior is a global consequence of patterns of neural activity, the genetic contribution to change in those patterns (e.g., development) can only be effected within individual cells. We cannot explain the role of genes in development by postulating direct effects on complex neural circuitry, any more than on the behavior that those circuits control. Our models of gene action must recognize that genetic activity occurs within the nuclei of individual cells. All other genetic influences are mediated by higher levels of the organism/environment system.

An important feature of the model depicted here is that all but the most transient effects of experience on behavior are mediated through molecular changes within individual nerve cells, including changes in gene activity. Notice that there is no arrow pointing directly from the box labeled *sensory stimulation* (the means by which experience is transduced by the nervous system) to that labeled *neural connectivity*, changes in which underlie the long-term, persistent changes in behavior involved in learning and development. In order for sensory experience to play a role in remodeling neural circuitry (e.g., by changing synaptic strengths), the activity in individual nerve cells that is produced by the experience must initiate biochemical events within the cell, including those events that lead to changes in protein synthesis and gene activity.

Not all changes in protein synthesis involve modification in gene activity in the nucleus. A number of studies (see reviews by Steward & Schuman, 2001; Wang & Tiedge, 2004) have shown that translation of mRNA into proteins can occur in dendrites, close to synaptic sites. Both the transport of specific mRNA molecules from the nucleus and the synthesis of proteins in the dendrite have been shown to be regulated by neural activity, providing highly differentiated local control of the protein synthesis necessary for synaptic modification (Steward & Schuman, 2001, p. 314ff). Connectionism relies very heavily on changes in synaptic strength induced by experience; such changes may be involved in development. Changes in synaptic strengths are produced by molecular events occurring both at the synapse and within the nucleus; developmental systems theory incorporates those events within an overall picture of developmental change.

As depicted in Figure 14.1, neural activity resulting from experience may also influence the transcription of genes in the cell nucleus. The finding that experience produces long-term effects on behavior by modifying gene activity is entirely consistent with the integrated perspective of developmental systems theory; however, it is contrary to any view that sets experiential and genetic influences on behavioral change in opposition. A class of genes known as immediate-early genes (IEGs) has been shown to play a ubiquitous role in translating experience into changes in gene expression. Transcription of IEGs within specific neural structures typically follows sensory stimulation within a few minutes. For example, presentation of song results in transcription of the IEG *ZENK* (an acronym for four earlier names given to the same gene: *zif-268*, *egr-1*, *NGFI-A*, and *Krox-24*) in parts of the songbird brain known to be involved in the perception and production of song (Ribeiro & Mello, 2000). *ZENK* has also been shown to be involved in a variety of examples of neural and behavioral plasticity, such as long-term potentiation and associative learning (Knapska & Kaczmarek, 2004). The molecular pathways subsequent to *ZENK* transcription are still being worked out, but its product has been shown to regulate the transcription of synapsins (Petersohn, Schoch, Brinkmann, & Thiel, 1995; Thiel, Schoch, & Petersohn, 1994), which are part of the molecular structure of synapses. Presumably, such regulation plays a role in the modification of synaptic efficiency, which in turn produces the developmental changes in both neural activity and behavior explicated by connectionist models.

## Social Contributions to Development

Developmental systems theory characterizes development as a situated, self-organizing, probabilistic process where pattern and order emerge and change as a result of complex interactions and relationships among developmental resources at multiple timescales. From this perspective, what is inherited across generations is a structured developmental system that includes components internal (e.g., genes, cytoplasm, cells) and external (e.g., diet, light cycles, conspecifics) to the organism. Developmental systems theory recognizes that the recurrence from generation to generation of the specific resources and interactions that make up an organism's physical, biological, and social environments are as causally informative to the development and transmission of phenotypic traits as are genes contained within this system. A central concern for developmental systems theory is thus the identification of the developmental resources that allow for the repeated assembly of constructive interactions both within and across generations.

In this light, an ever-present feature of life for nearly all vertebrates is that the developing organism is situated in and interacts with some form of a social environment. Across a diverse array of species and ecologies, the typical circumstances of prenatal and postnatal development include conspecifics, especially parents and siblings. Like other aspects of the developmental system, these social companions are experiential resources for the developing individual and have been shown to play a significant role in the achievement of a wide range of species-typical behaviors, including food choice, predator avoidance, mate choice, and species identification (Galef & Laland, 2005; Gottlieb, 1997; Lickliter, 2005).

For example, research with several precocial bird species has consistently demonstrated that social experience with conspecifics both before and after hatching can have a significant effect on chicks' emerging perceptual preferences and early filial behavior (reviewed in Lickliter, Dyer, & McBride, 1993). In particular, the type, amount, and timing of sensory stimulation provided by the embryo or hatchling's social environment can significantly influence the young bird's arousal levels, perceptual processing, and learning and memory. For example, bobwhite quail chicks denied physical contact with broodmates during their prenatal development by isolating them from contact with other eggs (and thereby decreasing normally available vestibular stimulation) or reared with quail chicks of another species following hatching failed to prefer species-specific maternal auditory and visual cues at ages when normally socially reared chicks reliably demonstrate such filial preferences (Lickliter & Lewkowicz, 1995; McBride & Lickliter, 1993). Further, simply modifying the amount of tactile, auditory, or visual stimulation available from siblings in the period following hatching delays the emergence of perceptual responsiveness to maternal auditory and visual cues in quail chicks (Columbus & Lickliter, 1998).

Mallard ducklings denied physical contact and interaction with siblings following hatching likewise show altered patterns of auditory and visual responsiveness to their maternal hen. Gottlieb (1993) found that by manipulating the social stimulation available to ducklings immediately following hatching, he could affect both their arousal levels and their ability to learn non-conspecific maternal calls (in this case, a chicken maternal call). Specifically, he found that the malleability that allows ducklings to learn to prefer the chicken maternal call over their own species-specific mallard maternal call was significantly affected by the presence or absence of tactile stimulation from siblings. Ducklings not permitted tactile contact with siblings slept less and were highly aroused before, during, and after exposure to recordings of the chicken call. In contrast, socially reared ducklings slept more and were less aroused than the isolates during exposure to the call. Ducklings that experienced normal levels of tactile contact with siblings appeared highly malleable, readily learning to prefer the non-conspecific chicken maternal call in the days following hatching, whereas those deprived of tactile contact with siblings did not. It is important to note that the presence of conspecifics is virtually guaranteed in the context of normal development for avian embryos and hatchlings, and early interaction with and stimulation from these social companions appears essential for the normal development of early perceptual and behavioral organization.

Additional evidence for the role of social factors as basic developmental resources is provided from research on the determinants of song learning in various species of songbirds. The presence of conspecifics that sing is a reliable and repeatable feature of the context of development for most songbirds and the opportunity to interact with social companions has been shown to influence the nature and course of song learning in a variety of songbird species. For example, West, King, and their colleagues have shown that juvenile and adult social experience with other cowbirds is a critical component of the development of male song patterns, female preference for male songs, and courtship and mating preferences in the brown-headed cowbird (e.g., West, King, & Arburg, 1988; West, King, & White, 2003; White, King, Cole, & West, 2002). Their work consistently shows that male and female cowbirds develop critical vocal and reproductive skills through recurring social interactions with their conspecifics. Further, male song patterns, female preference for songs, and courtship behaviors can be transmitted across generations (Freeberg, 1996). Young males that were captured from one geographic area and reared with adult males from another geographic area eventually served as models for their own offspring, inducing these offspring to behave like the males that had instructed their fathers. These types of changes in development, which result in novel behavior that recur across generations, can facilitate new organism–environment relationships. These new relationships can reveal latent possibilities for morphological, physiological, or further behavioral change. Changes in behavior driven by the physical or social environment can thus be the first step in creating new phenotypic variants on which natural selection can act (Gottlieb, 2002).

It is important to emphasize that social influences on development are not independent of the genetic influences discussed in the previous section. As already noted, exposure to song, a critical social factor in song development, also stimulates gene expression (e.g., Ribeiro & Mello, 2000), which is necessary for the changes in neural conductivity and connectivity that result in long-lasting behavioral change (i.e., learning of a species-typical song or song dialect). Indeed, we presume that any instance in which social experience produces a developmental change in behavior will involve changes in gene expression, as implied by Johnston and Edwards's (2002) model. Moffitt (2005) and Shanahan and Hofer (2005) provide overviews of recent studies that are beginning to identify the relationships between social experience and gene expression in human development and discuss various ways in which such studies could be extended.

## Relationship between Developmental and Evolutionary Change

In addition to including gene expression in explanations of development and broadening the domain of environmental influences on development to include social factors, developmental systems theory also addresses the relationship between developmental and evolutionary change. Developmental systems theory posits that development guides and constrains evolution in at least two major ways. First, it constrains phenotypic variation in such a way that the traits and characters presented to the filter of natural selection are not random. In other words, developmental processes and resources serve to limit the range of the possible in terms of phenotypic form. This is the *regulatory* function of development in evolution. Second, development provides variants in traits and characters that may lead to enduring alterations of the phenotype. This production of phenotypic novelties is the *generative* function of development and has significant implications for evolutionary change (Gottlieb, 2002; West-Eberhard, 2003). According to developmental systems theory, the regulatory and generative nature of development renders the natural selection of random mutations (the cornerstone of the neo-Darwinian view of evolutionary change) insufficient to account for evolution and its outcomes. Rather, the individual organism's entire developmental system is the source of both the stability *and* the variations in phenotype that researchers in evolutionary and developmental science seek to understand.

Given that all phenotypic traits arise during ontogeny as products of individual development, variations in the patterns and processes of development must be a primary basis for evolutionary change. Variations in morphologies, physiologies, and behaviors resulting from modification in developmental processes can place their possessors in different ecological relationships with their environments, and if these phenotypic variations provide slight advantages in survival and reproduction, then competitors without the novel phenotype will eventually decrease in frequency in the population. The novelty-generating aspects of the evolutionary process are thus the result of the developmental dynamics of living organisms, situated and competing in specific ecological contexts. From this view of evolution, genetic change is often a secondary or tertiary consequence of enduring transgenerational behavioral changes brought about by alterations in the organism's developmental system.

For example, a European passerine bird, the blackcap, has recently shown changes in its migratory behavior that appear to result in changes in mating behavior, size of egg clutches, and success at fledging young. At one time all European blackcaps migrated back and forth together, spending summers in northern Europe and the British Isles and winters in Portugal, Spain, and North Africa before gathering in mating grounds in southern Germany and Austria to breed. Over the last several decades, however, more and more blackcaps have remained north for the winter rather than heading south. Blackcaps were typically seen in the British Isles only during the summer months, but the number of them wintering in Britain and Ireland has soared over the last forty years. This change is likely due to the increased availability of winter provisioning provided by bird feeders, gardeners, landscapers, and other similar human activities. The resulting shift in migratory patterns has allowed northern wintering blackcaps to be exposed to the critical photoperiods that trigger migration and the onset of gonadal development some 10 days earlier than their southern-wintering counterparts.

Even though all blackcaps continue to gather each year at the same mating sites in Germany and Austria, northern blackcaps typically arrive earlier at the breeding grounds and establish territories and mate with other earlier arriving birds; southern-wintering blackcaps arrive at the same mating sites some two weeks later and are more likely to mate with each other, serving to reproductively isolate northern-wintering birds from the later-arriving southern-wintering population. Importantly, the shift in migratory pattern seems to confer an advantage to the northern blackcaps, who tend to lay about one more egg per season than do the late arrivals from the south (Bearhop et al., 2005).

The blackcap provides a compelling example of how a change in behavior (in this case, a change in migratory patterns brought on by changes in food availability) can lead to changes in the timing of breeding, which in turn can lead to the effective reproductive isolation of populations and ultimately, divergence and even sympatric speciation. Contrary to the assumptive base of the neo-Darwinian synthesis of the last century, the introduction of phenotypic variation upon which natural selection acts is not simply the result of random genetic mutations. Rather, variations in phenotypes and the resulting opportunities for evolutionary change are the result of a wide range of epigenetic processes occurring at different timescales and involving factors internal *and* external to the developing individual.

## Timescales of Change

The expanded perspective on development and evolution provided by developmental systems theory situates the issues addressed by DST and connectionism in a broader theoretical context and also highlights a persistent challenge for developmental theory—making sense of the relationship among the various timescales over which change occurs. These timescales include *real time*, the immediate experiences and encounters of the individual with the extant physical, biological, and social environments, *developmental time*, the continuing influence of prior experiences and encounters on the individual's ongoing interaction with varied environments, and *evolutionary time*, the transgenerational effects of the individual's experiences and activities during ontogeny. Characterizing how real time and developmental time combine to contribute to developmental change has been a primary focus of concern for both the DST and connectionist approaches represented in this volume. Developmental systems theory highlights the additional complexity of evolutionary time, raising the important issue of how real time and developmental time also combine with evolutionary time to contribute to evolutionary change.

As reviewed earlier, modifications in experience can initiate a range of physical and behavioral novelties, some of which can (given the availability of appropriate developmental conditions) persist across subsequent generations. Dramatic findings from developmental psychobiology indicate that modified early experience in one generation can predictably influence phenotypic outcomes in subsequent generations, even in the absence of the original experiential modification. For example, differences in physical measures (e.g., body weight, endocrine responses) and behavioral measures (e.g., fearfulness) have consistently been observed between groups of rats whose mothers (Denenberg & Whimby, 1963; Francis et al., 1999; Whimbey & Denenberg, 1967) or grandmothers (Denenberg & Rosenberg, 1967) were handled or not handled as pups. Despite their obvious importance to both developmental and evolutionary concerns, these types of transgenerational effects on both physiological responsiveness (in particular, the development of the hypothalamic adrenocortical system) and behavioral responsiveness (including curiosity, novelty seeking, and emotional regulation) remain poorly understood. Once again, we emphasize that both experiential and genetic factors are involved in mediating transgenerational effects such as these. Maternal behavior has been shown to influence gene expression, as measured by increases in mRNA coding for proteins involved in behaviors such as stress response and spatial learning (reviewed by Meaney, 2001). These findings support the message of developmental systems theory that experiential and genetic influences are co-actors in developmental change, including that which extends across generations. We now know that a wide range of supra-genetic and environmental factors are key participants in gene activity and gene expression, in some cases, well beyond the timescale of individual development.

Developmental systems theory thus enriches the accounts of development provided by DST and connectionism by emphasizing the importance of a transgenerational approach to the study of behavior, in which the development of one generation is studied in relation to the experiences and developmental resources of previous generations (Honeycutt, 2006). Knowledge of the dynamics of developmental processes at multiple timescales can lead to a deeper understanding of the ways and means of both developmental and evolutionary change.

## CONCLUSION

In this chapter our aim has been to indicate some aspects of development that are not currently addressed by either connectionism or dynamic systems theory but that will eventually need to be encompassed as our theories of development become more complete. Considering the number of different resources, factors, and interactions that must be taken into account in order to explain developmental change in behavior, and the variety of timescales over which that change occurs, it is not surprising that any theoretical explanation of development will lack either completeness or precision, and frequently both. Both connectionism and dynamic systems theory have gone further than most theories of development in providing quite precise explanations of certain developmental phenomena. In particular, these approaches have provided important insights into the dynamics of self-organization and the emergence of new skills and abilities at real-time and developmental timescales. Dynamic systems theory has concentrated on motor skills, emphasizing in particular the importance of recognizing that behavior involves physical systems whose components have mass and inertia and are subject to forces that both constrain and facilitate the course of development (e.g., Thelen, 1995). That insight finds an echo in the emphasis that developmental systems theory places on the physical constitution of the organism, rather than on construing development in the abstract terminology of information, plans, and blueprints. Connectionism offers detailed models of the changes in neural networks that underlie learning, although it has not included a place for the mechanisms of gene expression that regulate those changes in response to experience. Connectionism and dynamic systems theory provide explanations of behavioral change that emphasize particular aspects of development in the interest of increased precision. Their insights provide developmental systems theorists with a more precise index of the resources and processes that can contribute to the transgenerational stability *and* variability of phenotypic form and function. This chapter shows that developmental systems theory provides a broader theoretical perspective, entirely compatible with dynamic systems theory and connectionism, that can help inform and extend them both.

## ACKNOWLEDGMENTS

The writing of this chapter was supported in part by NICHD grant RO1 HD048423 awarded to RL.

## REFERENCES

Ast, G. (2005). The alternative genome. *Scientific American, 292*(4), 58–65.

Bateson, P. P. G. (1976). Specificity and the origins of behavior. *Advances in the Study of Behavior, 6*, 1–20.

Bearhop, S., Fiedler, W., Furness, R. W., Voter, S. C., Waldron, S., Newton, J., et al. (2005). Assortative mating as a mechanism for rapid evolution of a migratory divide. *Science, 310*(5747), 502–504.

Columbus, R. F., & Lickliter, R. (1998). Modified sensory features of social stimulation alter the perceptual responsiveness of bobwhite quail chicks (*Colinus virginianus*). *Journal of Comparative Psychology, 112*(2), 161–169.

Denenberg, V. H., & Rosenberg, K. M. (1967). Nongenetic transmission of information. *Nature, 216*(5115), 549–550.

Denenberg, V. H., & Whimbey, A. E. (1963). Behavior of adult rats is modified by the experiences their mothers had as infants. *Science, 142*(3596), 1192–1193.

Francis, D., Diorio, J., Liu, D., & Meaney, M. J. (1999). Nongenomic transmission across generations of maternal behavior and stress responses in the rat. *Science, 286*(5442), 1155–1158.

Freeberg, T. M. (1996). Assortative mating in captive cowbirds is predicted by social experience. *Animal Behaviour, 52*(6), 1129–1142.

Galef, B. G., Jr., & Laland, K. N. (2005). Social learning in animals: Empirical studies and theoretical models. *BioScience, 55*(6), 489–500.

Gottlieb, G. (1976). The roles of experience in the development of behavior and the nervous system. In Gottlieb, G. (Ed.), *Studies on the development of behavior and the nervous system: Vol. 3. Neural and behavioral specificity* (pp. 25–54). New York: Academic Press.

Gottlieb, G. (1993). Social induction of malleability in ducklings: Sensory basis and psychological mechanism. *Animal Behaviour, 45*(4), 707–719.

Gottlieb, G. (1995). Some conceptual deficiencies in 'developmental' behavior genetics. *Human Development. 38*(3), 131–141.

Gottlieb, G. (1997). *Synthesizing nature-nurture: Prenatal roots of instinctive behavior.* Mahwah, NJ: Erlbaum.

Gottlieb, G. (1998). Normally occurring environmental and behavioral influences on gene activity: From central dogma to probabilistic epigenesis. *Psychological Review, 105*(4), 792–802.

Gottlieb, G. (2002). Developmental-behavioral initiation of evolutionary change. *Psychological Review, 109*(2), 211–218.

Griffiths, P. E., & Gray, R. D. (1994). Developmental systems and evolutionary explanation. *Journal of Philosophy, 91*, 277–305.

Honeycutt, H. (2006). Studying evolution in action: Foundations for a transgenerational comparative psychology. *International Journal of Comparative Psychology, 19*(2), 170–184.

Johnston, T. D. (1985). Environmental constraints and the natural context of behavior: Grounds for an ecological approach to the study of infant perception. In G. Gottlieb & N. A. Krasnegor (Eds.), *Measurement of audition and vision in the first year of postnatal life: A methodological overview* (pp. 91–108). Norwood, NJ: Ablex.

Johnston, T. D. (1987). The persistence of dichotomies in the study of behavioral development. *Developmental Review, 7*(2), 149–182.

Johnston, T. D. (2001). Towards a systems view of development: An appraisal of Lehrman's critique of Lorenz. In S. Oyama, P. E. Griffiths, & R. D. Gray (Eds.), *Cycles of contingency: Developmental systems and evolution* (pp. 15–23). Cambridge, MA: MIT Press.

Johnston, T. D., & Edwards, L. (2002). Genes, interactions, and the development of behavior. *Psychological Review, 109*(1), 26–34.

Johnston, T. D., & Gottlieb, G. (1990). Neophenogenesis: A developmental theory of phenotypic evolution. *Journal of Theoretical Biology, 147*(4), 471–495.

Knapska, E., & Kaczmarek, L. (2004). A gene for neuronal plasticity in the mammalian brain: Zif268/Egr-1/NGFI-A/Krox-24/TIS8/ZENK? *Progress in Neurobiology, 74*(4), 183–211.

Lehrman, D. S. (1953). A critique of Konrad Lorenz's theory of instinctive behavior. *Quarterly Review of Biology, 28*(4), 337–363.

Lehrman, D. S. (1970). Semantic and conceptual issues in the nature-nurture problem. In L. R. Aronson, E. Tobach, D. S. Lehrman, & J. S. Rosenblatt (Eds.), *Development and evolution of behavior: Essays in Memory of T. C. Schneirla* (pp. 17–52). San Francisco: W. H. Freeman.

Lewontin, R. (2000). *The triple helix: Gene, organism, and environment.* Cambridge, MA: Harvard University Press.

Lickliter, R. (2005). Prenatal sensory ecology and experience: Implications for perceptual and behavioral development in precocial birds. In P. J. B. Slater, C. T. Snowden, T. J. Roper, H. J. Brockmann, & M. Naguib (Vol. Eds.), *Advances in the Study of Behavior, Vol. 35* (pp. 235–274). New York: Academic Press.

Lickliter, R., Dyer, A. B., & McBride, T. (1993). Perceptual consequences of early social experience in precocial birds. *Behavioral Processes, 30*(3), 185–200.

Lickliter, R., & Honeycutt, H. (2003). Developmental dynamics: Toward a biologically plausible evolutionary psychology. *Psychological Bulletin, 129*(6), 819–835.

Lickliter, R., & Lewkowicz, D. J. (1995). Intersensory experience and early perceptual development: Attenuated prenatal sensory stimulation affects postnatal auditory and visual responsiveness in bobwhite quail chicks. *Developmental Psychology, 31*(4), 609–618.

Liu, D., Diorio, J., Day, J. C., Francis, D. D., & Meaney, M. J. (2000). Maternal care, hippocampal synaptogenesis and cognitive development in rats. *Nature Neuroscience, 3*(8), 799–806.

Lorenz, K. Z. (1937). The companion in the bird's world. *The Auk, 54*(3):245–273.

Lorenz, K. Z. (1965). *Evolution and modification of behavior.* Chicago: University of Chicago Press.

McBride, T. C., & Lickliter, R. (1993). Social experience with siblings fosters species-specific responsiveness to maternal visual cues in bobwhite quail chicks (*Colinus virginianus*). *Journal of Comparative Psychology, 107*(3), 320–327.

Meaney, M. J. (2001). Maternal care, gene expression, and the transmission of individual differences in stress reactivity across generations. *Annual Review of Neuroscience, 24*, 1161–1192.

Moffitt, T. E. (2005). The new look of behavioral genetics in developmental psychopathology: Gene-environment interplay in antisocial behaviors. *Psychological Bulletin, 131*(4), 533–554.

Moore, D. S. (2001). *The dependent gene: The fallacy of nature vs. nurture.* New York: W. H. Freeman.

Morange, M. (2001). *The misunderstood gene* (M. Cobb, Trans.). Cambridge, MA: Harvard University Press.

Oyama, S. (1985). *The ontogeny of information: Developmental systems and evolution*. Durham, NC: Duke University Press.

Oyama, S. (2000). *The ontogeny of information: Developmental systems and evolution* (2nd ed.). Durham, NC: Duke University Press.

Oyama, S., Griffiths, P. E., & Gray, R. D. (Eds.). (2001). *Cycles of contingency: Developmental systems and evolution*. Cambridge, MA: MIT Press–Bradford Books.

Petersohn, D., Schoch, S., Brinkmann, D. R., & Thiel, G. (1995). The human synapsin II gene promotor: Possible role for the transcription factor zif268/egr-1, polyoma enhancer activator 3, and AP2. *Journal of Biological Chemistry, 270*(41), 24361–24369.

Ribeiro, S., & Mello, C. V. (2000). Gene expression and synaptic plasticity in the auditory forebrain of songbirds. *Learning and Memory, 7*(5), 235–243.

Robert, J. S. (2004). *Embryology, epigenesis, and evolution: Taking development seriously*. New York: Cambridge University Press.

Schneirla, T. C. (1956). Interrelationships of the "innate" and the "acquired" in instinctive behavior. In P. P. Grassé (Ed.), *L'Instinct dans le comportement des animaux et de l'homme*, (pp. 387–452). Paris: Masson.

Shanahan, M. J., & Hofer, S. M. (2005). Social context in gene-environment interactions: Retrospect and prospect [Special issue]. *Journals of Gerontology: Series B, Psychological Sciences and Social Sciences, 60B*, 65–76.

Steward, O., & Schuman, E. M. (2001). Protein synthesis at synaptic sites on dendrites. *Annual Review of Neuroscience, 24*, 299–325.

Thelen, E. (1995). Motor development: A new synthesis. *American Psychologist, 50*(2), 79–95.

Thiel, G., Schoch, S., & Petersohn, D. (1994). Regulation of synapsin I gene expression by the zinc finger transcription factor zif268/egr-1. *Journal of Biological Chemistry, 269*(2), 15294–15301.

Wang, H., & Tiedge, H. (2004). Translational control at the synapse. *Neuroscientist, 10*(5), 456–466.

West, M. J., & King, A. P. (1987). Settling nature and nurture into an ontogenetic niche. *Developmental Psychobiology, 20*(5), 549–562.

West, M. J., King, A. P., & Arberg, A. A. (1988). The inheritance of niches: The role of ecological legacies in ontogeny. In E. M. Blass (Ed.), *Handbook of behavioral neurobiology: Vol. 9. Developmental psychobiology and behavioral ecology* (pp. 41–62). New York: Plenum Press.

West, M. J., King, A. P., & White, D. J. (2003). The case for developmental ecology. *Animal Behaviour, 66*(4), 617–622.

West-Eberhard, M. J. (2003). *Developmental plasticity and evolution*. New York: Oxford University Press.

Whimbey, A. E., & Denenberg, V. H. (1967). Experimental programming of life histories: The factor structure underlying experimentally created individual differences. *Behaviour, 29*(2), 296–314.

White, D. J., King, A. P., Cole, A., & West, M. J. (2002). Opening the social gateway: Early vocal and social sensitivities in brown-headed cowbirds (*Molothrus ater*). *Ethology, 108*(1), 23–37.

# PART VI

# Reactions from the Inside

# CHAPTER 15

## Transitions in Cognitive Development: Prospects and Limitations of a Neural Dynamic Approach

*Han L. J. van der Maas and Maartje E. J. Raijmakers*

In this chapter we discuss old-fashioned cognitive development from the point of view of two modern approaches, connectionism and nonlinear dynamical systems theory. Old-fashioned cognitive development refers to the development of typical human cognitive capabilities, such as reasoning, planning, and abstract thinking. These cognitive abilities are distinctively human, in contrast to the abilities usually studied in connectionism and the dynamic systems approach, which generally focus on perception, pattern recognition, association, motor action, and simple memory. Such abilities are of interest, of course, but do not constitute the greatest challenge for researchers in cognitive development. Yet, connectionists and dynamic systems modelers claim that their models are relevant to higher cognitive functioning. This is apparent in their rejection of other approaches, such as the Piagetian and information-processing accounts of development. The main assertion in both connectionist and dynamic systems approaches is that higher cognitive functioning is largely based on nonsymbolic, graded, and dynamic properties, of which these same approaches provide the best account.

In this chapter, we criticize this claim. Both the connectionist and the dynamic systems approaches have made important contributions to current developmental psychology, including solutions to problems within Piagetian and information-processing accounts. Examples include the explanation of novel forms of thinking in the dynamic systems approach (van der Maas & Hopkins, 1998) and the integration of learning and development in connectionist approaches (Quartz, 1993). In our view, however, the claim concerning nonsymbolic higher order cognition is overstated. A critical examination of this claim might help in developing more realistic models.

To explain our position, we focus on sudden transitions in cognitive development. There is ample evidence for domain-specific transitions, for instance, in the Piagetian concrete operational and formal operational tasks (van der Maas & Hopkins, 1998). Our key example will be the famous balance scale task as applied to access proportional reasoning. A typical balance scale item is shown in Figure 15.1.

Children are asked to predict the movement of the scale when the blocks, as depicted in Figure 15.1, are removed. It is well known that children use different rules or strategies to solve such items (Jansen & van der Maas, 2002). Young children will focus on the weights and ignore the distance information. Older children and many adults understand that the weight and distance cues conflict in the item that is displayed in Figure 15.1, but only a small minority understand the correct torque rule that provides the solution to this conflict. The majority of people guess or use alternative rules, like the compensation rule, which is based on the sums of weight and distance on each side of the fulcrum. Siegler (1976, 1981) showed that children's progression on this task takes place by the use of increasingly more complex rules.

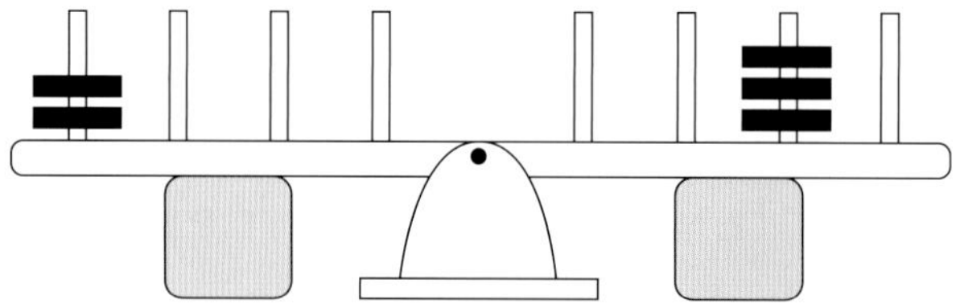

**Figure 15.1. Balance scale item**
What happens when the blocks are removed?

This is evident in children's responses and in their verbal explanations.

There are three reasons that this task constitutes an interesting case for our present discussion. First, the balance scale task appeals to proportional reasoning. The proportional reasoning that is required in the balance scale task clearly classifies as higher cognitive functioning. Interestingly, the data on this task show evidence for both graded, continuous processing and for symbolic, discrete processing (Jansen & van der Maas, 2002). Second, there exists a large body of empirical results on this task, starting with the work of Siegler (1976, 1981). A serious model for this task should explain the various well-established results, such as the diversity of rules, the order of rules, some deviations of rules, response time patterns, and transitions. Third, the balance scale task is used as a benchmark task in computational modeling. Specifically, in the last 25 years, a number of symbolic and connectionist models have been proposed to explain the empirical results (van Rijn, Van Someren, & van der Maas, 2003).

The main computational models for balance scale task performance are the connectionist model of McClelland (1989, 1995), the cascade correlation network model of Shultz, Mareschal and Schmidt (1994), and the ACT–R model of Van Rijn, Van Someren, and van der Maas (2003). The ACT–R model is based on symbolic/subsymbolic architecture. In view of the topics of this volume, we concentrate on the connectionist models, especially on their potential to explain rules and transitions between rules in development on the balance scale task. This will provide us with the means to evaluate the general claim that higher cognitive function can be understood in terms of graded, nonsymbolic, and dynamic properties of the brain.

However, there is one issue that has to be resolved first. We present the connectionist and the dynamic systems approach as unitary because they make the same general claim about cognition and they both reject the traditional symbolic approaches to cognition. In our view the two approaches are very similar indeed. Because not everyone will subscribe to this view (e.g., Beer, 2000), we first explain our position with respect to this.

## Relation between Dynamic Systems and Connectionist Models

A narrow definition of connectionism refers mainly to the typical multilayered, feed-forward back-propagation networks, as they were proposed in McClelland and Rumelhart (1986). We could broaden this view to include cascade correlation networks (Fahlman, 1988), recurrent networks (cf. Elman, 1990), and networks based on adaptive resonance theory (Grossberg, 1976). In fact, at present the field of neural network modeling includes a large number of network types, from the very low-level, single-neuron processes to the very high-level neuronal fields. The only common aspect of these models is that they are somehow models of the neural substrate of the brain.

A narrow definition of the dynamic systems approach refers to the well-known work of researchers such as Thelen and Smith (1994) and the dynamic field approach of Spencer and Schöner (2003). In this definition, the dynamic systems approach is a theory of development, in which actions in real time are modeled by means of differential equations. Most applications focus on infant behavioral development with more recent extensions that examine the development of spatial cognition, motor planning dynamics, and *embodied* cognition in autonomous robots. These researchers present this theory as a new theory of developmental processes that provides an alternative to both the standard theories in developmental psychology and to connectionism (Thelen & Smith, 1994; but see also Thelen & Bates, 2003). However, alternative accounts have been given. For instance, van Geert (1991) proposed nonlinear growth models to explain several developmental phenomena. van der Maas and Molenaar (1992) proposed the use of catastrophe theory to model and test transitions in development. For other examples, we refer to Lewis (2000). These accounts do not make the same theoretical claims about the relation between dynamic systems theory and other accounts of development.

Our own work is inspired by a broad definition of the dynamic systems approach (for recent examples see, Jansen & van der Maas, 2001; van der Maas et al., 2006; van der Maas & Raijmakers, 2000; Wagenmakers, Molenaar, Grasman, Hartelman, & van der Maas, 2005). In the broad sense, any application of concepts, methods, and models—from the mathematical study of nonlinear dynamical systems to developmental psychology—falls under the dynamical approach. In principle, any dynamical model with nonlinear interactions—developed from whatever theoretical perspective—is a nonlinear dynamical model. Any research on phase transitions or chaos theory is dynamical systems research.

In our view, there is no reason to prefer the narrow definitions of connectionism and the dynamical systems approach. The close similarity between the dynamical field model (Schutte, Spencer, & Schöner, 2003) and the connectionist model of the A-not-B task (Munakata, 1998) supports our view (see also, Thelen & Bates, 2003). Of course, the choice of definitions makes a large difference in thinking about the relations between the two approaches, as we discuss below.

## Neural Networks Are Dynamical Systems

An important and general conclusion that can be stated from the start is that connectionist models are a subset of the larger class of dynamical models, and that, therefore, conflict or competition between the approaches should be absent. Nonlinear dynamical models describe systems in which the new state depends nonlinearly on its old state. In addition, the state can depend on states of other processes (such as other neurons) and on parameters describing external influences. Such models can be formulated either in discrete time (with difference equations) or in continuous time (with differential equations). Mathematical study of such models has revealed very rich, complicated behaviors, such as catastrophic changes, chaotic attractors, and self-organization. It has been shown that many physical, chemical, and biological processes can be modeled this way (Wiggins, 1990).

Neural network models are dynamical models by definition. The state of neurons or neural systems depends on its earlier states, and most models account for this explicitly. These dependencies are generally nonlinear. Activations of neurons are usually a nonlinear function of the activity of connected neurons. This clearly applies to the many connectionist models developed and used in developmental psychology.

In this broad definition of nonlinear dynamical systems, it is not really possible to view the dynamic systems approach as a new theory of development. In the broad definition, the dynamic systems approach amounts to an application of concepts, techniques, and models from this field to developmental psychology. Such work can be carried out from any theoretical perspective.

We have argued in the past that many concepts from dynamical systems theory are surprisingly consistent with the old Piagetian theory of development (van der Maas, 1995). Application of nonlinear dynamical systems theory may lead to formalization of Piagetian concepts, such as disequilibrium and equilibration, and not actually lead to a new theory of development. It might also be fruitful to use nonlinear models to formalize Vygotskian concepts, such as the zone of proximal development, or information processing concepts, such as short-term memory. We do not see this consistency as a disadvantage. New tools to formalize and test developmental theories should be more than welcome. We propose to view dynamic systems accounts as a theory-free approaches, in the same way that statistical methods are largely theory-free approaches.

## Nonlinear Phenomena

Three phenomena of nonlinear dynamical systems are highly relevant to developmental psychology. A well-known phenomenon is *deterministic chaos*. Deterministic, nonlinear dynamical systems may show unpredictable chaotic behavior. One of the paradoxes of developmental psychology is that virtually all developmental psychologists agree that development during infancy is important for later development; but strong and convincing relations between developmental measures from infancy and measures of later development are hard to demonstrate (McCall & Carriger, 1993). In theory, deterministic chaos may explain this paradox, that is, why development, in certain respects, is so irregular and unpredictable. Unfortunately, it is very difficult to prove this conjecture; certainly the alternative, namely that our measurements are simply too noisy, remains compelling. Developmental data are relatively sparse and often quite unreliable. Attempts have been made to demonstrate chaos using psychophysiological time series data sets that are much more frequent and more reliable than developmental data; in general, these attempts have failed. The reason for this failure is than detection of chaos in time series with noisy measurements turns out to be extremely difficult. Demonstrating chaos in developmental data is, therefore, hardly feasible.

Another well-known phenomenon in nonlinear dynamical systems is self-organization. This refers to spontaneous, spatiotemporally coherent behavior of the parts of a system in the absence of some prespecified plan or algorithm that controls this behavior. Such behavior is purely a function of the dynamics of (nonlinear) interactions among the components of the system. Self-organization has been demonstrated in a variety of physical, chemical, and biological processes. The prototypical example is the Beluzhov-Zabotinsky (BZ) reaction, which displays temporal or spatial self-organization depending on whether the reactors are stirred (Winfree, 1980). As we describe elsewhere (van der Maas & Hopkins, 1998), we view self-organization as the most promising aspect of nonlinear dynamical systems theory. The reason is that self-organization could be a mechanism for structural developmental change, as it has been proposed not only in Piagetian theory but also in many other developmental accounts. Self-organization might be the definite resolution of Fodor's well-known learning paradox—that learning fundamentally new concepts seems impossible without composites of innate primitive concepts (Raijmakers & Molenaar, 2004).

However, like with chaos, to empirically investigate self-organization is no simple matter. Since neural networks are nonlinear dynamical systems, it should be possible to demonstrate self-organization in the learning process of certain types of neural networks. Demonstrating functional self-organization appears to be more difficult than many of us expected (Schlesinger, 2004). However, phase transitions are critical in self-organizing processes and, thus, present an indirect way to study self-organization.

Below we focus on this third phenomenon of catastrophes or phase transitions. There is a good deal of evidence for phase transitions in psychological processes, including development. We discuss this evidence and assess how well various computational models, including connectionist models, can account for such data. We also look at alternative neural network models, in which phase transition can be modeled. As it turns out, it is not simple to construct neural models that really show interesting nonlinear properties.

## Phase Transitions

Developmental psychologists entertain different definitions of discontinuous development. The intuitively simple definition focuses on a relatively large change in a relatively short period. This change can be either acceleration or a sudden jump. In the latter case, the large change occurs instantaneously. We have argued in the past (van der Maas & Molenaar, 1992) that the impasse in the research on discontinuous development (see Brainerd, 1978) is due to this informal definition.

First, the difference between acceleration and a sudden jump is not trivial. The definition should be clear with respect to this distinction. Accelerations in development are interesting phenomena, but, as explained below, are not discontinuous. Second, limiting the definition to sudden jumps helps, but an empirical problem remains. How can we detect sudden jumps? The fundamental measurement problem is that we can never be sure what happens between measurement sessions. Suppose we tested children every week and found that some children changed suddenly from one state to another. We cannot exclude the possibility that within the week an acceleration took place. This always remains a possibility, irrespective of the measurement frequency.

Here, the dynamic systems approach can help. As we explained above, phase transitions (e.g., bifurcations, catastrophes) are very important in the study of nonlinear dynamical systems. Phase transitions are well understood in this field. They are classified in distinct types (e.g., bifurcation types), and many techniques for the analysis and detection of these phenomena exist.

One branch of nonlinear dynamical systems theory is especially relevant for psychology and for the social sciences in general. This branch is known as *catastrophe theory* (Thom, 1975). From a mathematical point of view, the main ideas of catastrophe theory are now incorporated within bifurcation theory, but from our perspective, this approach is still very relevant. The reason is that catastrophe theory provides possible applications in cases where one does not have a mathematical model of a process at one's disposal. These applications involve archetypical models for phase transitions and strong criteria for phase transitions.

### Cusp Model

In catastrophe theory, discontinuities occur when system states become unstable. A catastrophe is a large, sudden jump to a new stable state (i.e., an attractor state) as a function of small, smooth changes in independent control variables. Such catastrophes can have different forms. The simplest form is the cusp catastrophe. Figure 15.2 illustrates the cusp catastrophe with a simple example (more detailed descriptions can be found: van der Maas, & Molenaar, 1992; Wagenmakers, van der Maas, & Molenaar, 2005; Zeeman, 1976).

### Flags

What makes the cusp model so useful is that we can derive a number of unique predictions of this model in terms of behavioral properties. Gilmore (1981) called these behavioral properties, of which he distinguished eight, catastrophe flags. We will explain five of them with the example in Figure 15.2. First, sudden jumps occur when we exert sufficient vertical pressure on the spoon, while increasing the horizontal force smoothly. At a given moment, the present stable state will become unstable, and the spoon will switch abruptly to the other side. In switching, it will pass the inaccessible state in between the two stable states. This inaccessibility, the second flag, makes this model fundamentally different from models that define discontinuities as accelerations. In such models, all intermediate states are stable. Third, for certain control values, two stable states exist, implying bimodality, which is another important flag. Fourth, when increasing the splitting factor starting at zero (for a = 0), the system must *choose* between two emerging states; this is called *divergence*. Finally, we have *hysteresis*. This refers to the fact that the sudden jumps from left to right and vise versa, do not take place at the same values of horizontal pressure; this is caused by *bistability*. You have to press relatively hard to force the spoon to jump. A well-known example is the transition from water to ice and vise versa. In shock-free conditions, water freezes at −4° Celsius, whereas ice melts at 0° Celsius.

## Case Study: The Balance Scale

We used catastrophe flags to study the transition between rules that children apply in the

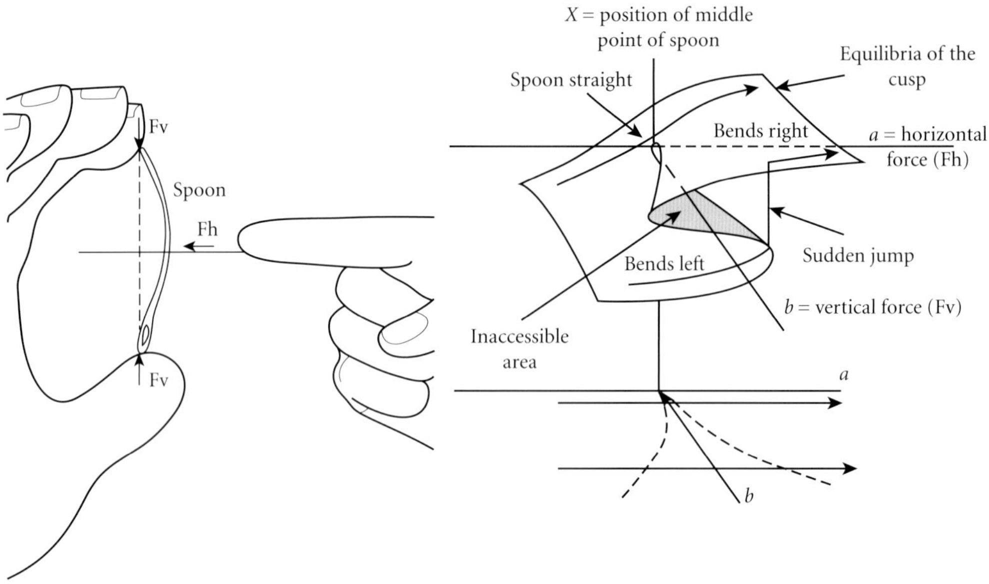


**Figure 15.2. Cusp catastrophe model**
On the left side an example of a catastrophic transition is shown. The stirring stick will jump from convex to concave or vice verse when pressed from the left or from the right side. Stronger jumps occur when the vertical pressure is increased. The cusp catastrophe (on the right) models this change in *X* (the position of the middle point of the spoon or stirring stick) as a function of changes in the factor *a* (horizontal force, Fh) and the splitting factor *b* (vertical pressure on the spoon, Fv). For certain values of *a* and *b*, the midpoint of the spoon will suddenly change from left to right or visa versa. When *a* is zero and *b* is high (e.g., there is only vertical pressure on the spoon), two attractor states exist separated by an inaccessible, unstable state.

balance scale task (Jansen & van der Maas, 2001). We focused on the transition from rule I to rule II because some prior support for this transition had been established. Children who apply either rule I or rule II respond differently to so-called distance items. These are items in which the weights are the same and only the distances to the midpoint of the fulcrum differ. Rule-I users think that balance scales with such a configuration of weights will remain in balance, whereas rule-II users correctly use the distance information. Rule-II users, like rule-I users, fail to do so when the number of weights are unequal. The initial evidence for a sudden transition concerns bimodality; scores on sets of distance items are usually strongly bimodally distributed.

Based on the work of Siegler and Chen (Siegler, 1976; Siegler & Chen, 1998), we hypothesized that encoding distance information is the relevant factor in this case. By changing the distance difference step by step, we tried to find sudden switches between incorrect and correct answers, as well as hysteresis. In our experiment, we varied the distance difference from 1 to 5 and back from 5 to 1. In control experiments, we reversed these series (with similar results) and checked whether irrelevant changes in the stimuli (i.e., color) also caused jumps and hysteresis (however, this was not the case).

The results were as follows. The large majority of children are stable rule-I or rule-II users. They give the same answer to all items. About 5% of children show hysteresis patterns. We tested this in several ways, and although small, this percentage could not be ascribed to chance (i.e., it is a statistically significant effect). In combination with the finding of bimodality, these results support the hypothesis of discontinuous development in the balance scale task. However, we did not find divergence. Also, a larger percentage of hysteresis patterns would have made our point stronger.

## Other Recent Results in Balance Scale Research

To evaluate the connectionist models of the balance scale task, we review some other recent results (see van der Maas, Jansen, & Raijmakers, 2004, for a more comprehensive review). Some doubts have been raised about the rules proposed by Siegler (1976, 1981). This does not pose a fundamental problem; additional rules may exist. Many children and adults seem to use compensation rules. They fail to understand the torque rule and, instead, compare the sums of weights and distances. More troublesome is that the classification of children into rule-using categories using the rule-assessment method of Siegler might be less robust than originally thought. For the rule-assessment method, it is essential that only the type of balance scale item matters and not the actual number of weights and the distances. Clearly the latter may vary within given item types. Rule-I users should fail all conflict distance items, and rule-II users should succeed on all distance items, irrespective of the actual values of the weights and distances.

Ferretti and Butterfield (1986) varied the torque differences (TD) in balance scale items and showed that these differences within item types influences rule classification. The idea is that a large difference in torque between the left and right side of the balance makes items easier. For the balance scale in Figure 15.1 the torque difference is only one (3x3 – 2x4). Compare this to a same balance scale but now with 5 instead of 3 weights on third peg on the right side. The torque difference is now 5x3 – 2x4 = 7. These are both conflict weight items, but the second is perhaps easier than the first. The empirical finding of an effect of TD on the probability of a correct answer greatly undermines the rule model and was used by the connectionists to distinguish their models (that show the TD effect) from symbol-oriented models and to argue for a more graded, nonsymbolic interpretation of the balance scale rules. However, Jansen and van der Maas (1997) offered two counter arguments. First they reanalyzed the data of Ferretti and Butterfield. Ferretti and Butterfield used four levels of torque difference (TD). The fourth level is very extreme, and Jansen and van der Maas demonstrated that the TD effect was solely due to this very extreme level, which never actually occurs in any known standard balance scale test (see Fig. 15.3). Second, they improved the rule assessment method by using latent class analysis. This statistical technique makes it possible to test the use of rules statistically. This technique can detect types or latent groups when data are categorical (Heinen, 1996; McCutcheon, 1987). The basic idea is to explain the dependencies between responses to items by latent classes so that within classes the responses are independent. In explanatory latent class analysis, no a priori hypotheses about the classes are required. In confirmatory analysis, specific hypotheses about the occurrence or type of latent classes can be tested. An important advantage of latent class analysis above other techniques is that the goodness of fit can be investigated with well-known statistical criteria. Jansen and van der Maas (2002) give a short introduction to this technique in the context of the balance scale task and discuss advantages and disadvantages of latent class analysis. Jansen and van der Maas (1997, 2002) analyzed responses for each item type of the balance scale test separately to test the homogeneity of items within item types. Homogeneity can be tested by using equality constraints between balance scale items of the same type. These constraints imply that the probability of an answer is equal between items. Jansen and van der Maas found that, in most cases, responses appeared to be homogeneous.

Thus, as van Rijn, van Someren, and van der Maas (2003) argue, any account of the data must address the question why children's behavior is homogenous for moderate torque difference levels and heterogeneous for extreme torque differences.

Finally, van der Maas and Jansen (2003) analyzed response times to balance scale items.

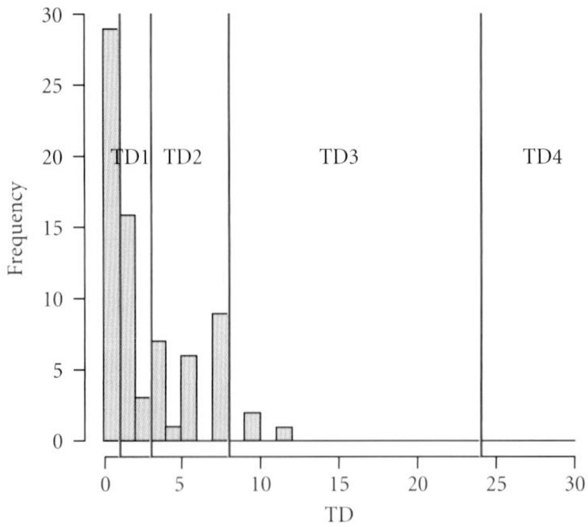


**Figure 15.3. Torque differences (TDs) in three balance scale tests**
Frequency of torque differences (TDs) in the set of items of three popular balance scale tests (Jansen & van der Maas, 2002; Siegler, 1981; Van Maanen, Been, & Sijtsma, 1989). The vertical lines show the TD levels used by Ferretti and Butterfield (1986). An effect of TD on the probability of a correct answer was shown only for level 4.

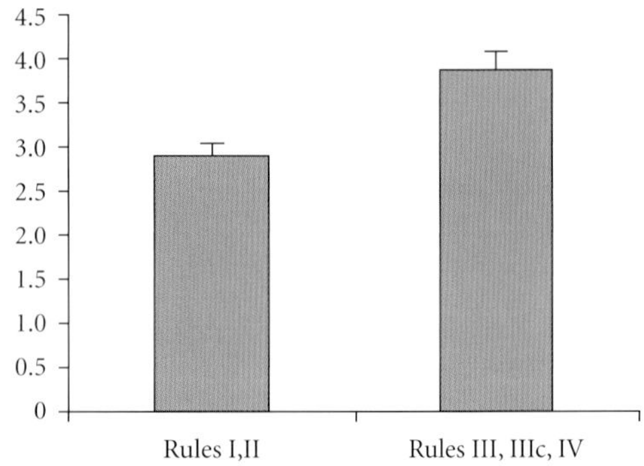


**Figure 15.4. Mean response times to balance scale items**
This figure depicts mean response times in seconds to balance scale items in which the side with the larger distance has more weights, the so-called weight distance items (van der Maas & Jansen, 2003). Here, responses are classified by which rule the subject appear to use in on the balance scale task (i.e., rules I, II, III, or IV); rule IIIc refers to the compensation rule, based on the sums of weight and distance on each side of the fulcrum. Error bars represent standard error (SEM). The sample included 28 Rule I and II-users and 40 Rule III, IIIc, and IV-users. (Further details can be found in van der Maas & Jansen, 2003.)

They collected data with a computerized test and analyzed the response times to items answered according to each rule with linear models (i.e., mixed effects models). They showed that the response time results are very well predicted using the rule model of Siegler. With respect to this, we give one example. We used one rather odd item type, which we call *weight distance items*. These are items with a larger number of weights on the side with the larger distance. These items are odd in that all rules should lead to the correct answer; thus, these items have no discriminatory value. However, if we look at response times, these items are of interest. According to the rule model of Siegler, subjects using rule I or II will only look at the weights. Since weights differ in the weight distance items, rule-I users and rule-II users immediately choose a response. Older subjects (including undergraduate students in our sample), using rule III and higher, make additional processing steps. They also first compare the number of weights, then they compare the distances, and finally they check whether the largest number of weights is on the side with the larger distance. So, they make two additional steps and are thus expected to be slower than the younger children who use rule I or II. As shown in Figure 15.4, younger children (rule-I and rule-II users), in spite of the fact that they process information slower, respond faster than older subjects (rule-III and rule-IV users).

## Do Connectionist Models Explain the Development of Proportional Reasoning?

van Rijn and colleagues (van Rijn et al., 2003) published an overview of computational models of the balance scale task. Between 1978 and 1996, several symbolic models were proposed; in the 1990s they were followed by connectionist models; and recently Van Rijn and colleagues (Van Rijn et al., 2003) proposed an *adaptive control of thought–rational* (ACT–R) model. We will not review the symbolic models, but rather focus on the connectionist models: the feed-forward back-propagation or *parallel distributed processing* (PDP) model of McClelland (1989, 1995) and the *cascade correlation* (CC) model of Shultz, Mareschal and Schmidt (1994). How do connectionist models address the development of proportional reasoning?

*The first issue concerns the application of rules.* The connectionists employ two arguments. First, when they simulated data with their models and subjected these data to the rule-assessment method of Siegler, they roughly found the same rules and also the same progression through the rules. Second, to explain the discrepancy between perfect rule use and the simulations, they used the TD effect of Ferretti and Butterfield. They argued that only the connectionist models were able to explain the TD effect. In addition, they used the TD effect to argue that information processing on this task is much more gradual than expected. But as we explained above, the TD effect is less convincing than suggested; it only exists for very extreme torque differences. Models should explain the absence of the TD effect for low and moderate TD levels, and the presence of a TD effect for very high levels of TD.

Jansen and van der Maas (1997) simulated data with the PDP model and analyzed the data with latent class analysis. They compared the results to data on children on exactly the same item set. They showed that the results were very different. Although the latent class models for children did show deviations from the rules, they generally confirm the use of rules. The latent class analysis of the PDP model data was a failure. There were no clear classes, and often they could not find a latent class model that fit the data. Also, the data on children appeared to be rather homogeneous within item types (i.e., no TD effects), whereas the data from PDP simulations were very heterogeneous.

Recently, Quinlan, van der Maas, Jansen, Booij, and Rendell (2007) reported a similar analysis of the CC network. Simulation results from the CC model more closely resemble the empirical results. Given data generated by these models, well-fitting latent class models were found, and some of the expected rules seem to be present. Support for rules I and II and weighted addition rules established. Human subjects, in contrast to the CC model, use unweighted addition rules and rule IV (i.e., the correct rule). In latent class analysis, the CC model was found to employ rules (e.g., always select the right side of the balance scale) that are not observed in empirical data. Especially troublesome is that even after extensive learning, the CC model, like the PDP model, fails to use the correct torque rule.[1]

*The second issue is discontinuous development.* Raijmakers, Van Koten, and Molenaar (1996) showed that the PDP model develops continuously. In the original analysis of the PDP model (and the CC model, as well), response patterns were first classified with the rule assessment method, and then these rules were plotted against time (i.e., epochs). Such a plot strongly suggests transitions and stages. Raijmakers, Van Koten, and Molenaar (1996) demonstrated that this is an artifact of the data-analytic procedure. The raw score on each item type should show discontinuities, but these turn out to be very gradual. Also, no evidence for bimodality could be found.

The CC model does a better job. In his book, *Computational Developmental Psychology*, Shultz (2003) showed that the CC model exhibits strong accelerations in performance. Since this model also showed evidence for rule I and II in the latent class analysis, it will exhibit bimodality on distance items. We did not perform a hysteresis experiment with either the PDP model or the CC model. One problem is that in our experiments, items are administered without feedback. But without feedback, CC and PDP models will not change their weights. These networks learn by back-propagation of error that depends on feedback of a trainer. For that reason, we do not see how connectionist

[1] The reason for this failure probably lies in the fact that the activation rules of these networks are additive. The networks have many free parameters so that in a limited item set they may mimic multiplication with weighted addition, but they will never generalize to all balance scale items (see also, Dawson & Zimmerman, 2003).

networks could show hysteresis. Furthermore, these networks do not show real-phase transitions or self-organization. The accelerations in the CC model are due to the addition of hidden nodes when learning falters. Such accelerations are unrelated to nonlinear self-organization. Note that the ACT–R model of Van Rijn and colleagues (Van Rijn et al., 2003) does provide an account for the hysteresis effect, although not in terms of self-organization.

New simulations might be in order to test hysteresis in connectionist networks. This is certainly the case with the response-time data reported by van der Maas and Jansen (2003). Both the ACT–R model and connectionist models are able to make predictions about response times. These response-time results constitute a new challenge for computational modeling of proportional reasoning.

*Our present conclusion about the connectionist models is that they are largely unconvincing as models of proportional reasoning and, more generally, as models of higher cognitive functioning.* Connectionists downplay the role of rules, reasoning, and representations, but at the same time demonstrate that they can approximately simulate the main empirical phenomena. Indeed, children are not perfect rule users, and some aspects of their reasoning (rule III, for instance) may be best described with statistical rules, as implemented in connectionist models. But overall these connectionist models are just too simple; they miss essential computational properties, and the evidence for genuine rule use by children is stronger than expected.

## Can Neural Networks Model Higher Cognitive Processes?

As we stated in the introduction, the real challenge for psychology lies in typically human abilities, such as reasoning and planning. The main successes of the connectionist approach do not lie here, but rather they can be found in other fields, such as language, memory and perception. As clear from the above, we are not impressed with the connectionist models of higher reasoning processes. Does this mean that we should abandon this approach?

We do not think so. First, neuronal processes evidently form the basis for all our information processing, including formal abstract thinking. Second, connectionist models have been subject to development over a period of 10–15 years. The field of neural network modeling has expanded in many directions, and at present, there are many types of networks to consider, and improvements are being made.

Below, we briefly discuss two important general lines of research. The first concerns the coupling of neural models with symbolic models, and the second concerns the utilization of the nonlinear properties of networks.

### Hybrid Models

The ACT–R model proposed by Van Rijn, Van Someren, and van der Maas (2003) has proven to be quite successful in simulating the main empirical phenomena in the area of the balance scale task. This model uses the ACT–R architecture, which combines symbolic and nonsymbolic aspects of information processing. More specifically, production rules are accorded activation values, which change in a gradual fashion like weights of network connections. The most important idea in the ACT–R model is that all balance scale rules can be based on one general rule: search for a difference between the right and left side of the scale, and if no difference can be found, the response is *balance*. The only difference young children apperceive is a difference in weights. They simply do not see the difference on distance items (i.e., items in which only the distances differ). Somewhat later in development, children also apperceive a difference in distance, a difference in sum, and finally, a difference in product of weight and distance. In the ACT–R model, there is a role for changes in knowledge *outside* the domain of proportional reasoning (e.g., learning to multiply, increases in capacity of short-term memory), as well as certain ACT–R mechanisms (e.g., production rule compilation).

The ACT–R model is not really a hybrid model; although it includes activation equations, it does not have neural properties. Hybrid models explicitly combine neural and symbolic properties. Examples have been described by Wermter and Sun (2000). Jacobsson (2005) reviewed a related line of research, which discusses techniques for rule extraction from neural networks. Another interesting line is the work on symbolic computation in stochastic neural networks (Tabor, 2002; see also Tabor, Chapter 8).

### Neural Networks with Functional Self-Organization

In the second general line of research, the focus is on constructive neural networks. Raijmakers and Molenaar (2004) discuss the learning paradox put forward by Fodor (1980). The point of this paradox is that learning new concepts necessarily involves composites of innate primitive concepts, since learning is a form of inductive inference involving hypothesis formation and confirmation. Learning fundamentally new concepts seems impossible. The addition of conceptual resources is the only way a system can become more powerful. If we want to present neural networks as viable models of higher cognition, we will have to show how such models can solve this paradox; thus, we will also show how such models can acquire more powerful concepts.

Generative neural networks, such as cascade correlation (CC) networks, are proposed as constructive neural models that escape the learning paradox (Quartz, 1993). Raijmakers (1997) criticized this claim. She argues that additions of new hidden units are discrete events (i.e., qualitative changes in the network), which represent additions of conceptual resources. It would be more convincing if a series of arbitrary and small quantitative changes in the network (e.g., a small change in learning rate) led to qualitative changes in the network.

As Raijmakers and Molenaar (2004) argue, this requires phase transitions in network behavior. Ideally, we would observe continuous changes in model parameters (learning rate, connection strengths, range of connections, etc.) leading to sudden transitions to new learning regimes, in which more powerful concepts are mastered (Molenaar & Raijmakers, 2000). Examples of work in this line can be found in Pollack (1991); Rodriguez, Wiles, and Elman (1999); and Van Ooyen, Van Pelt, and Corner (1995). In these three examples, the changes in model parameters are activity driven. The parameter changes in the model of Van Ooyen and colleagues (Van Ooyen et al., 1995) are considered to be typically maturational because they are only dependent on a global activity characteristic (i.e., the total amount of activity).

Raijmakers and Molenaar (2004) and Raijmakers, van der Maas, and Molenaar (1996) present evidence for phase transitions in a specific ART (adaptive resonance theory) network (see Grossberg, 1976). In this network, *Exact ART*, phase transitions occur between two learning regimes based on localized and distributed representations, as a function of small changes in the range of inhibitory and excitatory connections. As yet, the changes in these parameters are not activity dependent. These two types of representation in neural networks form the basis of very different types of computation (Page, 2000).

This is just one example of work in this general line of research. As neural networks are nonlinear dynamical systems, they will display all kinds of nonlinear phenomena (e.g., chaos, bifurcations, self-organization). The construction of networks in which such properties have a function in learning more powerful concepts is not easy (Tabor, 2002), but in our view, it is the most promising way to proceed.

## Conclusion: Connectionist and Dynamic Systems Approaches to Development—Are They the Cusp of a New Unified Theory or Are These Two Approaches Still too Distributed?

The subtitle of this concluding section is essentially the title of the symposium that led to this book. In this section we summarize our point of view. We, the authors of this chapter, adopt broad definitions of the connectionist and the dynamic approach. Connectionist models are dynamic systems models, and dynamical systems models are applicable to all kinds of theories and models.

In our view, the most promising part of dynamic systems theory concerns phase transitions and self-organization, which we study empirically and in simulation research. We are not impressed with the classical connectionist networks (a) because they fail to model higher cognitive processes and (b) because they do not show interesting dynamical behavior. The ultimate challenge is to construct neural networks with nonlinear properties that allow for functional self-organization via phase transitions. Clearly, the networks we propose also fail as models of higher cognitive processes, such as proportional reasoning. We can but admit this, and we do not exclude the possibility that, at least in the short run, more yields can be expected from symbolic/subsymbolic approaches, such as ACT–R and hybrid models.

We argued that there is no solution in downplaying the importance of higher cognitive functioning. Our studies of the balance scale showed us that children often use rules that strongly resemble the traditional production rules (e.g., if *equal weight* then *balance*). At the same time, there is evidence for graded processing (e.g., extreme TD effects, violations of rules, rule switching). The real challenge is to develop neural network models that capture both phenomena (i.e., traditional production rules *and* graded processing). We opt for a very fundamental research line in which we focus on self-organizing phase transitions in neural networks.

## REFERENCES

Beer, R. D. (2000). Dynamical approaches to cognitive science. *Trends in Cognitive Science, 4*(3), 91–99.

Brainerd, C. J. (1978). The stage question in cognitive-developmental theory. *Behavioral and Brain Sciences, 2*(1), 173–182, commentary 182–206, author's response 207–213.

Dawson, M. R. W., & Zimmerman, C. (2003). Interpreting the internal structure of a connectionist model of the balance scale task. *Brain and Mind, 4*(2), 129–149.

Elman, J. L. (1990). Finding structure in time. *Cognitive Science, 14*(2):179–211.

Fahlman, S. E. (1988). *An empirical study of learning speed in back-propagation networks* (Tech. Rep. No. CMU-CS-88-162). Pittsburgh, USA: Carnegie-Mellon University, Department of Computer Science. Retrieved October 27, 2008, from http://citeseerx.ist.psu.edu/viewdoc/summary?doi=10.1.1.38.549

Ferretti, R. P., & Butterfield, E. C. (1986). Are children's rule-assessment classifications invariant across instances of problem types? *Child Development, 57*(6), 1419–1428.

Fodor, J. A. (1980). On the impossibility of acquiring "more powerful" structures: Fixation of belief and concept acquisition. In J. Piaget, N. Chomsky, & M. Piattelli-Palmarini (Eds.), *Language and learning: The debate between Jean Piaget and Noam Chomsky,* pp. 142–162. Cambridge, MA: Harvard University Press.

Gilmore, R. (1981). *Catastrophe theory for scientists and engineers.* New York: Wiley.

Grossberg, S. (1976). Adaptive pattern classification and universal recoding. II. Feedback, expectation, olfaction and illusions. *Biological Cybernetics, 23*(4), 187–202.

Heinen, T. (1996). *Latent class and discrete latent trait models: Similarities and differences.* Thousand Oaks, CA: Sage.

Jacobsson, H. (2005). Rule extraction from recurrent neural networks: A taxonomy and review. *Neural Computation, 17*(6), pp. 1223–1263.

Jansen, B. R. J., & van der Maas, H. L. J. (1997). Statistical test of the rule assessment methodology by latent class analysis. *Developmental Review, 17*(3), 321–357.

Jansen, B. R. J., & van der Maas, H. L. J. (2001). Evidence for the phase transition from rule I to rule II on the balance scale task. *Developmental Review, 21*(4), 450–494.

Jansen, B. R. J., & van der Maas, H. L. J. (2002). The development of children's rule use on the balance scale task. *Journal of Experimental Child Psychology, 81*(4), 383–416.

Lewis, M. D. (2000). The promise of dynamic systems approaches for an integrated account of human development. *Child Development, 71*(1), 36–43.

McCall, R. B., & Carriger, M. S. (1993). A meta-analysis of infant habituation and recognition memory performance as predictors of later IQ. *Child Development, 64*(1), 57–79.

McClelland, J. L. (1989). Parallel distributed processing: Implications for cognition and development. In R. G. M. Morris (Ed.), *Parallel distributed processing: Implications for psychology and neurobiology* (pp. 8–45). New York: Oxford University Press.

McClelland, J. L. (1995). A connectionist perspective on knowledge and development. In T. J. Simon & G. S. Halford (Eds.), *Developing cognitive competence: New approaches to process modeling* (pp. 157–204). Hillsdale, NJ: Erlbaum.

McClelland, J. L., Rumelhart, D. E., & PDP Research Group. (1986). *Parallel distributed processing: Explorations in the microstructure of cognition: Vol. 2. Psychological and biological models.* Cambridge, MA: MIT Press.

McCutcheon, A. L. (1987). *Latent class analysis.* Newbury Park, CA: Sage.

Molenaar, P. C. M., & Raijmakers, M. E. J. (2000). A phase transition between localist and distributed representation. *Behavioral and Brain Sciences, 23*(4), 486.

Munakata, Y. (1998). Infant perseveration and implications for object permanence theories: A PDP model of the AB task. *Developmental Science, 1*(2), 161–184, commentary 184–204, author's response 205–211.

Page, M. (2000). Connectionist modelling in psychology: A localist manifesto. *Behavioral and Brain Sciences, 23*(4), 19, 443–467, commentary 467–496, author's response 496–505.

Pollack, J. B. (1991). The induction of dynamical recognizers. *Machine Learning, 7*(2–3), 227–252.

Quartz, S. R. (1993). Neural networks, nativism, and the plausibility of constructivism. *Cognition, 48*(3), 223–242.

Quinlan, P. T., van der Maas, H. L. J., Jansen, B. R. J., Booij, O, & Rendell, M. (2007). Re thinking stages of cognitive development: An appraisal of connectionist models of the balance scale task. *Cognition, 103*(3), 413–459.

Raijmakers, M. E. J. (1997). Is the learning paradox resolved? [Commentary] *Behavioral and Brain Sciences, 20*(4), 573–574.

Raijmakers, M. E. J., & Molenaar, P. C. M. (2004). Modeling developmental transitions in adaptive resonance theory [Special section/issue]. *Developmental Science, 7*(2), 149–157.

Raijmakers, M. E. J., van der Maas, H. L. J., & Molenaar, P. C. M. (1996). Numerical bifurcation analysis of distance-dependent on-center off-surround shunting neural networks. *Biological Cybernetics, 75*(6), 495–507.

Raijmakers, M. E. J., van Koten, S., & Molenaar, P. C. M. (1996). On the validity of simulating stagewise development by means of PDP networks: Application of catastrophe analysis and an experimental test of rule-like network performance. *Cognitive Science, 20*(1), 101–136.

Rodriguez, P., Wiles, J., & Elman, J. L. (1999). A recurrent neural network that learns to count. *Connection Science, 11*(1), 5–40.

Schutte, A. R., Spencer, J. P., & Schöner, G. (2003). Testing the dynamic field theory: Working memory for locations becomes more spatially precise over development. *Child Development, 74*(5), 1393–1417.

Schlesinger, M., Parisi, D., Sirois, S., Munakata, Y., Pfaffly, J., Raijmakers, M. E. J., et al. (2004). Beyond backprop: Emerging trends in connectionist models of development [Special issue/section]. *Developmental Science, 7*(2), 131–166.

Shultz, T. R. (2003). *Computational developmental psychology.* Cambridge, MA: MIT Press–Bradford Books.

Shultz, T. R., Mareschal, D., & Schmidt, W. C. (1994). Modeling cognitive development on balance scale phenomena. *Machine Learning, 16*(1–2), 57–86.

Siegler, R. S. (1976). Three aspects of cognitive development. *Cognitive Psychology, 8*(4), 481–520.

Siegler, R. S. (1981). Developmental sequences between and within concepts [Special issue]. *Monographs of the Society for Research in Child Development, No. 189, 46*(2).

Siegler, R. S., & Chen, Z. (1998). Developmental differences in rule learning: A microgenetic analysis. *Cognitive Psychology, 36*(3), 273–310.

Smith, L. B., & Thelen, E. (Eds.). (1993). *A dynamic systems approach to development: Applications.* Cambridge, MA: MIT Press–Bradford Books.

Spencer, J. P., & Schöner, G. (2003). Bridging the representational gap in the dynamic systems approach to development [Special section/issue]. *Developmental Science* 6, 392–412.

Tabor, W. (2002). The value of symbolic computation. *Ecological Psychology, 14*(1/2): 21–51.

Thelen, E., & Bates, E. (2003). Connectionism and dynamic systems: Are they really different? [Special section/issue]. *Developmental Science, 6*(4), 378–391.

Thelen, E., & Smith, L. B. (1994). *A dynamic systems approach to the development of cognition and action.* Cambridge, MA: MIT Press–Bradford Books.

Thom, R. (1975). *Structural stability and morphogenesis an outline of a general theory of models.* Reading, MA: Benjamin.

van der Maas, H. L. J. (1995). Beyond the metaphor? A review of L.B. Smith & E. Thelen (1993) a dynamic processes approach to development. *Cognitive Development*, 10, 621–642.

van der Maas, H. L. J., Dolan, C. V., Grasman, R. P. P. P. , Wicherts, J. M. , Huizenga, H. M., & Raijmakers, M. E. J. (2006). A dynamical model of general intelligence: The positive manifold of intelligence by mutualism. *Psychological Review, 113*(4), 842–861.

van der Maas, H. L. J., & Jansen, B. R. J. (2003). What response times tell of children's behavior on the balance scale task. *Journal of Experimental Child Psychology, 85*(2), 141–177.

van der Maas, H. L. J., & Hopkins, B. (1998). Developmental transitions: So what's new? *British Journal of Developmental Psychology, 16*(1), 1–13.

van der Maas, H. L. J., & Molenaar, P. C. M. (1992). Stagewise cognitive development: An application of catastrophe theory. *Psychological Review, 99*(3), 395–417.

van der Maas, H. L. J., & Raijmakers, M. E. J (2000). A phase transition model for mother-child interaction: Comment on Olthof et al., 2000. *Infant and Child Development, 9*(2), 75–83.

van der Maas, H. L. J., Jansen, B. R. J. & Raijmakers, M. E. J. (2004). Developmental patterns in proportional reasoning. In A. Demetriou & A. Raftopoulos (Eds.), *Cognitive Developmental Change Theories, Models and Measurement*, 118–156. Cambridge University Press.

van Geert, P. (1991). A dynamic systems model of cognitive and language growth. *Psychological Review, 98*(1), 3–53.

Van Maanen, L., Been, P., & Sijtsma, K. (1989). The linear logistic test model and heterogeneity of cognitive strategies. In E. E. Roskam (Ed.), *Mathematical psychology in progress* [Series: Recent research in psychology] (pp. 267–287). New York: Springer.

Van Ooyen, A., van Pelt, J., & Corner, M. A. (1995). Implications of activity dependent neurite outgrowth for neuronal morphology and network development. *Journal of Theoretical Biology, 172*(1), 63–82.

Van Rijn, H., van Someren, M., & van der Maas, H. L. J. (2003). Modeling developmental transitions on the balance scale task. *Cognitive Science, 27*(2), 227–257.

Wagenmakers, E.-J., Molenaar, P. C. M., Grasman, R. P. P. P., Hartelman, P. A. I., & van der Maas, H. L. J. (2005). Transformation invariant stochastic catastrophe theory. *Physica D: Nonlinear Phenomena, 211*(3–4), 263–276.

Wagenmakers, E.-J., van der Maas, H. L. J., & Molenaar, P. C. M. (2005). Fitting the cusp catastrophe model. In B. S. Everitt & D. C. Howell (Eds.), *Encyclopedia of Statistics in Behavioral Science: Vol. 1. A–D* (pp. 234–239). New York: Wiley.

Wiggins, S. (1990). *Introduction to applied nonlinear dynamical systems and chaos* [Series: Texts in applied mathematics, Vol. 2]. New York: Springer.

Winfree, A. T. (1980). *The geometry of biological time* [Series: Biomathematics, Vol. 8]. New York: Wiley.

Wermter, S., & Sun, R, (2000). *Hybrid neural systems* [Series: Lecture notes in computer science]. Berlin, Germany: Springer.

Zeeman, E. C. (1976). Catastrophe Theory. *Scientific American, 234*(4), 65–83.

# CHAPTER 16

## Dynamic Systems and the Quest for Individual-Based Models of Change and Development

*Paul van Geert and Kurt W. Fischer*

The aim of this chapter is to discuss the question of how dynamic systems theory can be fruitfully applied to the development of the kind of phenomena and variables that have interested the current authors for a long time. Examples of these phenomena are (a) the development of language, including the development of the lexicon and syntactic and grammatical knowledge and skill, (b) the development of cognition and thinking, including the emergence and acquisition of cognitive skills and knowledge in various domains, (c) the development of reflective judgment, including metacognition and social understanding, and (d) the development of social skills and behavior. Behind all these phenomena are the development of context-specific but overarching principles of skill formation, such as principles of relationships, systems of relationships, and so on.

Viewed in the context provided by the current volume, the answer to the question of how to apply dynamic systems theory to these phenomena is not trivial. Some contributors to the conference on which this book is based argued for a particular type of dynamic systems model, which both (a) is firmly rooted in a physical system, such as the body or the brain (the dynamics of which are, in principle, relatively well understood), and (b) describes a real-time interaction loop between an organism and its direct environment. Hence, psychologically relevant properties are not in the person alone, but in the person–environment dynamics.

We will argue that application of dynamic systems theory to presumed internal properties, such as the development of self-perception and identity, reflective judgment, or grammatical knowledge and skill, works with a different kind of dynamic systems model but is also built upon these two characteristics. The behaviors and cognitive properties being modeled are inherently embedded in a physical system and based on a real-time interaction loop between organism and environment. It will come as no surprise that the authors of the present chapter believe the dynamic systems view can be fruitfully applied to these subjects and topics—complex and high-end though they may be—in a way that is considerably more than metaphorical and illuminates development in important ways.

### Grand Old Views of Development and Learning

#### Etymological Analysis of *Development*

Before answering the question of how dynamic systems theory can be fruitfully applied to development in the broad sense of the word, it is good to remind ourselves of what the term *development* actually means and how it relates to associated terms such as *change*, *learning*, *acquisition*, *emergence*, *evolution*, and so on.

Etymologically, the term *development* means un-wrapping, that is, taking something out of its original wraps, thus uncovering something that is present at the beginning but concealed. There is a connotation of an inner logic in the term *development*, that is, the intrinsic necessity of a certain outcome. As a verb, *develop* appears in a transitive as well as in an intransitive, more precisely reflexive, form. For instance, you can develop a photo (people used to do these things before the digital camera) or develop a real estate project. However, it is a bit odd to say that a parent develops a child. One can say that the child develops grasping, for instance, or that the child develops. In the latter sense, *develop* is used as a reflexive verb, that is, a verb where the agent and patient are the same.

This historical and colloquial meaning of the term development is reflected in the structure of the classic grand theories of development, that of Piaget, for instance. These theories are implicitly based on a retrospective approach, that is, an approach that takes the properties of a final state (i.e., an outcome of development) as its conceptual starting point and reasons back, so to speak, to find the state or states that do not yet have these properties. These states must, by logical necessity, precede the final state. The properties that define each possible state are conditionally linked. For instance, Piaget's theory explains why you cannot think about formal contents if the thinking itself is not operational; hence the emergence of operational thinking precedes the emergence of formal thinking because it is a precondition to the former. These properties and their conditional relationships define a specific temporal arrangement of possible developmental states. The first author (van Geert, 1987a, 1987b, 1987c) has analyzed various classic developmental theories in the form of generative *rewrite rules* that very much resemble the principles of current symbol dynamics, which is a specific branch of (formal) dynamic systems theory (Dale & Spivey, 2005; Jaeger, 1999).

The classic developmental theories tried to explain the properties of the developing person and the conditional relationships among these properties by focusing on the major developmental mechanism(s). These mechanisms were in most cases internal, driven by the developing person himself. Thus, development is an internally driven process of growing toward an end state that is, in some way, potentially present from the beginning. This is not, in any way, a form of nativism, claiming that the end state (or at least some essential core of it) is present from the beginning. *Potential* is derived from the Latin word for *power*, thus meaning that something at the beginning has the power of developing into some particular final state through transformations that unwrap the potential by making it actual.

Several chapters in this book have already discussed the main properties of dynamic systems, and we can thus ask ourselves to what extent the classic view on development relates to the notion of a dynamic system. The answer, we believe, is that there is in fact a considerable overlap or similarity, in that *development is a typical example of a dynamic system*—if one is prepared to condone the terminological remnants of prescientific times that may still color the use of the term *development*.

### Dynamic Systems and the Grand Old View of Development

How does the dynamic systems approach relate to this *unwrapping* scheme and the notion of a reflexive activity (that is, something one imposes on oneself)? Dynamic systems theory has replaced this primitive notion of *unwrapping* by notions of emergence, self-organization, and attractors. Attractors are states to which the dynamic system is driven by internal necessity, that is, as the necessary consequence of a dynamic mechanism (the *growth term*) that governs the change of the system. In this sense, it has given a formal and precise meaning to the intuitive notion of potentiality—the power or possibility to move through a series of steps and transformations toward a state that has some permanence and stability. Dynamic systems theory has made the developmental notion of a self-imposed process explicit by specifying the system dynamics empirically and formally (for instance, by means of models).

The theoretical distance between the dynamic systems approach and the classic stage theories of development is perhaps smaller than that between dynamic systems and the majority of current empirical studies in development (van Geert, 1998b). As described in the preceding section, the stages of classic theories are not trivial qualifications but result from dynamic ordering principles, based on postulated, major developmental mechanisms. This ordering of stages thus defines a sort of crude yardstick for development, the tick marks of which are the stages as distinguished by the theory at issue. In a similar vein, a dynamic system describes the change of a developing property specified in a geometric space. In a physical system, the geometric space can be the real physical space or some property such as temperature or velocity. The simplest possible space is one-dimensional, in the sense that it specifies the change in one particular dimension, such as the child's lexicon, mathematical problem solving, or popularity in a peer group. High-dimensional geometric spaces, referring to basically every relevant dimension or variable useable for describing development, are also possible and useable for dynamic modeling (van Geert, 1998a). Current dynamic systems investigations in social, dyadic interaction (e.g., between mother and child) often employ two-dimensional categorical spaces, one dimension for each participant (Hollenstein & Lewis, 2006). An important challenge for dynamic systems theory in development is to specify this geometric space and its constituting dimensions in ways that make description, explanation, and research of developmental phenomena theoretically and empirically adequate. (We will come back to this issue in section The Metric of the Dynamics and Developmental Rulers, defining the notion of *developmental rulers.*)

## Development, Learning, and Teaching

Development has always been seen as different from, but at the same time intimately related to, learning and teaching. The grand old view of learning and teaching implies the notion of transmission. Thus a *teacher* (of whatever kind) transmits knowledge or skills to a learner. Transmission implies the bringing of something (e.g., a knowledge content) from one place to another (e.g., from an adult's brain to a child's brain). It requires the presence of a sender or transmitter and a receiver of the transmitted content. Connectionist models have considerably altered the transmission view on learning and have replaced it by a model of networks that self-organize in response to the input (as in supervised and unsupervised learning and statistical learning). However, the self-organizing formation of connections in a network is the way in which the transmission process is taking concrete shape. It is not literally transmission of content that takes place, but the formation or construction of an internal structure that in some way replicates certain structural properties or regularities of the environment.

For instance, a connectionist model of language learning explains the learning by means of networks that self-organize under the influence of linguistic inputs, resulting in a network organization that is capable of producing language itself, according to the rules that governed the language productions that were taken as input (for instance, see Plunkett & Juola, 1999). Thus, connectionist networks are capable of picking up structure from the environment, in addition to being able to build up any form of statistical regularity between the network's inputs and its outputs. Development, in its basic meaning, implies the unwrapping of an internal potential, not the mere appropriation—in whatever form—of an external content. The question is, of course, whether this distinction is anything more than an artifact. There is no process of development thinkable that does not also require the acquisition of external content. On the other hand, the initial architecture of a connectionist network greatly determines the way in which inputs affect the network's internal organization and its actual relating of inputs to outputs (for instance, see Elman et al., 1996). It is, thus, imaginable to constrain or specify a network in a way that the regularities it establishes between its actual inputs and outputs become relatively independent of the exact properties of its input and are, thus, *underdetermined* by the latter. This is functionally similar to saying that the network's output is *innate* or that inputs are no more than triggers of internally driven epigenetic processes (Chomsky, 1980; Fodor, 1975). However, in connectionist networks, such eventual nativism makes no reference to a representationalist form of innateness and does not imply that a representation (e.g., a grammar of one's language) is present at the start and then at some later time comes out in the form of explicit performance.

The basically constructive nature of development (i.e., using its inputs to construct its own trajectory and content), which connectionist networks imply, is also endorsed by Piagetian and neo-Piagetian theorists. They have argued that development occurs in the form of organizational/reorganizational and constructive processes, not just internally copying what is out there or incorporating what is transmitted from some outside source (Case, 1991; Fischer, 1980; Fischer & Bidell, 2006). For instance, when children start to understand the world in terms of relationships between concepts, it is not because they have somehow captured the notion of relationship as a thing present in the outside world, as if *relationship* were some sort of statistical regularity. In fact, the notion of relationship—whatever its actual psychological form—is constructed in an active attempt to understand and control the regularities of the world that the child encounters. In short, the distinction between development and learning is more gradual than discrete, given the fact that both learning and development rely on the working of a single underlying network structure, which is the brain. However, the gradualness of the distinction and the explanatory success of connectionist networks does not imply that it is therefore better to no longer make the distinction at all and to subsume all forms of learning and development under the same type of connectionist network operation. For instance, understanding teaching as a process of transmission requires not only an understanding of how the child processes the transmitted information, but also of how the teacher processes the information about the learning progress—or not—made by the child (van Geert & Steenbeek, 2005b). This transactional process between learner and teacher can be modeled by a dynamic system, linking the teaching of the teacher to the learning of the learner and also the other way round. The fact that the teacher and learner can be represented in the form of two interacting connectionist networks is no substitute for or reduction of the fact that teacher and learner constitute a long-term dynamic system, the properties of which can be modeled and understood (relatively) independent of the way the networks embody this dynamics.

## Dynamic Systems and Their Application to Developmental Processes

### A General Definition of a Dynamic System

A convenient definition of a dynamic system is "a means of describing how one state develops into another state over the course of time" (Weisstein, 1999, p. 501). It can be described by simple equations such as $x_{t+1} = f(x_t)$, with $f$ standing for some sort of function that yields a new value of $x$ given a value of $x$ at time $t$.[1] In order to have a dynamic systems model, one thus needs three things. The first is some sort of $x$ (i.e., a description of something that changes). The second is some sort of metric describing the set of all possible states of $x$. The third is an explicit description of the form of $f$, the function that relates a preceding state of $x$ to a succeeding state of $x$.

### The Dynamic Variable: What Develops during Development?

To begin with, $x$ can be anything that is sufficiently real to be the subject of a dynamic systems model. Thus, $x$ can be the number of carrots in a field, or the physical force exerted on a force plate, or a person's level of skill in making reflective judgments. That is, $x$ need not

[1] In the current volume, readers are likely to encounter two different ways of representing the form of a dynamic system. The present format uses a notation where time proceeds in incremental steps (+1), where (1) can represent any convenient duration. Another form uses the notation of differential equations, with time $t$ represented by real numbers and infinitesimally small increments in $t$. The first format is customarily called a *map*, the other a *flow*. For our current purposes, the distinction between these representational formats is a matter of taste.

be a physical thing and neither does it need to be directly reducible to known physical states of a system. Take reflective judgment as an example. We do not know what happens in the brain when a person makes reflective judgments, but we would certainly not deny that people do make reflective judgments and that some people are better at it than others. We can construct tests to measure reflective judgments, help students improve their reflective judgment, and so forth, all of which is sufficient support of the fact that reflective judgment is real and not a member of the family of unicorns and other fantastic creatures. Of course, the *real-ness* of reflective judgment is of a different kind than the *real-ness* of, say, temperature or a telephone booth, but the difference in kind does not imply that reflective judgment is a metaphor, something vague, or ill defined. We shall go further into this issue in our discussion of *unsolicited ontological claims* (section Unsolicited Ontological Claims).

### The Metric of the Dynamics and Developmental Rulers

To have a dynamic system, the metric—the second aspect of a dynamic systems model—must be something like a geometric dimension or space where the notion of distance between two states of $x$ can be given sufficiently concrete meaning. More precisely, the dimension should be sufficiently similar to a real-number line in order to describe measurable distances and meaningful movements along the distance axis at issue. Can a property such as reflective judgment, to name but an example, be described by means of a dimension that looks like a real-number line? The assumption that psychological properties like reflective judgment or intelligence can indeed be fruitfully described by (at least a certain stretch) of the line of real numbers (or integers, for that matter) is one of the main tenets of psychometrics and dates back to the beginning of intelligence measurement.[2] The point is not whether phenomena like reflective judgment or intelligence are of the kind of physical phenomena the magnitudes of which demonstrably vary in accordance with a real-number line—it is most likely they are not. What is at stake is whether variation in these phenomena can be sufficiently approximated by a real-number line to make the application of a dynamic systems model more than an empty exercise—it is most likely that this is indeed possible.

This dimension in the form of a real-number line is a mathematical object. It relates to empirically definable properties in the following way. First, it must relate to a dimension, number line, or series of ordered categorical distinctions that can be used to describe developmental processes. Dependent on one's theoretical goals, this dimension can be general and encompassing or highly context-specific. Whatever its scope, however, it must be possible to specify a person's developmental level—whatever is implied by *level*—as a region or point on the dimension at issue, and thus to specify a person's developmental trajectory as a temporally ordered series of such regions or points. The second author of this chapter has coined the term *developmental ruler* to describe such descriptive dimensions, which can function as yardsticks for specifying developmental levels (Fischer & Rose, 1999; Rose & Fischer, 1998). Developmental rulers must be linked with measurable or observable phenomena, for instance, results from a test, observations, or experiments, in some explicit way.

A developmental ruler corresponds with a sequencing of developmental stages. Instead of identifying the stage with an identifiable phase in a person's life (i.e., classic Piagetian interpretation of stage), stages should be identified as the major ticks on the developmental ruler. The question of whether these stages correspond with a real scale has been strongly supported by empirical research that shows discontinuities between levels (or stages) in both long-term development of optimal level performance and short-term variations in performance in one

[2] In the behavioral and developmental sciences, the relationship between measurements and an underlying mathematical object such as a real-number dimension is often a little vague, but such relationships can fruitfully be modeled by means of fuzzy-logic descriptions (van Geert, 2002, 2004; van Geert and van Dijk, 2003).

session (Dawson-Tunik, Commons, Wilson, & Fischer, 2005; Fischer, 2008; Fischer & Bidell, 2006). For instance, Rasch analysis shows the same scale marked by discontinuities in adult speech and writing that is evident in child and adolescent development.

In addition to using developmental rulers *sensu stricto*, one can use simple quantitative variables, such as the number of words counted in a child's lexicon, to specify the space of developmental dynamics, provided of course, that such quantitative variables can indeed be meaningfully linked to developmental processes.

### Dynamic Function or Growth Term and Its Level of Application

The third aspect of a dynamic systems model concerns the form of the *f* or *function* in the basic dynamic systems equation. This function is called the *growth term* and consists of a fixed principle that operates on the value of any $x$, as described through its descriptive dimension(s) or developmental ruler(s), and that produces another value of $x$, given the previous value of $x$ and any other influence that enters into the dynamic growth term. It must be a function that makes the value of $x$ (e.g., reflective judgment at a particular moment for a particular individual) move along a specific developmental ruler (e.g., the ruler for reflective judgment). Although the function itself may describe decrease, increase, or stagnation, it must be based on well-established growth mechanisms, which we will discuss later in this chapter (for instance, see, section An Example: A Dynamic Growth Model of Language Development).

The growth term applies to a specific timescale. Although we will discuss this issue further in this chapter (section Timescales and Levels of Aggregation), it is important to note that developmental issues usually invoke a distinction among the short-term timescale of real-time human action and the long-term timescale of development. These timescales are characterized by different types of phenomena and different types of dynamics. For instance, a dynamic model describing a playground interaction between two children, taking place over a short-term timescale (e.g., the course of several minutes), including processes of selection of the interaction partner (e.g., turn taking), and interacting (e.g., interpreting one's own and the other child's emotional expressions), differs quite considerably from a dynamic model that describes changes in peer selection and preference and emotional communication over the course of years (Steenbeek & van Geert, 2005, 2007, 2008). Events on the two timescales are related: short-term peer interactions contribute to long-term changes in, for instance, peer preference and status. Long-term changes in peer status and preference, in their turn, directly contribute to the properties and unfolding of short-term interactions (see also section Timescales and Levels of Aggregation).

Finally, although the nature of the dynamics (that is, the particular form of the growth term) depends on the timescale at issue, the use of a developmental ruler is not necessarily confined to the long-term developmental timescale. Across or within action sequences, especially those that involve more demanding problem solving, people—children and grown-ups alike—may fluctuate greatly in the developmental levels instantiated in their successive actions. For instance, they may fluctuate back and forth between a purely sensorimotor approach (with no reflection at all) and a highly abstract reflective approach. These short-term fluctuations may be described by means of the same developmental ruler or yardstick that is used to describe long-term development.

### Discussion Regarding Dynamic Systems

To summarize this discussion, in its most general form, a dynamic systems model describes the change in a state of a variable of interest, over a space of possible states, which are described by one or more dimensions that take the form of developmental rulers. We see this general definition as a starting point for the development of many kinds of specific dynamic systems models. Examples of such more specific models are *embodied-embedded action models*, which describe the immediate online emergence of behavior that takes place in a closed loop between individual and context, which is the sort of model put forward by Thelen, Smith, Schöner, Spencer, and others (for example, Schöner & Thelen, 2006; Smith, 2005; Smith & Thelen, 1993; Spencer, Smith, & Thelen, 2001; Thelen & Smith, 1994, 1998). Another example of a specific dynamic systems model is a connectionist model, whose description of the evolution law takes the form of connectionist networks operating under specific universes of input. Yet another type of specific dynamic systems model is the type of dynamic growth model developed by the current authors. Characteristic of this type of model is that it tries to capture the principles that govern the long-term change of any sort of variable that can be of interest to the researcher (general developmental level, size of the lexicon, popularity among peers, use of correct syntactic forms, etc.).

## Unsolicited Ontological Claims

### Terms and Potential Confusions

Sometimes the term *dynamic systems theory* is reserved only for the first type of model, that is, the embodied-embedded action model, with the second, connectionist type viewed as a useful extension of dynamic systems in order to explain long-term changes (for instance, see Thelen & Bates, 2003; see the special issue of *Developmental Science*, Spencer & Thelen, 2003). The third type, which for simplicity we refer to as the *dynamic growth model*, has been criticized as metaphorical, less grounded in actual behavior and empirical data (Spencer & Schöner, 2003; Thelen, 1996).

The most serious criticism is probably that the dynamic growth model is, in fact, implicitly representationalist and thus not a dynamic model at all (Thelen & Smith, 1998; van der Maas, 1995; van Geert, 1996; Witherington, 2007). The representationalist stance, also known as the *computational-representational understanding of mind*, or *information-processing theory*, states that thinking can best be understood in terms of representational structures in the mind and in terms of computational procedures that operate on these structures (Thagard, 1996). That is, in order to explain thinking, you need internal entities.

An important issue in Thelen and Smith's dynamic systems thinking is that it enables the researcher to get rid of the internal entities hypothesis that characterized representationalist and information-processing accounts (Smith & Thelen, 1993; Thelen & Smith, 1994, 1998). For instance, a baby's searching for hidden objects is not governed by internal computations on representations and does not involve a special representation called the *object concept* (Smith, Thelen, Titzer, & McLin, 1999). Such explanation, according to Thelen and Smith, is typical of Piaget's theory, or of information-processing theory (see also Smith, 2005). An important part of the discussions that took place during a workshop on dynamic systems hosted by the Netherlands Royal Academy of Sciences in 1997, focused on the criticism that the approach of Dutch investigators to dynamic systems, in particular growth models and catastrophe models, was in fact a version of representationalist, information-processing-like theory that revived the notion of internal representations or structures outside the actual space–time person–environment dynamics. As participants to that discussion, the current authors' view is that the criticism was based on an unnecessary identification of certain words (for instance the *object concept* or *representation*) with a particular meaning (an internal representation featured in some sort of internal information-processing machine). We feel that such an identification or interpretation is not warranted. The next two sections will provide some background for this.

### The Unsolicited Entity Claim

In his book, *The Concept of Mind* (Ryle, 1949/2002), the British ordinary-language philosopher Gilbert Ryle (1900–1976) warned against interpreting mental terms as referring to entities in the form of mental mechanisms. Such entities are almost automatically interpreted as little *machines* that cause or produce the behavior seen as an expression of these mental terms. Smith and colleagues (Smith et al., 1999) provide a detailed empirical account of the processes involved in retrieving hidden and replaced objects by babies and thus give a concrete example of the fact that the term *object concept* does indeed *not* refer to a little representational engine in the child's head, generating solutions to object concept problems in case the baby is confronted with one. What Ryle meant to say is that a mental term (such as *concept*, *desire*, and so on) tends to invoke an association with some internal entity. However, if you believe that using a mental term such as *concept* automatically makes you adhere to the unproved belief that it is some internal representational engine that manufactures behavior and problem solving, you are making an unsolicited ontological claim. It is a claim about the nature and existence (hence *ontological*) of concepts that have not been asked for; and that claim is not necessary for justifying the use of that term in natural or scientific language. (The claim is hence unsolicited.) That is, you can perfectly say that a child has a number concept, or not, without making the unsolicited ontological claim that what the child has in this case is some sort of causal entity roaming around in his mind. (Identifying a *concept* or *representation*, for that matter, with an internal entity operated upon by computational rules is an explicit theoretical choice, which one makes if one adheres to a computational representationalist approach, as in information-processing theory.) In its most fundamental meaning, *concept*, coming from the Latin *concipere*, implies an activity, namely to take in and hold, and is directly related to an Indo-European root meaning to grasp (which is still literally visible in the translation of *concept* in several Germanic languages). Hence, if a mental term such as *concept* implies any tacit ontological claim, it is more naturally related to the notion of carrying out some sort of activity with a particular result (i.e., *holding*) than to a notion of a causal–mental entity.

In short, colloquial words such as *concept* (or *representation*, for that matter) are relatively undefined and vague terms. In order to be used scientifically, they require further empirically supported description. One such description, the computational-representationalist choice, entails that they are like isolable entities that produce particular behaviors and thinking. Being such entities, they exist independent of the experiences and actions in which they serve.

Another description is that they are not entities, but processes, patterns of perception and action over time, and thus do not exist independent of the actions and experiences in which they are expressed. The latter definition is clearly the one favored by a dynamic systems approach.

However, the actual use of a term such as *concept* or *representation* in scientific discourse is in principle neutral with respect to the sort of definition or description that the user wishes to assign to the term at issue. Thus, one can very well talk about the development of *representations* and *concepts* without making the implication that *representations* or *concepts* are independent entities.

### The Unsolicited Internality Claim

Just like using mental terms such as *concept* and *representation* cannot be identified with implicitly making a claim that such terms refer to entities (see section Terms and Potential Confusions), the use of such terms can neither be identified with making the claim that they are entities of an internal nature, located in a person's mind (and eventually, by implication, represented in the form of a static structure in the brain). What is the nature of *concept*, *representation*, or comparable mental terms, if they are not defined as internal entities (fixed patterns, information-processing rules, etc.)?

Various chapters in this book argue that what we tend to see as mental entities—*concept*, *representation*—are in fact deeply contextual and thus not *internal* at all. We agree with that point of view, but not with the eventual implication that *concepts*, *representations*, and so on, as such, do not exist. What we wish to say is that by using mental terms, such as *concept* or *knowledge*, one does not implicitly make the ontological claim that these things are in essence internal and, in essence, independent of a context. In fact, all terms referring to aspects and properties of human actions, including the covert action of *thinking*, are, in their very origins, deeply contextual. Take for instance the meaning of *grasp* in "grasping a kilo of lead" and "grasping a kilo of feathers." The form of the grasping is entirely determined by the thing grasped, and the properties that make both movements an instance of *grasping* do not exist in the form of literal similarities but of resemblances in terms of form–function relationships and results. In its very origin, a mental term such as *concept* obeys the same implicit embedding of context. For instance, if you are saying that a child has a certain *concept*, you are implicitly referring to a property of a child–context system, with *context* being the child's characteristic environment. There is no intrinsic contradiction in saying that the development of a *concept* in a child is in fact the development of a child–context system. The context is not independent of the person, because the person seeks certain contexts actively, constructs certain contexts actively, and perceives certain contexts actively. (We will come back to this issue in section Can *Context* be an Independent Variable?) By the same token, the person is not independent of the context, since what the person does and can do, feels, perceives, and so on, depends on the person's current context.

The interplay between person and context is very intricate; it is a continuous dynamic loop (a "strange loop" in the words of Hofstadter, 2007), thus giving a clear ground to the conjecture that a person is in fact a person–context system. What we mean if we speak about the person *an sich* (*as such, in itself*) are certain similarities and regularities in the person's actions that we perceive over diverse but person-characteristic contexts and over sufficiently long stretches of time (van Geert, 2002, 2004). The question is whether the observation that development applies not to persons as isolated entities but to intricately interwoven person–context structures, also implies that there is no other way to study development than to study a particular kind of model of the dynamics of person–context interaction. The study of this person–interaction loop is the topic of dynamic systems theory as defined by Thelen, Smith, Schöner, Spencer, and others, and it makes an important contribution to our understanding of thinking, acting, and development. However, from a different perspective, it is also the topic of the dynamic growth models that the authors of this chapter have articulated, which explicitly deal with how speech, concepts, and skills involve a person interacting with contexts, based on an explicitly hierarchical view of levels of skill and levels of explanation.

## Complex Systems and the Superposition of Properties

We began section Unsolicited Ontological Claims with the quibbles among Dutch and Indiana adherents of dynamic systems theory, focusing on the issue of whether dynamic growth models are implicitly representationalist, thus invoking internal, independent mental entities, and in that sense, are either not true dynamical systems or primarily metaphorical. In the preceding sections, we have explained that the use of terms such as *concept, representation*, and the like does not logically imply the notion of internal, independent entities, and that they can just as well be used to refer to phenomena that are local, temporal processes of organism–context loops.

The question remains, however, whether this is an issue of either–or. As far as human experience and action is concerned, we are dealing with a very complex system of brain–body–environment dynamics. Such complex systems can have surprising properties, and although the matter is highly speculative, it is likely that they can be characterized by superposition of seemingly incompatible properties and duality (van Geert & Steenbeek, 2005a, 2005b, 2008). Although the comparison is certainly somewhat immodest, the entity-versus-process distinction in psychological phenomena might be comparable to the wave–particle distinction in physics. Physical objects show a wave–particle duality, for instance, properties of particles as well as of waves, which are seemingly incompatible conceptualizations. Maybe it is a good working hypothesis to conceive of the entity-versus-process debate as an inadequate conceptualization of the complexity of experience and action, which is better served by a view that presumes a form of superposition and duality of these two aspects that, at present, are not yet sufficiently understood (for alternative conceptualizations of this issue, see Spencer & Schöner, 2003). Accepting the assumption of such superposition, however, does not mean that one also accepts that all types of models—classic entity-based process accounts such as information-processing models, dynamics systems models, connectionist models—are thus, by definition, equally valid. Validation is a matter of empirical support in addition to theoretical coherence.

## Explaining Phenomena and Taking Loans from Other Explanations

### Hierarchical Levels and the Principle of Explanatory Loans

The phenomena that developmentalists deal with are hierarchically nested all the way down. Thus, the level of human biological evolution encloses the levels of historical developments of human cultures. Cultures encompass the sequences of individual life trajectories (which consist of ontogenetic processes that entail sequences of concrete actions). Life trajectories involve close person–context interactions, which in their turn are based on neurophysiological processes consisting of chemical and physical processes at atomic and subatomic scales. Any of these levels can be taken as the starting point of a dynamic model, that is, a model that describes the time evolution of some variable of interest that is characteristic of the level of aggregation chosen. These levels of aggregation, or levels of description, correspond with different timescales, and different types of dynamics. Any level of description takes a loan, so to speak, on any of the other levels. For instance, a dynamic model of a phenomenon playing on the level of historical time, takes for granted the proper working of mechanisms and phenomena at the timescale of biological evolution (the level above it) and at the timescale of ontogenesis (the level below it). Similarly, a dynamic model describing evolution of variables or properties at the ontogenetic timescale, takes for granted the proper working of phenomena at the timescale of real action (involving close interaction loops between persons and their environments) and at the historical timescale of the development of cultures and cultural artifacts. But whatever one's target level of description, there is always an implicit loan taken on other levels.

In many cases, the *loan* is guaranteed, in the sense that the link with adjacent levels is well understood. For instance, a biological model of population growth in a particular species is based on a good understanding of procreation mechanisms in individual members of the species. A model of particular person–context interaction (for instance, the infant's A-not-B error) is backed up by an understanding of the way in which the brain, in principle, processes the information involved (see for instance Schutte & Spencer, 2002; Schutte, Spencer, & Schöner, 2003; Spencer & Schöner, 2003). Models of the short-term dynamics of dyadic interaction in children (for instance, in play) can be related to long-term models of the development of social status and peer selection (Steenbeek & van Geert, 2005, 2007, 2008). Models of the long-term development of complicated thought processes (such as reflective judgment or a person's view on the self in relationships) are based on models of mechanisms in brain development (involving the formation of new neural networks for each developmental level; see Fig. 16.1; Fischer, 2008; Fischer & Bidell, 2006).

### An Example: A Dynamic Growth Model of Language Development

Although the availability of conceptual links to underlying (or, for that matter, overarching) levels of explanation, such as the brain level in models of action, significantly contributes to the explanatory value of the model at issue, the availability of such links is not a *sine qua non* for scientific explanation, as such. Take, for instance, the acquisition of words and the development of the lexicon. The first author of this chapter has formulated a model of long-term lexical development (and early language development, in general) based on dynamic principles of growth (for instance, see van Geert, 1991, 1993, 1994). These principles have been inspired by general biological principles of growth featured in population ecology, in which they have a firm biological basis (Kingsland, 1985). The application of these principles to developmental

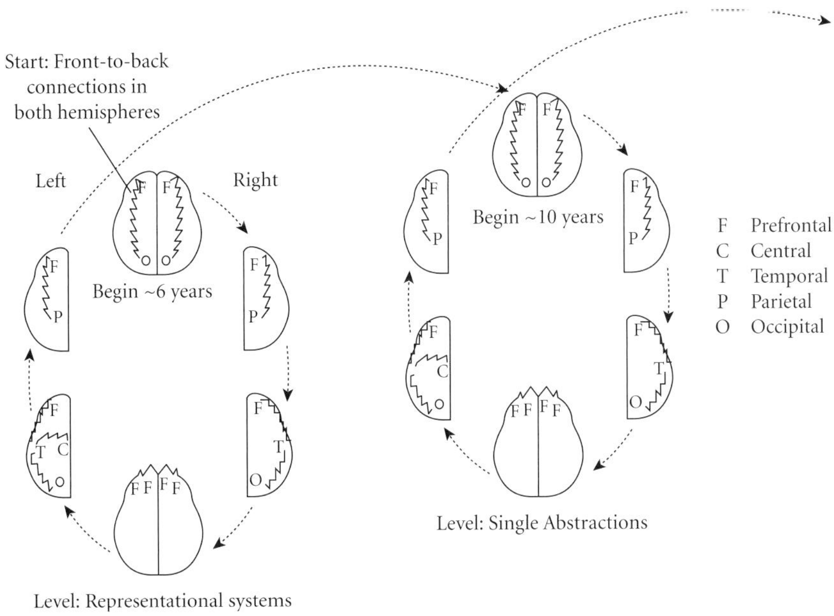


**Figure 16.1. Cortical network cycle for two successive cortical levels**
According to the neural-network model of cognitive development, connections between cortical areas are revised in a systematic growth cycle. For each level of skill development, connections between cortical areas grow systematically, with peak growth moving around the cortex in a generally clockwise pattern. For each repetition of the cycle, neural networks are reworked and retuned, creating the neural grounding for a new cognitive developmental level. The two cycles here are for the levels of (1) representational systems, which first emerge at approximately 6–7 years of age, and (2) single abstractions, which first emerge at approximately 10–12 years of age. Jagged-line connections mark the areas showing the leading edge of growth of connection, which begins with maximal growth of connections between the frontal and occipital cortexes in both hemispheres. Next, maximal growth moves through the right hemisphere and then through the left. For the left hemisphere, connections between the middle and back are more prevalent than for the right hemisphere; the temporal-central connection for the left hemisphere is shown as an example of that difference.

psychology is thus metaphorical in the literal sense of that word, meaning that these principles have been carried over from another field. To put it differently, there is a loan from another scientific discipline, population ecology. How can the loan be paid back?

We can pay our debts by showing that the growth dynamics provide a parsimonious, spare, generic, and justifiable theory of lexical and language growth. It is a simple or *parsimonious* theory in that it says that the change in some linguistic variable (e.g., the lexicon) over time, ΔL, is a simple function of the level already attained, L, and of a limited set of resources, C, that can be treated as a constant. In the case of the lexicon, a crucial aspect of the resource constant is the lexicon used in the environment to the extent that it is accessible to the child. Finally, ΔL depends on a set of coupled variables, {V}, that can be of any type and that can either favor or counter the growth of L.

The elementary format of the model can be written as $L_{t+1} = L_t + \Delta L_t$, with $\Delta L_t$ (i.e., the change in L) being equal to $f(L_t, C, \{V_t\})$ (with $f$ meaning *is a function of*).

In the simplified condition where there are no interacting variables, ΔLt is calculated based on the Verhulst growth equation, whioch takes the form $\Delta L_t = r\, L_t\, (1 - L_t/C)$. The effect of additional variables, e.g. variable V, can be expressed as m $L_t$ $V_t$. The parameters r and m are growth parameters, and by convention, r refers to the basic growth parameter affecting L (see van Geert, 1994; Fischer & Bidell, 2006).

It is a *generic* theory in that it offers an approach to a wide variety of phenomena (e.g., various kinds of linguistic and cognitive variables) that can be given a concrete form by specifying the variables at issue and the exact ways in which they relate to each other, by estimating the values of the interaction parameters.

Finally, it is a theory that can be *justified* in various ways. First, we can show that the general principle of L-C-{V} dependence is supported by the available theoretical and empirical literature, for instance, on lexical acquisition or the growth of sentence length (see for instance Bassano & van Geert, 2007; van Geert, 1991, 1994; van Geert & Steenbeek, 2005a, 2005b). Second, we can link these principles to an analysis of the underlying biology, that is, a model of how the brain accomplishes the task of learning new words (Fischer, 2008; Fischer & Bidell, 2006). Third, we can show that specific models based on these general principles of growth provide good predictions of qualitative and quantitative properties of developmental processes. For instance, nonlinear dynamic growth models have predicted developmentally discontinuous trajectories at high growth rates and smooth trajectories at low growth rates, consistent with the data (Fischer & Kennedy, 1997). Another model has predicted normative developmental functions for various capacities in a domain with socioeconomic differences in the functions (Case & Okamoto, 1996). Nonlinear growth models generate patterns of stepwise growth, including temporary regressions, which are consistent with data from various academic fields (Fischer & Bidell, 2006). In particular in the field of language development, dynamic growth models explain the quantitative forms of the empirical growth curves. Examples include early lexical development (van Geert, 1991, 1993,1994), lexical development in relation to growth of plurals (Robinson & Mervis, 1998), growth of closed-class words (Ruhland & van Geert, 1998), and pattern of growth and decline of sentences of various lengths (Bassano & van Geert, 2007; see Fig. 16.2).

By extending the growth model (which represents the level of a variable by a single vector) to a vector field model (which represents the level of a variable as a wave of levels), a variety of empirical developmental phenomena can be predicted, including discontinuous change, temporal multimodality, temporal increase in variability during a period of rapid transition, overlapping waves of development, and so forth. (For an explanation of the model and an extensive overview of the empirical data that the model explains, see van Geert, 1998b).

### Description, Explanation, and the Dynamic Systems Backpack-Traveler

A possible objection against dynamic growth models is that they are basically *descriptive*, in spite of their claims that they *explain* something, for instance, trajectories of developmental variables (see the preceding section). However, it is important to note that the term *explanation* always has a model-specific meaning: a dynamic growth model explains a developmental trajectory through the iterative application of a transformation function, entailing the principles of growth as specified above. Like any other model, it confines its explanatory concepts to an elementary set that it sees as fundamental to its particular form of explaining. By doing so, any model takes a loan on other forms of explanation that further ground its fundamental concepts; we have already expressed this point of view in section Hierarchical Levels and the Principle of Explanatory Loans. For instance, a dynamic growth model of lexical growth takes a loan on the possibility of connectionist models explaining its principles of growth by invoking networks processing inputs (an example of such a model is given in Davis, 2003). A connectionist network takes a loan on models explaining how the brain implements the network's connectionist structure and processes. This process of taking loans continues to other disciplines.

The discussion on whether a model is explanatory in a nontrivial sense often relates to at least one of the following assumptions. The first assumption is that it matters to which deductive principle or mechanism the model is referring. A statistical regression model explaining

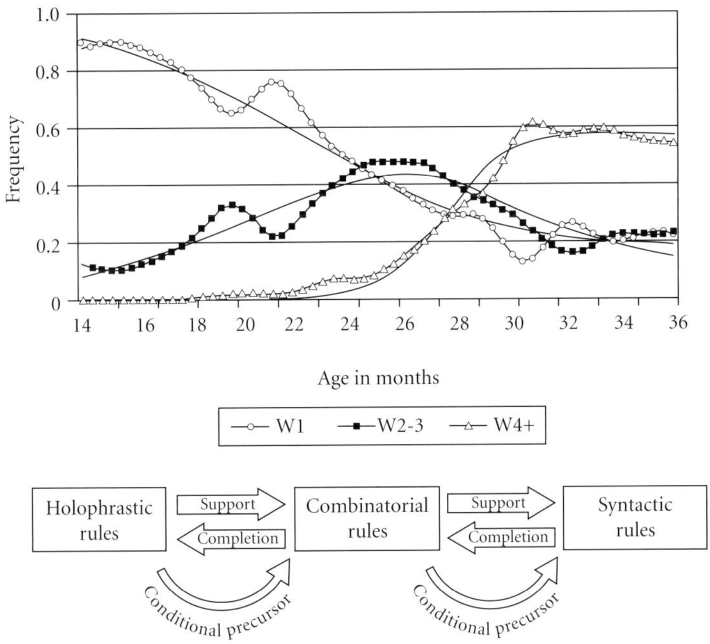


**Figure 16.2. Dynamic growth model: Changes in word length of spontaneous sentences**
This figure depicts changes in the frequencies of sentences of one, two, three, or more words in the spontaneous language production of a French-speaking girl, Pauline (after Bassano & van Geert, 2007). The frequencies relate to the emergence and eventual disappearance of three types of underlying linguistic rule systems (holophrastic, combinatorial, and syntactic rules). A dynamic growth model based on principles of conditional precursor, competitive, and supportive relationships has been fitted to the data (unmarked lines following smoothed data lines).

the development of some function by means of a polynomial of time is less theoretically convincing than a model explaining that same curve by reference to a mechanism that looks like a working brain, or, for that matter, a model that takes the preceding step as an explanation for the next step in the curve. (There are no well-established models that link pure temporal duration and the square of temporal duration directly and causally to word learning; duration matters in so far that it allows for a certain iteration of working principles, such as learning, changing of weights in a network, and so on.)

A second assumption refers to the fact that levels of explanation are hierarchically nested. (Such levels are often associated with different timescales, for instance, on the level of development, of real-time action, or of brain processes; see the preceding section and the following sections for further explanation.) According to basic reductionism, any primitive or unanalyzed element at a particular level of explanation is a thing to be explained by yet more primitive or elementary principles at some underlying level. A particularly appealing feature of the theory of complex dynamic systems, however, is its anti-reductionist nature. That is, any particular level of explanation has its own reality and cannot be reduced to the workings of lower levels. An understanding of some level of explanation—ontogeny, for instance—can be deepened, but not replaced, by an understanding of a lower level of explanation—the level of real-time action, for instance. In addition, levels are interconnected, in the sense that an understanding of one level—processes that take place on the timescale of ontogeny, for instance—is not only deepened by an understanding of a lower level—real-time action, for instance—but also by an understanding of a superordinate level—cultural history, for instance. In this sense, any explanation exists by virtue of leaving some things unexplained, namely things that are the topic of investigation for another level of explanation.

We have already seen that the principles featured in nonlinear dynamic growth models can be grounded on connectionist explanations. Why not replace such growth models by the connectionist models, then? The answer is that we see a dynamic systems model of long-term change and development as a traveler who travels lightly, with only the barest necessities in his small backpack. Such a person travels fast and may see a lot of the world, having the disadvantage that he can stay anywhere no longer than a little while. To put it more concretely, a dynamic systems model of long-term change may model the interactions among many components of a developing system, including the interactions and adaptations of educators and instructors. By doing so, the model builder uses a model that is relatively simple to construct to study the qualitative and quantitative properties of change, the nonlinearities and stabilities that emerge, and so forth. It is true that each and every component of that model can, in all likelihood, be replaced by a connectionist network that analyzes the working of each dynamical step in terms of changes in underlying network connections. However, building such a model will require hard labor; before doing this, the researcher might wish to study the general properties of the system through a much simpler dynamic model that captures the same general principles of change as those that emerge from the connectionist supra-model.

## Timescales and Levels of Aggregation

### Units of Developmental Description and the Forces That Hold Them Together

There is a wonderful book and movie called *Powers of Ten* (Morrison & Morrison, 1982; Office of Charles & Ray Eames, 1978; Office of Charles & Ray Eames, Demetrios, Mills, & Demetrios, 1990) that takes the reader on a journey through the universe, from the smallest possible levels of quantum-mechanical phenomena to the greatest possible astrophysical objects. As the reader turns the pages, as if turning a focus button of a giant microscope-telescope (or *mitescope*, pronounced as "mighty-scope"), several levels of aggregation pass by the reader/viewer. The reader/viewer see clusters of galaxies and separate galaxies, held together by the long-distance forces of gravity at the cosmological level, and then moves to distinct celestial bodies, for instance our own earth, on which one can observe separate layers of organization that become ever smaller and smaller as they zoom in to the subatomic level. What is interesting is that during our journey we discover aggregations or assemblies held together by specific interaction principles that apply to the level of magnification at issue. (Well-known physical interaction principles are the fundamental forces of gravity, electromagnetism, and strong and weak forces.) At one point as we move along the magnitude scale, we discover human cities, dwellings, and human beings themselves. In order to understand aggregations at the social and behavioral level, we must take into consideration that the aggregating forces are quite complicated emergent phenomena, such as the human laws, customs, artifacts, and so forth that hold people's activities and living spaces together.

However, the principle is that throughout our journey across the universe we encounter changeable assemblies that consist of ever smaller nested assemblies. The interactions that constitute the assembly in question (pick any one you like) vary from very strong to very weak (or soft, if one prefers that word); interactions are thus, to various extents, subject to external perturbations. What matters, however, is whether the assemblies in question are sufficiently stable across time and whether the reproduction of these assemblies occurs with sufficient stability and similarity. (Think of weak assemblies such as biological organisms; think of even weaker assemblies such as human action patterns, perhaps an infant acting in an A-not-B error situation.)

This basic issue raises a fundamental question about development and learning, which we have addressed in research on transient knowledge (a common state) and on how it can be transformed through dynamic growth processes into robust knowledge, which can be used with some stability across many contexts (Fischer, Bullock, Rotenberg, & Raya, 1993; Rappolt-Schlichtman, Tenenbaum, Koepke, & Fischer, 2007). In short, what emerges through development are similar to force–relationships, to put it somewhat metaphorically, that temporarily bind the components of characteristic assemblies (e.g., characteristic forms of problem solving, of acting, and so on) as they are formed in and through particular contexts.

As developmentalists, we can also look at the human life span and beyond by means of a *mitescope* that focuses on ever smaller (or greater) timescales, ranging from the timescales necessary to describe what happens at the level of neurons and synapses to the timescales necessary to describe what happens at the level of historical societies and human evolution. As we turn the focus button of this time-*mitescope*, we will encounter aggregations, or assemblies, based on specific interactions or forces that apply to the particular timescale we are focusing on currently. These assemblies are, to various degrees, strong or weak, but again the question is how successful the assembly or aggregation is in terms of duration or reproducibility. At the timescale of concrete action, we find soft but relatively durable and reproducible assemblies that take the form of particular actions, such as the much-investigated A-not-B error action–context pattern, holding a conversation with another person, or reading the abstract of a paper in an electronic database. At longer term timescales, we are able to observe certain regularities among the actions that people perform, regularities that we can describe by means of terms such as concept, reflective judgment, understanding of relationships, personality, and so forth. The meaning of these terms applies, of course, to the level of aggregation or assembly for which they are intended, and we should be aware of making any *category mistakes* regarding the use and meaning of these terms as discussed in the preceding sections.

### The Notion of Development as Relating to Person–Context Assemblies on Various Timescales and Levels of Aggregation

Several chapters in this book argue that psychological phenomena such as thinking and acting, if treated at the real-time level, can only be accounted for if the mutuality of person and context and the soft assembly of person–context aggregations is taken as a central feature (for a comparable discussion from the perspective of dynamic skill and dynamic growth models, see, for instance, Fischer et al., 1993; Fischer & Bidell, 2006; Van Geert, 2002). However, dynamic growth models of development focus on the growth of the lexicon, for instance, and not explicitly on *word use in context*. One may thus easily suspect that growth models refrain from incorporating the context by focusing on variables internal to the person. However, as we stated in the sections on unsolicited ontological claims, the implication that psychological variables such as lexical knowledge are internal entities is a theoretical choice that we do not make. Variables such as a child's level of lexical knowledge or of reflexive thinking always and by definition refer to person–context assemblies, that is, to aggregations of a person in his or her characteristic contexts.

If we then speak about the development of the child, we actually refer to long-term changes in the child and automatically also to the corresponding long-term changes in the child's contexts. These changes refer primarily to patterns of correlation over time (that is, in a single person's life trajectory) and not to identifiable entities. That is, development refers to a person–context assembly throughout the life span; contexts are no longer to be seen as independent variables or circumstances in which the person can be placed at pleasure, according to the whim of a researcher who treats such contexts as independent variables, the effect of which has to be estimated over many subjects.

The long-term assembly of person and context refers to the person in his *characteristic multitude of contexts* and not, as the short-term dynamics requires for its description, the here-and-now assembly of the person and a specific temporary context. The characteristic multitude of contexts differs for a young child and an adult, for instance. For the young child, the characteristic contexts will be the family, the play relationships with peers, and a particular type of school. (In underdeveloped countries, the characteristic context of many children will not include the school, but the context of child labor, for instance.) For an adult, the family context will be an entirely different one than for the child because the adult stands in the context as caretaker and provider. A characteristic context for the adult is the professional context. Persons may differ in their characteristic contexts as they may differ in their organismic characteristics, including intelligence, temperament, and so on. The person–contexts assemblies over the long term are also *soft assembled* in the same way person–context assemblies are soft assembled in the context of short-term individual actions and thinking processes.

## Can *Context* be an Independent Variable?

As stated earlier, the apparent softness or weakness of the association between a person and the person's characteristic contexts should not lure us into thinking that contexts are to be treated as an independent variable and not as genuine properties of the person himself. For instance, one might say that it is very easy to change a person's job (an *external context*) and very difficult, if not impossible, to change a person's temperament (an *internal property*). In reality we see that in both cases—jobs and temperaments—interactions and forces are at work that keep the associations relatively stable (although those forces are different for the job than for the temperament). Adults may change jobs, for instance, and children may change schools, but in many cases the change is relatively superficial; what remains is a very similar context as far as functionality and psychological meaning of the context for the person concerned. This idea brings us very close to the notion of *Umwelt* by Jakob von Uexküll (1864–1909; Von Uexküll, 1909/2001) and Thomas Sebeok (1920–2001; Sebeok, 2001; see also Kull, 2001a, 2001b, including the special issue of *Semiotics, 134*, 2001). Von Uexküll defined *Umwelt* (literally "surrounding world") as the organism's experienced environment, the qualities of which are codefined by the organism's senses and activities and the organism's interests and concerns. (The notion comes close to Lewin's notion of psychological life space (Lewin, 1936/1966). The notion of *Umwelt* implies that contexts and environments are inextricable entanglements of physical and psychological forces; and thus, context is never to be treated as independent of the organism or person.

With regard to specifying dynamic growth models, it is important to realize that persons and contexts (essentially, multitudes of contexts) are glued together in different ways. The person actively seeks certain types of contexts and by doing so actively extends the person–context association over the long term. Some persons may actively assign contexts to other persons, such as parents who send their children to a school of their choice. Contexts are dependent on the person: the context as it functions for the person is the perceived context. For instance, a book is a physical context with very special properties simply because the person has learned to read and has, thus, learned to see the physical symbols in the book as information of a very different kind than that included in the mere physicalness of the letters and words. People are bound to certain contexts through (legal) laws, ethics, customs, and so forth. People are bound to certain contexts because they need these contexts to achieve their goals. As an example, modern-day scientists have invested part of their memory into the hard disks of their computers and part of their intelligence into the processing capacities of statistical programs. As a further example, truck drivers have invested their physical carrying force into powerful trucks. Just as there are dynamic forces that keep our internal properties together—our intelligence, personality, and so on—there are dynamic forces that keep us and our characteristic contexts together. Such forces move us and our contexts along the path of development, stability, and change.

### Developmental Changes in Person–Context Assemblies

The person–multiple context assemblies that form the subject of developmental statements and theories change over developmental time, but never cease to exist as person–context assemblies (with *context* codefined by the person's senses, skills, interests, and so on). This observation has several implications for developmental research. One implication concerns the issue of assessment or measuring a person's developmental level (or any other psychological property, for that matter). For instance, if we wish to specify a person's level of thinking or understanding at a particular moment in the person's life, we cannot do so without specifying a particular context into which this level is assessed. By definition, the assessment will refer to the person–specific-testing–context assembly. However, the person is characterized by the fact that there exists a multitude of contexts in which his thinking skills can manifest themselves. Trivially speaking, the number of such contexts is quasi-infinite; however, the number of contextual distinctions that really matter (that is, that have functional significance) is relatively limited because they are constrained by organismic properties, such as the person's senses, skills, interests, brain, impairments, and so on. Thus, we can specify the person's developmental level of thinking in a state space that reckons with the contextual dimensions that really matter. In a developmental perspective, one contextual dimension involves the support of other persons (others helping the individual person solve a particular question, giving suggestions, etc.). Another contextual dimension involves the person making use of material and physical supports and contexts in a relatively autonomous way, without apparent help by others. If we see the person and the context as an assembly, a diagnostic assessment of the person's developmental level of thinking must thus be specified in terms of a state space which is at least two-dimensional, namely his functioning with help and without help (Fischer & Bidell, 2006).

A second implication is that the nature of the relationship between person and context changes as a result of development. We already mentioned the change from transient knowledge (that relies on contextual support) to robust knowledge (that remains stable across many contexts; Fischer et al., 1993; Rappolt-Schlichtman et al., 2007). In the transition from transient to robust knowledge, the need for contextual support (perhaps from a more competent person) will gradually diminish (e.g., through diminishing scaffolding, see van Geert & Steenbeek, 2005b). On the other hand, as a consequence of developing skills, the context may gradually become more supportive. For instance, as a child learns to read; written signs available in the context become accessible to the child, giving support to more complicated forms of thinking (e.g., the use of written schemes/texts). This aspect of development has been the focus of distributed cognition approaches (e.g., Clark, 2006; Clark & Chalmers, 1998).

In summary, we must accustom ourselves to reading phrases such as "the development of the child" as shorthand for statements about the development of [child]–[characteristic-contexts] assemblies and about the emergence of long-term correlational patterns in the life trajectory of individual persons and their *Umwelts* (e.g., their continuously and dynamically constructed life spaces). These statements refer to the long-term dynamics of such assemblies and consider these assemblies as the proper units of analysis or description.

## Developmental Rulers and Building Dynamic Systems Models of Developmental Change

### Developmental Rulers and Contexts of Assessment

In section Can *Context* be an Independent Variable?, we have seen that in order to describe a process as a dynamic system, one needs a metric or dimension to describe different states of a phenomenon that one is interested in. The best possible metric is formed by the line of real numbers; this enables one to describe a process in the form of Cartesian coordinate space. Cartesian coordinates are great for describing

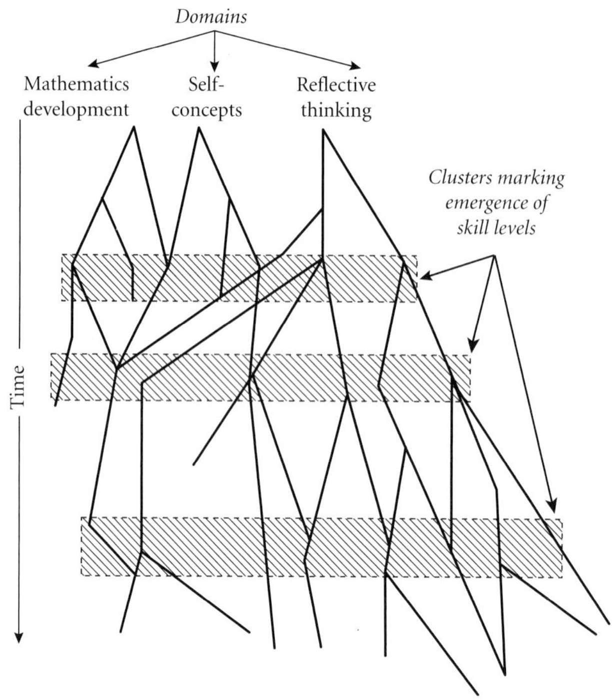


**Figure 16.3. Developmental web: Includes clusters of discontinuities across strands**

movement, but the most important characteristics of behavior in development and education involve cognition and emotion, not movement. A simple counting of correct or incorrect answers on a cognitive test, for instance, isn't appropriate. What we need is a developmental scale that works across tasks and domains, that is, a *developmental ruler*. Research must deal with scaling issues directly, such as by testing scales under several conditions. We can get at such developmental scales by studying developmental webs.

What do we mean by *developmental webs*? Development and learning take place along many parallel, independent strands that have similar properties but separate structures/contents. Examples of such strands are a child's skills in and understanding of social relationships, understanding of numerical problems, mastery of technological tools, ability to follow stories and story lines, and so forth. In a developmental web (Fig. 16.3), discontinuities may emerge that reflect new capacities emerging in common zones across strands. Clusters of discontinuities mark a universal developmental scale. That is, across disparate skills we may find one scale, for instance, a single underlying developmental ruler.

The scale or ruler not only applies to disparate skills, but also shows a developmental range, which depends on the context in which the skill at issue is put to work. As in Vygotsky's (1978) zone of proximal development, there are differences between unsupported and supported levels of skills. In that sense, there is no single competence, not even for a domain (see also section Developmental Changes in Person–Context Assemblies). A child's developmental level achieved under conditions of little or no external support (this we call the *functional level*) differs from the child's level assessed under conditions of support that primes complex behavior (this we call the *optimal level*) in the following ways. First, the child's level of growth is higher in the optimal/high support condition (Kitchener, Lynch, Fischer, & Wood, 1993). Second, optimal levels show qualitative properties that functional levels do not show, such as growth cycles and recurrent discontinuities. The distinction

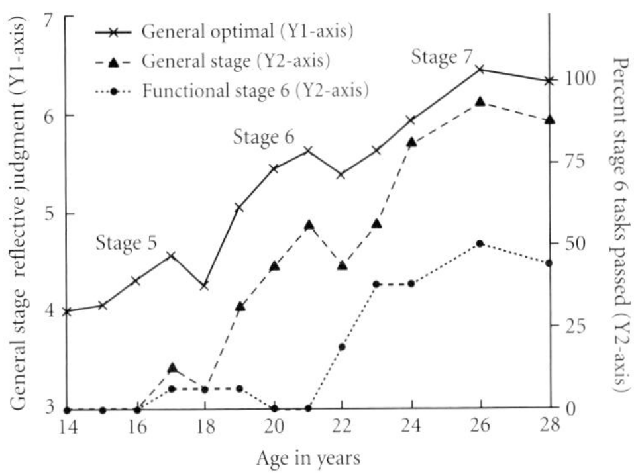


**Figure 16.4. Development of reflective judgment: Optimal and functional levels**
Performance at optimal level spurts on emergence of stages 5, 6, and 7 of reflective judgment. The top line (solid) shows the general score for optimal levels of reflective judgment across all tasks. The two dotted lines illustrate the difference between optimal and functional level, which shows slower, more gradual increase. The two dotted lines show percentage of correct performance for the subset of tasks assessing stage 6, which is the beginning of true reflective thinking. The upper dotted line shows optimal level for stage 6, and the lower line shows functional level.

between these functional and optimal levels, reflecting the sensitivity of the system to support, is a typical example of the person–context assembly as a complex system. For instance, a system may shows superposition of seemingly conflicting properties (e.g., being dependent on and independent of the context at the same time). In addition, the child in the system with high support has a tendency for performance to be more robust by trying to become less dependent on support.

The development of reflective judgment provides an example of the difference between optimal and functional levels (Fig. 16.4). Reflective judgment relates to the question of how we know that something is either true or false (given that much of our knowledge is intrinsically uncertain and incomplete). Reflective judgment is a typical example of the cognition- and knowledge-related aspects of development that requires a developmental ruler that goes far beyond a simple number line.

From 2 to 30 years of age, 7 stages in the development of reflective judgment can be observed (see Fig. 16.4 based on studies by; Kitchener & King, 1990; Kitchener et al., 1993). Briefly, the first stages entail absolute knowledge: a right or wrong judgment. Later stages involve relative knowledge, in which truth varies with perspective. Final stages imply true reflective judgment: truth depends on argument and evidence. These 7 stages are not stages in the classic developmental sense, that is, they are not identifiable discrete phases (such as being in stage 1 from age a to b, in stage 2 from age b to c, and so forth). Here, stages correspond to steps in the progression toward reflective thinking. For example, in stage 6 a person understands that although knowledge is uncertain, strong, justified conclusions can be made across domains viewpoints based on evidence and argument. In high support assessment conditions, where the subjects' more complex behavior is primed through suggestions (e.g., questions), some subjects will provide answers at stage 6, but without support, they exhibit behavior/arguments consistent with stages 4 or 5. The stage 6 knowledge is transient (that is, exists by virtue of the support given, but does not last if the support is withdrawn), while the stage 4 or 5 knowledge is more robust. Hence, the stage is not a fixed point that a person occupies, but is rather a point on a scale—a tick mark on the developmental ruler—along which the person's skills vary in both long-term development and short-term performance. (For an example regarding changes in linguistic level across days, see van Geert & van Dijk, 2002.)

### Developmental Rulers and the Space of Development

The relationship between the developmental steps the ticks on the developmental ruler—and the geometrization of developmental space needed for the application of dynamic systems models is not trivial. The stages amount to a hierarchical growth pattern and, consequently, the scale or ruler is a hierarchical complexity scale. It can be used in different ways, to show and underlie different types of developmental patterns.

First, as we explained earlier, stages can be used as tick marks on a single developmental ruler, thus measuring the successive developmental levels found with a particular type of assessment in a particular developmental domain (e.g., support assessment of reflective judgment). A nice example of a developmentally relevant pattern captured by employing the developmental ruler in this fashion is the *scalloping pattern* one can observe across a series of repeated activities, where a person (or a dyad) tries to build a skill for solving a particular type of problem (Fischer, Yan, & Stewart, 2003; Granott, 2002). The person moves back and forth between widely diverging developmental skill levels. What happens, functionally, is that a person is repeatedly moving down to a low level and then rebuilding skills. This repetition contributes to the emergence of a consolidated, general skill—robust knowledge—and provides a key to the plasticity of human behavior.

Why would one need regression in development? The problem is that developmental levels show hierarchical relationships and that higher levels depend on the organization of their lower level components; for instance, representations as higher order patterns require coordination of sensorimotor patterns. If a higher level pattern (e.g., a representation, or, at a still higher level, an abstraction) does not work in a given problem context, it needs to be reworked or reorganized; in so doing, one needs to be able to move down to the lower level component patterns and reconstruct the skill (Fischer et al., 1993). In fact, what is often viewed as a *critical* period (that is, a period beyond which certain developments are no longer possible), arises when we cannot move down and reorganize.

Second, the stages or skill levels that are used as tick marks on a developmental ruler can also be used as developmental dimensions in their own right. Each skill level acts as a *developmental grower*, increasing its quantitative level over time, eventually showing temporary regressions. A series of three successive skill levels, for instance, can be represented as three growth curves, modeled in the form of a coupled system (van Geert, 1991, 1994; Fischer & Kennedy, 1997; Fischer & Bidell, 2006). Alternatively, the developing system can be seen as a single point in a state space described by the three skill levels, which function as dimensions of the state space. Based on the simplified growth equation explained in section An Example: A Dynamic Growth Model of Language Development, the relationship between the growers can be described and modeled as a dynamic system of the form $x_{t+1} = f(x_t, C_x, \{y_t, z_t\})$, $y_{t+1} = g(y_t, C_y, \{x_t, z_t\})$, and $z_{t+1} = h(z_t, C_z, \{y_t, x_t\})$. The symbols $x$, $y$, and $z$ refer to the three hierarchical growers in question, and $C_x$, $C_y$, and $C_z$ the support or resource constants for $x$, $y$, and $z$, respectively. The functions $f$, $g$, and $h$ are the dynamic functions that describe how a next state of a grower, for instance, $x$, is affected by its preceding state and the preceding states of the coupled growers $y$ and $z$ (details have been explained in a variety of publications, see Fischer & Bidell, 2006; Fischer & Kennedy, 1997; van Geert, 1991, 1994, 2003; van Geert & Steenbeek, 2005a, 2005b). Basically, these dynamic functions imply (a) that the quantitative change in a grower over some time interval is proportional to the level already attained, (b) that it depends on limited resources and thus levels off as the resource level is approached, (c) that growers are either positively or negatively affected by the growth in other growers, and (d) that the onset of growth can depend on the presence of a precursor. These relationships can be mathematically modeled fairly easily if the growth of each variable is represented by a single vector. The modeling is more complicated if growth is represented by a vector field (van Geert, 1998a, 1998b), although the vector field models yield interesting discontinuous patterns in addition to continuous ones and can explain a host of empirical developmental phenomena.

## Conclusion: Evaluation of Dynamic Growth Models

The patterns generated by the dynamic growth models have been repeatedly found in a variety of skills and developmental domains as well as in development of brain electrical activity. In that sense, empirical data provide good support for the underlying dynamic models. As noted earlier, we have begun to create knowledge of how these growers—hierarchical skill levels—are physically instantiated; that is, how exactly they are represented in the brain and in the supportive, cultural contexts that make them possible. In this sense, we are moving toward founding the dynamic growth models from the bottom up; that is, we are working from the underlying physical principles up to the level of the model. In addition, the principles of support, competition, and precursors that are used in these models are firmly grounded in dynamic skill theory and various other theories of development and are, therefore, far from accidentally selected. Moreover, developments in brain research and neuroimaging provide further circumstantial evidence of the validity of the underlying principles.

In this context, dynamic growth models have one major advantage over some of their theoretical competitors, in that they are able to subsume a wide variety of developmental phenomena under a common general framework based on a small set of universal and simple principles of interaction among variables. For example, similar models explain growth patterns for reflective judgment concepts and brain activity (Fischer, 2008). As we stated earlier in this chapter, however, explanation is multileveled, and having an explanation on the level of, say, dynamic growth models, does not preclude the need for a different sort of explanation by means of connectionist models or neurocognitive models. The reality of development is too dynamic and nonlinear to be captured fully under a single theoretical framework. As with development itself, scientific understanding progresses by coordinating and by integrating perspectives, by decomposing and recomposing our insights in order to arrive at ever more powerful and richer explanations of the patterns of developmental change.

## References

Bassano, D., & van Geert, P. (2007). Modeling continuity and discontinuity in utterance length: A quantitative approach to changes, transitions and intra-individual variability in early grammatical development. *Developmental Science, 10*(5), 588–612.

Case, R. (Ed.). (1991). *The mind's staircase: Exploring the conceptual underpinnings of children's thought and knowledge.* Hillsdale, NJ: Erlbaum.

Case, R., & Okamoto, Y. (with Griffin, S., Siegler, R. S., & Keating, D. P). (1996). The role of central conceptual structures in the development of children's thought [Special issue]. *Monographs of the Society for Research in Child Development, No. 246, 61*(1–2).

Chomsky, N. (1980). *Rules and representations.* Oxford, United Kingdom: Blackwell.

Clark, A. (2006). Language, embodiment, and the cognitive niche. *Trends in Cognitive Sciences, 10*(8), 370–374.

Clark, A., & Chalmers, D. J. (1998). The extended mind. *Analysis, 58*(1), 7–19.

Dale, R., & Spivey, M. J. (2005). From apples and oranges to symbolic dynamics: A framework for conciliating notions of cognitive representation. *Journal of Experimental & Theoretical Artificial Intelligence, 17*(4), 317–342.

Davis, M. H. (2003). Connectionist modelling of lexical segmentation and vocabulary acquisition. In P. T. Quinlan (Ed.), *Studies in Developmental Psychology* [series]. *Connectionist models of development: Developmental processes in real and artificial neural networks* (pp. 151–187). London: Taylor & Francis–Psychology Press.

Dawson-Tunik, T. L., Commons, M., Wilson, M., & Fischer, K. W. (2005). The shape of development. *European Journal of Developmental Psychology, 2*(2), 163–195.

Office of Charles & Ray Eames (Creators). (1978). *Powers of ten: A film dealing with the relative size of things in the universe and the effect of adding another zero* [Motion picture (9 minutes)]. United States: Pyramid Films.

Office of Charles & Ray Eames (Creators), Demetrios, E., Mills, S. (Producers), & Demetrios, L. E. (Executive producer). (1990). *Powers of ten: A film dealing with the relative size of things in the universe and the effect of adding another zero* [Motion picture (9 minutes)]. United States: Pyramid Film & Video. Viewed on October 25, 2008, from http://powersof10.com/

Elman, J. L., Bates, E. A., Johnson, M. H., Karmiloff-Smith, A., Parisi, D., & Plunkett, K. (1996). *Rethinking innateness: A connectionist perspective on development.* Cambridge, MA: MIT Press–Bradford Books.

Fischer, K. W. (1980). A theory of cognitive development: The control and construction of hierarchies of skills. *Psychological Review, 87*(6), 477–531.

Fischer, K. W. (2008). Dynamic cycles of cognitive and brain development: Measuring growth in mind, brain, and education. In A. M. Battro, K. W. Fischer, & P. Léna (Eds.), *The educated brain: Essays in neuroeducation*. New York: Cambridge University Press.

Fischer, K. W., & Bidell, T. R. (2006). Dynamic development of action, thought, and emotion. In W. Damon & R. M. Lerner (Eds.), *Handbook of child psychology: Vol. 1. Theoretical models of human development* (6th ed., pp. 313–399). New York: Wiley.

Fischer, K. W., Bullock, D., Rotenberg, E. J., & Raya, P. (1993). The dynamics of competence: How context contributes directly to skill. In R. H. Wozniak & K. W. Fischer (Eds.), *The Jean Piaget Symposium series: Development in context: Acting and thinking in specific environments* (pp. 93–117). Hillsdale, NJ: Erlbaum.

Fischer, K. W., & Kennedy, B. (1997). Tools for analyzing the many shapes of development: The case of self-in-relationships in Korea. In E. Amsel & K. A. Renninger (Eds.), *The Jean Piaget Symposium series: Change and development: Issues of theory, method, and application* (pp. 117–152). Mahwah, NJ: Erlbaum.

Fischer, K. W., & Rose, S. P. (1999). Rulers, models, and nonlinear dynamics: Measurement and method in developmental research. In G. J. P. Savelsbergh, H. L. J. van der Maas, & P. L.C. van Geert (Eds.), *Nonlinear developmental processes* (pp. 197–212). Amsterdam: Royal Netherlands Academy of Arts and Sciences.

Fischer, K. W., Yan, Z., & Stewart, J. (2003). Adult cognitive development: Dynamics in the developmental web. In J. Valsiner & K. J. Connolly (Eds.), *Handbook of developmental psychology* (pp. 491–516). Thousand Oaks, CA: Sage.

Fodor, J. A. (1975). *The language of thought*. New York: Crowell.

Granott, N. (2002). How microdevelopment creates macrodevelopment: Reiterated sequences, backward transitions, and the zone of current development. In N. Granott & J. Parziale (Eds.), *Microdevelopment: Transition processes in development and learning* (pp. 213–242). New York: Cambridge University Press.

Hofstadter, D. R. (2007). *I am a strange loop*. New York : Basic Books.

Hollenstein, T., & Lewis, M. D. (2006). A state space analysis of emotion and flexibility in parent-child interactions. *Emotion, 6*(4), 656–662.

Jaeger, H. (1999). From continuous dynamics to symbols. In W. Tschacher & J.-P. Dauwalder (eds.), *Studies of nonlinear phenomena in life science: Vol. 8. Dynamics, synergetics, autonomous agents: Nonlinear systems approaches to cognitive psychology* (pp. 29–48). Singapore: World Scientific.

Kingsland, S. E. (1985). *Modeling nature: Episodes in the history of population ecology*. Chicago: University of Chicago Press.

Kitchener, K. S., & King, P. M. (1990). The reflective judgment model: Ten years of research. In M. L. Commons, C. Armon, L. Kohlberg, F. A. Richards, T. A. Grotzer, & J. D. Sinnott (Eds.), *Adult Development: Vol. 2: Models and methods in the study of adolescent and adult thought* (pp. 62–78). New York: Praeger.

Kitchener, K. S., Lynch, C. L., Fischer, K. W., & Wood, P. K. (1993). Developmental range of reflective judgment: The effect of contextual support and practice on developmental stage. *Developmental Psychology, 29*(5), 893–906.

Kull, K. (Ed.). (2001a). Jakob von Uexküll: A paradigm for biology and semiotics [Special issue]. *Semiotics, 134*(1/4).

Kull, K. (2001b). Jakob von Uexküll: An introduction. *Semiotica, 134*(1/4), 1–59.

Lewin, K. (1966). *Principles of topological psychology* (F. Heider & G. M. Heider, Trans.). New York: McGraw-Hill. (Original work published 1936)

Morrison, P., & Morrison, P. (1982). *Powers of ten: A book about the relative size of things in the universe and the effect of adding another zero*. Redding, CN: Scientific American Library.

Plunkett, K., & Juola, P. (1999). A connectionist model of English past tense and plural morphology. *Cognitive Science, 23*(4), 463–490.

Rappolt-Schlichtman, G., Tenenbaum, H. R., Koepke, M. F., & Fischer, K. W. (2007). Transient and robust knowledge: Contextual support and the dynamics of children's reasoning about density. *Mind, Brain, and Education, 1*(2), 98–108.

Robinson, B. F., & Mervis, C. B. (1998). Disentangling early language development: Modeling lexical and grammatical acquisition using an extension of case-study methodology. *Developmental Psychology, 34*(2), 363–375.

Rose, S. P., & Fischer, K. W. (1998). Models and rulers in dynamical development. *British Journal of Developmental Psychology, 16*(1), 123–131.

Ruhland, R., & van Geert, P. (1998). Jumping into syntax: Transitions in the development of closed class words. *British Journal of Developmental Psychology, 16*(1), 65–95.

Ryle, G. (2002). *The concept of mind.* Chicago: University of Chicago Press. (Original work published in 1949)

Schöner, G., & Thelen, E. (2006). Using dynamic field theory to rethink infant habituation. *Psychological Review, 113*(2), 273–299.

Schutte, A. R., & Spencer, J. P. (2002). Generalizing the dynamic field theory of the A-not-B error beyond infancy: Three-year-olds' delay- and experience-dependent location memory biases. *Child Development, 73*(2), 377–404.

Schutte, A. R., Spencer, J. P., & Schöner, G. (2003). Testing the dynamic field theory: Working memory for locations becomes more spatially precise over development. *Child Development, 74*(5), 1393–1417.

Sebeok, T. A. (2001). Biosemiotics: Its roots, proliferation, and prospects. *Semiotica, 134*(1/4), 61–78.

Smith, L. B. (2005). Cognition as a dynamic system: Principles from embodiment. *Developmental Review, 25*(3–4), 278–298.

Smith, L. B., & Thelen, E. (Eds.). (1993). *A dynamic systems approach to development: Applications.* Cambridge, MA: MIT Press–Bradford Books.

Smith, L. B., Thelen, E., Titzer, R., & McLin, D. (1999). Knowing in the context of acting: The task dynamics of the A-not-B error. *Psychological Review, 106*(2), 235–260.

Spencer, J. P., & Schöner, G. (2003). Bridging the representational gap in the dynamic systems approach to development [Special section/issue]. *Developmental Science, 6*(4), 392–412.

Spencer, J. P., Smith, L. B., & Thelen, E. (2001). Tests of a dynamic systems account of the A-not-B error: The influence of prior experience on the spatial memory abilities of two-year-olds. *Child Development, 72*(5), 1327–1346.

Spencer, J. P., & Thelen, E. (Section Eds.). (2003). Connectionist and dynamic systems approaches to development [Special section/issue]. *Developmental Science, 6*(4).

Steenbeek, H. W., & van Geert, P. (2005). A dynamic systems model of dyadic interaction during play of two children. *European Journal of Developmental Psychology, 2*(2), 105–145.

Steenbeek, H. W., & van Geert, P. L. C. (2007). A theory and dynamic model of dyadic interaction: Concerns, appraisals, and contagiousness in a developmental context. *Developmental Review, 27*(1), 1–40.

Steenbeek, H. W., & van Geert, P. (2008). An empirical validation of a dynamic systems model of interaction: Do children of different sociometric statuses differ in their dyadic play? *Developmental Science, 11*(2), 253–281.

Thagard, P. (1996). *Mind: Introduction to cognitive science.* Cambridge, MA: MIT Press.

Thelen, E. (1996). Dynamic systems for everyone. *PsycCRITIQUES*, 41, 1002–1003.

Thelen, E., & Bates, E. (2003). Connectionism and dynamic systems: Are they really different? [Special section/issue]. *Developmental Science, 6*(4), 378–391.

Thelen, E., & Smith, L. B. (1994). *A dynamic systems approach to the development of cognition and action.* Cambridge, MA: MIT Press–Bradford Books.

Thelen, E., & Smith, L. B. (1998). Dynamic systems theories. In W. Damon & R. M. Lerner (Eds.), *Handbook of child psychology: Vol. 1. Theoretical models of human development* (5th ed., pp. 563–634). New York: Wiley.

van der Maas, H. L. J. (1995). Beyond the metaphor? A review of L.B. Smith & E. Thelen (1993) a dynamic processes approach to development. *Cognitive Development*, 10, 621–642.

van Geert, P. (1987a). The structure of developmental theories. A generative approach. *Human Development,* 30(3), 160–177.

van Geert, P. (1987b). The structure of Erikson's model of the eight ages of man. A generative approach. *Human Development*, 30(5), 236–54.

van Geert, P. (1987c). The structure of Gal'perin's theory of the formation of mental acts. A generative approach. *Human Development,* 30(6), 355–381.

van Geert, P. (1991). A dynamic systems model of cognitive and language growth. *Psychological Review, 98*(1), 3–53.

van Geert, P. (1993). A dynamic systems model of cognitive growth: Competition and support under limited resource conditions. In L. B. Smith & E. Thelen (Eds.), *A dynamic systems approach to development: Applications* (pp. 265–331). Cambridge, MA: MIT Press.

van Geert, P. (1994). *Dynamic systems of development. Change between complexity and chaos.* New York: Harvester.

van Geert, P. (1996). The dynamics of Father Brown. Essay review of a dynamic systems approach to the development of cognition and action by E. Thelen and L. Smith. *Human Development*, 39 (2), 57–66.

van Geert, P. (1998a). A dynamic systems model of basic developmental mechanisms: Piaget, Vygotsky, and beyond. *Psychological Review, 105*(4), 634–677.

van Geert, P. (1998b). We almost had a great future behind us: The contribution of non-linear dynamics to developmental-science-in-the-making. *Developmental Science, 1*(1), 143–159.

van Geert, P. (2002). Developmental dynamics, intentional action and fuzzy sets. In N. Granott & J. Parziale (Eds.), *Microdevelopment: Transition processes in development and learning* (pp. 319–343). New York: Cambridge University Press.

van Geert, P. (2003). Dynamic systems approaches and modeling of developmental processes. In J. Valsiner & K. J. Connolly (Eds.), *Handbook of developmental psychology* (pp. 640–672). London: Sage.

van Geert, P. (2004). Dynamic modeling of cognitive development. Time, situatedness and variability. In A. Demetriou & A. Raftopoulos (Eds.). *Cognitive developmental change: Theories, models and measurement* (pp. 354–378). New York: Cambridge University Press.

van Geert, P., & Steenbeek, H. (2005a). Explaining after by before. Basic aspects of a dynamic systems approach to the study of development. *Developmental Review, 25*(3–4), 408–442.

van Geert, P., & Steenbeek, H. (2005b). The dynamics of scaffolding. *New Ideas in Psychology, 23*(3), 115–128.

van Geert, P., & Steenbeek, H. (2008). Understanding mind, brain, and education as a complex, dynamic developing system: Measurement, modeling, and research. In A. M. Battro, K. W. Fischer, & P. Léna (Eds.), *The educated brain: Essays in neuroeducation,* 71–94. New York: Cambridge University Press.

van Geert, P., & van Dijk, M. (2002). Focus on variability: New tools to study intra-individual variability in developmental data. *Infant Behavior & Development, 25*(4), 340–374.

van Geert, P., & van Dijk, M. (2003). Ambiguity in Child Language: The problem of inter-observer reliability in ambiguous observation data. *First Language, 23*(3):259–284

Von Uexküll, J. (2001). An introduction to Umwelt. *Semiotics, 134*(1/4), 107–110. (Original work published 1909)

Vygotsky, L. (1978). *Mind in society: The development of higher psychological processes.* Cambridge, MA: Harvard University Press.

Weisstein, E. W. (1999). *CRC concise encyclopedia of mathematics.* Boca Raton, FL: Chapman & Hall–CRC.

Witherington, D. C. (2007). The dynamic systems approach as metatheory for developmental psychology. *Human Development, 50*(2–3), 127–153.

# CHAPTER 17

## Dynamic and Connectionist Approaches to Development: Toward a Future of Mutually Beneficial Coevolution

*Michael S. C. Thomas, James L. McClelland, Fiona M. Richardson, Anna C. Schapiro, and Frank D. Baughman*

When modelers who exploit different approaches get together, there is a tendency to extol the virtues of one's own approach and try to promote it over the approaches of others. This can lead to a funny kind of *either-or* mentality—is it better to be a connectionist or a dynamical systems modeler? To us, this is truly the wrong question. We agree with Smith when she says (see Chapter 4), "[Which approach is better is not] an argument much worth having." Much more important are the insights that each approach offers toward an understanding of the nature of cognition and behavior and of the processes that underlie the development of cognitive and behavioral capacities. And of course, there are a number of reasons why any sort of either-or decision would be misguided. Neither school of thought is *stationary*—indeed, Schlesinger (Chapter 9) goes so far as to liken connectionism and dynamical systems theory (DST) to organisms that are constantly evolving over time and increasing their adaptive fit to the environment (of explanation); and neither school of thought is *unitary*—a cursory inspection within each field reveals separate groupings of researchers exploring different psychological phenomena with diverse (albeit related) modeling techniques and assumptions.

By way of illustration, within DST, one can distinguish (1) *dynamical field theory* (e.g., Thelen, Smith, Schöner, Spencer) focusing on behaviors unfolding in millisecond and second ranges, such as sensorimotor control in relation to objects; (2) *growth models* (e.g., van Geert, Fisher) focusing on phenomena occurring over days, weeks, months, and years, such as vocabulary development and exploiting linked growth functions; and (3) *catastrophe theory* (e.g., van der Maas, Raijmakers) focusing on development in reasoning occurring over minutes, hours, days, and weeks.

Within the connectionist modeling framework, one may also find a great deal of diversity. Some researchers use *feed-forward back-propagation network models* and target cognitive development over days, weeks, months, and years in domains such as language, semantics, and reasoning (e.g., McClelland, Plaut, Plunkett, Thomas). Others use constructivist networks (most notably, Schultz, Sirois, Westermann, and colleagues). Some researchers use *recurrent attractor models* and target both behavior unfolding in the moment and development over weeks, months, and years; they study domains such as grammar development and object-directed behavior (e.g., Elman, Mareschal, McClelland, Munakata, Plaut, Thomas). There are also those who use what might be called *neurocomputational models* targeting specific neural structures such as the prefrontal cortex, basal ganglia, or the hippocampus (e.g., Cohen, McClelland, O'Reilly). These choices of what to investigate are not based on doctrine and do not reflect doctrinal differences; rather they are generally pragmatic choices. The goal is to capture certain key features of human performance and human development—their experience-dependence, their partial but not complete tendency toward regularity, their graded sensitivities to different variables, and many other key features.

For us, the aim of juxtaposing connectionist and DST approaches within the same volume is to identify the common themes of (successful) models of developmental phenomena, so that future work can benefit from the combined insights of the two approaches. These themes will lie at the heart of any new theory of development, and their future evolution will, we suggest, be enhanced by maintained interaction among connectionist modelers and dynamical systems researchers.

It is true that, at times, a tension has existed between connectionism and DST, and it is instructive to consider why this should be the case. But in what follows, one should remember that the similarities between the two approaches far outweigh their differences. In this chapter, we argue that much of the tension, in fact, arises from a tenet that the two approaches share: both rely on the explicit, quantitative instantiation of ideas in mathematical or computational models. We argue that the use of such models is responsible for much of the theoretical progress generated by connectionism and DST beyond the theories of *good old-fashioned cognitive development* (GOFCD) (see Chapter 13 for an exposition of those theories). But we also argue that the use of explicit, quantitative models brings with it a new set of problems. In the next section, we discuss several consequences of the use of such models; some of these consequences are pertinent to any potential integration of dynamical and connectionist approaches. To illustrate these consequences, we then consider three points of apparent disagreement between connectionism and DST. We begin with a specific example: the nature of children's reasoning on the Piagetian balance scale task. We then consider two general concepts: the importance of embodiment and the role of stability. We finish with a brief summary of the themes that we imagine will feature in any future integration of connectionist and dynamical systems approaches. We believe that they are consistent with the current directions of both connectionist and DST theorizing.

## Explicit Quantitative Models and Cognitive Development

Several concepts have gained greater prominence in developmental theory through the work of connectionist and DST researchers. These include the idea of emergence; the demonstration that relatively sudden (apparently stagelike) transitions in behavior can arise from continuously changing underlying mechanisms; the idea that instability (or variability) is often associated with change; and the recognition that behavioral patterns may arise from competition between latent and active representations of knowledge.

Some have doubted that these ideas/concepts are genuinely new, arguing instead that connectionism and DST have simply served to shift the relative emphasis among preexisting ideas in our understanding of cognitive development (see Chapter 13). In some sense, it is not important to establish who thought of the idea first; in most cases, it happened a very long time ago. Instead, we argue that the core contribution of connectionism and DST has been their reliance on explicit, quantitative formulation. These methods have provided a new and sharper set of tools to drive forward theoretical progress in our field. In many scientific fields, explicit, quantitative methods have historically followed an earlier phase of exploratory data collection that was guided by informally specified theories. Progress via informally specified theories often asymptotes because such theories aren't sufficiently explicit to know exactly what they predict empirically. Moreover, their terminology frequently glosses over deeper conceptual problems. For example, a verbal theory may claim that different rules appear at different ages in children's reasoning on the balance scale task, but this theory hides the serious problem of specifying the nature of the experience-driven (or even maturational) mechanisms that can generate new rules. Ultimately, an explanatory theory must strive for mechanism, that is, a way in which behavior can be explained by the operation of the causes that shape it.

It is here that explicit, quantitative models of development—be they connectionist models, dynamical systems models, or some other kind—offer so many advantages. The advantages have been much discussed elsewhere (see, e.g., Elman et al., 1996; Mareschal & Thomas, 2007; McClelland, Rumelhart, & the PDP Research Group, 1986; Munakata & McClelland, 2003; Thomas & McClelland, 2008). Here, we simply allude to a few of them. Explicit, quantitative models necessitate that the theorist be much more specific about the causal entities in the theory—the same verbal term cannot be used (unwittingly) in subtly different ways. Unexpected behaviors may emerge from the complex interactions of many simple components, along with the structure of the problem domain. A formal model can test whether the theory as specified indeed generates the behavior it is supposed to explain at a quantitative level. Models can unify experimental data, for instance bringing together data from child development, adult performance, and breakdown in function at any stage with reference to a single well-specified system. Models can generate new predictions to be tested against quantitative data. Models can also produce general explanations by demonstrating how a small set of processing principles, when combined with the features of particular cognitive domains, can account for experimental data across a range of behaviors. Where new models are controversial, they stimulate further theoretically focused data collection, which in turn advances the field.

A key point about explicit, quantitative models is that they are not generally intended to provide a detailed account of all aspects of a situation or phenomenon. Rather, their role is to help us understand the consequences of certain constellations of assumptions. Central to the effort to achieve understanding is the role of simplification. All models make certain simplifications in order to focus on explaining the phenomenon of interest. This may mean focusing on a very restricted range of task situations and experiences relevant to them (for example, in the case of the balance scale task, there is a focus on experience with balance) and considering change only over a certain timescale. Finer timescales, believed to involve a graded and continuous real-time process, may be replaced with single computations that are essentially treated as occurring instantaneously. Simplifications will often be made in the way the environment is represented. For example, in the A-not-B task, the exact perceptual features of the objects that are manipulated in front of the child are not all held to be important to the phenomenon, so a model may provide only a single dimension for the presence or absence of each object.

The art of using explicit, quantitative models is first to make simplifications only in those aspects of the cognitive domain (e.g., regarding representation, process, or environment) that are not considered to be crucial in addressing the focal issues under consideration; and second to make simplifications that do not unduly narrow the focus of the model, so that it can still be generalized to new phenomena or novel experimental findings. Decisions about what those aspects are and which simplifications are best to address them are themselves reflections of scientific judgment; and they are subject to disagreement—not all researchers find the same aspect of a particular phenomenon to be its most central feature. Furthermore, there are natural differences among investigators in their perceptions of the factors that contribute most importantly to the explanation of a phenomenon. Thus, there is no single *correct* set of simplifications; and even if there were, there is no known algorithm for discovering what that set is. This is why explicit, quantitative modeling is best understood as an ongoing process of exploration.

The central role of simplification adds some complexity to theory development, in particular in evaluating the success and failure of particular models. When a model works, what does this mean for the underlying theoretical commitments from which it was derived? Often, it may not be fully clear what those commitments actually are. If a model succeeds in capturing the development of some target phenomenon, is this only because it has included unrealistic simplifications in its design? If a model fails, is it fatal for the underlying theoretical perspective that it attempts to embody? Assumptions that are introduced—either as simplifications or because the modeler did not choose to focus on a particular aspect of the phenomenon under consideration—may appear to others to be matters of theory or principle. The ensuing dialog is, in our view, a healthy process that, carried on over the course of several years, often leads to considerable progress.

It is here, however, that we come to the source of the tension that exists between connectionism and DST. Frequently, connectionist and DST models include different simplifications because they are targeting different issues. Common simplifications include the following:

- differences in the timescales over which developmental change is examined;
- relative emphasis on the role of learning (e.g., structural adaptations) versus priming/pre-shaping (e.g., temporally continuous activation states) in modulating behavior;
- simplifications regarding the importance of the structure of the problem domain in driving behavior (e.g., representations in connectionist models are typically of higher dimensionality than in DST models and place greater emphasis on the role of experience in the problem domain in shaping behavior);
- relative emphasis placed on the role of embodiment (connectionist models often omit constraints of embodiment);
- level of abstraction encoded in representations (in the past, connectionist models have focused on higher level phenomena); and
- roles of variability and stability in representational states (considered of greater importance in DST).

Indeed, in looking back across two decades of connectionist and DST models, it is striking how often these models have employed complementary simplifications. Critically, however, the differences in simplification that have emerged may not actually reflect any fundamental theoretical incompatibilities, but may instead reflect differences in the specific issues and phenomena that have been the focus of the modelers' attention.

The problem is not restricted to comparisons between connectionist and DST models. Model simplifications can vary as much within approaches as between them. One solution is to encourage researchers to be as clear as possible about what they take to be the core assumptions of their models and what they consider to be simplifications introduced only for the sake of tractability and transparency. Then it will become more apparent where actual theoretically important points of contention lie and which differences between models merely reflect differences in where the modelers have chosen to make simplifications. What might such a list look like? As an exercise, in Box 17.1 we list the simplifications made in one connectionist model of the development of syntax comprehension (Thomas & Redington, 2004); the simplifications are listed along with the rationale for making them. If more investigators provided detailed descriptions of their assumptions and simplifications, it would be easier for researchers to be clearer about points of principle and points of strategic simplification. This in turn should lead eventually to a clarification of exactly which aspects of a particular model are responsible for its successes and/or failures. Of course, even listing everything relevant in such a table is no easy task, since modelers may not always have in the focus of their attention all the factors that could potentially be relevant to their simulations. And knowing which factors are crucial for success and failure is even harder. In general, we adopt simplification for the sake of tractability—were the simplifications to be replaced, the model could become intractable, making it in fact very difficult to know just how important the role of the simplification is.

The complementarity between models frequently hinders a comparison between connectionism and DST. However, in support of our claim that these approaches have much in common, when the respective researchers have turned their minds to explaining *the same* developmental phenomenon, their models have tended to converge. The A-not-B error in infants provides one such case (see Chapters 4 and 7).

## Box 17.1 Example of explicit quantitative model of development

**"A Simple Recurrent Network model of the development of syntax comprehension"**

The Task:

- Participants listen to sentences and make a binary response to identify the agent of the sentence. Data for accuracy and speed of sentence classification have been collected for children and adults in various typical and atypical populations

**Phenomena to be captured:**

- Order of difficulty of accuracy for comprehending different sentence types (e.g., actives, passives); order of acquisition for the sentence types in children
- Changes in this pattern in adult breakdown and in different developmental disorders
- The type of information that is exploited in learning this task with an impoverished system (word specific, sequence specific)

**Assumed essential characteristics:**

- Stable, abstract representations of words (though not necessarily grammatical word classes)
- Other components of a language system (e.g., speech sound perception, lexicon, semantics)
- World that delivers examples of sentences where agent-patient knowledge is available to the system (the training set can be assumed to occur via episodes of experience rather than as an internally stored set)

**Representation of information in the model:**

- Localist input representations depict individual words in the sentence to be understood
- Localist output units (a) predict the next word in the sentence and (b) classify the sentence as *agent-precedes-patient* or *patient-precedes-agent*

**Simplifications:**

- System is not embodied: stable input representations and training signals (some auto-predictive) are delivered by an assumed external cognitive system, body, and world
- Learning algorithm: Back propagation of error signals as a proxy for some more plausible error-driven neural learning algorithm

**Timescale:**

- Real-time is simulated in discrete steps, each time step aligns with the presentation of the next word in the sentence. Network contains interval units. Activation is feedforward from the input but recurrent from the internal units (a copy of activation on the previous time step). No noise under normal conditions
- Target timescale is performance over seconds
- Shorter timescales simplified in activation dynamics of the model (steps of vector matrix calculation)
- Change over longer timescales (hours, months) assumed to be an accumulation of changes in the range of seconds

As we shall shortly see, both connectionist and DST models of this phenomenon employ unidimensional representations of objects and motor plans; both employ continuous activation and settling attractor states driven by recurrent connections; and both explain behaviors in terms of a competition between states induced by a sequence of previously encountered situations and the most recent event witnessed. Before examining this case, however, we consider another developmental phenomenon where connectionist and dynamical approaches have converged.

## Connectionist and Dynamical Modeling of Children's Development on the Balance Scale Task

As discussed in McClelland and Vallabha (see Chapter 1), connectionist models often focus on the overall time course of development, neglecting some of the details of shorter term processing. In the context of modeling the Piagetian balance scale task, this focus on overall developmental trends has, in part, led to a neglect of the question: exactly how well do these models capture the details of transitions between stages? van der Maas and Raijmakers (Chapter 15), researchers who use catastrophe theory as a framework for understanding stage transitions, have criticized connectionist models for not exhibiting the abrupt transitions between stages of the balance scale task that they claim are present in experimental data. In their view, these transitions exhibit several *catastrophe flags* that they see as indicative of underlying phase transitions in behavior. Looking for evidence of such catastrophe flags in the McClelland (1989, 1995) balance scale model, they argue that these flags are not exhibited in the model's behavior (Raijmakers, van Koten, & Molenaar, 1996).

There have been many interesting contributions made in the application of catastrophe theory to the balance scale task and in the related effort to analyze the connectionist model's ability to account for these effects (see also Jansen & van der Maas, 1997; 2001; 2002; Quinlan et al., 2007). Indeed, the McClelland (1989, 1995) model does have some shortcomings in accounting for several aspects of the relevant experimental data. However, these shortcomings arise from simplifications in the model in relation to its initial focus on the longer timescale over which developmental change occurs in tasks such as the balance scale task and not from deficiencies in the underlying theory embodied in the model. In the following paragraphs, we take some time to consider in detail which simplifications produced the shortcomings in the original McClelland (1989, 1995) model. We then demonstrate how new work has extended the model to address shorter term dynamics in behavior (Schapiro & McClelland, in press). This extended model exhibits the *catastrophe flags* that van der Maas and Raijmakers (Chapter 15) take to be indicators of underlying phase transitions in development.

First, let us consider the relevant empirical data, taken from experiment 1 of Jansen and van der Maas (2001). A paper-and-pencil version of the balance scale task was administered to over 300 children between 6 and 10 years of age. The study contained a pretest and posttest to assess children's performance on several items of various problem types. Between the pretest and posttest the investigators inserted a *hysteresis test*, a special series of items progressing stepwise from a minimum to a maximum difference in the distance of the weights from the fulcrum on the two sides of the scale, then stepping back down from the maximum to the minimum again. Key findings from the study were (1) a tendency toward bimodality in test scores (most children got all or none of the distance problems correct on the pretest and most got all or none of the distance problems correct on the posttest); (2) an upward shift in the distributions of scores from pretest to posttest (i.e., testing itself seemed to improve behavior); and (3) the presence of transitions in performance during the hysteresis test.

The original McClelland model exhibited some degree of bimodality in test scores, but no shift in the distributions of scores from pretest to posttest and no transitions during the hysteresis test. This was by virtue of its focus on longer term behavior change. Trial-by-trial changes, particularly those consequent on the system's previous behavior, were outside the scope of the original model. However, in new work, Schapiro and McClelland (in press) added three extensions to the model to allow it to address the Jansen and van der Maas (2001) data.

The first extension was to allow the model's own output to serve as a teaching signal that could lead to a change in behavior. The lack of a mechanism addressing how change can occur without a teaching signal was not explicitly recognized as a simplification in the initial formulation of the model; the focus was on how experience with situations involving weight and distance might lead to progress in understanding balance scale task behavior and the apparent succession of developmental stages. However, it is clear that when we behave, the responses we make can lead to changes in our behavior, and any satisfactory model should provide a mechanism that makes this possible. To address this, Schapiro and McClelland (in press) used the network's output for a given problem as the basis for assigning it a corresponding teaching signal. This assignment was based on the same discrete categorization of the network's continuous output that was used to assign one of the three possible responses (*left side down*, *right side down*, or *balance*). For example, if the activation of the left output unit was 0.333 or more greater than the activation of the right output unit, the output was scored *left side down* and a teaching signal of 1 for the left output unit and 0 for the right output unit was assigned.

The second extension was the use of the teaching signal to modulate the *gain* on the inputs to the hidden layers in the model during network testing. The adjustment of gain has been proposed as one way of implementing an attention-like mechanism in connectionist networks (Kruschke, 1992; Kruschke & Movellan, 1991). The gain is simply a scalar multiplier that scales the net input to the hidden units in the model. An increase in gain thus corresponds to an increased sensitivity to the inputs coming to the affected units. Following Kruschke (1992), we adopted the idea that dimensional attention, operationalized as an adjustment to a dimension-specific gain parameter, can be adjusted using the back-propagated error signal. In this case, the adjustment is guided by the teaching signal generated by the network's own output. Adjustment to the gain variable provides one possible mechanism that may explain changes in children's sensitivity to the distance dimension during testing.

The third extension was the incorporation of noise, or intrinsic variability, into network processing. This was motivated by earlier work (McClelland, 1991, 1993; Movellan & McClelland, 2001; Usher & McClelland, 2001) indicating the importance of including intrinsic variability in the activations of units throughout the network. Based on this work, a sample of normally distributed zero-mean Gaussian noise was added to a given unit's net input before its activation was calculated. The addition of noise is an example of a more detailed level of modeling that can be quite important but missed in a model that focuses only on overall developmental patterns.

Together, the three extensions were aimed at increasing the temporal granularity of the model from coarse developmental stages to trial-by-trial performance during testing. However, the modified model still contains simplifications. In other work, we and many other connectionists have assumed that processing within connectionist networks is itself a real-time, continuous process; indeed, it was with this specific assumption that McClelland (1979) first began to explore connectionist models, and this assumption is part of the set of basic processing principles that McClelland (1993) later proposed. But Schapiro and McClelland (in press) have maintained the simplification of relying on a single-pass feed-forward computation in the present extension of the McClelland (1989) balance scale model.

Schapiro and McClelland (in press) demonstrated that the addition of a self-generated teacher, along with the use of gain and noise, appeared sufficient to produce all the evidence of catastrophe-like transitions in behavior found in the data from Jansen and van der Maas (2001). First, the model now showed a pattern of bimodality quite similar to that seen in the Jansen and van der Maas data. Second, it showed a progression to higher scores from pretest to posttest. Third, it exhibited a pattern of transitions in the hysteresis test quite similar to that observed by Jansen and van der Maas. This included the pattern those authors viewed as most clearly demonstrative of a catastrophe, the so-called delay pattern. In the delay pattern, the participant switches from making an incorrect *balance* response to making a correct distance-based response at some point during a problem sequence with increasing distances; the participant then persists in making the correct response on the way back down through a problem sequence with decreasing distances past the point of the switchover during the earlier increasing sequence.

It is interesting to consider the consequences of the model's ability to account for the presence of these catastrophe flags. Is there really a phase transition happening in the model? Our inclination is to say no; the delay patterns displayed by the model (and many of the so-called sudden-jump patterns; for discussion, see Schapiro & McClelland, in press) appear to be the result of relatively small changes in unit activations.

In fact, there seem to be several indications in the Jansen and van der Maas (2001) data that the transitions exhibited by many of the children tested are actually more continuous than the presence of catastrophe flags might at first suggest. First, both on the pretest and the posttest, there are many children who get some but not all distance problems correct. The distance problems these children are most likely to get correct involve larger variations of distance. In other work, Jansen and van der Maas (2002) actually identified such a pattern as characterizing a group of children, and treated it, as we do, as a developmental phase that lies between the more categorical patterns associated with what Siegler (1976) called *rule 1* (relying only on weight) and *rule 2* (relying on distance only when weights are equal). Furthermore, about half of the transitions to the categorical rule 2 pattern on the posttest are from intermediate patterns on the pretest, and most of the transitions from the rule 1 pattern on the pretest are to an intermediate pattern on the posttest. These results suggest that many children have a graded sensitivity to the distance cue which increases over the course of the experiment, instead of a sudden realization that distance is important prior to showing no previous awareness of the distance dimension. Such a change in an underlying graded pattern is, of course, the core claim of the connectionist model, distinguishing it from approaches based on the actual representation and use of categorical rules, as in models such as those of Van Rijn, Van Someren, and van der Maas (2003) or Klahr and Siegler (1978).

There do seem to be a small number of children who really do make a bigger jump, moving from taking only weight into account on all the problems on the pretest to consistently relying on distance in the posttest when the weight on both sides is equal. This pattern, as well as some large jump patterns seen in the hysteresis test, is sometimes seen even in the Schapiro and McClelland model. However, the pattern is also consistent with the possibility that more categorical or stagelike change may be occurring in a small number of children. Granting this, there still appears to be a persistent tendency among researchers to see performance in the balance scale task as essentially more rule-like, and transitions as more discrete or categorical, than the data actually warrant. Several phenomena we have considered, and others that are not reviewed here (see McClelland, 1995, for more discussion) point to an underlying continuity, especially around points of transition, at least for a substantial proportion of children.

To return to our broader argument, what is of fundamental importance for the present analysis is the observation that, by replacing simplifying assumptions (such as noise-free processing) with assumptions that are considered crucial in dynamical systems research (such as intrinsic variability), we have increased the extent to which the connectionist models can be successful. The absence of intrinsic variability and short-term modulation of behavior in the McClelland (1989) model were not points of principle, but only of simplification. The successful outcome of the model's extensions, prompted by findings offered by proponents of the dynamical systems approach, points toward a future convergence in which the principles that both approaches have focused are seen as part of an integrated and improved approach that exploits the key insights of both.

## Embodiment

Corbetta (Chapter 3) illustrates the emphasis that DST frequently places on embodiment: "the body, with its physical properties, is the vital liaison between the mind and the outer world; this liaison is constant throughout the life span as we grow and interact with the environment." In this view, we cannot escape the fact that the mind is encapsulated within a body. Moreover, this body undergoes a series of changes throughout life—particularly during early development as we learn to carry out basic actions such as reaching, grasping, and walking. The extent to which our physical status constrains the formulation of our intentions to act upon the outer world is a topic of debate. The solution to this problem is intimately linked to the nature or indeed actual existence of internal representations.

There is a range of different perspectives on the nature of embodiment. One perspective, perhaps the middle of the road between extreme representationalist and antirepresentationalist approaches, is that embodiment acts as an additional constraint on cognitive processing (Mareschal et al., 2007). This stance recognizes the contribution of our physical status, while at the same time not rejecting out of hand or devaluing the significance of internal representations—a historical point of contention between connectionist and DST approaches to cognition. This perspective is consistent with the ideas of Glenberg (1997), who argues that representations held in memory reflect the structure of the environment making them analogical or *embodied*—because they are mapped to the outside world. This allows the representation of perceptual states to become meaningful in themselves through their use in interpreting the environment, making our own actions central to our understanding.

Evidence for physical constraints in our internal representations can be found in studies investigating the correlation between real rotary movements and mental rotation. These studies concluded that the same laws of motion govern mental rotation as actual movement (Decety, 1996; Georgopoulos & Pellizzer, 1995; Jeannerod, 1995; Wexler, Kosslyn, & Berthoz, 1998). The role of embodiment can be found even under conditions that do not directly invoke any physical aspects of task performance. For example, when Van den Bergh, Vrana, and Eelen (1990) presented letter pairs to typists and nontypists and asked them to express preferences between competing pairs, they found that typists preferred letter pairs that were typed with two different fingers to letter pairs that could be typed with the same finger. The non-typists showed no such preference. Judgments were implicit, in the sense that the typists were unable to verbalize an explanation for their preference. Van den Bergh and colleagues concluded that motor program information is encoded within the representations for letter pairs, which then influences the selection of the preferred pair for typists.

Consideration of embodiment can certainly lead to elegant explanations of phenomena that must be explained very differently (and perhaps incorrectly) from a cognitive or neurocomputational viewpoint. For example, Thelen and Fisher's (Thelen & Fisher, 1982) embodied account of the disappearance of infants' stepping reflex is that as the legs grow heavy with subcutaneous fat during development, the infant does not have the strength to lift them when supported upright. This contrasts with an alternative explanation that the disappearance reflects a process of cortical inhibition of the reflex. Thelen (1986) later found evidence against the inhibition theory by demonstrating that the infants could still make stepping movements under certain conditions, such as when placed on a treadmill.

Given the potential insights that embodied accounts offer, why have connectionists so often chosen to simplify their models by excluding the constraints of embodiment? Why have they instead construed developmental problems in terms of learning transformations between abstract (disembodied) representational states?

In many cases, connectionists would argue that this is because the phenomena they are targeting are those in which embodiment is less relevant—for example, in the study of language acquisition. Indeed, when DST researchers turned their attention to the sensorimotor basis of learning object labels in infancy, they too found that the infants' application of these labels generalized beyond the sensorimotor circumstances of acquisition. Object labels therefore seem to require a more abstract level of encoding than sensorimotor links (see Chapter 4). For instance, in the section Beyond Fields to Association fields, Smith postulates a new, more abstract level of representation in learning object labels. The *association field* in this account has no direct sensory input, is not a plan for action, and serves to free the mapping of word and object from spatially directed action plans.

The role of embodiment in constraining the design of explicit quantitative models becomes more apparent when we compare connectionist and DST models of precisely the same phenomenon. For this, we turn to the A-not-B task. The A-not-B task is a classic Piagetian task, in which infants demonstrate perseverative reaching behavior. The typical task set up consists of two covered hiding locations (A and B, respectively). During A-trials, the experimenter waves a toy near location A and hides the toy under the cover in that location. The infant then reaches for the toy. This procedure is repeated for several trials at location A. Following these trials, the toy is then waved near and hidden in location B. Typically, 8- to 11-month-old infants will (erroneously) continue to reach to location A to retrieve the toy on these B-trials (Piaget 1954; Diamond, 1985).

The connectionist account of perseverative reaching (Munakata, 1997, 1998; see Chapter 7) focuses on a competition between active and latent internal representations that link object locations to actions such as reaching and gazing. Active representations correspond to sustained neuronal firing for current events, implemented through recurrent connectivity. By contrast, latent representations correspond to a longer term memory of previous events, implemented through experience-dependent change to connection weights. During A-trials, the model learns that objects will be at the location A, thereby building up a latent representation that biases interest to that location and produces correct reaching performance on the A-trials. On the B-trial, the system must overcome its bias to reach to location A based on the observation that the object is now at location B. During early development, the strength of recurrent connections for maintaining active representations of current events is low. If there is a delay between the observation and the opportunity to reach for the object, infants may be unable to overcome the latent bias of location A on B-trials—resulting in perseverative reaching. Across development, the strength of recurrent connectivity is increased, allowing active representations to be maintained in memory with a sufficient strength to override the bias of latent representations—resulting in a decrease in perseverative reaching and correct reaching to location B.

The DST account, by contrast, explains perseverative reaching in the A-not-B task in terms of the infant's ability to sustain the cued location despite a *motor habit* of reaching towards location A on B-trials. The crucial difference is the claim that infants develop a *motor memory*—a long term memory trace generated by populations of motor planning field neurons. Infants develop this motor memory during reaching on A-trials and keep on using it even when no longer appropriate. This DST account is supported by evidence that the A-not-B error can occur even when infants are simply reaching to visible covers over empty containers (Smith, McLin, Titzer, & Thelen, 1995). Moreover, the influence of this motor memory on reaching dynamics can be eliminated by altering the infant's body position between the A-trials and the B-trial (Smith, Thelen, Titzer, & McLin, 1999; see also Chapter 4).

The DST model focuses on the evolution of activity in a dynamic field representation of motor memory over time. The dynamic field is influenced by the current sensory input, the most recent event, and long-term memories of previous reaches. When the field's activity exceeds threshold, the location of an activation peak drives a reach to a certain location in space (either location A or location B, in this case). Perseverative reaching occurs when the long-term memory comes to dominate the persisting activation from the last event (the B-trial). Errors are overcome across development through a change to an external control parameter, *h*, which modifies the excitability of the populations of neurons that encode the motor memory, increasing the probability that the model will enter a self-sustaining activation state, and in doing so retain the memory of location B during short delays (Thelen, Schöner, Scheier, & Smith, 2001).

These accounts do place a different emphasis on the role of embodiment in how the activation states of the models are characterized, but *underneath the hood*, how different are they? Both formalizations consist of two forms of memory that may be placed in competition; both involve an in-the-moment memory system that utilizes cycling activation in a recurrent computational circuit; both employ (externally applied) parameter changes that affect the strength of active representations of the cue on the B-trial in order to capture developmental change in levels of perseveration; both simplify the encoding of objects, spatial locations, and motor actions to unidimensional variables (representing, for example, a reach to location A). Also similarly, the connectionist model includes a learning mechanism for building up the latent representations of A-trials, while in the DST model the build-up accumulates in a *pre-shaped* or *long-term memory* field. However, the dynamics of the DST model enable it to account for the degree of random activity shown by infants (Thelen et al., 2001), while the connectionist model does not.

It appears the primary role of embodiment in the DST model is in its characterization of the dynamic field as the formulation of a motor plan, while the connectionist account includes more abstract internal representational states that intervene between sensory systems and motor behavior. Neither model actually incorporates any biomechanical aspects of reaching. Is the difference between the models just skin deep, then? Are these similar computational systems merely labeled in different ways, with no body in sight? Perhaps. But the difference in labeling nevertheless reflects the theoretical concerns of each set of researchers and it results in real consequences for the interface between model, theory, and empirical data. It leads the DST researchers to focus on bodily manipulations to the infant, on manipulations to the sensory properties of the objects, and on situations in which the role of motor habits may be adaptive in learning, as avenues of further research of the A-not-B phenomenon (see Chapter 4). DST also motivates researchers to probe the role of embodiment through the use of autonomous robots (see Chapters 2 and 18 for examples).

By contrast, connectionist researchers have focused on the graded nature of the internal memory representations without particular regard for their content, and therefore their potential to drive different behaviors (e.g., comparing reaching behavior versus gaze behavior) and to account for perseveration in other tasks (such as children's rule-guided behavior in card-sorting and speech interpretation tasks; see Chapter 7). In the example of A-not-B errors, then, the common developmental phenomenon has led connectionist and DST researchers to include many of the same assumptions and simplifications into their explicit quantitative models, while retaining subtly different emphases in the theories that these models are claimed to instantiate. It seems likely that a synthesis of the two would lead to a more complete model than either of the current models taken on their own. For instance, a more comprehensive model necessitates the integration of timescales, accounting for both trial-by-trial events (as in the DST model) and the longer term developmental picture (as in the connectionist account). Ultimately, such a model also requires fully embracing the principle of embodiment through the implementation of effector-specific encoding, putting both connectionist and DST equally to the test (see Chapter 18 for steps in this direction).

## STABILITY

Many connectionist models of development to date have explored the ability of neural networks to learn associations between representations encoding information relevant to generating behavior in particular cognitive domains. For example, infamously, Rumelhart and McClelland (1986) trained a network to learn the relationship between phonological representations of the present and past tense of English verbs, so that the system could output the correct past tense form when required. Rumelhart and McClelland explored whether the network would go through the same stages of development that children exhibit when learning this feature of language. Other examples can be found in models of reasoning, memory, and category/concept formation (see Elman et al., 1996).

If one puts the issue of embodiment to one side, DST researchers have expressed further reservations about models of this type. First, as Schlesinger points out, connectionist models have rarely investigated timescales at the fast end of real time (i.e., milliseconds), despite being inspired by the concept of neural processing that operates over such timescales (Chapter 9). Moreover, it is far from obvious that the abstract, stable representations employed in some connectionist models of development are realistic starting assumptions. (For example, connectionist models of language acquisition tend to assume that developmental processes operate over stable representations of word meanings, word sounds, and written word forms.) The real cognitive system, however, is in a continuous state of flux—the world usually offers a continually shifting stream of sensory data, much of it a consequence of the individual's own actions in the world. Perhaps in downplaying sensorimotor contributions to cognition, connectionists have created artificial, neat-and-tidy, abstract problems for their networks to solve, problems that are nothing like those faced by the child embedded in his or her own, continuously unfolding, subjective world. Indeed, perhaps the leisurely timescales over which these connectionist models operate are simply too blunt to reveal the key phenomena that characterize developmental change.

Stability is a concept that is central to dynamic systems theory. Schöner (Chapter 2) describes how cycling activation in recurrent circuits can produce representational states that are stable over time, both in being self-sustaining and (potentially, where appropriate) robust to perturbations. Given a gradual change in external input, the system can appear stable up until a certain point, when it may flip into another stable pattern of behavior. Changes in control parameters in the system may have similar effects, leading the system to change qualitatively in nature despite the quantitative (and perhaps linear) change in the control parameter. Moreover, before a flip takes place, it may be anticipated by a period of increased instability. Stability and instability in behavior, therefore, themselves become a focus of investigation in the study of developmental change. These arguments are often illustrated with the example of motor control but are also viewed as pertinent to the development of higher cognition (see, e.g., van der Maas & Raijmaker's analysis of children's reasoning, Chapter 15).

As before, our interest in this chapter is to consider whether there is a fundamental difference between connectionism and DST on the issue of stability, or whether the difference arises from model simplifications made in the service of explaining divergent developmental phenomena. Our sense here is very much the latter, and perhaps even that connectionist models are *better situated* to address issues of stability in development.

Mareschal, Leech, and Cooper (Chapter 10) convincingly argue that much of the disagreement on the centrality of stability lies in the different historical origins of connectionism and DST. Connectionism arose from the study of neural memory systems, where the objective is the retrieval of a stable representation of a memory given an appropriate cue. By contrast, DST arose from the study of motor control, where the task involves a continuous computational loop of motor commands given the goal and the unfolding sensory information that is (in part) the consequence of previous motor commands. In this domain, the adjustments are continuous.

It is true that connectionist models have tended to examine developmental change over longer time periods, simplifying questions of change over shorter time ranges. Thus, the model of sentence comprehension described in Box 17.1 assumes the existence of distinct representations of individual words and has a temporal dimension specified by the rate at which words arrive. It is not immediately clear what would be gained in such a model by including the millisecond range given the nature of the target behavioral phenomena to be explained, other than to force attention onto the issues of phoneme recognition and word recognition and away from those touching on sentence comprehension. But of course, it is always possible that new findings might emerge from extending the model to include just such constraints.

Importantly, however, connectionist models can and do operate at finer timescales. Simple recurrent models of the type proposed by Elman (1991) allow the researcher to study developmental change over months as the model adapts to the training set. They also allow the researcher to study the online recurrent processing dynamics as each subsequent word is processed in the sentence. Even simple feed-forward networks can be treated in this way, by allowing activation to build up in a cascading fashion rather than to be computed in a single pass (Cohen, Dunbar, & McClelland, 1990; McClelland, 1979). In these models, the chosen timescale can be arbitrarily small. With very small timescales, mathematically, the difference equations used in connectionist models approach the differential equations of DST. The temporally extended versions of connectionist models have allowed researchers to examine the consequences of persisting activation states in networks, for example to explain short-term priming effects in word recognition (Thomas, 1997). Mareschal and colleagues (Chapter 10) use precisely this approach in an attractor network to model the development of analogical reasoning in children. If the model is given an initial pair (*cat* is to *kitten* as...), cycling activation causes the network to settle into a state that encodes the implicit relation (*parent of*). When a new first term is applied to the network's input units (...*dog* is to...), this input combines with the persisting activation state to settle into the solution to the analogy at the output (*puppy*).

Crucially, the model of Mareschal and colleagues demonstrates how the settling activation states (*attractors*) exhibited by a recurrent connectionist network are created by a longer term, experience-dependent, development process. The Mareschal model is trained on the relationship between pairs of terms ("cat is parent of kitten"). As it develops this conceptual knowledge, the nature of the analogies it can draw *in the moment* alters. The model exhibits a developmental phenomenon known as the *relational shift*, in which its analogies move from being driven by perceptual similarities to relational similarities as a function of the knowledge that has been acquired. In the same vein, Thomas (1997) demonstrated how short-term and long-term, word priming effects can be reconciled by considering the first to be a consequence of persisting activation states in the word recognition system and the second a consequence of experience-dependent structural (weight) change produced by recognizing words (that is, an accumulation of short-term events).

DST examines how the attractor states that it builds into its equations impact on behavior. Connectionism is able to show how these dynamic properties arise as a consequence of learning over a longer time period—or as McClelland and Vallabha (see Chapter 1) put it, how new macroscopic behavioral properties emerge from microscopic mechanistic changes within adaptive systems. Connectionist models, therefore, offer a potential bridge between timescales—even if connectionist researchers have often focused their attention on developmental changes occurring at longer timescales. As previously discussed, a similar bridging of timescales has also been achieved by Schapiro and McClelland's augmented version of the earlier McClelland (1989) balance scale model.

Does the issue of stability solely revolve around the question of incompatible model simplifications? We would argue that there are a number of areas where it does not. First, as in the case of embodiment, empirical data about stability—of input, of representations—are additional constraints that must influence the construction of models. For the computation of syntactic relations in sentences, it may be reasonable to assume some prior availability of stable, word-level information. But in other domains, particularly those closer to the senses or to motor interfaces, an assumption of stability may be more questionable.

Second, connectionism usually commits to a richer representation of knowledge than the unidimensional variables present in DST models. This has led connectionism to face what is called the *stability-plasticity dilemma*, that is, how new knowledge may be incorporated into an information processing system while preserving existing knowledge (see Richardson & Thomas, 2008, for discussion). The stability–plasticity dilemma has particular importance where the individual's environment is nonstationary—that is, where the information content of experience tends to change over time. In models employing distributed representations, the stability of knowledge may be especially problematic and necessitate intermediate memory systems to *damp* the changes on long-term knowledge wrought by fleeting, in-the-moment experiences (see, for example, McClelland, McNaughton, & O'Reilly, 1995).

Lastly, some connectionist theorists have taken the issue of stability very seriously and argued that the presence or absence of stability in a dynamic representational state may have real consequences for the experiential states of the organism. For example, O'Brien and Opie (1999) proposed that stable, explicit, neural representations are the only states that contribute to the contents of consciousness. These authors further proposed that the connectivity within parallel distributed processing systems provides a set of *potentially explicit* representations that may influence future behavior, an idea similar to that of latent representations discussed previously in the context of the A-not-B models (see Chapter 7).

## Conclusion

We believe a constructive integration of connectionist and DST approaches is not only possible but desirable. It will be driven beyond the borders of GOFCD by the use of explicit quantitative models championed by connectionism and DST. We have argued that such models may appear to exaggerate the differences between DST and connectionist theories of development by virtue of their different simplifications in service of explaining different empirical phenomena. We have illustrated this point via the examples of the balance scale task, the role of embodiment, and the role of stability. Equally, we could have considered other points of debate, such as the level of abstraction or the appropriate dimensionality of representational states employed in explicit quantitative models, and we would have drawn similar conclusions.

Connectionism and DST share the greater part of their vision of cognitive development, a vision that is conditioned by the neurocomputational substrate that delivers cognition. The nature of (what we believe will be) an eventual convergence is as yet hazy on the horizon, but already some of its features can be discerned. We might speculate that such a unified theory of development would include concepts such as distributed and graded knowledge, experience-dependent change, attractor dynamics, partial representations, soft assembly, and the constraints of embodiment. The concept of cognitive development itself may have to expand to embrace constraints from as low as the genome, from as high as society, and from as wide as evolution (see Chapter 14). But, crucially, we also see the future in pluralistic terms. No one single set of assumptions makes sense for all models, and researchers within and between the two converging approaches will continue to exploit a range of different simplifications appropriate to the specific focus of their interests and the demands of the tasks and issues under consideration. This convergent, but still pluralistic, activity will continue to depend crucially on explicit quantitative models of cognition and behavior, and of the mechanisms of change. It is for this reason that developmentalists should care about this book and its consideration of debates arising between connectionists and dynamical systems theorists. The great challenge (and opportunity) that lies ahead is to apply such explicit quantitative models across the full range of developmental phenomena that constitute the richness of human cognition.

## Acknowledgments

This work was supported by UK Medical Research Council Grant G0300188 and European Commission grant NEST-029088 to Michael Thomas, and National Institute of Mental Health Centre Grant P50 MH64445, James L. McClelland, Director.

## References

Cohen, J. D., Dunbar, K., & McClelland, J. L. (1990). On the control of automatic processes: A parallel distributed processing account of the Stroop effect. *Psychological Review, 97*(3), 332–361.

Decety, J. (1996). The neurophysiological basis of motor imagery. *Behavioral Brain Research, 77*(1–2), 45–52.

Diamond, A. (1985). Development of the ability to use recall to guide action, as indicated by infants' performance on AB. *Child Development, 56*(4), 868–883.

Elman, J. L. (1991). Distributed representations, simple recurrent networks, and grammatical structure. *Machine Learning, 7*(2–3), 195–225.

Elman, J. L., Bates, E. A., Johnson, M. H., Karmiloff-Smith, A., Parisi, D., & Plunkett, K. (1996). *Rethinking innateness: A connectionist perspective on development*. Cambridge, MA: MIT Press–Bradford Books.

Georgopoulos, A. P., & Pellizzer, G. (1995). The mental and the neural: Psychological and neural studies of mental rotation and memory scanning. *Neuropsychologia, 33*, 1531–1547.

Glenberg, A. M. (1997). What memory is for: Creating meaning in the service of action. *Behavioral and Brain Sciences, 20*(1), 1–19, commentary 20–40, author's response 41–55.

Jansen, B. R. J., & van der Maas, H. L. J. (1997). Statistical test of the rule assessment methodology by latent class analysis. *Developmental Review, 17*(3), 321–357.

Jansen, B. R. J., & van der Maas, H. L. J. (2001). Evidence for the phase transition from rule I to rule II on the balance scale task. *Developmental Review, 21*(4), 450–494.

Jansen, B. R. J., & van der Maas, H. L. J. (2002). The development of children's rule use on the balance scale task. *Journal of Experimental Child Psychology, 81*(4), 383–416.

Jeannerod, M. (1995). Mental imagery in the motor cortex. *Neuropsychologia, 33*(11), 1419–1432.

Klahr, D., & Siegler, R. S. (1978). The representation of children's knowledge. In H. W. Reese & L. P. Lipsitt (Eds.), *Advances in child development and behavior* (Vol. 12, pp. 61–116. New York: Academic Press.

Kruschke, J. K. (1992). ALCOVE: An exemplar-based connectionist model of category learning. *Psychological Review, 99*(1), 22–44.

Kruschke, J. K., & Movellan, J. R. (1991). Benefits of gain: Speeded learning and minimal hidden layers in back-propagation networks. *IEEE Transactions on Systems, Man and Cybernetics, 21*(1), 273–280.

Mareschal, D., Johnson, M. H., Sirois, S., Spratling, M., Thomas, M. S. C., & Westermann, G. (2007). *Neuroconstructivism, Vol. 1: How the brain constructs cognition*. New York: Oxford University Press.

Mareschal, D., & Thomas, M. S. C. (2007). Computational modeling in developmental psychology. *IEEE Transactions on Evolutionary Computation* [Special issue on autonomous mental development], *11*(2), 137–150.

McClelland, J. L. (1979). On the time relations of mental processes: An examination of systems of processes in cascade. *Psychological Review, 86*(4), 287–330.

McClelland, J. L. (1989). Parallel distributed processing: Implications for cognition and development. In R. G. M. Morris (Ed.), *Parallel distributed processing: Implications for psychology and neurobiology* (pp. 8–45). New York: Oxford University Press.

McClelland, J. L. (1991). Stochastic interactive processes and the effect of context on perception. *Cognitive Psychology, 23*(1), 1–44.

McClelland, J. L. (1993). Toward a theory of information processing in graded, random, and interactive networks. In D. E. Meyer & S. Kornblum (Eds.), *Attention and performance: Vol. XIV. Synergies in experimental psychology, artificial intelligence, and cognitive neuroscience* (pp. 655–688). Cambridge, MA: MIT Press.

McClelland, J. L. (1995). A connectionist perspective on knowledge and development. In T. J. Simon & G. S. Halford (Eds.), *Developing cognitive competence: New approaches to process modeling* (pp. 157–204). Hillsdale, NJ: Erlbaum.

McClelland, J. L., McNaughton, B. L., & O'Reilly, R. C. (1995). Why there are complementary learning systems in the hippocampus and neocortex: Insights from the successes and failures of connectionist models of learning and memory. *Psychological Review, 102*(3), 419–457.

McClelland, J. L., Rumelhart, D. E., & PDP Research Group. (1986). *Parallel distributed processing: Explorations in the microstructure of cognition: Vol. 2. Psychological and biological models.* Cambridge, MA: MIT Press.

Movellan, J. R., & McClelland, J. L. (2001). The Morton-Massaro law of information integration: Implications for models of perception. *Psychological Review, 108*(1), 113–148.

Munakata, Y. (1997). Perseverative reaching in infancy: The roles of hidden toys and motor history in the AB task. *Infant Behavior and Development, 20*(3), 405–416.

Munakata, Y. (1998). Infant perseveration and implications for object permanence theories: A PDP model of the AB task. *Developmental Science, 1*(2), 161–184, commentary 184–204, author's response 205–211.

Munakata, Y., & McClelland, J. L. (2003). Connectionist models of development. *Developmental Science, 6*(4), 413–429.

O'Brien, G., & Opie, J. (1999). A connectionist theory of phenomenal experience. *Behavioral and Brain Sciences, 22*(1), 127–148, commentary 148–175, authors' response 175–192.

Piaget, J. (1954). *The construction of reality in the child* (M. Cook, Trans.). New York: Basic Books.

Quinlan, P. T., van der Maas, H. L. J., Jansen, B. R. J., Booij, O., & Rendell, M. (2007). Re-thinking stages of cognitive development: An appraisal of connectionist models of the balance scale task. *Cognition, 103*(3), 413–459.

Raijmakers, M. E. J., van Koten, S., & Molenaar, P. C. M. (1996). On the validity of simulating stagewise development by means of PDP networks: Application of catastrophe analysis and experimental test of rule-like network performance. *Cognitive Science, 20*(1), 101–136.

Richardson, F. M., & Thomas, M. S. C. (2008). Critical periods and catastrophic interference effects in the development of self-organizing feature maps. *Developmental Science, 11*(3), 371–389.

Rumelhart, D. E., & McClelland, J. L. (1986). On learning the past tense of English verbs. In J. L. McClelland, D. E. Rumelhart, & PDP Research Group (Eds.), *Parallel distributed processing: Explorations in the microstructure of cognition. Vol. 2: Psychological and biological models* (pp. 216–271). Cambridge, MA: MIT Press–Bradford Books.

Rumelhart, D. E., McClelland, J. L., & PDP Research Group. (1986). *Parallel distributed processing: Explorations in the microstructure of cognition: Vol. 1. Foundations.* Cambridge, MA: MIT Press–Bradford Books.

Schapiro, A. C., & McClelland, J. L. (in press). A connectionist model of a continuous developmental transition in the balance scale task. Cognition.

Siegler, R. S. (1976). Three aspects of cognitive development. *Cognitive Psychology, 8*(4), 481–520.

Smith, L. B., Thelen, E., Titzer, R., & McLin, D. (1999). Knowing in the context of acting: The task dynamics of the A-not-B error. *Psychological Review, 106*(2), 235–260.

Smith, L. B., McLin, D., Titzer, B., & Thelen, E. (1995). The dynamics of the A-not-B error. Paper presented at the meeting of the Society for Research in Child Development, Indianapolis, IN.

Thelen, E. (1986). Treadmill-elicited stepping in seven-month-old infants. *Child development, 57*(6), 1498–1506.

Thelen, E., & Bates, E. (2003). Connectionism and dynamic systems: Are they really different? [Special section/issue]. *Developmental Science,* 6(4), 378–391.

Thelen, E., & Fisher, D. M. (1982). Newborn stepping: An explanation for a "disappearing reflex". *Developmental Psychology, 18*, 760–775.

Thelen, E., Schöner, G., Scheier, C., & Smith, L. B. (2001). The dynamics of embodiment: A field theory of infant perseverative reaching. *Behavioral and Brain Sciences, 24,* 1–34, commentary 34–70, authors' reply 70–80.

Thomas, M. S. C., & Redington, M. (2004). Modelling atypical syntax processing. In W. Sakas (Ed.), *Proceedings of the First Workshop on Psycho-Computational Models of Human Language Acquisition at the 20th International Conference on Computational Linguistics* (pp. 85–92). Retrieved October 17, 2008, from http://www.colag.cs.hunter.cuny.edu/psychocomp/PsychoCompProceedings.pdf

Thomas, M. S. C. (1997). *Connectionist networks and knowledge representation: The case of bilingual lexical processing.* Unpublished doctoral dissertation, Oxford University, United Kingdom.

Thomas, M. S. C., & McClelland, J. L. (2008). Connectionist models of cognition. In R. Sun (Ed.), *Cambridge handbook of computational cognitive modeling.* New York: Cambridge University Press.

Usher, M., & McClelland, J. L. (2001). The time course of perceptual choice: The leaky, competing accumulator model. *Psychological Review, 108*(3), 550–592.

van den Bergh, O., Vrana, S., & Eelen, P. (1990). Letters from the heart: Affective categorization of letter combinations in typists and nontypists. *Journal of Experimental Psychology: Learning, Memory, and Cognition, 16*(6), 1153–1161.

Van Rijn, H., van Someren, M., & van der Maas, H. L. J. (2003). Modeling developmental transitions on the balance scale task. *Cognitive Science, 27*(2), 227–257.

Wexler, M., Kosslyn, S. M., & Berthoz, A. (1998). Motor processes in mental rotation. *Cognition, 68*(1), 77–94.

# CHAPTER 18

## Moving toward a Unified Theory While Valuing the Importance of the Initial Conditions

*John P. Spencer, Evelina Dineva, and Gregor Schöner*

Dynamical systems and connectionist approaches to development are fundamentally complementary (Smith & Samuelson, 2003). The two communities that have grown up around these approaches share a basic outlook on development, share a conception of what understanding development entails, and are committed to joint principles. These principles include the following: a firm grounding in neural function; an awareness of the role structured environments play during development; a mistrust of static architectural accounts, which both communities view as incompatible with learning/development; and a deep appreciation for the step-by-step, emergent, and self-organizing nature of developmental processes.

Our impression of convergence of these two approaches was not seriously destabilized at the conference from whence this volume originated, even though disagreements were openly and sometimes controversially discussed. Indeed, in several instances, we felt the disagreements that did arise were based on misperceptions rather than deep differences. That said, several *real* differences did arise, and both groups articulated their positions on issues that ultimately require clarity and resolution. We also discovered that sometimes there are subtle differences in outlook that are not readily perceived—differences in the role that theory and models play that, in the extreme case, can lead people to use the same mathematical equations to quite different ends. Such differences reside at a metatheoretical level and have practical consequences for what we communicate to the field of developmental science more generally.

We left the conference inspired and with the sense that the stuff of *real* development is well within grasp as our two communities work to understand development, to understand each other, and to ultimately capitalize on our complementary strengths. As we look toward this integrative future, it will be critical to establish a deep appreciation for the initial conditions, that is, to articulate and value the current strengths of each approach. This will foster a deep sense of integration that will push both ways of thinking to new levels. In this spirit, our so-called reactions chapter works to articulate and clarify cases of perceived differences between approaches that, in our view, are *not* real, as well as cases of perceived differences that *are* real. We conclude by discussing the implications of efforts to integrate the two approaches for developmental science more generally. In our view, clarifying similarities/differences between approaches offers far more that just technical clarity for co-called modeler types; it truly offers a vision of a new, integrative, developmental theory.

### Perceived Differences That Are Not Real Differences

In this section, we discuss two perceived differences that, in our estimation, are not real differences between dynamical systems theory (DST) and the connectionist (CN) approach: the neural commitment of each approach and the degree to which each approach addresses the processes that underlie learning.

### How Neural Are CN and DST?

Connectionism is explicitly based on a neuronal framework. The very definition of the connectionist program refers to this: parallel, distributed, processing neural units that integrate inputs across synaptic weights, architectures, and connectivity patterns that determine neural function, and so on.

Less clear is whether DST shares this neural/neuronal commitment. This is understandable given that several *flavors* of DST have emphasized the *collective* behavior of organisms, as in the cases of movement coordination (Schöner & Kelso, 1988), motor development (Spencer et al., 2006), early language development (van Geert, 1991, 1993), and educational theory (Fischer, 2008). Such efforts to understand the dynamics of collective behavior often over a developmental timescale have resulted in abstract dynamical models that capture, for instance, the rate of change in relative phase as people oscillate their fingers (Schöner, Haken, & Kelso, 1986) or the complex behavior of systems consisting of multiple *growers* that change in complex ways over developmental time (van Geert, 1998).

Although it is clear that not all flavors of dynamic systems theory have been firmly grounded in neural principles, some have. Our dynamic field theory (DFT) is an example. The dynamic field framework was originally developed to capture the dynamics of neural activation in the visual cortex (Amari, 1977). More recently, this framework has been extended to account for the processes that underlie saccadic eye movements (Kopecz & Schöner, 1995; Wilimzig, Schneider, & Schöner, 2006), motor planning (Erlhagen & Schöner, 2002; Schutte & Spencer, 2007), the dynamics of neural activation in the visual cortex and the premotor cortex (Bastian, Riehle, Erlhagen, & Schöner, 1998; Bastian, Schöner, & Riehle, 2003), the behavior of autonomous robots (Bicho, Mallet, & Schöner, 2000; Iossifidis & Schöner, 2006; Steinhage & Schöner, 1998), infants' performance in Piaget's A-not-B task (Thelen, Schöner, Scheier, & Smith, 2001), and developmental changes in visuospatial cognition (Simmering, Schutte, & Spencer, 2008; Spencer, Simmering, Schutte, & Schöner, 2007).

Reviews of the dynamic field approach can be found in the chapters by Schöner (Chapter 2) and Spencer (Chapter 5). For the purposes of the present issue—the neural commitment of DST—we focus on one particular example.

Erlhagen and Schöner (2002) proposed a dynamic field theory of movement preparation. One of the central concepts in this theory is that movement preparation can be *pre-shaped* by either perceptual *pre-cues* or by a *memory trace* of past motor decisions (a concept which also underlies performance in other tasks in early development, see Spencer & Hund, 2002, 2003; Thelen et al., 2001). For instance, in the case of perceptual pre-cues, dynamic field theory predicted that patterns of neural activation at the population level would be centered around pre-cued locations. If the pre-cued pattern was relatively precise, this would form a sharp distribution of activation associated with the average, pre-cued location. When the target cue was given, this sharp distribution could quickly shift in the direction of the target, leading to fast reaction times and accurate performance with sufficient planning time. Distributed pre-cues, however, would lead to a distributed activation pattern with consequently slower reaction times and lower accuracy under certain testing conditions (for successful tests of these claims, see Schutte & Spencer, 2007).

Although Erlhagen and Schöner (2002) demonstrated that DFT could capture behavioral patterns from a variety of paradigms, more direct support for the pre-shape concept came from neurophysiological evidence. In particular, time-dependent changes in neural activation in DFT were compared to single-unit neural activity in the premotor cortex of Rhesus monkeys measured in a pre-cue paradigm using population coding techniques (Bastian et al., 1998; Bastian et al., 2003; Erlhagen, Bastian, Jancke, Riehle, & Schöner, 1999). The first step in making this comparison was to map the responses of neurons to basic stimuli and create a continuous field by ordering the neurons based on their *preferred* stimulus directions. This was followed by a behavioral pre-cuing task that probed whether neural activation was pre-shaped as predicted by DFT (Bastian et al., 1998; Bastian et al., 2003; Erlhagen & Schöner, 2002). Results confirmed this prediction, showing a robust relationship between the dynamics of neural activation in real and simulated cortical fields.

Tests of DFT 's account of movement preparation clearly show that dynamic systems theories can be tightly linked to neural data. And there are many such examples from our own group (Bastian et al., 2003; Erlhagen et al., 1999; Jancke et al., 1999) as well as groups using related theoretical concepts (Pouget, Dayan, & Zemel, 2000; Rolls, Stringer, & Trappenberg, 2002). We also note that links between dynamic fields and neural data are not confined to single-unit or multiunit neurophysiology. For instance, a DFT account of movement preparation has been tested using ERP techniques (McDowell, Jeka, Schöner, & Hatfield, 2002). Moreover, recent ties between our dynamic field approach and more biophysical approaches to working memory suggest that neural measurements can also be used to test theories of developmental change (see Simmering, Schutte, et al., 2008).

DFT shows that dynamical systems approaches can have a strong neural commitment. Given this, where does DFT sit in the broad collection of dynamic systems approaches? Put differently, what is the link between DFT and other dynamical systems approaches that emphasize the collective behavior of organisms?

Central to DFT are the attractor states in which patterns of neural activation live. Much of our understanding of these attractor states comes from initial work by Amari (1977) who provided a quasi-formal analysis of the attractor layout of dynamic neural fields. Our work on spatial cognition, for example, has utilized Amari's analysis, emphasizing three particular attractors: a resting state, a self-sustaining state, and a self-stabilized state where peaks of activation remain stable only in the continued presence of input (see Spencer & Schöner, 2003; Spencer et al., 2007). We have a firm grasp on the parameters of the model that move it into and out of these states; we understand the bifurcations that arise when one state is destabilized and another state is created; and we have demonstrated how these attractor states give rise to observable behavioral/neural properties.

Importantly, this emphasis on attractors, bifurcations, stability/instability, and so on, has real consequences for the behavior of the systems we develop. As an example, we contend that stability is a fundamental requirement for *any autonomous, real-time behavioral system*. If this is the case, then understanding the attractor states of a neural network becomes a fundamental requirement for any theory of behavior. We discuss this claim in greater detail below.

Our emphasis on concepts of dynamical systems also connects DFT at a deep level with other dynamical systems approaches that emphasize these concepts at other levels of analysis and at longer timescales (Newell & Molenaar, 1998; van der Maas & Molenaar, 1992; Van Geert, 1998). In this context, we emphasize that not all approaches—DST, CN, or otherwise—need to be tightly linked to the neural level of analysis. Rather, the level of analysis should depend on the questions being asked and the phenomena under study (see van Geert & Fischer, Chapter 16).

Critically, however, dynamical systems concepts provide a link among these different behavioral and measurement levels. Our work on the Piagetian A-not-B error and related phenomena illustrates how this link can be quite generative. The dynamical systems approach to the A-not-B error was motivated largely from a conceptual analysis of infants' performance in the classic Piagetian task using the concepts of DST (see Smith & Thelen, 1993; Thelen & Smith, 1994). This conceptual analysis led to a host of novel behavioral predictions (Smith, Thelen, Titzer, & McLin, 1999) which were collected and formalized in the DFT account of infant perseverative reaching (Dineva, 2005; Thelen et al., 2001). This theory, in turn, enabled strong ties between work in infancy and work in later development on children's spatial working memory abilities (Simmering & Spencer, 2008). Additionally, there may be ways to test the developmental hypotheses that have emerged from this body of work in the near future using fMRI (see Edin, Macoveanu, Olesen, Tegnér, & Klingberg, 2007).

What, ultimately, will this move toward neural data buy us? In our view, moving to this level provides more constraints for theory and novel ways to test theoretical claims. Work by Bastian and colleagues (Bastian et al., 1998; Bastian et al., 2003; Erlhagen et al., 1999) described above provides an example of this use of neurophysiology—this measurement level provides direct evidence for the pre-shape concept, testing in precise detail whether our population dynamics account offers an accurate view of the brain's role in the behavioral pattern under study.

Thus far, we have described how DST can have a strong neural commitment. In this context, we can also ask how committed connectionism is to neural principles. Real neurons in the higher nervous system typically have 10,000 synapses, not the limited number that most connectionist models can accept. Moreover, real neural networks are overwhelmingly recurrent. Braitenberg (Braitenberg & Schüz, 1991) estimated that more than 95% of all neuronal connections are part of recurrent loops rather than part of feed-forward paths. The simplifications commonly used in connectionist networks—a small number of neurons with 50 or so connections—seem like reasonable simplifications of this complexity until one considers that well-known learning algorithms do not generalize to more complex networks with high degrees of interconnectivity (Braitenberg & Schüz, 1991; Douglas & Martin, 2004). The status of these simplifications is important in determining what CN tells us about how brains really work.

Furthermore, in contrast to the demonstration of tight links between neurophysiological recording and activation in dynamic neural fields (Bastian et al., 1998; Bastian et al., 2003; Erlhagen et al., 1999), connectionist networks have been linked to neural data primarily at the macroscopic level, where a few units of computation stand for activation in entire brain areas. The connectivity of the network then reflects the macro-connectivity between such areas (Cohen, Braver, & O'Reilly, 1996; Cohen & Servan-Schreiber, 1992; O'Reilly, Braver, & Cohen, 1999). This is a useful approach and a timely one, given the enormous amount of data emerging from neuroimaging laboratories. But this approach highlights that connectionist networks may provide more abstract descriptions of neurophysiology than variants of dynamical systems thinking that are grounded in neurology/neurophysiology.

In summary, although there is a sense in the literature that CN has a stronger neural commitment than DST, this is not the case. There are differences in how both approaches have been linked to neural data, but both approaches have a shared commitment to neural function. It will be useful to clarify the differential strengths/weaknesses of each approach in this regard as we collectively try to integrate brains, behavior, and development.

Critically, however, we must place this future work in the correct context. Such work is geared toward refinement of theoretical concepts, tighter/richer links to neural data, and so on. It is not about determining whether one group is more committed to neural function than the other. We feel that the commitment to a grounding of perceptual, motor, and cognitive function in principles of neuronal function is a shared and deep conviction between the CN and DST approaches. We are much closer here than it first appears. In fact, when contrasted with the still largely dominant information processing approach and good old-fashioned cognitive development (see Chapter 13), our differences appear negligible.

### How Much Do CN and DST Address Learning?

As with claims about neural grounding, there exist misconceptions regarding learning. In particular, DST thinking is viewed as relatively mute on learning while CN models are assumed to confront learning head on.

We certainly agree that the CN approach has led to major advances in our understanding of particular classes of learning. For instance, the CN approach has fostered a revolution in thinking about statistical learning and the ability of neural systems to extract statistical patterns from complex and rich inputs. This work has had a particularly strong impact within the domain of language (e.g., Altmann, 2002; Elman, 1990; MacDonald & Christiansen, 2002), and stands in opposition to nativist claims about language acquisition (Elman et al., 1996; Spencer et al., in press).

But this is only one aspect of learning. As we highlight below, dynamical systems approaches to learning have much to offer that is complementary to insights revealed by the CN approach. In particular, DST offers a path toward process models of learning that explain how learning is constrained by an organism's own activity and ability to determine its own stimulation. By this view, learning is not solely about the statistics of input; learning is also about the statistics of experience, of internal states, and of the systems' own autonomous decisions, all occurring within the context of the systems' intrinsic dynamics.

Dynamical systems thinkers have examined learning within a variety of domains. We highlight two here: motor learning and spatial cognition. Within the domain of motor learning, there have been at least two dynamical systems approaches to learning. The first is summarized by Newell and colleagues in Chapter 6; thus, we focus solely on the second DST approach to learning, which emerged from the study of coordination dynamics by Schöner, Zanone, and Kelso (1992).

Schöner and colleagues examined learning within the *finger twiddling* paradigm where participants move their fingers at a particular relative phase (e.g., in-phase [0°] when the index fingers move toward the palm in synchrony, or anti-phase [180°] when one finger moves toward the palm while the other moves away). A series of empirical and theoretical analyses of these coordination patterns revealed that they showed key *signatures* of a dynamical system with attractors at 0° and 180° (Scholz & Kelso, 1989; Schöner & Kelso, 1988). Within this context, the learning question was simple: what happens as people learn a new pattern of coordination, for instance, 90° relative phase?

To examine this question, Schöner, Zanone, and Kelso (1992) first had to specify what it is that is being learned as people acquire a new coordination pattern. Schöner (1989) proposed that learning can be conceptualized as a dynamic process that operates on the coordination dynamics but at a slower timescale. This means the people should be learning about the *dynamics* of the coordination pattern and not simply about the specific, target coordination pattern (e.g., 90° relative phase). Moreover, because the dynamics at the learning timescale have their own stability properties, there can be dramatic, qualitative transitions over learning which lead to large fluctuations at specific times during the learning process.

Empirical observations confirmed these insights (Zanone & Kelso, 1992, 1997). As people learn the new coordination pattern (90° relative phase), the entire landscape of coordination patterns changes: the anti-phase pattern (180°) becomes unstable and the attractor landscape shows a qualitative shift toward the new pattern (90°). This qualitative shift is accompanied by large fluctuations that are distinct from fluctuations at the faster timescale of real-time behavior. Thus, learning is about changing the dynamics of the system rather than learning the details of each individual pattern; as a new attractor is created, others can be de-stabilized (see also Schöner & Dineva, 2007).

A second dynamical systems approach to learning has emerged within the framework of DFT. This approach to learning emerged from a fundamental observation. Organisms determine their own stimulation by exploring their surrounds, looking at different things, moving through the world, and so on. These activities strongly constrain the external stimulation they receive. But the organisms' own *internal* activity—the second-to-second decisions they make about what to attend to, what is task relevant, what must be remembered—also places strong constraints on what is learned.

The DFT approach captures this sense of active learning as our work on infants' perseverative reaching illustrates. Based on an analysis of data from hundreds of infants who participated in the Piagetian A-not-B task, Dineva and colleagues (Dineva, 2005; Dineva, Schöner, & Thelen, 2008, in preparation) have demonstrated that the trial-by-trial decisions infants make in this task have a profound impact on their performance on the critical B-trial. In particular, infants who spontaneously reach to location B on an A trial have a significantly weaker tendency to reach back to location A on a subsequent B-trial. Thus, it is the infants' own spontaneous decisions—and not just the statistics of input—that structure what is learned; the infants' decisions affect their ability to make subsequent decisions.

Within the dynamic field framework, we have captured these findings using a simple form of Hebbian learning implemented using memory traces. In particular, when a peak of activation emerges in, say, a motor planning field that makes decisions about which lid to pick up in the A-not-B task, it leaves a trace of activation in a pre-shape or long-term memory (LTM) field (Dineva et al., 2008, in preparation; Simmering & Spencer, 2008). Such traces build up over a slower timescale than the activation dynamics within the planning field; they can become quite focused as similarly positioned peaks form from trial to trial. Thus, if the system builds a peak associated with location A over and over, a robust LTM trace associated with location A will also form.

Critically, activation in LTM fields is coupled back into the field of origin. In this way, LTM traces can bias which decisions/peaks are formed in the motor planning field on subsequent trials. This is the mechanism that underlies infants' perseverative reaching in the A-not-B task. As 8- to 10-month-old infants repeatedly reach to location A, they build up stronger memory traces, which make it more likely that they will reach to location A again on subsequent trials; that is, they form habits. When the hiding location is switched to location B, however, the strong traces associated with location A can lead to perseverative reaching if infants are not able to sustain activation associated with responding to location B. Note that this is less likely to occur if the memory traces associated with location A are weaker and there is some trace associated with location B—precisely what occurs when infants make spontaneous errors (i.e., reaches to location B on an A trial) in the A-not-B task.

The tendency to perseveratively reach to location A in the canonical A-not-B task disappears around 12 months of age as infants develop the ability to sustain an active or *working* memory for location B in the context of the perceptual cues available in the task. Nevertheless, the same LTM mechanism leads to systematic trial-to-trial distortions between the ages of 2 and 6 years when toys are hidden in a sandbox instead of beneath visibly marked hiding locations (Schutte, Spencer, & Schöner, 2003; Spencer, Smith, & Thelen, 2001). In this case, children's reaches to a B location in a sandbox are biased in the direction of a nearby A location after a short delay.

Given that our LTM fields implement a form of Hebbian learning, what is the link between this learning mechanism and Hebbian learning in connectionist networks? In many cases, these different forms of Hebbian learning lead to the same outcomes. For instance, just as strengthening synapses in a connectionist network makes it more likely that a stimulated neuron will excite other neurons to which it is strongly connected, LTM traces in DFT boost activation within a local region of a field, making it more likely that a local group of neurons will excite one another to, for instance, form a stabilized activation peak.

Similarly, just as Hebbian learning can lead to the extraction of statistical patterns from input, memory traces in DFT can reflect the statistics of past decisions (i.e., past peaks). For instance, we have shown that such traces produce a bias toward the average responded-to location in spatial planning and memory tasks (Schutte & Spencer, 2007; Spencer & Hund, 2002; Spencer et al., 2007). Traces are sensitive to frequency information as well. If, for instance, one location in a set of three close locations is more frequent than the others, spatial recall responses will not be distorted toward the average location; rather, recall responses will be pulled toward the frequent location (Spencer & Hund, 2002). Note that there are some differences between our implementation of Hebbian learning and Hebbian learning in connectionist networks. These differences are discussed in the section that follows.

In conclusion, the examples of DST approaches to learning highlighted in this section illustrate two points. First, we believe there is a strong, shared commitment to understanding learning in both the CN and the DST approaches. Second, we think both approaches offer unique insights into learning. With regard to the DST approach, dynamical systems concepts offer insights into the process of learning, usefully confronting classes of learning that extend beyond the boundaries of statistical learning. We think such efforts are needed to capture aspects of *active learning* that most certainly play an important role over development.

## Perceived Differences That Are Real Differences

Although there are shared commitments to the issues of neural grounding and learning, the DST and CN approaches do differ in important ways. We discuss six real differences below, highlighting issues that we think must be addressed head-on as we collectively build on the differential strengths of the two approaches.

### Learning the Statistics of Decisions Versus Inputs

Although the learning mechanism we use in DFT implements a form of Hebbian learning because this learning is a direct result of field dynamics (e.g., the formation of an activation peak) it is not identical to the type of Hebbian learning implemented in most connectionist networks. Put simply, learning in DFT is a reflection of *past decisions* rather than *past inputs*. This is not a trivial distinction.

Consider the dynamic field model of the A-not-B error described above. There are two central points of contrast between this model and the PDP model proposed by Munakata (Munakata, 1998) with regard to learning. First, the way the output of the PDP model is mapped to infants' performance in the A-not-B situation is to interpret the level of activation of each response node as a probability of responding. Thus, the PDP model does not make a response on each trial per se. Second, the responses generated on each trial do not feedback and affect synaptic weights in the network. Because of these features in the model, it cannot account for spontaneous errors and the effect such errors have on perseveration.

This contrasts with the dynamic field model proposed by Dineva and colleagues (Dineva, 2005; Dineva et al., 2008, in preparation) where an active decision—a stabilized peak of activation—is generated at the end of each trial. This difference in read-out has consequences: because DFT forms a stabilized peak of activation corresponding to location A or B at the end of each trial, the model generates a robust trace of each decision. Such traces are necessary to explain why spontaneous errors in the A-not-B task have such a dramatic effect on infants' perseverative reaching. The alternative—accumulating small fluctuations in the activation of response nodes from trial-to-trial—does not create enough of a bias toward location B to account for the data and, in some cases, to override the statistics of input (see Dineva et al., 2008, in preparation).

A second example from spatial cognition highlights another way in which Hebbian learning in DFT and CN models differ. Spencer and colleagues have shown that spatial recall responses are biased away from perceived reference frames during memory delays. If, for instance, people are asked to remember a location to the left of a visible line, they exaggerate the *leftness* of the location after a short memory delay (see Simmering & Spencer, 2007). Critically, several lines of evidence suggest that these spatial recall biases reflect repulsion away from perceived reference axes (rather than attraction toward spatial prototypes as proposed by Huttenlocher, Hedges, & Duncan, 1991; see Spencer et al., 2007 for discussion). This type of bias creates an interesting case study because working memory is *drifting in a direction with no clear input.* Importantly, results have shown that these drift effects are superimposed on LTM biases. For instance, biases over learning in spatial recall tasks are not precisely toward the average remembered location. Rather, biases over learning can be systematically distorted away from reference axes in the task space (Schutte et al., 2003). This provides empirical evidence that the LTM traces learned by participants are a reflection of distortions that occur in working memory rather than in the statistics of the input distribution.

A final example supports another aspect of our account of learning—that memory traces can directly affect the dynamics of decision making in dynamic fields, leading to nonlinear changes over learning. As memory traces build in DFT, they effectively elevate the overall level of excitation within a local region of the field. Such increases in activation can make it more likely to form self-sustaining activation peaks, that is, to move from one state (an input-driven state) to a qualitatively different state (the self-sustaining or *working memory state*). This has been empirically demonstrated recently by Dineva and Smith (2008, in preparation) who used distributed memory traces to make it more likely that infants would be able to sustain a decision to reach to location B in the A-not-B task. The result: with distributed practice, 8- to 10-month-old infants were able to sustain a memory of location B during a 3–5-second delay—a delay that typically produces perseverative responses. Note that this example parallels work on learning and coordination dynamics by Schöner (1989) in that learning in this case directly affects the dynamics of the system (see also Schöner & Dineva, 2007).

### Autonomy and Stability

One generally perceived difference between the DST and CN approaches is that dynamicists have focused more on the real-time behavior of organisms in time-varying and structured environments. We think that this perception is correct.

Organisms do not sit around waiting to receive a stimulus to give a response. They actively generate behavior and, thus, control their own sensory stimulation. The resulting continuous flow of behavior, which we like to think of as an expression of the fundamental autonomy of organisms, is not simply understandable as a form of stimulus–response chaining. Instead, the current behavioral state and recent behavioral history both matter and impact on upcoming behavioral decisions. The dynamical systems approach is centrally concerned with such autonomy and the associated real-time and closed-loop coupling of an organism to its environment (for related concerns, see Turvey, 1990; Warren, 2006).

Any system that controls its own sensory information must have stability properties that ensure that small changes in sensory information lead to small changes in behavior and vice versa. We have known this in the technical domain since the inception of control theory (Mees, 1981). Nevertheless, this fundamental constraint is often overlooked in models that address brain–behavior relations.

An insightful example comes from our own work. In the original dynamic field model of perseverative reaching in young infants (Thelen et al., 2001), activation evolved under the influence of various inputs (e.g., the presentation of a toy, vision of lids in the task space). To generate a response at the end of the delay, the location in the field with maximal activation was mapped onto the reaching target that was closest to that field location. This is a standard read-out procedure used in many other approaches that involve piercing a threshold of activation to make a response (Schall, 2004).

When we embodied this model on an autonomous robot, however (something pushed for by the ever-insightful Esther Thelen), we discovered that our read-out procedure was missing something fundamental—stability! Because the robot was continuously linked to perceptual input and motor output, the representation of the movement direction after the delay—that is, the maximal activation value in the field that specified whether the robot should turn left or right to acquire the target—was pushed around by all sorts of things. The result: the robot would change course erratically, wiggling its way somewhere in-between the two targets, and not reaching either.

The solution was simple: we needed to stay true to the dynamical systems concepts we had proposed. In particular, rather than reading out the maximal activation in the field after the delay—something that the robotic implementation showed was intrinsically unstable—we needed to have the model generate a *stable* decision to reach to location A or B. This led to the next iteration of our account of infant perseverative reaching (Dineva et al., 2008, in preparation). As described previously, this new model generates a stable decision—a stabilized activation peak—at the end of each trial due to a boost in the inputs (which corresponds to pushing the box toward the infant in the A-not-B task at the end of each trial). Recent work with an autonomous robot shows that this works: our robotic implementation now effectively captures the details of decision making in an A-not-B situation (Dineva, Faubel, & Schöner, 2007).

Note that there was an important corollary of this work with regard to learning: stable decisions also enable organisms to build up different behavioral histories depending on the choices they make. As described previously, small fluctuations in the activation of response nodes do not generate sufficient differences across trials to explain how infants' own decisions impact subsequent behavioral events. Stable decisions, by contrast, are an effective way to structure trial-to-trial learning, and simulations of the new DFT explain how the behavioral history of individuals in the A-not-B task predicts the extent to which they perseverate (for related results, see Clearfield, Diedrich, Smith, & Thelen, 2006; Dineva et al., 2008, in preparation; Schöner & Dineva, 2007).

### Embodiment

A prerequisite for autonomy is that cognitive systems can be interfaced with the sensory and motor surfaces of a body and that cognitive processes can be acted out by such an embodied system in a real, structured environment. We believe that DST and CN differ in the extent to which they enable and are committed to embodiment.

The best way to prove that a model may be embodied and situated is to implement it on an actual body—a robotic device—with real-time sensory and motor systems, and to have the robotic system generate behavior in a natural environment. This has been done extensively within the realm of DST. In fact, an approach to autonomous robotics inspired by the principles of DST has been developed and demonstrated in robotic navigation tasks (Bicho et al., 2000; Schöner & Dose, 1992; Schöner, Dose, & Engels, 1995; Steinhage & Schöner, 1997), in tasks involving robotic cooperation (Large, Christensen, & Bajcsy, 1999; Soares, Bicho, Machado, & Erlhagen, 2007), in tasks involving autonomous reaching by robot arms (Iossifidis & Schöner, 2006), and in tasks that involve fast learning and recognition of objects in the work space of a robotic arm (Faubel & Schöner, 2008). This approach literally employs the same attractor dynamics and instabilities used to model the behavior of human infants, children, and adults to generate robotic behavior acted out using conventional robotic hardware in real-world situations.

Connectionist models have proved more difficult to implement in robotic devices (but see Schlesinger, Chapter 9; Lovett & Scassellati, 2004; Sirois, 2005). One reason is that connectionist models commonly are interfaced with sensory and motor systems in relatively abstract ways. For instance, Morton and Munakata (Chapter 7) use a discrete set of input neurons to represent the details of lids and toys in their model of the A-not-B task. Similarly, they use a discrete set of input neurons to represent objects such as trucks or flowers in their model of the Wisconsin card sorting task. As another example, the categorization model of McClelland and Vallabha (Chapter 1) has input units representing plants or animals. These interfaces essentially presuppose that an object recognition system extracts the appropriate categorical information from raw sensory input to represent these objects. The interface between model and sensory surface, thus, presupposes that fundamental problems of vision such as segmentation and invariant object recognition are somehow solved. By contrast, we have demonstrated that dynamic neural field models can work with relatively *raw* sensory inputs, even in cases where some form of *higher level* learning must take place such as learning the names for objects (Faubel & Schöner, 2008).

At the other end, single neurons in CN models sometimes represent planned actions. When the activity of these neurons exceeds a threshold, the response is assumed to be generated. But as our robotic implementation of the A-not-B task showed, this form of read-out falls far short of the temporal and spatial dynamics required to control an actual effector system. To realize the real-world control of an effector, one must tackle the problems of initiating action, stabilizing decisions (even when the sensory information that led to the decision is lost during movement), controlling the effector, and updating motion plans in closed loop.

Although simple input and output neurons might be a reasonable simplification for an abstract model of behavior, one cannot overlook such problems if the demonstration of embodiment is a goal (which, to be clear, is not an explicit goal of many CN approaches). Interestingly, when connectionist models have been implemented in robotic systems, they start to look remarkably like dynamical systems models (see, for example, Sirois, 2005). For instance, Schlesinger and colleagues (Schlesinger, Amso, & Johnson, 2007) studied visual selective attention in infants and proposed a model that was applied to real-world images. The raw image stream was first filtered to produce feature maps (reflecting intensity, oriented edges, motion, and opponent-color contrast). In each feature dimension, short-range excitatory and long-range inhibitory connections worked for several iteration steps. Then, the sum of feature intensities was used to form a saliency map. For read-out, a stochastic process mapped saliency to a probability distribution for fixation. Given that neural representations in this model operate over continuous feature dimensions via short-range excitation and long-range inhibition, these models are similar to the types of dynamic neural fields used in our own work. Thus, this example highlights how there can be clear convergence between the two approaches when there is a shared commitment to embodiment.

Although convergence is possible, embodiment has certainly been more of a focus within DST than in the CN approach. The critical question is whether this difference matters. Put differently, is it acceptable to use abstract input–output nodes for the sake of simplification, or does this simplification miss something critical? We suspect that this simplification does hide some important aspects of cognitive and neural systems, even when applied to issues in higher level cognition. This suspicion comes from our work developing a formal account of *embodied* cognition within DFT.

Our use of stabilized activation peaks as a form of working memory provides one example (Spencer et al., 2007). This form of working memory differs in several ways from CN instantiations of working or active memory. Active memory traces are transient in nature, slowly returning to a resting level. Such transients have limited stability and will tend to blend inputs presented to the network provided that sufficient time elapses. Working memory peaks in DFT, by contrast, are in a stable attractor state capable of suppressing distracting inputs via lateral inhibition or maintaining separate peaks simultaneously in working memory. This mechanism has behavioral consequences that can be measured by experiment (e.g., Johnson & Spencer, 2008). For instance, we recently developed a dynamic neural field model of multi-item working memory and change detection (Johnson, Spencer, Luck, & Schöner, in press). This model made the counterintuitive prediction that visual working memory for similar non-spatial features (e.g., two similar colors; two similar orientations) would be enhanced relative to memory for dissimilar items. This effect arises in the model because close peaks share lateral inhibition, effectively sharpening locally excitatory interactions. Most theories of working memory predict the opposite—that there should be interference when similar items are held in memory. Behavioral results confirmed that adults are, in fact, better at detecting changes when they hold similar items in visual working memory.

In what sense is this prediction linked to embodiment? We have explicitly linked the working memory construct to self-sustaining peaks in DFT based on our past work implementing neural fields on autonomous robots.

This work led to the insight that working memory must be stable to effectively drive real behavior. Importantly, there is a conceptual bridge between working memory in DFT and classic ideas discussed by Baddeley (1986). Baddeley defines working memory as a type of memory used in the service of other cognitive operations—it does some *work* for the cognitive system. In our view, such work requires stability as a prerequisite. Just as stability is needed when a movement decision is required to drive a robotic arm in real time, we suspect stability is needed when working memory must help drive a cognitive operation in real time.

We believe that the pressure to examine the embodiment and situatedness of cognition will ultimately be a force of convergence between our two approaches. As models by Schlesinger and colleagues (Schlesinger et al., 2007; see also Schlesinger & Parisi, 2007) and Sirois (2005) attest, embodiment is not outside the purview of the CN approach. We encourage connectionists to take a closer look at this issue.

### Higher Level Cognition

The previous two sections highlighted how the DST approach has been used to examine the issues of autonomy and embodiment, issues that have been explored less thoroughly by connectionists. The state of affairs is reversed in the case of higher level cognition: the CN approach has been much more thoroughly applied to problems in higher level cognition. Multiple examples in this book bear this out, including work on the acquisition of grammar (Tabor, Chapter 8), analogy (Mareschal et al., Chapter 10), and language development (McMurray et al., Chapter 11).

Although the application of DST to issues in higher level cognition has been limited, there are some notable efforts in this direction. Van Geert and colleagues (Van Geert & Van Dijk, 2002; see also Van Geert & Fischer, Chapter 16), for example, have used dynamical concepts to examine individual trajectories in language development. Similarly, Van der Maas and colleagues (Jansen & Van der Maas, 2002; see also van der Maas & Raijmakers, Chapter 15) have used dynamical concepts to understand developmental changes in performance on the balance scale task. And more recent work within the dynamic field framework has examined several issues in early word learning (see Samuelson, Schutte, & Horst, 2008; Smith, Chapter 4), including an embodied approach to the fast acquisition of novel words in an autonomous robotics setting (Faubel & Schöner, 2008).

Considered in the context of the issues of autonomy and embodiment, one can readily see the complementary nature of the CN and DST approaches; each approach has its respective strengths. A fundamental goal for the future is to clarify these strengths and work to build on them to achieve systems that can develop higher level forms of cognition yet stay grounded in the real-time details of sensorimotor systems and real-world environments.

### DST and CN Have Different Scientific Histories

Although we have described the CN and DST approaches as complementary, the presence of differences can be confusing, particularly given overlap in the concepts and mathematical equations we use. In a formal sense, for instance, all recurrent neural networks are discrete time dynamical systems. Similarly, there are instances of connectionist models which are formulated all the way through as continuous time dynamical systems, sometimes using the same equations that dynamists claim for themselves (e.g., McClelland & Vallabha, Chapter 1; Usher & McClelland, 2001). Conversely, dynamic neural fields can be viewed as a special class of recurrent neuronal networks (Amari, 1977; Wilson & Cowan, 1973). If the equations are the same, then why are there differences between these approaches? One reason is that, due to our unique scientific histories, models play a different role in the two lines of thinking.

Connectionism emerged within the tradition of mathematical psychology and the information-processing approach to cognition and development. This tradition places a heavy emphasis on models of cognition and is concerned with the identification of the model with the best fit that is also compatible with the broadest set of experimental facts. Note that connectionism made a fundamental contribution to mathematical psychology in that it expanded the range of relevant evidence to include neuroscientific observations at multiple levels. To illustrate this scientific tradition, consider the model of perceptual decision making proposed by Usher and McClelland (2001; see Chapter 1). The mathematics of this model are, essentially, a special case of the generic dynamic field model of Amari (1977) in which only two locations of a field are modeled—the two associated with one perceptual choice or another. The perceptual decision is conceptualized as an attractor. And the stability of the attractor is varied depending on, for instance, the strength of available sensory input.

To test this account, the model has been set in competition with a set of alternative models, including diffusion models. The central question here is which model fits reaction time distributions more effectively and how closely each model maps onto neural observations. A somewhat biased analysis by Smith and Ratcliff (2004) provides an example of this style of thinking and arrives at the conclusion that a diffusion account fits more facts than the competitive accumulator model of Usher and McClelland (2001).

Although this is a valid way to test competing theories, the tradition within DST is to take a more qualitative view of theoretical comparisons. For instance, the competitive accumulator model has some obvious conceptual strengths that are unmatched by the diffusion account. Unlike the decision boundaries invoked in diffusion models, it is easy to see how choices in the competitive accumulator model could reflect learned neuronal representations. Moreover, the diffusion model is inherently nonautonomous (e.g., how is the diffusion process reset to prepare for the next decision? what neuronal processes amplify small changes at a decision boundary into a macroscopic behavioral act?) The competitive accumulator model, in contrast, could easily be extended to address the dynamics of continuous, online behavior.

Importantly, this emphasis on qualitative comparisons between classes of models has deep roots in the application of dynamical systems concepts within psychology. In dynamical systems thinking, the identification of a specific model is a relevant goal only in a limited sense. Providing quantitative fits between a dynamical model and a set of experimental data is primarily an existence proof, showing that there is at least one model that is consistent with all known facts. Typically, however, such fits do not provide a way of uniquely identifying a single model as the only plausible formulation for a given system. Why? The problem is that dynamical systems are strongly nonlinear and multicausal. A whole equivalence class of models may bring about a particular state, and different members of the equivalence class may be used to invariantly realize some function, depending on the details of the environmental conditions or behavioral history. This makes it difficult (perhaps impossible) to uniquely identify the system in question.

This is one of the reasons why DST has emphasized theoretical concepts rather than models (see, for example, Schöner, Chapter 2). Indeed, when we think of the new "Unified Theory" mentioned in the title of this volume, we think of the shared theoretical concepts that have emerged from the CN and DST approaches rather than an effort to develop a new "Unified Model" of development.

This is also one of the reasons why much of the early work within the DST framework was explicitly aimed against information processing and the identification of the modules and stages of processing that underlie cognition (Port & van Gelder, 1995). In contrast, emphasis was placed on evaluating, for instance, the concept of stability and loss of stability as a necessary ingredient of understanding behavioral patterns (van der Maas et al., 2006; van der Maas & Molenaar, 1992). DST predicts instability and certain signatures of instability that investigators should be able to observe, such as enhanced fluctuations and critical slowing down. Some of the predicted signatures have been measured empirically. For instance, data suggest that switches in coordination patterns are driven by instabilities because prior to coordination changes we observe large fluctuations in behavioral patterns (Schöner & Kelso, 1988). Although these data have been modeled quantitatively (Schöner et al., 1986), it is also true that *any* model consistent with the qualitative theory of dynamical systems leads to these predictions. The predictions are, therefore, universal within the dynamical systems framework (van der Maas et al., 2006; van der Maas & Molenaar, 1992). This is a strength in that it provides evidence for the framework. This also means, however, that the observation of instabilities cannot be used to identify any particular model of the coordination system.

Given that DST and CN have emerged from different scientific traditions, are they ultimately incompatible? We, of course, do not think this is the case. For instance, the mathematical psychology tradition can be quite informative. We can ask, for instance, whether the dynamic field model of infants' perseverative reaching provides a better quantitative fit to empirical observations than Munakata's PDP model (Munakata, 1998). Similarly, we can ask which framework has been generalized to other observations more effectively—the PDP model to different forms of perseveration (e.g., perseveration in the Wisconsin card sorting task; see Morton & Munakata, Chapter 7) or DFT which has been generalized to the development of visuospatial working memory (Spencer et al., 2007) and language development (Lipinski, Spencer, & Samuelson, in press; Samuelson et al., 2008). Finally, we can ask which set of concepts maps more effectively onto neural reality (Cohen et al., 1996; O'Reilly et al., 1999; Simmering, Schutte, et al., 2008).

Although these are all valid questions, we also need to be cautious about how seriously we take these comparisons and how deeply we allow ourselves to fall into the information processing attractor, so to speak. In our view, the nonlinear concepts we are collectively working with are simply not compatible with strong forms of information processing that embrace modules, serial stages, and so on. Thus, we must stay true to our scientific traditions, but we must also look toward the future and acknowledge that we are pursuing qualitatively new ways of thinking.

### Is Learning the Same as Development?

We end this section with a topic that is more food for thought than a well-articulated case of a real difference between approaches. As we discussed previously, both CN and DST have a shared commitment to understanding the processes that underlie learning, although the CN approach has taken this issue as central. Interestingly, however, the issue of learning raises a second fundamental issue: is learning the same thing as development?

Although there is not a clear answer to this question at present, our sense is that dynamists tend to answer this question *no*, while connectionists tend to answer *yes*. If learning and development are not the same thing, then what is development? One place dynamists have looked for development is in patterns of neural reorganization that reliably occur at different times/stages across the life span (Fischer, 1980; Fischer & Bidell, 2006). Dynamists have also looked for development in the body—changes in motor coordination patterns related to, for instance, muscle growth and physical development (Thelen, 1985; Thelen, Fisher, & Ridley-Johnson, 1984). And in our dynamic field approach, we have made a distinction between learning that operates over a trial-to-trial timescale and more general changes at the level of the tuning of cortical maps that occurs over a slower timescale (see also Kohonen, 1982; Miikkulainen, 1991; Sirosh & Miikkulainen, 1994). Although changes at these two levels can both arise due to Hebbian processes (Abbott & Regehr, 2004), our empirical evidence suggests that these processes occur over at least two different timescales (Spencer et al., 2007).

Clearly the issue of whether learning and development are the same thing is fundamental to our science. And we look forward to future discussions about this topic. Critically, however, we contend that the only way to examine this issue effectively is to use the full power of computational modeling and formal theory. We take up this issue in greater detail in our concluding section below.

## Implications for Development: Where Next and Why Should Developmentalists Care?

Some of the issues we have raised in this reactions chapter are quite technical in nature: how does one evaluate different types of models? What do different forms of Hebbian learning produce? In this context, why should developmentalists who are not firmly rooted in either the *CN camp* or the *DST camp* care about either of these two approaches? Is this book merely about two groups squabbling over technical issues? Or do the issues raised in this volume pose deeper problems that require decisions (and actions!) on the part of those who don't do modeling? In our view, the issues at stake here are important, not just for those of us enmeshed in the world of modeling and formal theories. These issues matter for all developmental scientists. Here's why.

First off, real thinking about learning and development is happening in both approaches, and both approaches have adopted a process view of development at their core. This view contrasts with talk about innateness, modules, and encapsulated, evolutionary adaptations. Thus, in the grand scheme, there is a paradigm shift happening. We contend that it is finally time to leave the old framing of our science behind. And, critically, this argument is not based on belief in one set of concepts over another. It is based on empirical facts and contemporary developmental theory (see, for example, Spencer et al., in press). We had a tangible sense at the conference in 2005 that the group assembled was close to a deep understanding of developmental processes that can make this paradigm shift complete. The discussions were thoughtful, the work was innovative, and we left more energized than from any other scientific gathering.

Second, the conceptual issues that have emerged from the in-depth exploration of DST and CN approaches are important to all researchers in development. Indeed, we contend that a process-based account of development must address *all* of the central themes we have raised in this chapter and others that form the backbone of this volume: the autonomy of behavior; the processes that underlie statistical learning as well as other classes of learning; the embodiment of cognitive systems that are situated in real environments; the emergent, multicausal nature of development that yields individual pathways over the life span; the emergence and soft assembly of behavioral functionality; the integration of processes over multiple timescales; the real time and developmental relation between *lower level* and *higher level* forms of cognition; and the interface between brain and behavior as we seek to understand the processes that underlie development from neurons to neighborhoods. As examples from this volume illustrate, DST and CN have begun to address all of these challenges, but we are only just beginning to integrate the insights from our diverse endeavors. Both approaches must make steps to achieve more. As they do, both will change and, we contend, ultimately fuse.

Are these concepts unique to DST and CN? No. They have been invoked in other forms within other theoretical traditions (e.g., Piagetian theory; Vygotsky's theory; the Gibsonian tradition; good old-fashioned cognitive development, see Chapter 13). That said, CN and DST have taken these concepts to new levels of clarity. Indeed, we will go so far as to state that *developmental science cannot make progress without the type of computational modeling and formal theories pioneered within the dynamical systems theory and connectionist approaches*. In our view, the problems are simply too hard (see also, spencer & plumert, 2007). This means that the differences between connectionist and dynamical systems theory approaches are not merely technical. They are fundamental to our science.

What is the way forward? We must first appreciate our strengths. But we must also appreciate our differences and the initial conditions that brought us to this point. Bringing these issues into the foreground reveals our different commitments as scientists, our different points of emphasis, and our different styles. We hope this volume has taken an important first step in this direction. If we stay grounded in the concepts of our science and the shared theoretical ideas that are the foundation for our different approaches, we will ultimately be stronger and more cohesive. We agree with Linda B. Smith and Larissa Samuelson (2003; see also Thomas et al., Chapter 17) that diversity is good, complementarity is good. There is no rush to fuse our approaches. But we do think a new unified theory of development is in the offing.

Critically, however, an ungrounded unified theory will do us absolutely no good. Just as we need strong theory to tackle the complexity of development, we need tight theory–experiment links to understand processes that operate at multiple levels and across multiple timescales. In this context, *we need to forge strong ties to experimentalists*. This is no small feat. There is a real mistrust of modeling (sometimes, rightly so!) by some researchers in the developmental community; this mistrust is often expressed by claiming that *anything* can be modeled. The reality of modeling is, of course, quite the opposite. It is generally quite difficult to come up with a single model that is consistent with all the constraints known experimentally. Thus, we need to find ways to bridge the gap between theoreticians and experimentalists, to forge a shared sense of trust, and to deeply value both approaches (see Simmering, Triesch, Deàk, & Spencer, 2008). Clearly, this is possible as this volume—rich with theory and data alike—attests.

Surveying the content at the conference in 2005 and surveying the content of this book, it is clear to us that there is no other community making more progress on discovering the reality of developing systems. We are eager to push forward and do the hard work of understanding what William James, Kurt Lewin, Jean Piaget, Lev Vogotsky, Eleanor Gibson, Elizabeth Bates, Esther Thelen, and countless other scholars all saw as the path to understanding the details of developing systems. Let us take up their charge, move forward smartly, and do the hard work of understanding development.

## Acknowledgments

Preparation of this manuscript was supported by NSF HSD 0527698 awarded to John P. Spencer, as well as support from the Obermann Center for Advanced Studies at the University of Iowa awarded to John P. Spencer and Gregor Schöner. Correspondence concerning this article should be addressed to John P. Spencer, Department of Psychology, University of Iowa, E11 Seashore Hall, Iowa City, Iowa 52242. E-mail: john-spencer@uiowa.edu.

## References

Abbott, L. F., & Regehr, W. G. (2004). Synaptic computation. *Nature, 431*(7010), 796–803.

Altmann, G. T. (2002). Learning and development in neural networks—the importance of prior experience. *Cognition, 85*(2), B43–B50.

Amari, S.-I. (1977). Dynamics of pattern formation in lateral-inhibition type neural fields. *Biological Cybernetics, 27*(2), 77–87.

Baddeley, A. D. (1986). *Working memory.* New York: Oxford University Press.

Bastian, A., Riehle, A., Erlhagen, W., & Schöner, G. (1998). Prior information preshapes the population representation of movement direction in motor cortex. *NeuroReport, 9*(2), 315–319.

Bastian, A., Schöner, G., & Riehle, A. (2003). Preshaping and continuous evolution of motor cortical representations during movement preparation. *European Journal of Neuroscience, 18*(7), 2047–2058.

Bicho, E., Mallet, P., & Schöner, G. (2000). Target representation on an autonomous vehicle with low-level sensors. *International Journal of Robotics Research, 19*(5), 424–447.

Braitenberg, V., & Schüz, A. (1991). *Anatomy of the cortex.* New York: Springer.

Clearfield, M. W., Diedrich, F. J., Smith, L. B., & Thelen, E. (2006). Young infants reach correctly in A-not-B tasks: On the development of stability and perseveration. *Infant Behavior and Development, 29*(3), 435–444.

Cohen, J. D., Braver, T. S., & O'Reilly, R. C. (1996). A computational approach to prefrontal cortex, cognitive control, and schizophrenia: Recent developments and current challenges. *Philosophical Transactions of the Royal Society of London. Series B, Biological Sciences, 351*(1346), 1515–1527.

Cohen, J. D., & Servan-Schreiber, D. (1992). Context, cortex, and dopamine: A connectionist approach to behavior and biology in schizophrenia. *Psychological Review, 99*(1), 45–77.

Dineva, E. (2005). *Dynamical field theory of infant reaching and its dependence on behavioral history and context.* Unpublished doctoral dissertation, Ruhr-Universität Bochum, Bochum, Germany.

Dineva, E., Faubel, C., & Schöner, G. (2007, July). *Dynamic field theory and embodiment: Robotic demonstration of infant perseveration.* Paper presented at the 14th International Conference on Perception and Action, Yokohama, Japan.

Dineva, E., Schöner, G., & Thelen, E. (2008). Behavioral history matters: A dynamic field theory of spontaneous and perseverative errors in infant reaching. Manuscript in preparation.

Dineva, E., & Smith, L. B. (2008). Dynamical field approach to memory guided reaching: The role of behavioral history for stabilizing decisions. Manuscript in preparation.

Douglas, R. J., & Martin, K. A. C. (2004). Neuronal circuits of the neocortex. *Annual Review of Neuroscience, 27,* 419–451.

Edin, F., Macoveanu, J., Olesen, P., Tegnér , J., & Klingberg, T. (2007). Stronger synaptic connectivity as a mechanism behind development of working memory-related brain activity during childhood. *Journal of Cognitive Neuroscience, 19*(5), 750–760.

Elman, J. L. (1990). Finding structure in time. *Cognitive Science, 14*(2), 179–211.

Elman, J. L., Bates, E. A., Johnson, M. H., Karmiloff-Smith, A., Parisi, D., & Plunkett, K. (1996). *Rethinking innateness: A connectionist perspective on development.* Cambridge, MA: MIT Press–Bradford Books.

Erlhagen, W., Bastian, A., Jancke, D., Riehle, A., & Schöner, G. (1999). The distribution of neuronal population activation (DPA) as a tool to study interaction and integration in cortical representations. *Journal of Neuroscience Methods, 94*(1), 53–66.

Erlhagen, W., & Schöner, G. (2002). Dynamic field theory of movement preparation. *Psychological Review, 109*(3), 545–572.

Faubel, C., & Schöner, G. (2008). Learning to recognize objects on the fly: A neurally based dynamic field approach. *Neural Networks, 21*(4), 562–576.

Fischer, K. W. (1980). A theory of cognitive development: The control and construction of hierarchies of skills. *Psychological Review, 87*(6), 477–531.

Fischer, K. W. (2008). Dynamic cycles of cognitive and brain development: Measuring growth in mind, brain, and education. In A. M. Battro, K. W. Fischer, & P. Léna (Eds.), *The educated brain: Essays in neuroeducation.* New York: Cambridge University Press.

Fischer, K. W., & Bidell, T. R. (2006). Dynamic development of action, thought, and emotion. In W. Damon & R. M. Lerner (Eds.), *Handbook of child psychology: Vol. 1. Theoretical models of human development* (6th ed., pp. 313–399). New York: Wiley.

Huttenlocher, J., Hedges, L. V., & Duncan, S. (1991). Categories and particulars: Prototype effects in estimating spatial location. *Psychological Review, 98*(3), 352–376.

Iossifidis, I., & Schöner, G. (2006, October). *Dynamical systems approach for the autonomous avoidance of obstacles and joint-limits for a redundant robot arm.* Paper presented at the IEEE International Conference on Intelligent Robots and Systems (IROS), Beijing, China.

Jancke, D., Erlhagen, W., Dinse, H. R., Akhavan, A. C., Giese, M., Steinhage, A., et al. (1999). Parametric population representation of retinal location: Neuronal interaction dynamics in cat primary visual cortex. *Journal of Neuroscience, 19*(20), 9016–9028.

Jansen, B. R. J., & van der Maas, H. L. J. (2002). The development of children's rule use on the balance scale task. *Journal of Experimental Child Psychology, 81*(4), 383–416.

Johnson, J. S., & Spencer, J. P. (2008). Working memory for location is attracted toward the focus of attention. Manuscript submitted for publication.

Johnson, J. S., Spencer, J. P., Luck, S. J., & Schöner, G. (in press). A dynamic neural field model of visual working memory and change detection. *Psychological Science.*

Kohonen, T. (1982). Self-organized formation of topologically correct feature maps. *Biological Cybernetics, 43*(1), 59–69.

Kopecz, K., & Schöner, G. (1995). Saccadic motor planning by integrating visual information and pre-information on neural, dynamic fields. *Biological Cybernetics, 73*(1), 49–60.

Large, E. W., Christensen, H. I., & Bajcsy, R. (1999). Scaling the dynamical approach to path planning and control: Competition among behavioral constraints. *International Journal of Robotics Research, 18*(1), 37–58.

Lipinski, J., Spencer, J. P., & Samuelson, L. K. (in press). It's in the eye of the beholder: Spatial language and spatial memory use the same perceptual reference frames. In L. B. Smith, M. Gasser, & K. Mix (Eds.), *The spatial foundations of language*. New York: Oxford University Press.

Lovett, A., & Scassellati, B. (2004). Using a robot to reexamine looking time experiments. *Proceedings of the Third International Conference on Development and Learning* (pp. 284–291). Cambridge, MA: IEEE Press. Retrieved October 27, 2008, from http://www.cs.yale.edu/homes/scaz/papers/Lovett-ICDL-03.pdf

MacDonald, M. C., & Christiansen, M. H. (2002). Reassessing working memory: Comment on Just and Carpenter (1992) and Waters and Caplan (1996). *Psychological Review, 109*(1), 35–54, discussion 55–74.

McDowell, K., Jeka, J. J., Schöner, G., & Hatfield, B. D. (2002). Behavioral and electrocortical evidence of an interaction between probability and task metrics in movement preparation. *Experimental Brain Research, 144*(3), 303–313.

Mees, A. I. (1981). *Dynamics of feedback systems*. New York: Wiley.

Miikkulainen, R. (1991). Self-organizing process based on lateral inhibition and synaptic resource redistribution. In T. Kohonen, K. Mäkisara, O. Simula, & J. Kangas (Eds.), *Proceedings of the International Conference on Artificial Neural Network* (pp. 415–420). New York: Elsevier.

Munakata, Y. (1998). Infant perseveration and implications for object permanence theories: A PDP model of the AB task. *Developmental Science, 1*(2), 161–184, commentary 184–204, author's response 205–211.

Newell, K. M., & Molenaar, P. C. M. (Eds.). (1998). *Applications of nonlinear dynamics to developmental process modeling*. Mahwah, NJ: Erlbaum.

O'Reilly, R. C., Braver, T. S., & Cohen, J. D. (1999). A biologically-based computational model of working memory. In A. Miyake & P. Shah (Eds.), *Models of working memory: Mechanisms of active maintenance and executive control* (pp. 375–411). New York: Cambridge University Press.

Port, R. F., & van Gelder, T. (Eds.). (1995). *Mind as motion: Explorations in the dynamics of cognition*. Cambridge, MA: MIT Press–Bradford Books.

Pouget, A., Dayan, P., & Zemel, R. (2000). Information processing with population codes. *Nature Reviews. Neuroscience, 1*(2), 125–132.

Rolls, E. T., Stringer, S. M., & Trappenberg, T. P. (2002). A unified model of spatial and episodic memory. *Proceedings of Biological Science, 269*(1496), 1087–1093.

Samuelson, L. K., Schutte, A. R., & Horst, J. S. (2008). The dynamic nature of knowledge: Insights from a dynamic field model of children's novel noun generalizations. Manuscript submitted for publication.

Schall, J. D. (2004). On building a bridge between brain and behavior. *Annual Review of Psychology, 55*, 23–50.

Schlesinger, M., Amso, D., & Johnson, S. P. (2007). The neural basis for visual selective attention in young infants: A computational account. *Adaptive Behavior, 15*(2), 135–148.

Schlesinger, M., & Parisi, D. (2007). Connectionism in an artificial life perspective: Simulating motor, cognitive, and language development. In D. Mareschal, S. Sirois, & G. Westermann (Eds.), *Neuroconstructivism: Vol. 2. Prospectives and prospects* (pp. 129–158). New York: Oxford University Press.

Scholz, J. P., & Kelso, J. A. S. (1989). A quantitative approach to understanding the formation and change of coordinated movement patterns. *Journal of Motor Behavior, 21*(2), 122–144.

Schöner, G. (1989). Learning and recall in a dynamic theory of coordination patterns. *Biological Cybernetics, 62*(1), 39–54.

Schöner, G., & Dineva, E. (2007). Dynamic instabilities as mechanisms for emergence. *Developmental Science, 10*(1), 69–74.

Schöner, G., & Dose, M. (1992). A dynamical systems approach to task-level system integration used to plan and control autonomous vehicle motion. *Robots and Autonomous Systems, 10*(4), 253–267.

Schöner, G., Dose, M., & Engels, C. (1995). Dynamics of behavior: Theory and applications for autonomous robot architectures. *Robotics and Autonomous Systems, 16*(2–4), 213–245.

Schöner, G., Haken, H., & Kelso, J. A. S. (1986). A stochastic theory of phase transitions in human hand movement. *Biological Cybernetics, 53*(4), 247–257.

Schöner, G., & Kelso, J. A. S. (1988). Dynamic pattern generation in behavioral and neural systems. *Science, 239*(4847), 1513–1520.

Schöner, G., Zanone, P. G., & Kelso, J. A. S. (1992). Learning as change of coordination dynamics: Theory and experiment. *Journal of Motor Behavior, 24*(1), 29–48.

Schutte, A. R., & Spencer, J. P. (2007). Planning "discrete" movements using a continuous system: Insights from a dynamic field theory of movement preparation. *Motor Control, 11*(2), 166–208.

Schutte, A. R., Spencer, J. P., & Schöner, G. (2003). Testing the dynamic field theory: Working memory for locations becomes more spatially precise over development. *Child Development, 74*(5), 1393–1417.

Simmering, V. R., Schutte, A. R., & Spencer, J. P. (2008). Generalizing the dynamic field theory of spatial cognition across real and developmental time scales. In S. Becker (Ed.), Computational cognitive neuroscience [special issue]. *Brain Research, 1202*(C), 68–86.

Simmering, V. R., & Spencer, J. P. (2007). Carving up space at imaginary joints: Can people mentally impose arbitrary spatial category boundaries? *Journal of Experimental Psychology: Human Perception and Performance, 33*(4), 871–894.

Simmering, V. R., & Spencer, J. P. (2008). Generality with specificity: The dynamic field theory generalizes across tasks and time scales. *Developmental Science, 11*(4), 541–555.

Simmering, V. R., Triesch, J., Deák, G., & Spencer, J. P. (2009). To model or not to model? A dialogue on the role of computational modeling in developmental science. Manuscript submitted for publication.

Sirois, S. (2005). *Hebbian motor control in a robot-embedded model of habituation.* Paper presented at the International Joint Conference on Neural Networks (IJCNN).

Sirosh, J., & Miikkulainen, R. (1994). Cooperative self-organization of afferent and lateral connections in cortical maps. *Biological Cybernetics, 71*(1), 65–78.

Smith, L. B., & Samuelson, L. K. (2003). Different is good: Connectionism and dynamic systems theory are complementary emergentist approaches to development. *Developmental Science, 6*(4), 434–439.

Smith, L. B., & Thelen, E. (Eds.). (1993). *A dynamic systems approach to development: Applications.* Cambridge, MA: MIT Press–Bradford Books.

Smith, L. B., Thelen, E., Titzer, R., & McLin, D. (1999). Knowing in the context of acting: The task dynamics of the A-not-B error. *Psychological Review, 106*(2), 235–260.

Smith, P. L., & Ratcliff, R. (2004). Psychology and neurobiology of simple decisions. *Trends in Neurosciences, 24*(3), 161–168.

Soares, R., Bicho, E., Machado, T., & Erlhagen, W. (2007). Object transportation by multiple mobile robots controlled by attractor dynamics: Theory and implementation, *International Conference on Intelligent Robots and Systems IROS 2007, IEEE/RSJ* (pp. 937–944). San Diego, CA.

Spencer, J. P., Blumberg, M. S., McMurray, B., Robinson, S. R., Samuelson, L. K., & Tomblin, J. B. (in press). Short arms and talking eggs: Why we should no longer abide the nativist-empiricist debate. *Child Development Perspectives.*

Spencer, J. P., Corbetta, D., Buchanan, P., Clearfield, M. W., Ulrich, B. D., & Schöner, G. (2006). Moving toward a grand theory of development: In memory of Esther Thelen. *Child Development, 77*(6), 1521–1538.

Spencer, J. P., & Hund, A. M. (2002). Prototypes and particulars: Geometric and experience-dependent spatial categories. *Journal of Experimental Psychology: General, 131*(1), 16–37.

Spencer, J. P., & Hund, A. M. (2003). Developmental continuity in the processes that underlie spatial recall. *Cognitive Psychology, 47*(4), 432–480.

Spencer, J. P., & Plumert, J. M. (2007). What makes thinking about development so hard? In J. M. Plumert & J. P. Spencer (Eds.), *The emerging spatial mind* (pp. 375–386). New York: Oxford University Press.

Spencer, J. P., & Schöner, G. (2003). Bridging the representational gap in the dynamical systems approach to development. *Developmental Science, 6*(4), 392–412.

Spencer, J. P., Simmering, V. R., Schutte, A. R., & Schöner, G. (2007). What does theoretical neuroscience have to offer the study of behavioral development? Insights from a dynamic field theory of spatial cognition. In J. M. Plumert & J. P. Spencer (Eds.), *The emerging spatial mind* (pp. 320–361). New York: Oxford University Press.

Spencer, J. P., Smith, L. B., & Thelen, E. (2001). Tests of a dynamic systems account of the A-not-B error: The influence of prior experience on the spatial memory abilities of two-year-olds. *Child Development, 72*(5), 1327–1346.

Steinhage, A., & Schöner, G. (1997). Self-calibration based on invariant view recognition: Dynamic approach to navigation. *Robotics and Autonomous Systems, 20*(2–4), 133–156.

Steinhage, A., & Schöner, G. (1998). Dynamical systems for the behavioral organization of autonomous robot navigation. In P. S. Schenker & G. T. McKee (Eds.), *Proceedings of SPIE: Vol. 3523. Sensor fusion and decentralized control in robot systems* (pp. 160–180). Bellingham, WA: SPIE Publishing.

Thelen, E. (1985). Developmental origins of motor coordination: Leg movements in human infants. *Developmental Psychobiology, 18*(1), 1–22.

Thelen, E., Fisher, D. M., & Ridley-Johnson, R. (1984). The relationship between physical growth and a newborn reflex. *Infant Behavior and Development, 7*(4), 479–493.

Thelen, E., Schöner, G., Scheier, C., & Smith, L. B. (2001). The dynamics of embodiment: A field theory of infant perseverative reaching. *Behavioral and Brain Sciences, 24*(1), 1–34, commentary 34–70, authors' response 70–80.

Thelen, E., & Smith, L. B. (1994). *A dynamic systems approach to the development of cognition and action.* Cambridge, MA: MIT Press–Bradford Books.

Turvey, M. T. (1990). Coordination. *American Psychologist, 45*(8), 938–953.

Usher, M., & McClelland, J. L. (2001). The time course of perceptual choice: The leaky, competing accumulator model. *Psychological Review, 108*(3), 550–592.

Van der Maas, H. L. J., Dolan, C. V., Grasman, R. P. P. P., Wicherts, J. M., Huizenga, H. M., & Raijmakers, M. E. J. (2006). A dynamical model of general intelligence: The positive manifold of intelligence by mutualism. *Psychological Review, 113*(4), 842–861.

Van der Maas, H. L. J., & Molenaar, P. C. M. (1992). Stagewise cognitive development: An application of catastrophe theory. *Psychological Review, 99*(3), 395–417.

Van Geert, P. (1991). A dynamic systems model of cognitive and language growth. *Psychological Review, 98*(1), 3–53.

Van Geert, P. (1993). A dynamic systems model of cognitive growth: Competition and support under limited resource conditions. In L. B. Smith & E. Thelen (Eds.), *A dynamic systems approach to development: Applications* (pp. 265–331). Cambridge, MA: MIT Press.

Van Geert, P. (1998). A dynamic systems model of basic developmental mechanisms: Piaget, Vygotsky, and beyond. *Psychological Review, 105*(4), 634–677.

Van Geert, P., & van Dijk, M. (2002). Focus on variability: New tools to study intra-individual variability in developmental data. *Infant Behavior and Development, 25*(4), 340–374.

Warren, W. H. (2006). The dynamics of perception and action. *Psychological Review, 113*(2), 358–389.

Wilimzig, C., Schneider, S., & Schöner, G. (2006). The time course of saccadic decision making: Dynamic field theory. *Neural Networks, 19*(8), 1059–1074.

Wilson, H. R., & Cowan, J. D. (1973). A mathematical theory of the functional dynamics of cortical and thalamic nervous tissue. *Kybernetik, 13*(2), 55–80.

Zanone, P. G., & Kelso, J. A. S. (1992). Evolution of behavioral attractors with learning: Nonequilibrium phase transitions. *Journal of Experimental Psychology: Human Perception and Performance, 18*(2), 403–421.

Zanone, P. G., & Kelso, J. A. S. (1997). Coordination dynamics of learning and transfer: collective and component levels. *Journal of Experimental Psychology: Human Perception and Performance, 23*(5), 1454–1480.

# AUTHOR INDEX

# SUBJECT INDEX

Note: Page numbers with "f" denote figures; "t" denote tables